MW00862376

Handbook of Natural Colorants

Wiley Series in Renewable Resources

Series Editor

Christian V. Stevens, Department of Organic Chemistry, Ghent University, Belgium

Titles in the Series

Wood Modification: Chemical, Thermal and Other Processes
Callum A. S. Hill

Renewables-Based Technology: Sustainability Assessment
Jo Dewulf and Herman Van Langenhove (Editors)

Introduction to Chemicals from Biomass
James H. Clark and Fabien E. I. Deswarte (Editors)

Biofuels
Wim Soetaert and Erick Vandamme (Editors)

Handbook of Natural Colorants
Thomas Bechtold and Rita Mussak (Editors)

Forthcoming Titles

Starch Biology, Structure and Functionality
Anton Huber and Werner Praznik

Industrial Application of Natural Fibres: Structure, Properties and Technical Applications
Jörg Müssig (Editor)

Surfactants from Renewable Resources
Mikael Kjellin and Ingegärd Johansson (Editors)

Thermochemical Processing of Biomass
Robert C. Brown (Editor)

Bio-based Polymers
Martin Peter and Telma Franco (Editors)

Handbook of Natural Colorants

Edited by

THOMAS BECHTOLD

and

RITA MUSSAK

Leopold-Franzens University, Austria

A John Wiley and Sons, Ltd., Publication

For details of our global editorial offices, for customer services and for information about how to apply for permission to reuse the copyright material in this book please see our website at www.wiley.com.

Reprinted February 2010

Library of Congress Cataloging-in-Publication Data

Bechtold, Thomas.
Handbook of natural colorants / Thomas Bechtold, Rita Mussak.
p. cm.
Includes bibliographical references and index.
ISBN 978-0-470-51199-2 (cloth : alk. paper) 1. Dyes and dyeing. 2. Dye plants.
3. Dyes and dyeing—Chemistry. I. Mussak, Rita. II. Title.
TP919.B43 2009
667′.26—dc22

2008053906

A catalogue record for this book is available from the British Library.

ISBN 978-0-470-511992 (H/B)

Set in 10/12pt Times by Integra Software Services Pvt. Ltd, Pondicherry, India
Printed and bound in Great Britain by CPI Antony Rowe, Chippenham, Wiltshire

Contents

List of Contributors

Harby Ezzeldeen Ahmed Department of Conservation, Faculty of Archeology, Cairo University, Giza, Egypt. Present address: Tomes IV, Biotechnology–Chemical Engineering School, National Technical University of Athens, 9 Iroon Polytechniou, 15780, Zografou, Athens, Greece

Claude Andary Laboratoire de Botanique, Phytochimie et Mycologie, Faculte de Pharmacie, UMR 5175 (CEFE)*, 15, Ave Charles Flahault, FR-34093 Montpellier Cedex 5, France

Luciana Gabriella Angelini Dipartimento di Agronomia e Gestione dell' Agroecosistema, University of Pisa, Via S. Michele degli Scalzi, 2 I-56127 Pisa, Italy

Thomas Bechtold Research Institute for Textile Chemistry and Textile Physics, University Innsbruck, Hoechsterstrasse 73, A-6850 Dornbirn, Austria

Andrea Biertümpfel Thüringer Landesanstalt für Landwirtschaft, Apoldaer Straße 4, D-07778 Dornburg, Germany

Daniela Borrmann Department of Food and Experimental Nutrition, Faculty of Pharmaceutical Science, University of São Paulo, Av. Prof Lineu Prestes, 580, Bloco 14, 05508-900 São Paulo, SP, Brazil

Dominique Cardon Le Vert, 30460 Colognac, France

U. Gamage Chandrika Department of Biochemistry, Faculty of Medical Sciences, University of Sri Jayewardenepura, Gangodawila, Nugegoda, Sri Lanka

Erika Ganglberger Austrian Society for Environment and Technology, Hollandstrasse 10/46, 1020 Vienna, Austria

Susanne Geissler University of Applied Sciences Wiener Neustadt FHWN, Wieselburg, Zeiselgraben 4, A-3250 Wieselburg, Austria

M. Monica Giusti The Ohio State University, Parker Food Science and Technology, Room 110, 2015 Fyffe Road, Columbus, OH 43210, USA

Michael Grabner BOKU – University of Natural Resources and Applied Life Sciences Vienna, Peter Jordan Str. 82, A-1190 Vienna, Austria

Hely Häggman University of Oulu, Department of Biology, PO Box 3000, FIN-90014 Oulu, Finland

Christian Hansmann Wood Kplus, Competence Center for Wood Composites and Wood Chemistry, St. Peter Str. 25, A-4021 Linz, Austria c/o BOKU – University of Natural Resources and Applied Life Sciences Vienna, Peter Jordan Str. 82, A-1190 Vienna, Austria

Philip John School of Biological Sciences, Harborne Building, University of Reading, Whiteknights, Reading RG6 6AS, UK

Riita Julkunen-Tiitto University of Joensuu, Faculty of Biosciences, PO Box 111, FIN-80101 Joensuu, Finland

Andreas Kandelbauer BOKU – University of Natural Resources and Applied Life Sciences Vienna, Peter Jordan Str. 82, A-1190 Vienna, Austria c/o Wood Carinthian Competence Center, Klagenfurter Str. 87-89, A-9300 St. Veit/Glan, Austria

Hoang Thi Linh Faculty of Textile–Garment Technology and Fashion Design, Hanoi University of Technology, 1-Dai co Viet, Hanoi, Vietnam

Ursula Maria Lanfer Marquez Department of Food and Experimental Nutrition, Faculty of Pharmaceutical Science, University of São Paulo, Av. Prof Lineu Prestes, 580, Bloco 14, 05508-900 São Paulo, SP, Brazil

Maria J. Melo Department of Conservation and Restoration, Requimte and CQFB, Faculty of Sciences and Technology, New University of Lisbon, Campus Caparica, 2829-516 monte da Caparica, Portugal

Adriana Zerlotti Mercadante Department of Food Science, Faculty of Food Engineering, University of Campinas – UNICAMP, CP: 6121, 13083-862, Campinas, Brazil

Ulrich Müller Wood Kplus, Competence Center for Wood Composites and Wood Chemistry, St. Peter Str. 25, A-4021 Linz, Austria c/o BOKU – University of Natural Resources and Applied Life Sciences Vienna, Peter Jordan Str. 82, A-1190 Vienna, Austria

Rita A. M. Mussak Institute for Textile Chemistry and Textile Physics, University of Innsbruck, Hoechsterstrasse. 73, A-6850 Dornbirn, Austria

Fernando Pina REQUIMTE, CQFB, Chemistry Department, Faculty of Sciences and Technology, New University of Lisbon, 2829-516 Monte da Caparica, Portugal

Johannes Pöckl Wood Kplus, Competence Center for Wood Composites and Wood Chemistry, St. Peter Str. 25, A-4021 Linz, Austria c/o BOKU – University of Natural Resources and Applied Life Sciences Vienna, Peter Jordan Str. 82, A-1190 Vienna, Austria

Riikka Räisänen Department of Home Economics and Craft Science, PO Box 8, University of Helsinki, Fin-00014 Helsinki, Finland

Veridiana Vera de Rosso Department of Health Science, Federal University of São Paulo – UNI FESP, 11030-400, Santos, Brazil

Taylor C. Wallace The Ohio State University, Parker Food Science and Technology, Room 110, 2015 Fyffe Road, Columbus, OH 43210, USA

Martin Weigl Wood Kplus, Competence Center for Wood Composites and Wood Chemistry, St. Peter Str. 25, A-4021 Linz, Austria c/o BOKU – University of Natural Resources and Applied Life Sciences Vienna, Peter Jordan Str. 82, A-1190 Vienna, Austria

Günter Wurl Thüringer Landesanstalt für Landwirtschaft, Apoldaer Straße 4, 07778 Dornburg, Germany

Series Preface

Renewable resources, their use and modification are involved in a multitude of important processes with a major influence on our everyday lives. Applications can be found in the energy sector, chemistry, pharmacy, the textile industry, paints and coatings, to name but a few.

The area interconnects several scientific disciplines (agriculture, biochemistry, chemistry, technology, environmental sciences, forestry, . . .), which makes it very difficult to have an expert view on the complicated interaction. Therefore, the idea to create a series of scientific books, focusing on specific topics concerning renewable resources, has been very opportune and can help to clarify some of the underlying connections in this area.

In a very fast changing world, trends are not only characteristic for fashion and political standpoints but science is also not free from hypes and buzzwords. The use of renewable resources is again more important nowadays; however, it is not part of a hype or a fashion. As the lively discussions among scientists continue about how many years we will still be able to use fossil fuels, opinions ranging from 50 years to 500 years, they do agree that the reserve is limited and that it is essential not only to search for new energy carriers but also for new material sources.

In this respect, renewable resources are a crucial area in the search for alternatives for fossil-based raw materials and energy. In the field of the energy supply, biomass and renewable-based resources will be part of the solution alongside other alternatives such as solar energy, wind energy, hydraulic power, hydrogen technology and nuclear energy.

In the field of material sciences, the impact of renewable resources will probably be even bigger. Integral utilization of crops and the use of waste streams in certain industries will grow in importance, leading to a more sustainable way of producing materials.

Although our society was much more (almost exclusively) based on renewable resources centuries ago, this disappeared in the Western world in the 19th century. Now it is time to focus again on this field of research. However, it should not mean a 'retour à la nature', but it should be a multidisciplinary effort on a highly technological level to perform research towards new opportunities, to develop new crops and products from renewable resources. This will be essential to guarantee a level of comfort for a growing number of people living on our planet. It is 'the' challenge for the coming generations of scientists to develop more sustainable ways to create prosperity and to fight poverty and hunger in the world. A global approach is certainly favoured.

This challenge can only be dealt with if scientists are attracted to this area and are recognized for their efforts in this interdisciplinary field. It is therefore also essential that consumers recognize the fate of renewable resources in a number of products.

Furthermore, scientists do need to communicate and discuss the relevance of their work. The use and modification of renewable resources may not follow the path of the genetic engineering concept in view of consumer acceptance in Europe. Related to this aspect, the series will certainly help to increase the visibility of the importance of renewable resources.

Being convinced of the value of the renewables approach for the industrial world, as well as for developing countries, I was myself delighted to collaborate on this series of books focusing on different aspects of renewable resources. I hope that readers become aware of the complexity, the interaction and interconnections, and the challenges of this field and that they will help to communicate on the importance of renewable resources.

I certainly want to thank the people of John Wiley & Sons, Ltd from the Chichester office, especially David Hughes, Jenny Cossham and Lyn Roberts, in seeing the need for such a series of books on renewable resources, for initiating and supporting it and for helping to carry the project to the end.

Last, but not least, I want to thank my family, especially my wife Hilde and children Paulien and Pieter-Jan for their patience and for giving me the time to work on the series when other activities seemed to be more inviting.

Christian V. Stevens
Faculty of Bioscience Engineering
Ghent University
Belgium
June 2005

Preface

Looking out of the window on a bright and colourful autumn day we can recognize that nature provides us with a firework of yellow, red and green colours, inspiring mankind to bring more colour into the products of daily life. However, we have known for a long time that access to the colours of nature is coupled with laborious procedures and a high number of restrictions.

The invention of synthetic organic chemistry and the desire for more bright and stable colorants can be seen as one of the strong driving forces in the historical development of natural sciences. In the 20th century synthetic colorants dominated almost every field of possible application, such as mass coloration of plastics, textiles, paints, cosmetics and food.

For almost 100 years research on natural colorants was continued by only a few groups who were enthusiastic enough to persist against the straightforward arguments for the use of synthetic dyes, relying on cost, performance, colour strength and brilliance, which can easily be achieved using artificial dyes.

During the last decade more and more new aspects were integrated into the assessment of any chemical product used. Interestingly, every new argument that had to be considered also added to the argument strengthening the position of natural colorants. Increased awareness of product safety and higher attention to the possible adverse impact of a chemical product on human health brought changes in the regulations for the use of colorants in food and cosmetics.

Concentration on renewable resources, sustainability and replacement of oil-based products are driving forces to reassess the potential of natural resources including natural colorants, at least for application in very specific fields. Growing consumer interest in purchasing 'green' products, which exhibit an improved environmental profile, can be seen as the breakthrough force for reintroducing natural colorants into the modern markets.

During our own scientific work on natural dyes for textiles and hair dyeing we learned that knowledge about natural colorants and their possible application, at present, is quite fragmented. There are collections of knowledge about natural colorants like Schweppe's *Handbuch der Naturfarbstoffe*, that summarizes properties and sources of natural dyes from a chemical and more historical aspect. However, for the demands of the future development of natural colorants into applications of the present, there is no useful source of information available that could help to give an overview of the state of the research and knowledge in the field of natural colorants.

The search for scientists working on natural colorants who were able and willing to write a contribution for this book was the major challenge in editing this book. The interdisciplinary range of content that should be covered by the different authors made our work particularly difficult, but is understandingly one of the key aspects of this book. The introduction of natural colorants into modern products is an interdisciplinary task that has to consider farming, dyestuff extraction, analysis, properties and application at the same time. Success will only be achieved if integrative concepts are presented that consider all stages of production at the same time.

The organization of the different chapters follows this order. In the first chapters a short review about plant sources and applications of natural colorants in historical times is given. Aspects of farming crops and product processing are then summarized for the different chemical classes of dyes. In the more application-oriented chapters the use of natural colorants in, for example, food, wood, textile and hair dyeing is presented. Sustainability and consumer aspects are discussed in the final chapters of the book.

We would like to thank all authors for their contributions. Their expertise in their particular field, covering a whole array of specialized knowledge, makes the book a unique source of information, which summarizes the present knowledge about natural colorants in depth.

We are aware that every collection of information will be incomplete and further aspects could have been introduced and considered in more detail. However, we are convinced that the book as it stands will be a useful instrument to overview the fragmented situation of natural colorants and will support a rapid and efficient entry of new researchers into this emerging field of sustainable chemistry. From this point of view we are also convinced that the book will strengthen the position of natural colorants in the future, by facilitating access to information and thereby indirectly helping the revival of natural colorants to gain momentum.

Thomas Bechtold and Rita A. M. Mussak
Dornbirn/Linz
2008

Part I
Historical Aspects

1

History of Natural Dyes in the Ancient Mediterranean World

Maria J. Melo

The colours used on textiles and artifacts, their social significance and the scope of their trade, are part and parcel of a people's overall history.

Jenny Balfour-Paul, in *Indigo*, British Museum Press, 2000

1.1 Introduction

1.1.1 Ancient Mediterranean World

The build-up of *Mare Nostrum* probably began much earlier than the 6th–5th millennium BC and there is material evidence pointing to such activity as early as the 12th–11th millennium BC [1]. *Mare Nostrum*, the Roman name for the Mediterranean Sea, was to become the home for a global market that expanded beyond its natural borders in the 1st millennium BC. The Phoenicians, the Etruscans, the Greeks and finally the Romans shaped *Mare Nostrum*, a geographic as well as a cultural domain. It was also home for the first global dye, Tyrian purple, which was traded by the ingenious and industrious Phoenicians. The purple of Tyre was famous, as were the textiles dyed and produced by the Phoenicians. It is said that the Greeks named the Phoenicians after *Phoinikes*, the ancient Greek word for 'red colour', probably as a result of their famous purple trade.

By the time of the founding of the Mediterranean civilizations, what we would consider the classical palette for natural dyes had already been established, and the most valued colours were indigo for the blues, anthraquinone-based chromophores for the reds and 6,6′-dibromoindigo for purple. These colours were traded all over the Mediterranean,

Handbook of Natural Colorants Edited by Thomas Bechtold and Rita Mussak

regardless of distance to be travelled or the price to be paid. The natural sources for yellows were much more diverse, so yellows could generally be obtained locally. For dyeing, with the exception of some browns, all other colours and shades, including green and orange, could be obtained with these blue, red, purple and yellow dyes. This classical palette was preserved over centuries, if not millennia. The first adjustment resulted from the loss of Tyrian purple following the fall of Constantinople and the subsequent collapse of the Roman social and commercial web. This was followed by a new entry, cochineal red, brought by the Spanish from the New World [2]. However, even with the introduction of cochineal the chemical nature of the classical pallete was maintained, as carminic acid is still a substituted 1,2-dihydroxy anthraquinone. This classical palette was only challenged by the audacity of chemists, who created new molecules, and colours never seen before, from the mid-19th century on [3].

1.1.2 Dyes from Antiquity

Natural dyes, discovered through the ingenuity and persistence of our ancestors, can resist brightly for centuries or millennia and may be found hidden in such diverse places as the roots of a plant, a parasitic insect and the secretions of a sea snail. By contrast, the bright colours that we see in the green of a valley, the red of a poppy, the purple of mauve or the blue of cornflower are less stable. Natural dyes were used to colour a fibre or to paint. It is useful to distinguish between dyes and pigments based on their solubility in the media used to apply the colour; dyes are generally organic compounds that are soluble in a solvent, whereas pigments, used in painting, are usually inorganic compounds or minerals that are insoluble in the paint medium (oil, water, etc.) and are dispersed in the matrix. A lake pigment is a pigment formed by precipitation, namely by complexation with a metal ion, forming a dye on the surface of an inorganic substance.

Dyeing, in red, blue, purple or yellow, is a complex task that requires skill and knowledge [4]; this is true now and has been for several millennia. Colour is obtained by applying a chemical compound called a chromophore or chromogen, something that brings or creates colour. When used as a textile dye, the chromophore must also be captured as strongly as possible into the fibres; i.e. it must be resistant to washing. Dyes can bind to the surface of the fibre or be trapped within them. They are often bound to the textile with the aid of metallic ions known as mordants, which can also play an important role in the final colour obtained. Alum, as a source of the aluminium ion, is an important historical mordant and was widely used in the past. Other important mordants used in the past were iron, copper and tin ions [4,5]. Dyes, like indigo, which are trapped in the fibres due to an oxidation–reduction reaction, without the aid of a mordant, are known as vat dyes.

Natural dyes, as lake pigments, have been widely applied in painting. For example, anthraquinones and their hydroxy derivatives have been used as red dyes and pigment lakes from prehistoric times, and we can find written accounts of the use of anthraquinone reds and purples as dyes in ancient Egypt [5, 6]; anthraquinone lakes (e.g. madder red) were also very popular with Impressionist painters, including Vincent van Gogh. Lake pigments can be prepared by precipitating the dye extract with aluminium or other inorganic salts, such as alum [7]. Pure dyes such as indigo were also used as painting materials, e.g. in medieval illuminations (Figure 1.1).

Figure 1.1 *Medieval Portuguese illumination, dating from a 12–13th century, Lorvão 15, fl. 50 kept at Torre do Tombo (Lisbon). Dark blues were painted with indigo, whereas for the backgound the inorganic and precious pigment lapis-lazuli was used (See Colour Plate 1)*

These *eternal* colours will be described in more detail in the following sections, after a brief account of the analytical techniques used to reveal the secrets of these ancient materials. The natural colorants will be organized according to the colour: first the anthraquinone reds, followed by the blues and purple, where indigo and its bromo derivatives play a major role. Yellows will close this historic overview.

1.1.3 Unveiling the Secrets of Ancient Dyes with Modern Science

Identifying the dyes and dye sources used in the past has only been possible with the development, in the past two decades or so, of sensitive new microanalytical techniques [8,9]. Chromophores are extracted, then separated chromatographically and characterized by UV-Vis spectrophotometry or mass spectrometry; whenever possible comparison with authentic references is performed. Currently, the use of HPLC-DAD (high-performance liquid chromatography with diode array detection) enables dyestuff characterization from as little as 0.1 mg of thread. For unknown components, or those not characterized before, analysis by HPLC-MS (HPLC with mass spectrometric detection) may provide further information. Recently developed mild extraction methods allow more detailed chemical information to be obtained on the historical natural dyes, and as a consequence it is sometimes possible to identify the natural sources [10, 11].

Mordant analysis can also provide relevant information about the dyeing method or process used. The metal ions can be quantified by inductively coupled plasma separation with atomic emission spectrometric (ICP-AES) or mass spectrometric detection (ICP-MS) of samples (*ca.* 0.25 mg textile strands) previously digested with nitric acid solutions [12, 13]. Before the sample analysis, calibration curves must be constructed using standards. Concentration linearity in the range of ppb to ppm (or higher) can be achieved.

1.2 Ancient Reds

1.2.1 Anthraquinone Reds

The most stable reds used in antiquity are based on the 1,2-dihydroxy anthraquinone chromophore (Figure 1.2), also known as alizarin. Dyes containing anthraquinone and its derivatives are among the most resistant to light-induced fading [5]. These

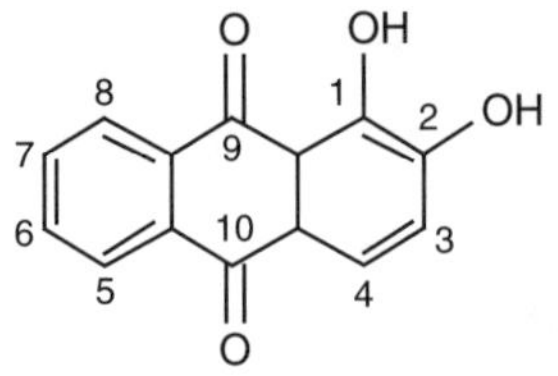

Figure 1.2 Alizarin; 1,2-dihydroxy anthraquinone

dyes were obtained from parasitic insects, such as the famous *Kermes vermilio*, or from the roots of plants belonging to the Rubiaceae or madder family, and were among the reds that dominated the dye markets of Europe [2, 14]. Alizarin and purpurin are the main chromophores in *Rubia tinctorum*, the most important species of the family Rubiaceae. In Persia and India, other red dyes – of animal origin – were also used. These dyes were imported, or sometimes found locally, and were considered luxury goods. Well-known examples are the reds based on the laccaic acids, kermesic acid and carminic acid (Table 1.1), from the parasite insects, lac, kermes and cochineal, respectively [2]. The female lac insect secretes a red resin, stick-lac, from which are obtained both the lac dye and the shellac resin. Common or Indian lac, *Kerria lacca* (= *Laccifer lacca*, *Carteria lacca*, *Tachardia lacca* and *Lakshadia lacca*) and *Kerria chinensis* are examples of species that have been widely exploited [5]. In both cochineal (*Dactylopius coccus*) and kermes (*Kermes vermilio*)

Table 1.1 Chemical structures for anthraquinone reds

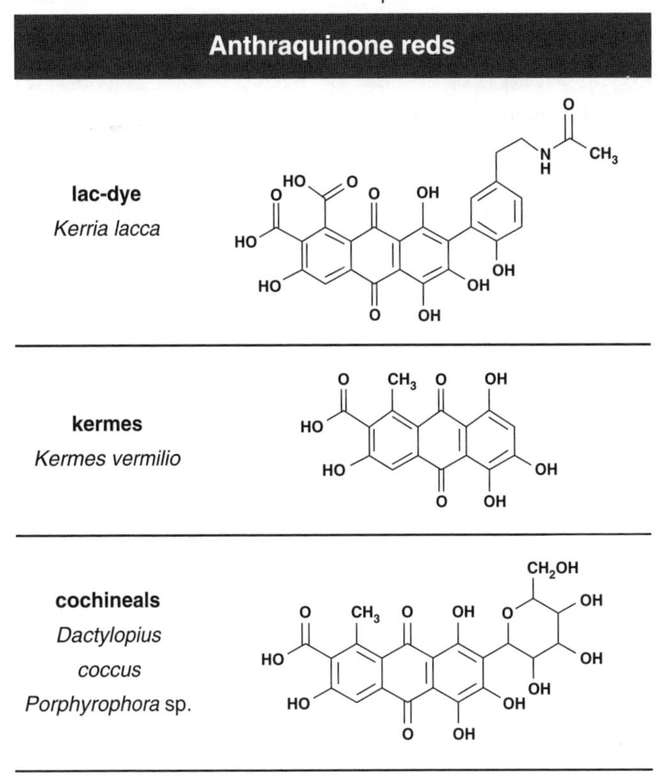

the red dye is obtained from the eggs of the female insect, and therefore there were harvest seasons that corresponded to the phase where the eggs displayed the highest dye content. It was at this stage that the impregnated female insects were cropped before depositing their eggs. *Kermes vermilio* is probably the most famous of the European insect parasites and produced a dye 'brighter than madder and faster than cochineal' [2, 7a], and was already described in the works of Theophrastus and Pliny. The word for worm in many latin languages is based on *verme-* (or *vermiculu* = small worm) and thus is the root of the word for red in several modern European languages, namely vermejo, vermelho, vermeil and vermilion in Spanish, Portuguese, French and English, respectively [15].

Other important historical insect sources of red or scarlet dyes derived from species of *Porphyrophora*, e.g. Polish cochineal (*P. polonica*) and Armenian cochineal (*P. hamelii*). In Europe, most of these sources were replaced in the 16th century by the American cochineal (*Dactylopius coccus*), which had earlier been carefully domesticated and cultivated by the indigenous peoples of the New World, and was commercialized by the Spanish empire. Although the various species of *Porphyrophora* also contain carminic acid, dried specimens of *Dactylopius coccus* have a much higher content (15–20%) of the dye [2, 5], compared with only 0.8% and 0.6% for the Armenian and Polish ones, respectively [5].

1.2.2 Redwoods

Redwoods, also known in antiquity as brasil [4, 16], were used as sources of dyes and pigment lakes [5, 7a]. It is said that the country Brasil was named after those redwoods, this possibly being the only country in the world named after a tree. The name the Portuguese had in mind was *Terra da Vera Cruz* (Land of the Holy Cross). The reds obtained from the bark of the tree are not as stable as the previously described anthraquinone reds, but they were much more affordable and were widely used for dyeing and in medieval miniature painting [17, 7a] as well as in cosmetics. In his book [16], Cennino Cennini recommends the use of a good *verzino* (brazilwood) to highlight the reddish-blue tone of purified lapis lazuli. This good *verzino* should be obtained from a young lady, who, using it for lipstick and other cosmetics, would take care that she had obtained a well-prepared product of a nice colour.

The main colorant in *Caesalpinia brasiliensis, C. echinata, C. sappan, C. violacea* and *Haematoxylum brasiletto* is brazilin. In *Haematoxylum campechianum* the main colorant is haematoxylum. Through oxidation, brazilin and haematoxylum are converted to the darker red compounds, brazilein and haematein, respectively (Figure 1.3) [18].

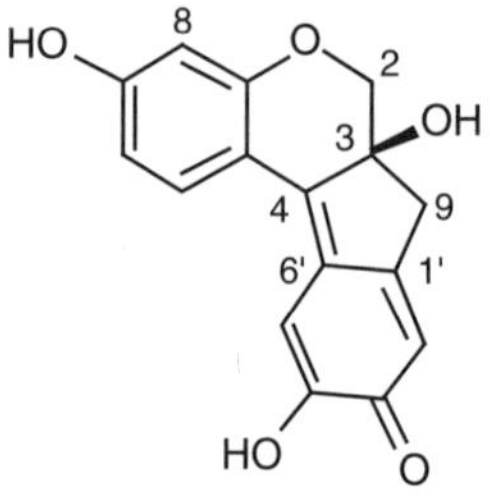

Figure 1.3 Brazilein; haematein is 8-hydroxy brazilein

1.2.3 Flavylium/Anthocyanin Reds

Anthocyanins are ubiquitous water-soluble colorants responsible for the impressive red and blue colours of flowers and fruits (Figure 1.4). Natural flavylium reds can be considered their aglycone ancestors. There are several references to the use of these

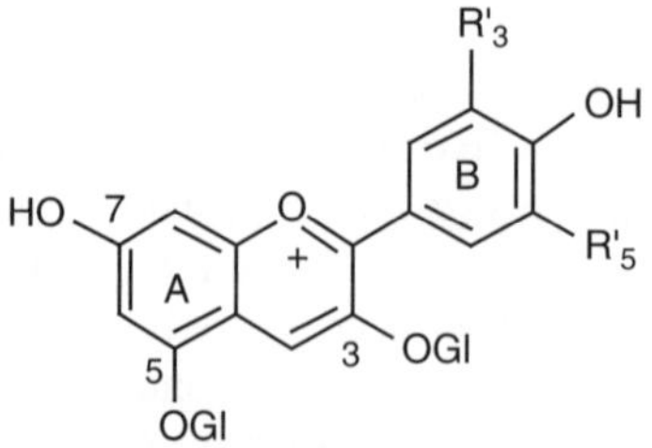

Figure 1.4 In the basic structure of anthocyanins, a hydroxyl group is present at positions 4′ and 7, and a sugar at position 3 (monoglycoside) or positions 3 and 5 (diglycoside)

compounds to substitute for unavailable inorganic pigments or to give special lighting effects in ancient illustrations. The use of these materials ranges from the Roman Empire (described by the famous architect Vitruvius) to paintings of the Maya civilization [19]. Their use as 'watercolours' has been described in several treatises or recipe books on illumination painting, including the Strasbourg manuscript, an 'Old Portuguese work on manuscript illumination: the book on how one makes colours of all shades' and 'De Arte Illuminandi' [20]. Anthocyanins were used to produce clothlets/watercolours [16], as described by Cennino Cennini in the 15th century, and to dye textiles [5, 21, 22]. With anthocyanins the colour domain ranges from red to blue, but with natural flavylium dyes it is limited to the yellow-red. One of the most famous examples of a natural flavylium red can be found in dragon's blood resins.

Dragon's blood is a natural resin, having a deep, rich red colour, which is obtained from various trees, namely from *Dracaena draco* and *Dracaena cinnabari*, which belong to the Liliaceae family. The resin appears in injured areas of the plant and has been used for centuries for diverse medical and artistic purposes [23]. These resins contain not only the red chromophores but also additional flavonoids and steroids; Dragon's blood resins have been used in traditional Chinese medicine. The molecules responsible for the red colour of the resin obtained from the palm tree *Daemonorops draco* (*Calamus draco* was used in the past) have been characterized by Brockmann *et al.* and named dracorubin and dracorhodin (Table 1.2) [24]. Brockman *et al.* concluded that dracorhodin was a natural flavylium chromophore belonging to the Anthocyanin family. More recently, other natural flavylium chromophores, such as dracoflavylium, have been identified in *Dracaena draco, Dracaena cinnabari* and other Dracaenaceae [23] (Table 1.2).

Table 1.2 *Chemical structures responsible for the red colour in Dragon's blood resins. The structures correspond to the quinoid bases (A)*

Dracorhodin	Nordracorhodin	Dracorubin	Dracoflavylium
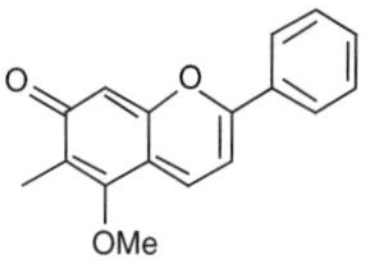	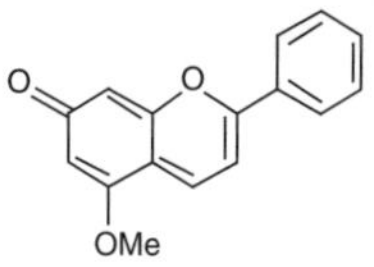	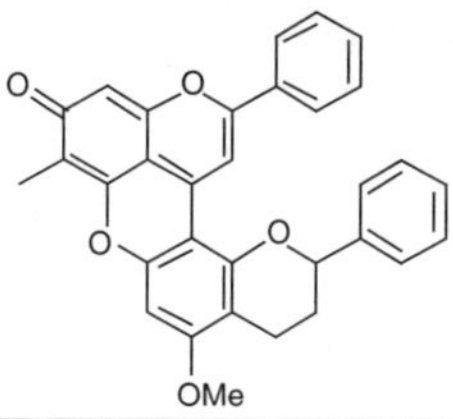	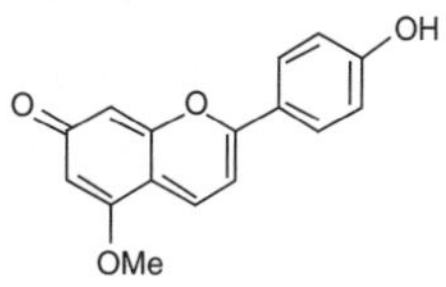

1.2.3.1 Equilibria in Solution

As stated above, dracorhodin and dracofalvylium are natural flavylium reds related to anthocyanins. Anthocyanins are characterized by a hydroxyl group in position 4′ and 7, and a sugar in position 3 (monoglycosides) or 3 and 5 (diglycosides). On the other hand, in anthocyanidins the hydroxyl groups take the positions of the glycosides, leading to unstable structures in solution. In contrast, the so-called deoxyanthocyanidins correspond to 'anthocyanidins' lacking the hydroxyl in position 3 (but bearing a hydroxyl in position 5), and are quite stable in solution [19, 25].

In the 1970s, it was firmly established by Dubois and Brouillard (anthocyanins) [26a] and McClelland (synthetic flavylium salts) [26b] that both families of compounds undergo multiple structural transformations in aqueous solution, following the same basic mechanisms [22, 27] (Figure 1.5). The flavylium cation (**AH+**) is the dominant species in very acidic solutions, but with increasing pH a series of more or less reversible chemical reactions take place: (1) proton transfer leading to the quinoidal base (**A**), (2) hydration of the flavylium cation giving rise to the colourless hemicetal (**B**), (3) a tautomerization reaction responsible for ring opening, to give the pale yellow Z-chalcone, form (**Cc**), and, finally, (4) *cis–trans* isomerization to form the pale yellow E-chalcone (**Ct**). Furthermore, at a higher pH, and depending on the number of hydroxyl groups, further deprotonated species are found, such as $\mathbf{Ct^{n-}}$ and $\mathbf{A^{n-}}$. The relevant contributers to colour are $\mathbf{AH^{+}}$ and the quinoid bases, **A** and $\mathbf{A^{-}}$. The red colour of dragon's blood resin was found to be due to the red quinoid bases of the respective yellow flavylium cations [23].

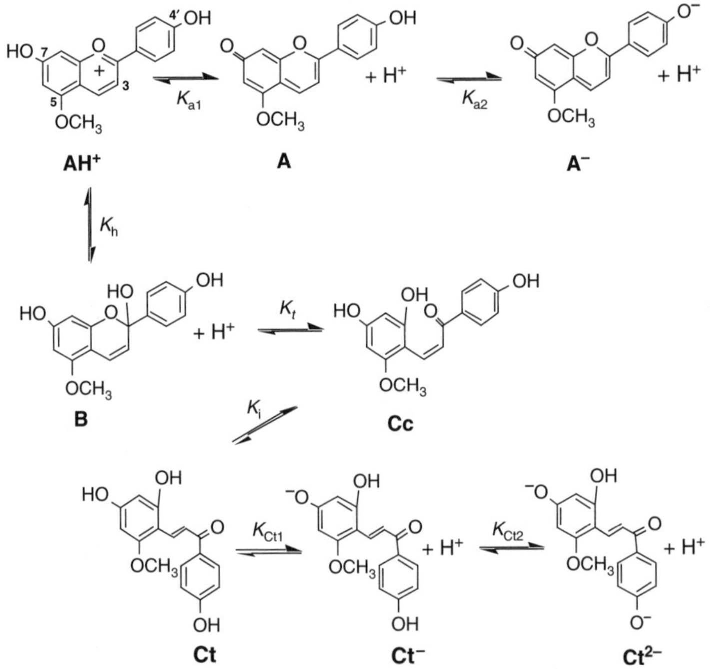

Figure 1.5 *Scheme of chemical reactions for dracoflavylium. Reprinted with permission from Reference [23], Melo, M. J., et al., Chem. Eur. J.,* ***13****, 1417–1422 (2007). © 2007, Wiley-VCH*

1.3 Ancient Blues

1.3.1 Indigo Blues

Indigo blue was one of the earliest and most popular dyestuffs known to man (Figure 1.6). It is still a universally used colour, as the worldwide use of indigo-dyed blue jeans can

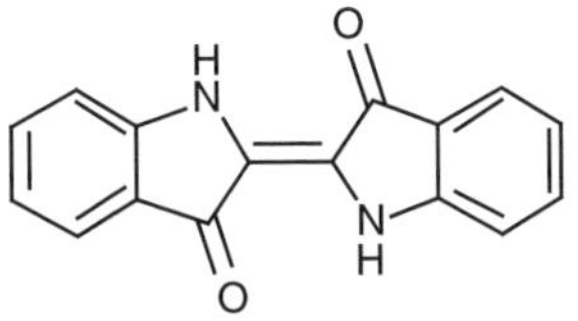

Figure 1.6 Indigo (=indigotin)

attest. Indigo, as the name implies, has its origins in ancient India. In the great civilizations of Egypt, Greece and Rome, it was prized for its quality as a dye even though transportation costs meant it was very expensive [28]. Indigo sources are found all over the world, and several plants were most probably used in antiquity. Julius Caesar, in his *De bello Gallico*, describes the warrior skin paintings of his Gaulish adversaries as being obtained from a blue juice; the warriors' faces were frightening and they believed that dyeing the skin would protect them, making them invulnerable!

Isatis tintoria was grown in Europe, but it is known that indigo from *Indigofera tinctoria* was traded to Europeans, namely from Persia to Muslim Spain, and from there distributed to other European countries [29]. Species of *Indigofera* produced a high-quality indigo, which was used in medieval illuminations (Figure 1.1) [30].

Indigo is also one of the most light-stable organic dyes, a characteristic that explains not only its wide use in antiquity and the pre-modern area but also its longevity as a colorant [31]. The stability of indigo is also the reason why it was used in medieval illuminations and by some of the great masters of the 17th and 18th centuries, such as Rubens [32]. Usually, with the exception of some necessary and almost irreplaceable colour lakes, organic compounds were avoided in oil painting, as it was known that they were far less stable to light than the inorganic available pigments. Highly pure indigo was an exception.

In 1865 Bayer started his work on indigo, and some years later proposed the chemical structure as well as a possible synthesis of indigo. At the time, indigo was an important molecule due to its commercial value as a dye; synthetic indigo paved the way not only for the development of 'big' German chemical and pharmaceutical industries but also helped to end the colonial production of indigo, which until then had come from natural sources in British, French and Iberian colonies. This particularly affected the British Indian colony, which was the major producer at that time. Even at the beginning of the 21st century, indigo is still a surprisingly important dye. Inasmuch as indigo can be obtained from natural sources using microbial fermentation, its production using 'green' chemistry is of considerable interest [33].

The precursors of indigo (= indigotin) can be extracted entirely from plants. The extract, a solution of the water-soluble indoxyl glucoside, indican, is hydrolysed via fermentation to give indoxyl (Figure 1.7), which will further react with another indoxyl molecule producing indigo or with isatin resulting in indirubin, a reddish dye [5, 34]. The indigo derivatives are generally known as vat dyes and, in their oxidized forms, are insoluble in water [5].

A common characteristic of vat dyes is the presence of one or more carbonyl groups, which, when treated with a reducing agent in the presence of an alkali, form a water-soluble dye known as the leuco species. The process of dyeing a textile with these species involves

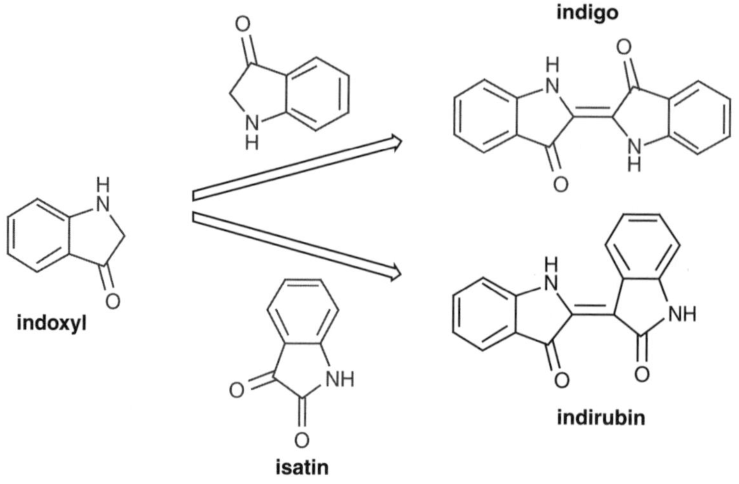

Figure 1.7 Indigo dye bath

the use of the vat dye in a reducing media, leading to its leuco form [5, 33]. The explanation of the blue colour of indigo and its derivatives has also been an intriguing and fascinating subject, and was explored with great detail and ingenuity during the 1970s and 1980s of the 20th century [35]. It has been discovered that the fundamental chromophore of the indigo dyes includes the central double bond (connecting the two rings) together with the nitrogen and carbonyl groups [35a–c]. Substitution in different ring positions of indigo is likely to promote significant shifts in both the visible long-wavelength and UV bands [35d].

Indirubin can be formed during the dyeing with indigo, namely in the interface where more oxygen is available; this is the reason why it is also known as the red shade of indigo [36]. As already mentioned, indirubin is formed through the reaction of indoxyl anion with isatin, a side reaction in the reaction of indoxyl anion with itself to give indigo (Figure 1.7). It can also be present in the Muricidae (see Section 1.4 on ancient purple). As for dragon's blood, its renaissance has been a consequence of its pharmaceutical activity [36]. Marine indirubines are contained in the purple pigment derived from marine selfish of the Muricaidae and Thaisidae families.

1.3.2 Anthocyanin Blues

As reported in Section 1.2.3, there are several ancient references to the use of anthocyanin watercolours for painting, particularly for illuminations, e.g. as described in the Strasbourg Manuscript [20a]. In the past, extracting and capturing an anthocyanin blue was not an easy task; these blues are complex, self-assembled, supramolecular structures, bound by non-covalent interactions. The overall supramolecular complex must be preserved, not just the monomeric subunits, otherwise the colour is lost. For example, to obtain a 'cornflower' blue, the supramolecular assembly must be maintained within a certain pH range. Apparently this was achieved by following the recipes described in ancient treatises such as the Strasbourg Manuscript [19, 20]. The structure of the self-assembled supramolecular pigment from the blue cornflower (*Centaurea cyanus*) was elucidated by Kondo *et al.* [37a]; the supramolecular components are succinylcyanin (Sucy), a cyanidin chromophore with a sugar in position 3, malonylflavone (Mafl), a flavone with a sugar in position 3, and the metal ions Mg^{2+} and Fe^{3+}, the exact composition being $[Sucy_6Mafl_6Fe^{3+}Mg^{2+}]$. These authors proposed that the supramolecular complex should be similar to that of commelinin (Figure 1.8). The metal centre organizes the geometry of the complex, which is held together by weak hydrophobic interactions, such as π–π stacking of the aromatic rings and hydrogen bonding through the sugar moieties [37].

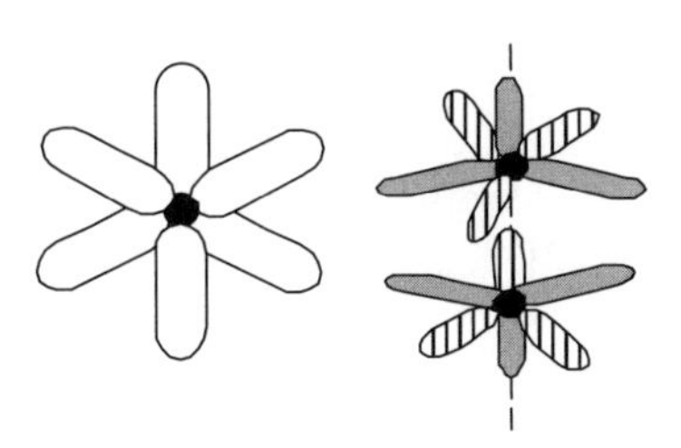

Figure 1.8 Schematic representation of the blue supramolecular complex involving anthocyanins. The two metals (dark spheres) are at the centre of two planes; a top view of one of these planes is represented on the left. Each metal coordinates three anthocyanin chromophores (vertical lines); three flavones (grey) are intercalated with the three anthocyanins, filling the void spaces

In both cornflower blue and commelin, the blue colour is only stable in concentrated solutions. When diluted, the solution quickly becomes colourless due to dissociation of the components of the complex.

1.4 Ancient Purple (Tyrian Purple)

The high status of the wearer of purple-dyed textiles in the past was never to be surpassed by any other colour, natural or synthetic. Purple has a commercial and economic history but also a cultural one, with profound political and religious impact [5, 38, 39]. Purple was a status symbol for the Roman emperors as well as, in later years, for the powerful representatives of the emergent power, the Catholic Church. It was the Persian monarchs who, as early as the 9th century BC, made purple the official mark of royalty. The kings of Israel also wore purple. Imperial or divine purple was used for powerful emotional moments, where authority was conveyed, in rituals of high visual impact. Purple was also the colour of the mantle Christ used in His sacrifice. Thus it is the symbol of His sacrifice but also of His glory, the glory He found when sacrificing Himself for mankind [38]. With it, purple achieved the status of a sacred colour. The original colour was the most expensive dye in the market, but even its imitations (as purple obtained by a combination of indigo with madder or other similar red) could maintain its status symbol [40]. In our present times, it would be hard to find an example that could convey the impact of purple in antiquity.

The first true purple dye, Tyrian purple, was obtained from Mediterranean shellfish of the genera *Purpura* (Figure 1.9) [41]. The Mediterranean purple molluscs, *Murex brandaris*, *Murex trunculus* and *Purpura haemastoma*, have recently been renamed *Bolinus brandaris*, *Hexaplex trunculus* and *Stramonita haemastoma*, respectively, with the result that *Murex* and *Purpura* have disappeared from the name, and with it the memory of their importance as historic dyes was cancelled [39]. As already mentioned in the introduction, it was one of the luxury goods traded by the Phoenicians, and hence the common name 'Tyrian purple', named after the Phoenician city of Tyre. This purple dye was produced by extraction of the secretions of the hypobranchial glands of Mediterranean gastropod molluscs, probably since the mid-2nd millennium BC. Piles of purple-yielding mollusc shells on Crete and along the Levantine coast, and elsewhere in the Mediterranean, have provided the primary archaeological evidence for the extent of this industry [41b]. Pigments from these invertebrates were used in the Akrotiri frescos, on Santorini in Greece, which date back to the 17th century BC [42].

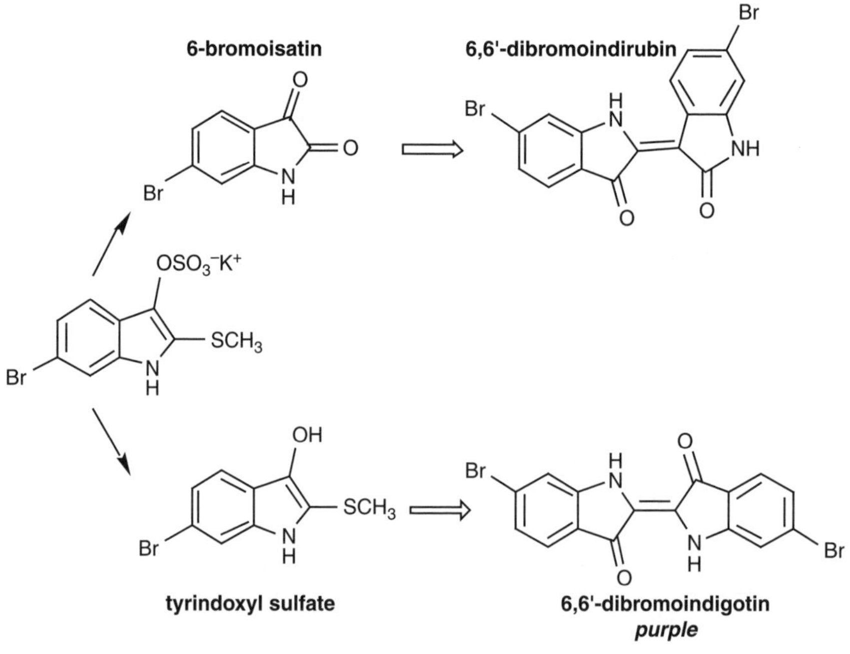

Figure 1.9 *Percursors and schematic pathways for the chromophores obtained from the muricids*

The percursors of Tyrian purple and derivatives such as monobromoindigo are stored in the hypobranchial glands of live muricids and can be released after death or when the mollusc is pressed and rubbed. The entire biosynthetic pathway for the production of dye precursors is yet to be elucidated, but the compounds generated in the pathway are biologically active and include choline esters, which may be employed as muscle relaxants and indole derivatives with cytotoxic (anticancer activities) and microbial properties [43]. Tyrian purple and its derivatives also occur naturally in the egg masses of the Muricidae.

In Figure 1.9 are depicted the coloured compounds that can be obtained as final products from the glandular secretions: true purple (6,6′-dibromoindigotin), the monobrominated indigo derivatives (6- and 6′-bromoindigotin), brominated indirubin derivatives (6,6′-dibromoindirubin, 6- and 6′-bromoindirubins), indirubin and indigo [44]. It is worth pointing out that indigo is blue, whereas its brominated derivatives are violet and indirubin is reddish. The shade of purple obtained depends on the species of mollusc and on the dyeing process used. McGovern and Michel report the chemical distribution of these insoluble dyes to be the following: *M. brandaris* and *P. haemastoma* produce mainly 6,6′-dibromoindigotin (purple), while *M. trunculus* gives also indigotin (blue), i.e. indigo in addition to the purple. The first steps of the synthesis are believed to involve an enzymatic process resulting in the production of tyrindoxyl sulfate (Figure 1.9), which is followed by chemical processes involving light and oxygen. Because the hypobranchial glands of the species that produces indigo, *M. trunculus*, contain a more efficient enzyme, *M. trunculus* was mixed with other species for purple vat dyeing in the Mediterranean. Hence the final colour obtained ranged between reddish purple and bluish violet [41b].

The story of purple was to be continued in the 19th century with the discovery of mauve by Perkin and the rebirth of the ancient *purple mania* as *mauve mania* [45]. The synthesis of mauveine in 1856 [46] – subsequently known as aniline purple (1857), Tyrian purple (1858) and mauve dye (from 1859) – is a story that demonstrates how a well-prepared mind can change history. The synthesis of the iconic mauveine is a major landmark in the history of science and technology, as it led to the establishment of the synthetic dye industry. In this case, the colour purple did not denote power or status, but opened a new era for chemistry.

1.5 Ancient Yellows

1.5.1 Flavonoid Yellows

Contrary to the situation for the reds, the sources for yellow dyes is enormous, as indicated in Table 1.3. Knowledge of many of these natural sources for yellow and how to dye with them has been lost, and therefore many of these species are not described or well documented in the literature. In addition, the yellow dyes are considerably less resistant to fading [8, 47, 48] than the reddish or bluish colours, and therefore it may be expected that the actual yellow colours observed today are different from the originals. Also, the reason why many old tapestries look blue (and red) is that all the yellows have faded, leaving only indigo blue (and some anthraquinone reds). The yellow chromophores based on the flavone chromophore can be divided in two large groups, those with and those without a hydroxyl group in the 3 position (Figure 1.10 and Table 1.3) [8, 49].

To the first group belong important chromophores such as apigenin and luteolin. These appear in the plant as glycosides, i.e. with sugar residues attached to one or more of the hydroxyl positions shown in Figure 1.10. Luteolin and its sugar derivatives are considered to be among the most stable yellows, and thus were widely used for dyeing. Perhaps the most used plant source for luteolin derivatives was weld, *Reseda luteola* (*lírio-dos-tintureiros* in Portuguese, *arzica* in Italian and *gaude* in French) (Table 1.3) [7a].

Table 1.3 *Chemical structures for flavone and flavonol yellows*

Flavones and flavonol yellows	
Weld *Reseda luteola*	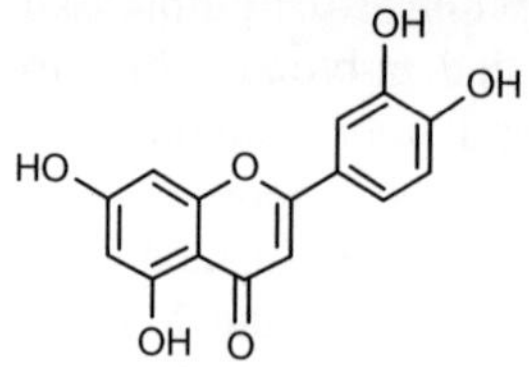 Luteolin 5, 7, 3', 4'-OH-flavone
Persian berries *Rhamnus* spp.	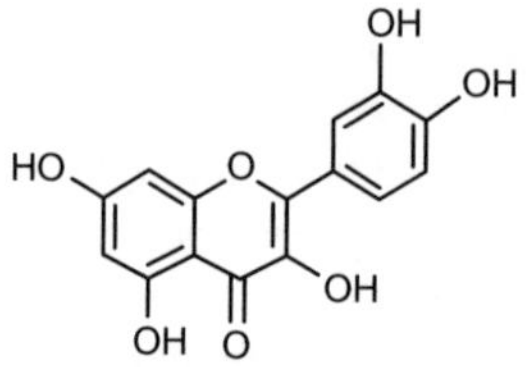 Quercitin 3, 5, 7, 3', 4'-OH-flavone
Old fustic *Chlorophora tinctoria*	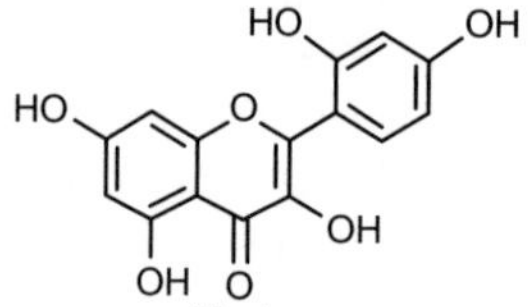 Morin 3, 5, 7, 2', 4'-OH-flavone

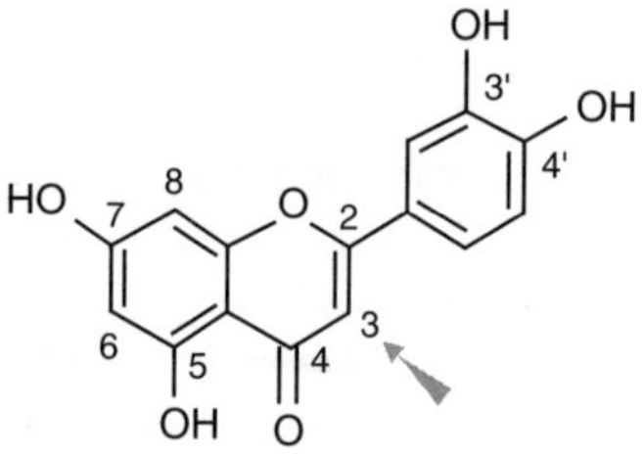

Figure 1.10 *Luteolin (flavone)*

The second group, the flavonols (3-hydroxy flavones), contains flavonoids, widely known for their antioxidant properties, e.g. quercetin, kaemferol and rhamnetin, and also morin, myrecetin and fisetin, just to name a few. These yellows are found in a variety of plants, including Persian berries (*Rhamnus* spp.), young fustic (*Cotinus coggygria*), old fustic (*Chlorophora tinctoria*) and yellow wood (*Solidago virgaurea*). Other sources are onion skins (*Allium cepa*), sawwort (*Serratula tinctoria*), dyer's greenweed (*Genista tinctoria*), marigold (*Chrysanthemum* spp.), dyer's chamomile (*Anthemis tinctoria*), flax-leaved daphne (*Daphne gnidium*), etc. [5].

1.5.2 Carotenoid Yellows

Carotenoids are present in our everyday diet, in the red of a tomato, in the orange of a carrot or, less frequently, in the golden yellow of 'Riso alla Milanese'. The traditional yellow used in this Italian receipe for rice is obtained from the stigmas of saffron, *Crocus sativus*, a beautiful flower, 10 cm high, and has been used for dyeing since antiquity. The main chromophore is the aglycone, crocetin, which can be found in the form of the glycoside, crocin (Figure 1.11) or with other substituted glycoside patterns. Other carotenoid yellows were used in the past, but the most famous was the golden saffron yellow.

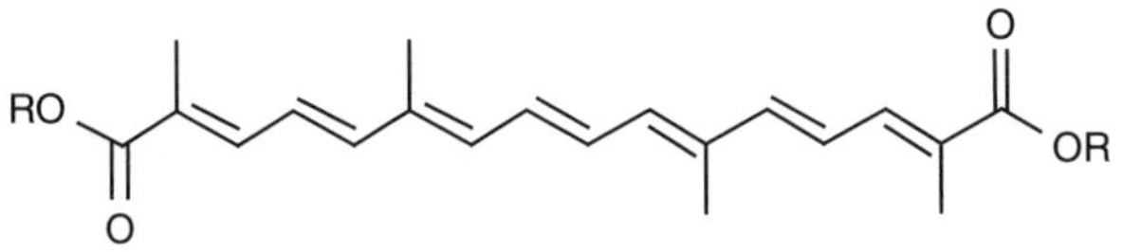

Figure 1.11 *Crocetin, R=H, and crocin, R=β-D-gentiobiosyl (a diglycoside), are all-trans carotenoids found in saffron*

1.5.3 Chalcone and Aurone Yellows

Like saffron, these chromophores absorb light at longer wavelengths and have more golden/orange hues. Chalcone yellows have been recently identified in ancient Andean textiles, as marein and other glycosides of okanin (2′,3′,4′,3′,4-pentahydroxychalcone), possibly obtained from a *Coreopsis* species (Figure 1.12) [50, 51].

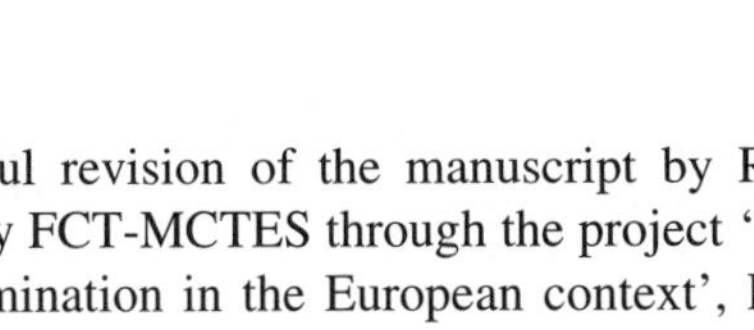

Figure 1.12 *Marein, okanin 4′-O-glucoside (chalcone)*

Acknowledgement

The patient and thoughtful revision of the manuscript by Richard Laursen is kindly acknowledged. Funding by FCT-MCTES through the project 'The identity of Portuguese medieval manuscript illumination in the European context', PTDC/EAT/65445/2006 is acknowledged.

References

1. F. Braudel, *The Mediterranean in the Ancient World*, Penguin Books, 2002.
2. (a) R. A. Donkin, Spanish red. An ethnogeographical study of cochineal and the Opuntia cactus, *Transactions of the American Philosophical Society*, New Series, **67**(5), 1–84 (1977). (b) R. A. Donkin, The insect dyes of western and west-central Asia, *Anthropos*, **72**, 647–680 (1977).
3. P. Ball, *Bright Earth: Art and the Invention of Color*, Farrar, Straus and Giroux, New York, 2002.

4. F. Brunello, *L'Arte della tintura nella Storia dell'Aumanità*, Neri Pozza, Vicenza, 1968.
5. (a) D. Cardon, *Le Monde des Teintures Naturelles*, Éditions Belin, Paris, 2003. (b) D. Cardon, *Natural Dyes. Sources, Tradition, Technology and Science*, Archetype Publications, 2007.
6. R. Halleux (ed.), *Les Alchimistes Grecs: Papyrus de Leyde, Papyrus de Stockholm, Recettes*, Les Belles Lettres, Paris, 2002.
7. (a) J. Kirby, Paints, pigments, dyes, in T. Glick, S. J. Livesey and F. Wallis (eds), *Medieval Science, Technology, and Medicine – an Encyclopedia*, Routledge, 2005. (b) A. Burnstock, I. Lanfear, K. J. van der Berg, L. Carlyle, M. Clarke, E. Hendriks and J. Kirby, Comparison of the fading and surface deterioration of red lake pigments in six paintings by Vincent van Gogh with artificially aged paint reconstructions, in Preprints of the 14th Triennial Meeting of the ICOM Committee for Conservation, Vol. 1, 2005, pp. 459–466.
8. E. Ferreira, A. Hulme, H. McNab and A. Quye, The natural constituents of historical textile dyes, *Chem. Soc. Rev.*, **33**, 329–336 (2004).
9. (a) J. Wouters and A. Verhecken, The coccid insect dyes: HPLC and computerized diode-array analysis of dyed yarns, *Studies in Conservation*, **34**, 189–200 (1989). (b) J. Wouters and A. Verhecken, The scale insect dyes (*Homoptera*: Coccoidea). Species recognition by HPLC and dyode-array analysis of the dyestuffs, *Annales de la Société Entomologique de France*, **25**(4), 393–410 (1989).
10. X. Zhang and R. A. Laursen, Development of mild extraction methods for the analysis of natural dyes in textiles of historical interest using LC-diode array detector-MS, *Anal. Chem.*, **77**(7), 2022–2025 (2005).
11. P. Guinot and C. Andary, Molecules involved in the dyeing process with flavonoids, in 25th Meeting on *Dyes in History and Archaeology*, Suceava, Romania, 21–22 September 2006 (forthcoming in *Archetype*).
12. D. A. Peggie, *The Development and Application of Analytical Methods for the Identification of Dyes on Historical Textiles*, PhD Dissertation, The University of Edinburgh, 2006.
13. M. J. Melo, A. Aguiar-Ricardo and P. Cruz, A green approach to antique textile cleaning, in Preprints of the 14th Triennial Meeting of the ICOM Committee for Conservation, Vol. 2, 2005, pp. 944–954, James & James, 2005.
14. R. Chenciner, *Madder Red. A History of Luxury and Trade*, Curzon Press, 2000.
15. H. Schweppe and H. Roosen-Runge, Carmine-cochineal carmine and kermes carmine, in Robert L. Feller (ed.), *Artists' Pigments – A Handbook of Their History and Characteristics*, Vol. 1, Cambridge University Press, Cambridge, 1986.
16. (a) Cennino Cennini, in F. Brunello (ed.), *Il Libbro dell'Arte,* Neri Pozza Editore, Vicenza, 1997. (b) Clothlets, translated from the Italian *pezzuole*; small pieces of cloth imbibed with the dye, that would be applied as a watercolour.
17. P. Roger, I. Villela-Petit and S. Vandroy, Les laques de brésil dans l'enluminure médiévale: reconstituition à partir de recettes anciennes, *Stud. Conserv.*, **48**, 155–170 (2003).
18. D. Cardon with C. Higgitt, Appendix: Chemical structures of the dyestuff groups, in Reference [5b].
19. M. J. Melo, Missal blue: anthocyanins in nature and art, in 21st Meeting on *Dyes in History and Archaeology*, Avignon, France, 10–12 October 2002 (forthcoming in *Archetype*).
20. (a) V. and R. Borradaile, *The Strasburg Manuscript – A Medieval Painters' Handbook*, Transatlantic Arts, New York, 1966. (b) S. Blondheim, An old Portuguese work on manuscript illumination, *JQR*, **XIX**, 97–135 (1928).
21. F. Brunello, *De Arte Illuminandi e altri Trattati sulla Miniatura Medievale*, Neri Pozza Editore, Vicenza, 1992.
22. A. Roquero, *Tintes y Tintoreros de America*, Ministerio de Cultura, Madrid, 2006.
23. M. J. Melo, M. Sousa, A. J. Parola, J. S. Seixas de Melo, F. Catarino, J. Marçalo and F. Pina, Identification of 7,4′-dihydroxy-5-methoxyflavylium in 'Dragon's blood'. To be or not to be an anthocyanin, *Chem. Eur. J.*, **13**, 1417–1422 (2007).
24. (a) H. Brockmann and H. Junge, Die Konstitution des Dracorhodins, eines neuen Farbstoffes aus dem 'Drachenblut', *Ber. Dtsch. Chem. Ges. B*, **76**, 751–763 (1943) (b) forenames>H. Brockmann and R. Haase, Über Dracorubin, den roten Farbstoff des 'Drachenblutes', *Ber.*, **69**, 1950–1954 (1936).

25. M. J. Melo, M. Moncada and F. Pina, On the red colour of raspberry (*Rubus idaeus*), *Tetrahedron Lett.*, **41**, 1987–1991 (2000).
26. (a) J. R. Brouillard and J. E. Dubois, Mechanism of structural transformations of anthocyanins in acidic media, *J. Am. Chem. Soc.*, **99**, 1359 (1977). (b) R. A. McClelland and S. Gedge, Hydration of the flavylium ion, *J. Am. Chem. Soc.*, **102**, 5838–5848 (1980).
27. (a) F. Pina, Thermodynamics and kinetics of flavylium salts – malvin revisited, *Chem. Soc., Faraday Trans.*, **94**, 2109–2116 (1998). (b) F. Pina, M. Maestri and V. Balzani, in H. S. Nalwa (ed.), *Handbook of Photochemistry and Photobiology*, Vol. 3, *Supramolecular Photochemistry*, American Scientific Publishers, 2003.
28. J. Balfour-Paul, *Indigo*, British Museum Press, 2000.
29. O. R. Constable, *Trade and Traders in Muslim Spain*, Cambridge University Press, Cambridge, 2003.
30. M. Clarke, Anglo-Saxon manuscript pigments, *Stud. Conserv.*, **49**, 231–244 (2004).
31. (a) J. Seixas de Melo, A. P. Moura and M. J. Melo, Photophysical and spectroscopic studies of indigo derivatives in their keto and leuco forms, *J. Phys. Chem. A*, **108**, 6975–6981 (2004). (b) Micaela M. Sousa, Catarina Miguel, Isa Rodrigues, A. Jorge Parola, Fernando Pina, J. Sérgio Seixas de Melo and Maria J. Melo, A photochemical study on the blue dye indigo: from solution to ancient Andean textiles, *Phtochem. Photobiol. Sci.*, **7**, 1353–1359 (2008).
32. M. van Eikema Hommes, *Discoloration in Renaissance and Baroque Oil Paintings*, Archetype Publications, 2004.
33. (a) A. Berry, T. C. Dodge, M. Pepsin and W. Weyler, Application of metabolic engineering to improve both the production and use of biotech indigo, *J. Ind. Microbiol. Biotech.*, **28**, 127–133 (2002). (b) A. N. Padden, V. M. Dillon, P. John, J. Edmonds, M. D. Collins and N. Alvarez, Clostridium *used in mediaeval dyeing,* Nature, **396**, 225–225 (1998).
34. (a) R. J. H. Clark, C. J. Cooksey, M. A. M. Daniels and R. Withnall, Indigo-red, white and blue, *Educ. Chem.*, 16–19 (1996). (b) C. J. Cooksey, Tyrian purple: 6, 6′-dibromoindigo and related compunds, *Molecules*, **6**, 736 (2001).
35. (a) M. Klessinger, Captodative substituent effects and the chromophoric system of indigo, *Angew. Chem. Int. Ed. Engl.*, **19**, 908–909 (1980). (b) E. Wille and W. Lüttke, 4,4,4′,4′-Tetramethyl-δ2,2′-bipyrrolidine-3,3′-dione, a compound having the basic chromophore system of indigo, *Angew. Chem. Int. Ed. Engl.*, **10**, 803–804 (1971). (c) G. Pfeifer, W. Otting and H. Bauer, Octahydroindigo, Angew. Chem. Int. Ed. Engl., 15, 52 (1976). (d) P. W. Sadler, Absorption spectra of indigoid dyes, *J. Org. Chem.*, **21**, 316–319 (1956). (e) G. M. Wyman, Reminiscences of an accidental photochemist, *EPA Newsletter*, **50**, 9–13 (1994).
36. L. Meijer, N. Guyard, L. A. Skaltsounis and G. Eisenbrand (eds), *Indirubin, the Red Shade of Indigo*, Life in Progress, Roscoff, France, 2006.
37. (a) T. Kondo, M. Ueda, M. Isobe and T. Goto, A new molecular mechanism of blue color development with protocyanin, a supramolecular pigment from cornflower, Centaurea cyanus, Tetrahedron Lett., **39**, 8307–8310 (1998). (b) T. Kondo, K. Yoshida, A. Nakagawa, T. Kawai, H. Tamura and T. Goto, Structural basis of a blue-colour development in flower petals from Commelina communis, Nature, **358**, 515–518 (1992). (c) T. Goto and T. Kondo, Structure and molecular stacking of anthocyanins – flower color variation, *Angewandte Chemie Int. Ed. Engl.*, **30**(1), 17–33 (1991).
38. Guglielmo Cavallo, La porpora tra scienze e culture. Una introduzione, in Oddone Longo (ed.), *La Porpora, Realtà e Immaginario di un Colore Simbolico*, Istituto Veneto di Scienze, Lettere ed Arti, Venezia, 1998.
39. R. Haubrichs, Natural history and iconography of purple shells, in Reference [36].
40. D. Cardon (ed.), *Teintures Précieuses de la Méditerranée: Poupre, Kermès, Pastel*, Musée des Beaux-Arts de Carcassonne – Centre de Documentació I Museu Tèxtil de Terrassa, Carcassonne, 1999.
41. (a) P. E. McGovern and R. H. Michel, Royal purple dye: the chemical reconstruction of the ancient Mediterranean industry, *Acc. Chem. Res.*, **23**, 152–158 (1990). (b) P. E. McGovern and R. H. Michel, Royal purple and the pre-Phoenician dye industry of Lebanon, *Masca Journal*, **3**, 67–70 (1984).
42. S. Sotiropolou and I. Karapanagiotis, Conchlian purple investigation in prehistoric wall paintings of the Aegean area, in Reference [36].

43. C. B. Westley, K. L. Vine and K. Benkendorff, A proposed functional role for indole derivatives in reproduction and defense Muricidae (Neogastropoda: Mollusca), in Reference [36].
44. (a) Z. C. Koren, High-performance liquid-chromatographic analysis of an ancient tyrian-purple dyeing vat from Israel, *Is. J. Chem.*, **35**, 117–124 (1995). (b) Zvi C. Koren, Historico-chemical analysis of plant dyestuffs used in textiles from ancient Israel, inM. V. Orna (ed.), *Archaeological Chemistry: Organic, Inorganic and Biochemistry Analysis*, ACS Symposium Series 625, 1996.
45. (a) A. S. Travis, *The Rainbow Makers – The Origins of the Synthetic Dyestuffs Industry in Western Europe*, Lehigh University Press, Bethlehem, Pennsylvania, 1993. (b) S. Garfield, *MAUVE. How One Man Invented a Color That Changed the World*, W. W. Norton & Company, 2002.
46. J. Seixas de Melo, S. Takato, M. Sousa, M. J. Melo and A. J. Parola, Revisiting Perkin's dye(s): the spectroscopy and photophysics of two new mauveine compounds (B2 and C), *Chem. Commun.*, 2624–2626 (2007).
47. (a) M. Heitor, M. Sousa, M. J. Melo, J. Hallett and M. C. Oliveira, The colours of the carpets, in T. Pacheco Pereira and J. Hallett (eds), *The Oriental Carpet in Portugal. Carpets and Paintings, 15th–18th Centuries*, Museu Nacional de Arte Antiga, 2007. (b) J. S. de Melo, M. J. Melo and A. Claro, As moléculas da cor na arte e na natureza, *Química-Boletim da Sociedade Portuguesa de Química*, **101**, 44–55 (2006).
48. H. Böhmer, *Koekboya, Natural Dyes and Textiles: a Colour Journey from Turkey to India and Beyond*, Remhob-Verlag, Ganderkesee, 2002.
49. (a) E. Haslam, *Practical Polyphenols. From Structure to Molecular Recognition and Physiological Action*, Cambridge University Press, Cambridge, 1998. (b) J. B. Harbone, *The Flavonoids – Advances in Research since 1986*, Chapman & Hall, London, 1994.
50. X. Zhang, R. Boytner, J. Cabrera and R. Laursen, Identification of yellow dye types in pre-Columbian Andean textiles, *Anal. Chem.*, **79**, 1575–1582 (2007).
51. M. J. Melo, I. Rodrigues, A. Claro, M. Montague and R. Newman, The color of Andean textiles from the MFA-Boston collection, in 26th Meeting on *Dyes in History and Archaeology*, Vienna, Austria, 7–10 November 2007 (forthcoming in *Archetype*).

2

Colours in Civilizations of the World and Natural Colorants: History under Tension

Dominique Cardon

2.1 Introduction

Before the invention of the first artificial and synthetic dyes – the start of a new type of intellectual and technological adventure – all civilizations in the world had managed to live in colours: they dyed human and animal skins, hair, teeth, bones, all sorts of vegetable fibres and woods, their drinks and food, in the whole range of colours of the rainbow [1]. This, however, implied a constant quest of the human mind and energy to discover, simultaneously, the amazingly rich colouring resources contained in living organisms (plants mainly and a few invertebrate animals) and the appropriate ways to use them in order to obtain the desired colour and make it last on each selected substrate. It is under this constant tension between the generosity of Nature and the strictness of its laws that ancient dyers developed their art.

Museums all over the world offer magnificent illustrations of this art, since all the colours we admire there on any artefact dated earlier than the second half of the 19th century are necessarily due to natural colorants or to the natural colour of the material they were made of. In the textile industry, some natural dyes went on being used well into the 20th century and in some countries their use for traditional textiles has survived until our time. They were favoured by prestigious textile designers like Mariano Fortuny or Issey Miyake, and the current worldwide success of *bogolan* cloth from Mali – handwoven cotton textiles decorated by a dyeing technique associating tannin plants and iron-rich mud – offers a striking instance of their appeal to contemporary aesthetics.

Handbook of Natural Colorants Edited by Thomas Bechtold and Rita Mussak

This induces us, therefore, to reflect upon the complex relationship between natural and synthetic colorants, the historical reasons for the (temporary?) decline of the former and the triumph of the latter, the differences in their systems of production and in their compositions, and the implications of these differences from an aesthetic and socioeconomic point of view. This may best be done by reconsidering some key periods in the history of natural dyes.

2.2 The Triumph of Mauvein: Synthetic Fulfilment of the Antique Purplemania

If the invention of mauvein by William Henry Perkin in 1856 represents such a strong symbol of the beginning of a new era – the age of synthetic dyes – it is largely because of the capital of prestige accumulated by the colour purple since remote antiquity. In the ancient Mediterranean world, the most beautiful, fast and luxurious purple/violet/mauve colours were obtained by fishing, sacrificing and vat-processing millions of marine sea-shells belonging to only four different mollusc species of the Muricidae family: the spiny dye murex, *Bolinus brandaris* (Linnaeus, 1758), the banded dye murex, *Hexaplex trunculus* (Linnaeus, 1758), the red-mouthed rock-shell, *Stramonita haemastoma* (Linnaeus, 1766) and, to a lesser extent, the sting winkle or oyster drill, *Ocenebra erinaceus* (Linnaeus, 1758). Less than 1 mg of the colourless product that will eventually give the purple dye – a complex mixture of indigoid colorants – is present in the hypobranchial gland of each animal. About 100 kg of glands were used per vat, according to a recipe in Pliny the Elder's *Natural History*. Therefore, the precious dye-baths were used until completely exhausted, to give colours ranging from the dark, intense purple called *blatta* ('coagulated blood') to pale mauves, the 'conchylian colours' [2].

Given the exclusive status of these true purple dyes, it is no wonder that a huge industry gradually developed in the ancient world to imitate their different shades, using different ingredients and processes. The best of these imitation purples was already described centuries BC, in Babylonian and Egyptian texts: fast purple shades were obtained by dyeing wool in two baths, first in blue with woad, *Isatis tinctoria* L. (Cruciferae), or indigo, *Indigofera tinctoria* L. (Leguminosae), and then, after mordanting with alum, in a red bath composed either of madder (*Rubia* spp., Rubiaceae) or of dye-insects (*Kermes* or *Porphyrophora* spp., Homoptera, Coccoidea). Not only are such combinations of dyes most commonly identified as the purple dyes of Coptic textiles, but they survived among European dyers more than a thousand years after the production of true mollusc purples disappeared. Many samples of such 'purple' dyes on wool-cloth, combining woad/indigo and cochineal or madder, figure in dyers' recipe books of the 18th–19th centuries preserved in European archives [3].

According to Egyptian alchemical sources, other plants were also used to imitate purple: alkanet or dyer's bugloss, *Alkanna tinctoria* (L.) Tausch (Boraginaceae), the root of which contains a purple dye identical to that of the traditional Japanese dye of *murasaki*, *Lithospermum erythrorhizon* Siebold and Zuccarini, from the same botanical family [4]. As late as the first half of the 19th century, when the industry of printed cotton had taken such economic importance in Europe, a fashion for textile designs of small flowers on 'purple alkannet ground' triumphed in France for some years.

Since antiquity, orchil (also called lichen purple) was the other common way to imitate purple. It was already a semi-synthetic dye, produced by macerating Mediterranean sea-coast lichens (*Roccella* spp., Roccellaceae) in the ammonia of putrid urine, with the addition of oxygen, introduced by frequent steering of the fermenting mass. The rich purples of orchil, an easy-to-apply, direct dye, were still very popular in Europe in the 19th century, when various species of lichens, collected in mountainous regions of Europe as well as on the coasts of all continents (Canary Islands, continental Africa and Madagascar, India, and even Baja California on the west coast of Central America), arrived by thousands of tons each year to the orchil factories of France (in Clermont-Ferrand, Lyons and Paris) and the United Kingdom (mostly near Edinburgh, in Scotland) [5].

The same year as mauvein was invented, in 1856, a group of French chemists even managed to overcome the main technical drawback of orchil, its poor light-fastness, patenting an improved fabrication process for an orchil dye they – significantly – named 'purple': 'pourpre française'... Too late! The times were ripe for the dawn of synthetic dyes and the 'democratization', not only of purple but of all colours.

2.3 Blue: from Kingly Regional to Globally Democratic

All over the world, the best sources of fast blue dyes are indigo-giving plants, scattered in many different botanical families, present in many different kinds of natural environments [6]. For European civilizations, from prehistoric to medieval times, there was no choice: to obtain a solid blue dye, the only locally available indigo plant was woad, *Isatis tinctoria* L. As previously mentioned, a woad blue was used as ground to imitate purple, the most prestigious of antique colours. One of the most prestigious of medieval colours, *brunette*, a perfect, fast black, was also based on a blue ground, but darker than for purple, followed by a mordant bath of alum and a red dye (most commonly madder but sometime kermes). Even darker than the ground of *brunette*, the blackish blue called 'perse' was so costly in expensive dyestuff and skilled labour that it was only available to the mighty and wealthy elite. Each of the degrees of this stepped gradation of woad blues, followed by a second dye bath with a yellow dye plant, gave a corresponding, standardized shade of green, from 'gay green' and 'grass green' to 'emerald green'. As the vat was used until exhaustion without further addition of woad, the strong clear blues of the new vats were gradually getting paler and greyish; only then were the inexpensive clothes of the lower class dipped into it.

All this technical organization obviously rested upon a complex production chain. This favoured the development in medieval Europe of what can already be called an industrial form of agriculture, involving whole regions. Woad was cultivated on hundreds of hectares, the leaves had to be picked, brought to mills, crushed into a pulp, shaped into balls and put to dry in huge sheds. Then, a second fermentation of the crushed balls, done either locally or in the big textile centres, was required to concentrate the indigo further in the granulous dyestuff, couched woad, about 150 kg of which were put into every new vat. Thousands of workers were employed in this production on a massive scale and huge fortunes were built by a few businessmen able to develop closely knit international connections [7].

This highly developed economic system paved the way for a further degree of globalization, as soon as new routes across the oceans, first to Asia, then to America, were opened and colonial methods of exploitation allowed massive imports of cheaper exotic indigo. Although already imported into medieval Europe, this indigo extracted from the indigo shrubs, *Indigofera* spp., had so far only been used to dye vegetable fibres and silk. Now, it was also added in ever-increasing proportions into the huge woad-vats of the wool-cloth industry and thousands of tons were also used in the cotton-printing centres scattered all over Europe, until, at the very end of the 19th century, the discovery of an economically viable synthesis of indigo abruptly ruined the agricultural production of indigo plants worldwide [8].

This double revolution, the triumph of exotic, then of synthetic, indigo, truly brought about a democratization of saturated blue and green colours in fashion, at least in industrialized countries, that would have been inconceivable in the era of the medieval woad-vat.

2.4 Red and Yellow: from Micro to Macro Scales

As outstanding as mollusc purple in preciousness and symbolic power were the scarlet and crimson dyes produced by a few tiny insects from Europe and Asia, parasites of trees or herbs: dyer's kermes, *Kermes vermilio* (Planchon, 1864), lac insects (*Kerria* spp.) and several species of *Porphyrophora* including Polish and Armenian crimson-dyeing scale insects [9]. As in the case of purple, their prestige largely derived from the astronomic numbers of animals needed: 1 gram of kermes represents 60 to 80 dry females, while ancient recipes recommended using at least an equal weight of this dyestuff and of textile to be dyed! Dyers'madder, *Rubia tinctorum* L., largely cultivated in several regions of Asia, the Middle East and Europe, was therefore the most popular source of red tones and also served to obtain pinks, oranges, purples, greys, browns and the most precious black, *brunette*.

Yellow dyes, present in many herbs, are abundant all over the world in nearly all types of environments. In medieval to 19th century Europe, to supply the textile industry, some yellow dye plants were cultivated on a large scale, like weld, *Reseda luteola* L. (Resedaceae), others were massively collected in wild extenses of land, like dyer's broom, *Genista tinctoria* L. (Leguminosae), sawwort, *Serratula tinctoria* L. (Compositae), flax-leaved daphne, *Daphne gnidium* L. (Thymelaeaceae); gathering, drying and carrying such light but bulky dyestuffs represented a branch of economic activity that has been largely overlooked by historians until recently [10].

The Age of Great Voyages brought dramatic changes in the range of red and yellow dyes that suddenly became available to European dyers. Within less than a century, all the insect dyes formerly used were replaced by a cochineal, *Dactylopius coccus* (O. Costa, 1835) that the ancient Indian civilizations of Central and South America had succeeded in domesticating on a cactus species. The new dye, made very abundant by their careful breeding techniques, was also much richer in red colorants (at least 10 times more than kermes) and – last but not least – unbeatably cheaper as a result of the colonial system of exploitation established by the Spaniards in their American empire [11].

The biggest change of scale, however, came with the adoption of exotic dye-woods, arriving in boats loaded with trunks and logs, full of red or yellow colouring matter. From

Asia came sappanwood, *Caesalpinia sappan* L., sandalwood, *Pterocarpus santalinus* L., and narrawood, *Pterocarpus indicus* L.; from Africa, camwood, *Baphia nitida* Lodd., and barwood, *Pterocarpus soyauxii* Taubb.; from America came the red brazilwood, *Caesalpinia echinata* Lamarck, and brasiletto, *Haematoxylum brasiletto* Karsten – all these red dyewoods belonging to the same botanical family of Leguminosae. New yellow dyestuffs were also provided by American tree species and imported massively into Europe: old fustic or dyer's mulberry, *Maclura tinctoria* (L.) D. Don. (Moraceae), and quercitron or black oak, *Quercus velutina* Lam. (Fagaceae). With each of these new dye sources, not only new shades of reds and yellows became available but also new recipes and combinations of dyestuffs soon followed: three to four different dye-woods figured in some recipes of maroon, beige, grey and black dyes for wool-cloth and printed cotton, for instance.

For centuries, these imported dyestuffs kept increasing the possibilities of creation for dyers and of choice for customers. By the mid-19th century, however, as the big European dye industries still completely depended on natural dyestuffs, dye crops like indigo and madder were already occupying millions of hectares all over the world; the dye-wood forests, over-exploited, were fast disappearing. Industrial society was facing a crucial problem that may rise again in our time: to keep constantly diversifying and increasing their supply in raw colorant materials to answer an ever growing demand.

2.5 What Future for Natural Colorants in the Dawning Era of Renewable Resources?

Whether, and how, a renewed massive use of natural colorants could address the above-mentioned problems of supply, in the present global context, without adding further threats to biodiversity, is certainly the most crucial question raised in this volume. The dawn of synthetic dyes opportunely corresponded with a time when industrialized societies were enthusiastically exploring the pathways opened by the exploitation of fossil resources: first coal, then petrol. The most emblematic of these dyes, manmade out of black tar from black coal or black petrol, was aniline black. One dye-bath, composed of a definite amount of one and the same molecule, could now produce a colour that, in the past, had always required complex combinations of dyestuffs and mordants.

Cheap and easy to apply, synthetic dyes and pigments have produced a major cultural revolution that has irreversibly changed the whole world. People everywhere are now accustomed to take colours for granted, to be surrounded by them in most circumstances of their lives without necessarily paying conscious attention to them.

As fossil resources are becoming less and less abundant, it is one of the challenges of the 21st century to keep colours available and affordable for the huge human masses of today and the even bigger numbers expected in the future. The alternative will be: either to the obliged to resort to a white or noncoloured environment because of a sheer scarcity of colorants or, through intensified scientific research, keep our use of colours an aesthetic choice.

Colorants of the future will probably result from an optimized combination of all the different types of resources of our globe, including recycled ones. They will most likely include again some natural dyestuffs, already known or still to be discovered, necessarily produced and processed by improved techniques. Research into natural colorants, therefore, as presented in this volume, is a major issue of our epoch.

Acknowledgement

The brief analysis of the long history of uses of natural dyes presented above has been much inspired by the opportunity I was offered a few years ago to collaborate in the preparation of the beautiful exhibition *Fashion in Colors*, organized by The Kyoto Costume Institute at the Museum of Modern Art in Kyoto, Japan, in Spring 2004. My sincere thanks go to the Curator of KCI, Ms Akiko Fukai, and to her enthusiastic and efficient team of collaborators.

References

1. (a) D. Cardon, *Natural Dyes – Sources, Tradition, Technology and Science*, Archetype Publications, London, 2007. (b) D. Cardon, *Le Monde des Teintures Naturelles*, Belin, Paris, 2003.
2. (a) D. Cardon, as Reference [1a], Chapter 11, pp. 553–606 and extensive bibliography. (b) D. Cardon (ed.), *Teintures précieuses de la Méditerranée: Pourpre, Kermès, Pastel/Tintes Preciosos del Mediterráneo: Púrpura, Quermes, Pastel*, Musée des Beaux-Arts Carcassonne, France, and Centre de Documentació i Museu Tèxtil, Terrassa, Spain, 1999/2000.
3. D. Cardon, as Reference [1a], Chapter 4, pp. 114–115; Chapter 8, p. 361, 374.
4. D. Cardon, as Reference [1a], Chapter 3, pp. 60–70.
5. D. Cardon, as Reference [1a], Chapter 10, pp. 487–514.
6. D. Cardon, as Reference [1a], Chapter 8, pp. 335–408.
7. (a) D. Cardon, as Reference [1a]., pp. 367–377. (b) J. Balfour-Paul, *Indigo*, Archetype Publications, London, 2006.
8. J. Balfour-Paul, as Reference [7b], Chapter 3, pp. 41–87.
9. D. Cardon, as Reference [1a], Chapter 12, pp. 609–618, 635–666.
10. (a) D. Cardon, as Reference [1a], Chapter 6, pp. 169–186. (b) D. Cardon, Dye crops, in *The Oxford Encyclopedia of Economic History,* Vol. 2, Oxford University Press, Oxford, London, New York, 2003, pp. 116–118.
11. (a) A. Roquero, *Tintes y Tintoreros de América – Catálogo de Materias Primas y Registro Etnográfico de México, Centroamérica, Andes Centrales y Selva Amazónica*, Instituto del Patrimonio Histórico Español, Madrid, 2006, pp. 143–9. (b) D. Cardon, as Reference [1a], Chapter 12, pp. 619–635.

3

History of Natural Dyes in North Africa 'Egypt'

Harby Ezzeldeen Ahmed

3.1 Introduction

The Egyptians were conscious that they excelled in weaving for many inscriptions extol the garments of the gods and the bandages for the dead. The preparation of clothes was considered as a rule to be women's work, for truly the great goddesses Isis and Nephthys had spun, woven and bleached clothes for their brother and husband Osiris. Under the old Empire this work fell to the household salves and in later times to wives belonging to the great departments [1].

Fine linen remained the most used fabric throughout all the eras of ancient Egypt. Because it breathes, linen worked well in the warm Egyptian climate where insulation was needed. A second advantage was that it could be loosely woven into gauze that allowed air to circulate around the skin. Cotton did not arrive in Egypt until Roman times. Wool clothing existed, but since it was forbidden in temples and tombs and since most surviving clothing comes from burials, actual articles of wool material are rare.

Spinning, which straightened and strengthened the fibres, meant exactly what the word says. Egyptian spindles consisted of a wooden shaft, four inches or more in length, through which a whorl – a two-to-four inch disk – was attached nearer the top than the bottom. The fibre was tied to a groove at the top of the shaft (later to a hook); then the spindle was rolled vigorously down the spinner's thigh, which set it spinning, stretching the fibre by its weight while twisting it for strength. This produced thread of a consistent diameter. Generally, Egyptian thread was two-ply: the ends of two fibres spun once were attached to the spindle and twisted together. Single Egyptian threads were spun counterclockwise, the natural

Handbook of Natural Colorants Edited by Thomas Bechtold and Rita Mussak

rotation of flax, while added plies were spun in the opposite direction to prevent the unwinding of individual threads [2, 3].

After the thread had been produced, it could be woven. The original Egyptian looms, dating from prehistoric times, were horizontal and simply two wooden beams anchored to the floor by short pegs. Working such a loom required kneeling, and the length of the fabric that could be woven was limited by the weaver's reach. By the New Kingdom, vertical looms consisting of two beams anchored to the end of the tall, upright wooden frame had come into wide use. These looms could be worked while sitting or standing to produce a longer length of cloth with less discomfort for the weaver, and allowed such techniques as tapestry weaves [2, 4].

3.2 Natural Dyes in Pharaonic Textiles

Plain white linen was preferred during most Egyptian eras, but only a limited number of people could afford high-quality linen. Colouring the fabric was a common method used during all eras to disguise its cheapness and lower quality. Of course, coloured fabric required coloured threads [1].

The use of dyed threads or dyed cloth can tentatively be traced back to the First Dynasty via a brownish piece of linen found at Tarkhan. Greater confidence about dyed cloth can be assigned to the late Third or Early Fourth Dynasty, based on a red cloth fragment from the site of the medium. In general, starting with the New Kingdom (18th Dynasty – 1550 BC), cloth woven with coloured threads was used more frequently. The ancient Egyptian dyestuffs can be divided into two basic types: ocher and plant dyes. Ocher is an earth that consists of hydrated oxide of iron (rust) mixed into clay. With heat, yellow iron oxide can be transformed gradually into red iron oxide; thus ocher can be used to create yellow, yellow-brown and red colours. The dyeing of linen with iron oxide has a long tradition in Egypt, which may date back to the early dynastic period (as, for example, the First Dynasty Tarkhan textile mentioned above). Linen that was coloured red from iron oxide dyes was also found at various places later, including the workmen's village of Tell el-Amarna [5, 6].

Two Greek papyri, dating from about the 3rd or 4th century AD, which were found in Egypt, probably at Thebes, describe the process of dyeing and nature of the colours used during that period. These are papyrus X, now in Leyden, translated by Berthelot, and papyrus Holm, now in Stockholm, published by Lagercrantz. These two papyri, in so far as they deal with dyes and dyeing, have been made the subject of a special study by Pfister [7].

Two methods of using colour in pharaonic textiles were identified: firstly, fully dyed cloth, usually of a shade of red, and, secondly, fabrics that included dyed parts. Coloured bands and stripes in combinations of red and blue were found [8].

3.3 Dyeing Techniques

The use of the natural dyes in ancient Egypt, both individual and in combination, indicated that at least three different dyeing techniques were available:

1. Direct dyeing. This method applies direct dye which requires no mordents or metallic salts in order to fix the colour. An ambient temperature bath is used.
2. Substantive dyeing. Madder, safflower and henna require mordents in order to fix the dye. The dye bath is usually prepared before the yarn or cloth is immersed. Both yarn and cloth could be dyed using this technique.
3. Double dyeing or over-dyeing. This technique consists of more than one dyeing step. One well-known example for this method is producing colours such as purple. Yarn or a piece of cloth was dyed twice, first in the blue dye bath and then in the red [6, 8].

3.4 Dye Sources

Five principal dyes are mentioned, which have been identified as:

(a) archil (orchil), a purple colour derived from certain marine algae found on rocks in the Mediterranean Sea;
(b) alkanet, a red colour prepared from the root of *Alkanna tinctoria*;
(c) *Rubia tinctorum*, which generates red coloured products;
(d) woad (*Isatis tinctoria*), a blue colour obtained by a process of fermentation from the leaves;
(e) and finally indigo from the leaves of the *Indigofera* species.

Alkanna and *Rubia* are common in the Mediterranean region according to Muschler, both having been found growing in Egypt, while Oliver mentions *Alkanna tinctoria* as growing in North Africa and south–east Europe. The ancient Egyptian names for alkanet and madder have been identified with some degree of certainty by Loret. The different dyes may conveniently be considered in alphabetical order of colour [5, 6, 7].

The ancient Egyptian blue has always been called indigo, from *Indigofera tinctoria* imported from India a hundred years ago. Thomson and Herapath identified it on ancient Egyptian fabrics, though unfortunately the date of the material is not stated. Indigo, however, is produced from a great variety of plants. One of most famous example is *Isatis tinctoria*. The colouring of different indigo-delivering plants is not absolutely identical, basically because of the presence of different side components. Therefore a distinction between different sources might be possible. The dye does not occur already formed in the plant, but is obtained by artificial fermentation of the leaves, which contain a precursor that becomes converted into indigo.

Indigo was cultivated in Egypt during the last century, but this cultivation probably does not date back earlier than the Middle Ages. Maqrisi (14th century AD) states that indigo was cultivated in Egypt in his day. The locally made dye has now been replaced by imported artificial indigo. The former cultivated indigo is *Indigofera argentea*, a species that grows wild in Nubia, Kordofan, Sennar and Abyssinia, though sometimes it is said to have been the Indian indigo *Indigofera tinctoria* [6, 7].

3.4.1 Woad

Woad was certainly cultivated in the Fayum province of Egypt in early Christian times, which is from the 1st to the 4th century AD, and probably earlier. Therefore, it has been assumed that the blue coloured ancient Egyptian fabrics of these days may have been woad. Vitruvius (1st century BC) mentions the scarcity of indigo and the use of woad.

Pfister examined a large number of dyed woven woollen fabrics, chiefly from Arsinoe in Upper Egypt, ranging in date from the 3rd century to the 7th century AD, on which he identified the blue colour as woad, which, however, he calls indigo. Schunk also identified what he call indigo on fabric Gurob, dating from 400 to 500 AD. Winlock, writing of a blue dye of the late Eighteenth Dynasty, says that it was probably the juice of the sunt berry (*Acacia nilotica*), but the evidence is not given and the sunt has a pod with seeds and not berries [5, 7, 9].

3.4.2 Indigo

Indigo is one of the oldest dyes used by man for dyeing. Indigo is found in various dye plants in the form of indoxyl glucoside, an intermediate for the blue indigo dye. Indigo is obtained from indoxyl glucoside by fermenting the leaves of the indigo plant, thus producing a greenish-yellow liquid which turns red when air-oxidized [10].

3.4.3 Red

The red colour that could be identified on the Antinoe fabric was generally madder, but occasionally kermes. Although Pfister called it 'cochineal' or sometimes 'Persin cochineal', this can be questioned since cochineal originally came from Mexico and was not known in Egypt at the time. A 'full deep red' and a 'dull chestnut brown' were found by Schunk to be madder, and a red-brown colour on one of the fabrics from the tomb of Tutankhamun was identified as madder by Pfister . An orange-red colour on mummy wrappings of the Twenty-first Dynasty Pfister found to be originated by henna, probably mixed with a red colour obtained from the flowers of *Carthamus tinctorius*.

This plant grew abundantly in Ancient Egypt and is still plentiful, and from the flowers both a red and a yellow dye are obtained. Thomson thought that a red on some fabrics examined by him was due to safflower [7,11.12].

3.4.4 Yellow

Yellow is domiciled in the whole area reaching from central Asia to the Mediterranean region and contains mainly two dyes: the safflower yellow, which dissolves in cold water and the red dye carthamin [10].

Thomson suggested that the yellow dye of the ancient Egyptians was derived from the safflower but was unable to prove this. Though it has been established definitely by Hubner, who identified it on fabrics of the Twelfth Dynasty. Hubner also found that another and slightly different shade of yellow of the same date was iron buff [7, 13].

3.4.5 Black

Although on several of the dyed fabrics from the tomb of Thutmos IV (Eighteenth Dynasty) there is a colour that appears to be black, recent careful examination of these fabrics indicates that the original colour may have been dark brown. The nature of the colour was not determined, but it possibly may have been made by imposing red on blue, similarly to the suggestion of Schunck, who investigated similar colours and suggested a superposition of madder on indigo [7, 13].

3.4.6 Brown

Pfister suggested that perhaps the brown colour on some of the Antinoe fabrics may be catechu (cutch), which is prepared from the *Mimosa catechu* grown in India, but this seems most unlikely.

3.4.7 Green

In one instance Pfister found a green colour to be due to indigo (woad), together with a yellow colour, and a similar green was also examined by Schunk. Lucks found that a green colour on thin plaster on a stick from the tomb of Tutankhamun consisted of a mixture of blue (blue firt) and a yellow.

3.4.8 Purple

Pfister found that the purple colour on the Antinoe fabrics was madder on indigo (woad) and Schunk recorded a purple made by mixing red and blue threads [5, 6, 7, 13].

3.5 Dyeing in Coptic Textiles

One of the first problems to arise is the question what is a 'Coptic textile' and how can it be defined? This is an old problem, which is addressed in most of the catalogues of other textile collections. However, it is worthwhile summarizing some of the aspects. The word 'Copt' is probably an abbreviation from the Greek *Aigyptios*. In order to avoid the problem of definition some authors have decided to use various terms to describe these textiles, for instance Greco-Roman, Late Antique, Late Roman, Early Byzantine or more cumbersome combinations such as Late Roman and Early Byzantine textiles [8].

The term 'Coptic textile' is commonly used to describe a large number of textiles from Egyptian findings dating from the Late Roman period into Islamic time. Coptic textiles usually comprise whole garments, such as tunics, or parts of garments, such as tunic ornaments or sections of cloaks; fragmentary or whole hangings; and textiles used to cover cushions or otherwise in household decoration. The patterns of Coptic textiles are purely ornamental or representational, sometimes including only animals and plants and sometimes humans as well as the main focus of interest. In general, most of the patterns, ornamental and representational, have a background in late classical art and indicate that its themes remained a strong source of decorative inspiration long after other forms of worship – Christianity and even Islam had converted large numbers away from paganism in the first instance and Christianity in the second [14].

Identification of natural dyes on Coptic textiles can be helpful in their dating. For example, kermes *(Kermes vermilio)* started to be widely used in Egypt only in Islamic times, which means in the second half of the 7th century . Reversed-phase HPLC with diode-array UV–VIS spectrophotometric detection has been used for identification of natural dyes in extracts from wool and silk fibres from archeological textiles. The examined objects originate from the 4th to the 12th century in Egypt and belong to the collection of Early Christian Art at the National Museum in Warsaw. The main individual chemical components of natural dyes, anthraquinone, indigoid and flavonoid dyes, including alizarin, purpurin, luteolin, apigenin, carminic acid, ellagic acid, gallic acid, laccaic acids A and B and indigotin, were identified. In particular, yellow dye was found to derive from weld (*Reseda luteola*), red dye originated from madder and other shades were identified as henna, lac dye, indigo, Tyrian purple and cochineal. The presence of gallic and ellagic acids indicates the use of tannins as organic mordants or weighting components [12, 15, 16]. Note the different colours of the Coptic textiles shown in Figure 3.1.

Figure 3.1 *Textile object 2045 from the Coptic period in the Ismaillia Museum, Egypt (photo picked up by the author) (See Colour Plate 2)*

3.6 Wool Dyed Fabric with Natural Dye

Woollen garments have been found in graves of the early Christian period and the use of coloured wool at this date for the decoration of fabrics is fairly common. The few findings of wool of earlier dates that can be trace included, in chronological order, one predynastic example of 'brown and woollen knitted stuff', a woollen cloth wrapped around the remains of a male skeleton of the First Dynasty, a specimen found in the pyramid of Menkaure at Giza, recorded as 'part of a skeleton enveloped in coarse woollen cloth of yellow colour', which was almost certainly an intruded burial of a much later date than the pyramid, and a specimen from the Twelfth Dynasty found by Petrie, who says regarding it, 'Wool was also spun; a handful of weaver's waste is mainly made up of blue worsted ends and blue wool, with some red and some green ends.' Brunton found yellow wool from the second intermediate period, and a sack of woven goat's hair excavated at El Amarna contained a mass of goat's wool, five large balls of wool each consisting of several hundred matters of spun wool and a square garment of woven wool. Braulik records a woollen textile from Saqqura, dating from not earlier than 300 BC and there is a woollen turban of pre-Christian Roman date, with reference to which, Winlock, the finder, says: 'Apparently it was the style in Thebes just before the Christian era to swathe the hair in fine linen veils until the head was twice its natural size, and then over that to pull such a brown and red turban, tied behind with drawing-strings.' Other woollen fabrics were found by Brunton at Mostagedda dated to the early Roman, later Roman and Coptic periods [6, 7, 12].

3.7 Dyes in Islamic Textiles

The attention that different Islamic governments paid to the textile industry had a great effect on its growth during the Middle Ages. Private factories also helped spread the textile arts, which were used for political purposes by rulers. The rulers wore these fabrics and gave them as gifts to princes and friends. The rulers bestowed on a select few of their subjects robes, called Khola, or pieces of cloth. Considering the importance of such an industry, the government established public factories called 'Dar al-Tiraz' or textile factory. These factories had two sections, one to produce fabric for the masses and another to make garments for the elite. Just like state workshops, private workshops were also closely monitored by the government, which supervised the factories' production and supplied them with raw materials. The government also tested the products and gave them quality stamps, ensuring there was no cheating with the proportions of raw materials used. Islamic dyers professionally used natural dyes on a large scale for different fabrics [17].

The basic information on dyes and dyestuffs used in Ottoman times derives from lists of colorants and dyed fabrics in Ottoman inventories and other documents; from the actual colours of the surviving fabrics, which may often clarify the sense of otherwise obscure colour words; and from Muslim manuals of dyes and dyeing prepared by guild masters [18].

The main Ottoman source of blue colour was indigo from *Indigofera tinctoria*, which mainly came with the spices from India via Egypt or the Hijaz. It was also used with henna, double-dyed to give a strong and lustrous black, and was particularly in demand as a hair dye. Indigo was also important in dyeing crimson and, with lac, kermes or cochineal, could also give a range of deep purples.

The most common yellow dye was obtained from safflower (*Carthamus tinctorius*), which was grown in Egypt and Iran as a commercial crop and which gives colours ranging from orange to a vivid golden yellow. Equally fine, though much more expensive, was saffron, which was mostly used for silks. Other dyestuffs like turmeric, buckthorn (*Rhamnus infectorius*) and the dyer's oak (*Quercus infectoria* and *Q. lusitanica*) are mentioned as other dyestuffs in the documents.

Green dyes could be obtained from the berries of two other buckthorns (*Rhamnus chlorophorus* and *Rh. utilis*) under the name 'Persian berries'. Greens, however, were mostly the product of double dyeing, for most of the yellow plant dyes went well over an indigo base. By far the largest range of dyestuffs available in Ottoman, however, lay in the brown-red purple part of the spectrum. Madder was of immense importance in the dyeing of wool for carpets, where the bright red known as 'Turkey red' was particularly favoured. By far the most important source of red crimson (which was named according to the Turkish word *Kirmizi*) was a series of aphids and scale insects, the dried bodies of which gave a rich, expensive, glowing colour to velvets and satins called *grana*. The most widespread one was Kermes (*Kermococcus vermilio*, formerly *Cocus ilicis*) which infests the branches of the Kermes oak (*Quercus coccifera*), a species that grows all round the Mediterranean. In the Ottoman period the favourite crimson seems to have been lac crimson, imported in enormous quantities from India, where the lac insect infests species of *Ficus* and produces shellac. Cochineal dye was very probably being imported in small quantities by the late 16th century [18].

Textile objects date back to the Ottoman age, such as case 12014 found in the Islamic Art Museum in Cairo (see Figure 3.2), which measures 116 by 69.7 cm. It contains many decorations, such as plant decorations (flowers and leaves), written decorations inside the lamp hanging from the arch and geometric decorations (columns and arches). Different colours were used, such as red, blue, green and yellow. The author, Harby E. Ahmed, analysed the dyes, took small samples of the different colours and investigated them using Fourier transform infrared (FT-IR) spectroscopy. Then all the dyes that gave these colours were obtained and investigated. The original samples were then compared with the new dyes, giving the following findings: the red colour is cochineal dye, the yellow colour is safflower dye, the blue colour is indigo dye and the green colour is a mixture of indigo and tumeric dyes [19].

The author studied the effect of different ageing procedures, such as light, thermal and chemical ageing, on silk dyeings with the safflower and cochineal as the natural dyes mordanted with different mordants, e.g. $A_{12}(NH_4)_2(SO_4).24H_2$ (alum), FeC_{12}, $CuSO_4$ and SbC_{12} (tartaric acid). This study was carried out to establish the standard conditions of light and temperature at which the archaeological textiles can be maintained without any deterioration. The CIE-Lab values of the dyeings were measured with a double-beam Optimatch spectrophotometer (Macbeth-UK, with a sample diameter of 10 mm) [20, 21].

3.8 Mordants

Certain dye sources, such as indigo and murex purple, produce colourfast dyes all by themselves, although the processes of reduction by fermentation and later oxidation make the 'vat dye' procedure long, complicated and far from obvious. Most of the vegetable dyes, however, require some sort of mordant to set permanently in any fibre. A mordant is a separate chemical that combines with the dye in such a way as to attach the colouring matter to the fibre by increasing affinity and/or strengthened interactions in some cases via a lasting chemical bond (mordant means 'biting in'), thereby making the colour stand fast against light and washing [3].

Figure 3.2 *Textile Ottoman 12014 in the Islamic Art Museum, Cairo, Egypt (photo picked up by the author) (See Colour Plate 3)*

Some mordants will also change the hue of certain dyes (different mordants on the same dye may darken, brighten or drastically alter the colour). What we know for certain about ancient mordants and developers is even less than what we know about dyes. The chemicals found by Heebner in his analysis of the colourfast yellow in two of his Twelfth Dynasty mummy linens indicated that colour came from iron buff together with a calcium compound probably used as a developer. The largish amounts of aluminium and calcium in the coloured linens from Tutankhamen's tomb that have been analysed suggest the use of aluminium and calcium salts as mordants [3].

The principal one was almost certainly alum, which occurs in Egypt and was worked in ancient times. No certain instance of its use can be quoted, though it has been suggested that it was employed on fabrics in the Twelfth Dynasty, and there is textual evidence that may indicate the use of alum as a mordant with madder during the New Kingdom. According to the two papyri already mentioned, the mordants used in Egypt in early Christian times included alum, but also salts of iron, such as the acetate, specially prepared from iron and vinegar, and the sulfate, which occurs frequently as an impurity in alum [7].

Generally mordants that were used over time were divided into two types, acid and basic, where acid mordants are used to bond acid dyes and basic mordants to bond basic dyes.

Acid mordants have generally been derived from tannin, readily available from oak balls or bark; occasionally they are vegetable oils. Basic mordants, however, come from the salts of various metals, particularly aluminium, chromium, iron, copper, zinc or tin.

References

1. A. Erman, J. M. White and H. M. Tirard, *Life in Ancient Egypt*, Dover Publications Inc., New York, 1971.
2. B. Brief and H. Hobbs, *Daily Life of the Ancient Egyptians,* Greenwood Press, London, 1990.
3. E. J. W. Barber, *Prehistoric Textiles: The Development of Cloth in the Neolithic and Bronze Ages*, Princeton University Press, UK, 1991.
4. E. Riefstahl, *Patterned Textiles in Pharaonic Egypt,* The Brooklyn Museum and the Brooklyn Institute of Arts and Sciences, Brooklyn, New York, 1945.
5. D. B. Redford, *The Oxford Encyclopaedia of Ancient Egypt*, Vol. 3, The American University in Cairo Press, Cairo, 2001.
6. T. C. Golyon, Le lin et sa teinture en Egypte des procedes ancestraux aux pratiques importee. Aspects de l' artisanat du textile dans le monde Mediterraneen, *VII Siecle av. J.C.a l'Epoque Recente*, 13–25 (1996).
7. A. Lukas, *Ancient Egyptian Materials and Industries*, History and Mysteries of Man, Ltd, London, 1989.
8. P. M.Van T. Hooft, M. J. Raven, E. H. C. Van Rooij and G. M. Vogelsang-Eastwood, *Pharaonic and Early Medieval Egyptian Textiles,* RIJKS Museum van Oudheden, Leiden, The Netherlands, 1994.
9. G. V. Eastwood, *Pharaonic Egyptian Clothing, Studies in Textile and Costume History*, Vol. 2, Brill Academic Publishers, Leiden, The Netherlands,1993.
10. H. Schweppe, Practical hints on dying with natural dyes: production of comparative dyeings for the identification of dyes on historic textile materials, Sponsored by the Conservation Analytical Laboratory of the Smithsonian Institution, Washington DC, USA, 1986.
11. G. M. Crowfoot and N. D. E. G. Davies, The tunic of Tutankhamun, *The Journal of Egyptian Archaeology*, **27**, 113–130 (1941).
12. D. T. Jenkins, *The Cambridge History of Western Textiles,* Cambridge University Press, Cambridge, 2003.
13. B. J. Kemp, G. V. Eastwood, A. Boyce, H. G. Farbrother, G. Owen and P. Rose, *The Ancient Textile Industry at Amarna,* Egypt Exploration Society, London, 2001.
14. D. Thompson, *Coptic Textiles, The Brooklyn Museum, Brooklyn, New York, 1971.*
15. I. Surowiec, J. Orska-Gawryś, M. Biesaga, M. Trojanowicz, M. Hutta, R. Halko and K. Urbaniak-Walczak, Identification of natural dyestuff in archaeological Coptic textiles by HPLC with fluorescence detection, *Analytical Letters*, **36**(6), 1211–1229 (2003).
16. J. Orska-Gawrys, I. Surowiec, J. Kehl, H. K. Urbaniak-Walczak and M. Trojanowicz, Identification of natural dyes in archaeological Coptic textiles by liquid chromatography with diode array detection, *Journal of Chromatography A*, **989**, 239–248 (2003).
17. P. L. Baker, Islamic Textiles, British Museum Press, London, 1995.
18. J. M. Regers and S. Delibas, *The Topkapi Saray Museum: Costumes, Embroideries, and Other Textiles*, Little Brown and Company, Boston, 1986.
19. H. E. Ahmed and Y. Zidan, The conservation treatment of a silk textile in the Islamic Art Museum, Cairo. Recent preoccupations concerning textiles, leather, legislation, in ICOM-CC Working Groups, Athens, Greece, 21–24 April 2004.
20. H. E. Ahmed, Y. Zidan and K. Elnagar, Ageing behavior of silk dyed fabric with safflower dye, in 26th Meeting on *Dye in Historical Textiles and Archaeology*, Vienna, Austria, 7–10 November 2007.
21. H. E. Ahmed, Y. Zidan and K. Elnagar, Studies on dyeing with cochineal and ageing of silk dyed fabric, in *Scientific Analysis of Ancient and Historic Textiles: Informing Preservation, Display and Interpretation*, AHRC Research Center for Textile Conservation and Textile Studies, First Annual Conference, UK, 13–15 July 2004.

Part II

Regional Aspects of Availability of Plant Sources

4

Dye Plants in Europe

Andrea Biertümpfel and Günter Wurl

4.1 Introduction

Dyeing with plant dyestuffs has a long tradition in Europe. The people of the Stone Age used dyes of plant origin, e.g. for body painting. For textile dyeing plants were used systematically first in antique Greece and later in the Roman Empire. The zenith of plant dyeing was reached in the 19th century. At this time the dyes of more than 30 plant species were used, most of them imports (von Wiesner, 1927). Only madder (*Rubia tinctorum* L.), weld (*Reseda luteola* L.), woad (*Isatis tinctoria* L.) and ai (*Polygonum tinctorium* Ait.) were cultivated to a large extent in Europe. With the discovery of synthetic dyes at the end of the 19th century natural dyes and their cultivation disappeared nearly completely.

For some time the health hazards of many synthetic dyes have been known, so consumers and the industry have been looking for more ecologically friendly and toxicologically safe products. Therefore plant dyes have become interesting and European agriculture could have the possibility to cultivate dye plants again. In order to achieve an efficient production of the raw materials the development of modern cultivation methods for important European dye plants is necessary.

4.2 Potential European Dye Plants

In Europe more than 100 plant species containing dyestuffs exist (Roth *et al.*, 1992). These are trees, shrubs and herbs, as well as mushrooms and lichens. The plant components used for dyeing are also very different. It can be the whole plant (e.g. weld), the leaves (e.g. woad), the roots (e.g. madder), the flowers (e.g. dyer's chamomile), the fruits (e.g. common buckthorn), the bark (e.g. oaks), the semen shell (e.g. Persian nut) or the skin (e.g. onion) (Cardon and du Chatenet, 1990).

Handbook of Natural Colorants Edited by Thomas Bechtold and Rita Mussak

Because of the low fastness of the dye some of them, like carrots, elderberries or nettles, can be used as dyestuffs only for food, but not for textiles. Only a limited number of plant species exhibit the potential for large-scale production (Vetter, 1997; Vetter *et al.*, 1997a, 1997b; Wurl, 1997). The most promising dye plants for agricultural cultivation are listed in Table 4.1. For all these plants modern cultivation recommendations have been worked out with the financial support of national institutions and the Commission of the European Union (EU).

Table 4.1 *Promising dye plants for cultivation under European soil and climatic conditions*

Plant species – botanical name	*Plant species – common name*	*TKW (g)*[a]	*Dyeing part of the plant*	*Colour*[b]
Isatis tinctoria L.	Woad	2.0	Leaves	Blue
Polygonum tinctorium Ait.	Ai	3.0	Leaves	Blue
Rubia tinctorum L.	Madder	20.0	Roots	Red
Reseda luteola L.	Weld	0.2	Whole plant	Yellow
Anthemis tinctoria L.	Dyer's chamomile	0.4	Flowers	Yellow
Solidago canadensis L.	Canadian golden rod	0.03–0.06	Whole plant	Yellow
Serratula tinctoria L.	Sawwort	3.5	Whole plant	Yellow
Carthamus tinctorius L.	Safflower	30.0–60.0	Leaves/flowers	Yellow/red
Genista tinctoria L.	Dyer's greenweed	3.5	Whole plant	Yellow
Centaurea jacea L.	Rayed knapweed	1.0	Whole plant	Yellow
Tagetes erecta L.	Big marigold	3.0	Flowers	Yellow
Verbascum spp.	Mullein	0.2	Whole plant	Yellow
Origanum vulgare L.	Wild marjoram	0.1	Whole plant	Brown
Chelidonium majus L.	Greater celandine	0.15	Whole plant	Brown
Tanacetum vulgare L.	Common tansy	0.75	Whole plant	Brown
Achillea millefolium L.	Milfoil	0.15	Whole plant	Brown
Alchemilla vulgaris L.	Lady's mantle	0.3	Whole plant	Brown

[a] Weight of a thousand seeds.
[b] With alum as mordant (except blue).

4.3 Cultivation of Dye Plants Yesterday and Now

The cultivation of dye plants largely took place in the Middle Ages. It was combined with an enormous amount of manual work that cannot be applied in modern agriculture; e.g. the siliquated seeds of woad were sown per hand. Nowadays only husked seeds are drilled. Also the post-harvest process of the harvested material is different. To minimize the dye lost from the woad and ai the water-soluble indigo precursors must be extracted out of the leaves by water immediately after harvesting. In the past they were fermented on a large scale and then dried. The yellow and red dyeing plants must nowadays be dried by technical means to get a high quality of the material. The dye content of the harvested material is only slightly lowered due to quick drying at high temperature (40–60 °C). In the past air drying of the weld was the only way to dry the plants, causing dyestuff losses of nearly 50% (Figure 4.1).

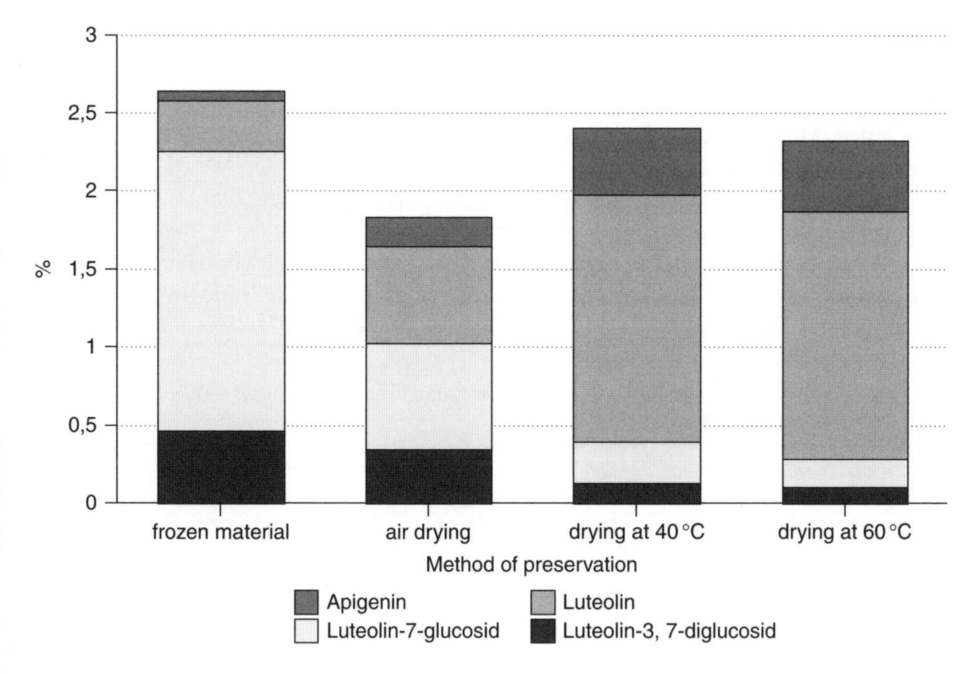

Figure 4.1 *Influence of the preservation method on the dye content and the dye compounds of weld (Rohrbach, 1996)*

Generally, all steps involved in the cultivation and processing of the material up to the dyed end-product must be mechanizable and practicable using available technical equipment in order to keep costs for natural dyes as low as possible (Biertümpfel *et al.*, 2000).

4.4 Modern Cultivation Methods for Important European Dye Plants

4.4.1 General Facts

Prerequisites to the successful cultivation of dyes are:

- deep ploughing in autumn;
- moisture-saving seed bed preparation in spring;
- fine crumbly and stabilized seed bed on the surface;
- shallow drilling (1–3 cm).

A general problem with the production of minor crops like dye plants is efficient weed control. Registered herbicides for such plants do not exist. In Germany, for example, the application of a herbicide, a pesticide or a fungicide is only permitted for crops for which the remedies are registered. For dye plants no permissions have been granted up to now. The farmer has to pay attention to these facts if he intends to cultivate dye crops.

Short cultivation recommendations of the most important European dye plants will now be given.

4.4.2 Blue Dyeing Plants

The number of blue dyeing plant species is very limited worldwide. Only two species are suitable for cultivation in Europe: woad (*Isatis tinctoria* L.) and ai (*Polygonum tinctorium* Ait.). Woad as well as ai needs high nitrogen fertilization for high yields of biomass and high dyestuff contents. An example for ai is given in Table 4.2. Table 4.3 contains the agricultural recommendations for the two blue dyeing plants.

Table 4.2 *Influence of different amounts of nitrogen fertilization on dry mass yield and indigo yield of ai in Mitscherlich pots (four replications/variant)*

N fertilization (g N/pot)	*Dry mass yield (g/pot)*	*Indigo content (% in the dry mass)*	*Indigo yield (g/pot)*
0.5	135.6	1.29	1.77
1.0	130.1	1.35	1.75
1.5	129.0	1.90	2.48

Table 4.3 *Cultivation instructions for woad and ai*

	Woad (Isatis tinctoria L.)	*Ai (Polygonum tinctorium* Ait.)
Botanical family	Cruciferae	Polygonaceae
Origin	East Mediterranean area, West Asia, Caucasus	Japan, where nowadays it is cultivated for the dyeing of traditional clothes
Duration of vegetation	Biannual, winter annual, develops in the first year a leaf rosette, in the second year 1–1.80 m high stems with thousands of yellow flowers	Annual
Demands on climatic conditions	Fully winterhard, continuous water supply during vegetation and high temperatures are positive for yield and dye content	Frost susceptible, continuous water supply during vegetation, high temperatures are positive for yield and dye content
Demands on the soil	Prefers deep grounded, humous soils	Prefers deep grounded, humous soils
Crop rotation	No demands on the preceding crops, summer cereals as the following crop	No demands on the preceding crops, all crops can follow
Sowing	Drilling with usual drill for fine seeds	Drilling with usual drill for fine seeds
Sowing time	In autumn: not earlier than the end of October In spring: as soon as possible	Middle of April–beginning of May
Sowing strength	4–5 kg of pure seeds per hectare	5 kg per hectare
Sowing depth	1–2 cm	2–3 cm

Table 4.3 *(Continued)*

	Woad (*Isatis tinctoria* L.)	Ai (*Polygonum tinctorium* Ait.)
Germination duration	2–3 weeks	2–3 weeks
Row distance	15–30 cm	20–30 cm
Fertilization	N 150–200 kg/ha Recommendation: 120 kg available N/ha[a] at the beginning of the vegetation, 50 kg mineral N/ha after each cut, except the last cut P, K and Mg after the withdrawal: P 50–60 kg/ha, K 250–300 kg/ha, Mg 15–20 kg/ha	N 150–200 kg/ha Recommendation: 120 kg available N/ha[a] at the beginning of vegetation, 40 kg mineral N/ha after the first cut P, K, Mg: low demand, fertilization in the frame of crop rotation is sufficient
Weed control	Hoeing with a machine is recommended; for the situation of herbicide application see above	
Harvest	2–4 cuts, beginning at the end of June in intervals of 5–7 weeks to the following cuts; the harvest is done with deep cutting machines like a green fodder harvester	Two cuts, end of July and end of September; the harvest is done with deep cutting machines like a green fodder harvester
Yields	20–30 to fresh mass/ha (=3–4 to dry mass/ha)[b]	20–30 to fresh mass/ha (=2.5–3.5 to dry leaves/ha[c])
Post-harvest treatment	For dye and also fermented juice[d] production the fresh leaves must be processed immediately after the harvest	The water-soluble dye precursor must be extracted immediately after the harvest. The dye content in ai leaves is about fivefold higher than in woad (Figure 4.2)
Seed production	Harvest of the siliquas with a usual reap thresher about 6 weeks after flowering of the plant. The siliquas must be dehulled by a clover huller	Ai is a strict short-day plant and does not begin to flower before the middle of August under Central European conditions. Seed production should be done in southern countries
Seed yield	0.2–1.0 ton/ha pure seeds	1.5 ton/ha pure seeds in a good year

[a] Soluble N in the soil in 0–60 cm depth + mineral N.
[b] This is the yield that is reached under Middle European conditions. In Italy, for example, with a warmer climate, the yield can be 9–9.5 ton/ha (Voltolina and Valeriani, 1996).
[c] Only the leaves contain the dye precursor indican.
[d] Fermented woad juice is used for the conservation of wood, stones and paper and also for fire prevention.

4.4.3 Red Dyeing Plants

The single European cultivatable plant containing red dye (in the roots) is madder (*Rubia tinctorum* L.). The dye content of *Lithospermum erythrorhizon* Sieb. and Zucc. is too low and *Galium* species do not develop enough biomass. The red dye won from the safflower flowers has too low a fastness to light and its production is too expensive. Old cultivation instructions are correct in recommending that the madder roots should be uprooted only after a three-year cultivation period. The dye content does not change during this time, but

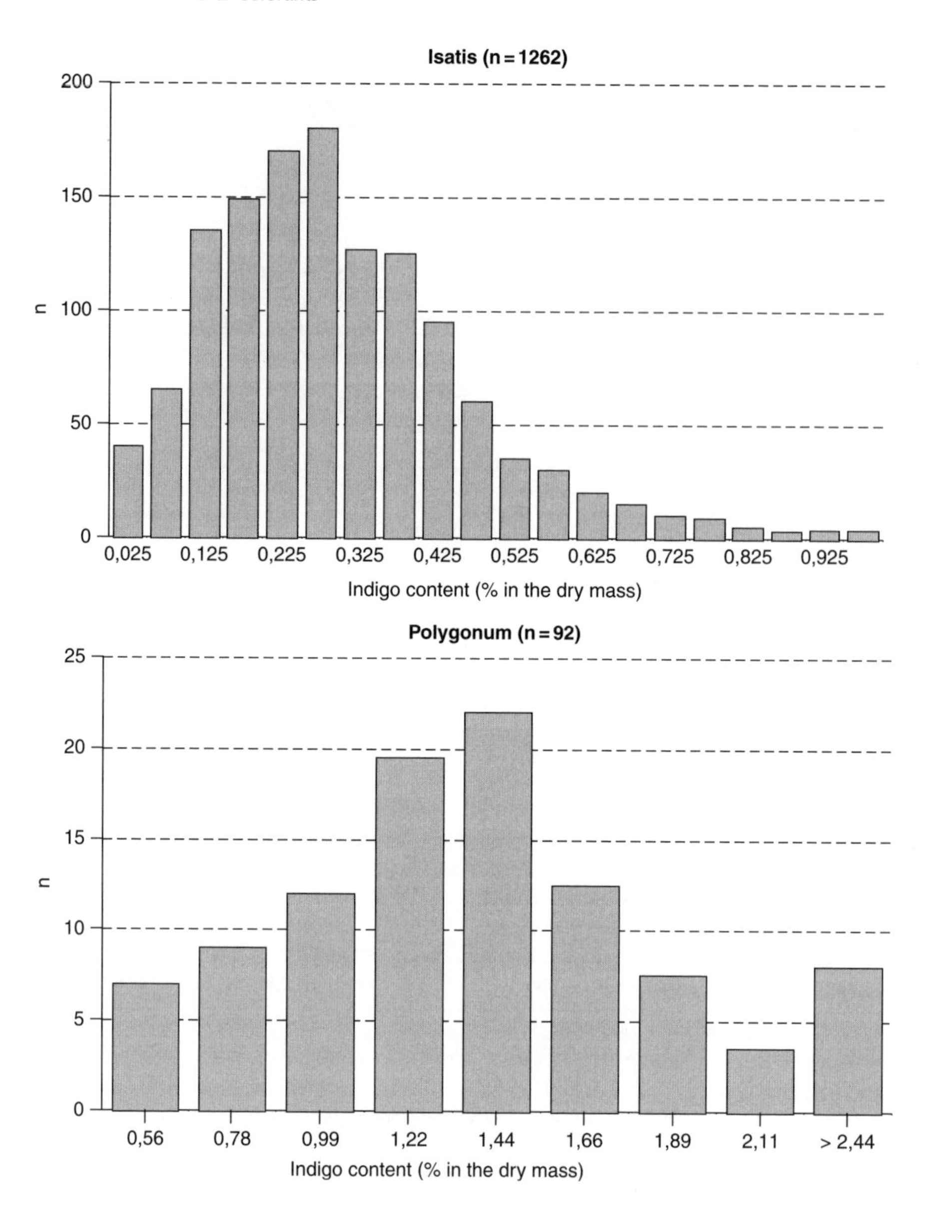

Figure 4.2 *Variability of the indigo content in Isatis and Polygonum*

the dry mass yield increases continually from year to year (Figure 4.3). Therefore, harvesting after three years is useful. Table 4.4 contains cultivation instructions for madder.

4.4.4 Yellow Dyeing Plants

Most European dye plants contain yellow dyestuffs. The dyeing compounds are flavones and flavonoles. The amount of nitrogen fertilization has a significant

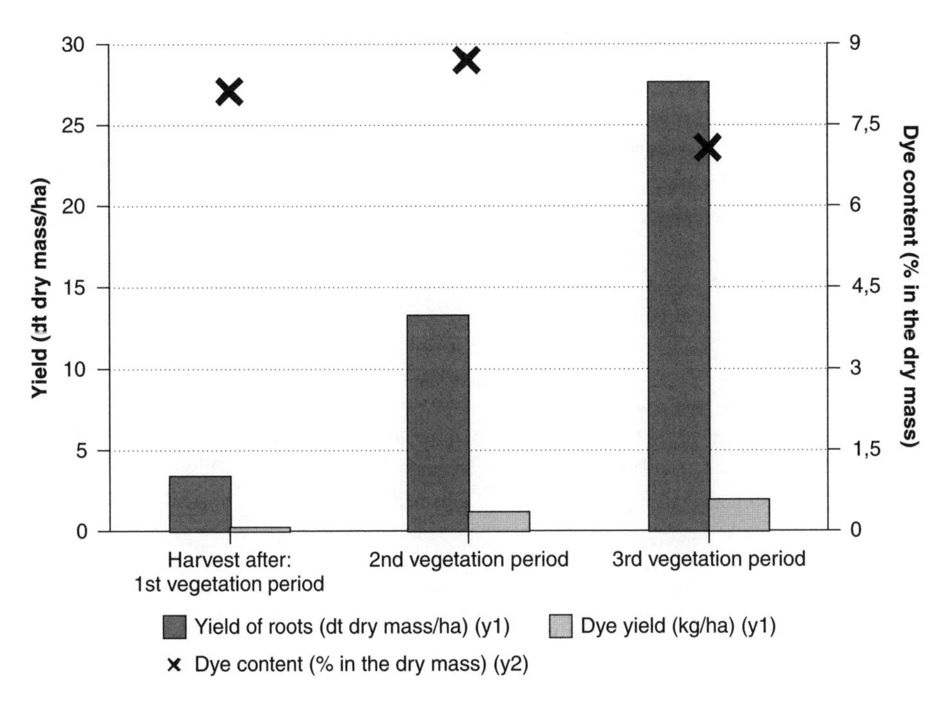

Figure 4.3 *Influence of the harvest time on the yields of roots and dye and the dye content of madder (Rohrbach, 1997)*

Table 4.4 *Cultivation instructions for madder*

Botanical family	Rubiaceae
Origin	Mediterranean area
Duration of vegetation	Perennial, at the cultivation a utilization of 3 years
Demands on the climatic conditions	No demands on the climate, grows overall in Europe, frosthardiness is high and overwinters without problems
Demands on the soil	Sandy humus loams are optimal for madder; on heavy soils the clearing and cleaning of the roots is difficult
Crop rotation	Well-fertilized leaf fruits, like legumes, potatoes, sugarbeet or rape, are especially suitable as preceding crops; following crops should be cereals
Sowing	Can be sown by drilling
Sowing time	In spring as early as possible (March–April)
Sowing strength	5–7 kg/ha at a germination ability of 50–60 % (aimed number of plants/m^2: 15)
Sowing depth	3–5 cm
Germination duration	2–4 weeks
Row distance	20–30 cm, on heavy soils the cultivation of madder on ridges like potatoes seems to be better, ridge distance depends on the machine (50–75 cm)
Planting	More regular stands by planting out of advanced plants or root runners in spring or autumn
Fertilization	The demand on nutrients is high. The yearly gift of N, P and K should be: N 120–160 kg/ha P 40 kg/ha K 150 kg/ha

(continued)

Table 4.4 *(Continued)*

Weed control	The young plants of madder growing slowly. Therefore, an effective weed control in the first vegetation year is absolutely necessary. Hoeing with a machine is the best method for weed control. At the cultivation on ridges the hilling of the crops can be used for weed control. From the second year the madder plants cover the soil nearly completely and suppress the weeds
Harvest	Rooting out the madder roots with a potato harvester after a cultivation duration of 3 years (sown madder), respectively 2 years (planted madder)
Yields	1.5–3[a] to dry mass/ha
Post-harvest treatment	The roots must be washed and cut into strikes of about 1 cm and then dried at 40 °C immediately after the harvest. The dyeing compounds of madder are di- and trihydroxyanthraquinones, especially glucosides of alizarin (= 1.2-dihydrogenanthraquinone). Their content is 5–7 % in the dry madder roots
Seed production	The seed set of madder depends on the climatic conditions during the flowering of the crop. Berries from the flowers develop only in dry warm weather. The seed production is difficult because of the unequal ripening of the fruits. Therefore, cutting the whole plants followed by drying and threshing seems to be the best method
Seed yield	1.5 ton/ha under suitable climatic conditions

[a] Under Italian conditions yields up to 8 ton/ha are possible (Angelini *et al.*, 1997; Angelini, 1999).

influence on the quality of the material for all yellow dyeing species. In a study of its effect on weld, it was found that an increase in the amount of nitrogen fertilization induced a decrease in the dye content (Figure 4.4). At the same time the dry mass yield increased only slightly.

For dyer's chamomile nitrogen fertilization of more than 80 kg of available N/ha increased the vegetative growth, but delayed the formation of flowers. A higher amount of nitrogen fertilization increases the danger of lodging and therefore the harvest bacomes complicated. Besides nitrogen fertilization the harvest time plays an important role with regard to the dye content; e.g. for weld the dye content is at a maximum 10 days after the beginning of flowering (Figure 4.5).

In the following the cultivation methods for weld and dyer's chamomile are described (Table 4.5). Weld is the traditional yellow dyeing plant in Europe while dyer's chamomile yields the most brilliant yellow tint, but its fastness to light is high only on textiles of plant origin.

4.4.5 Brown Dyeing Plants

The dyestuffs in brown dyeing plants are flavonoides and also tannic acids. In most cases they are pharmaceutical plants like greater celandine (*Chelidonium majus* L.), liverwort (*Agrimonia eupatoria* L.), common tansy (*Tanacetum vulgare* L.) or wild marjoram (*Origanum vulgare* L.). Like the yellow dyeing plants their demand on the nutrient content in the soil is low. Table 4.6 contains the cultivation methods for wild marjoram.

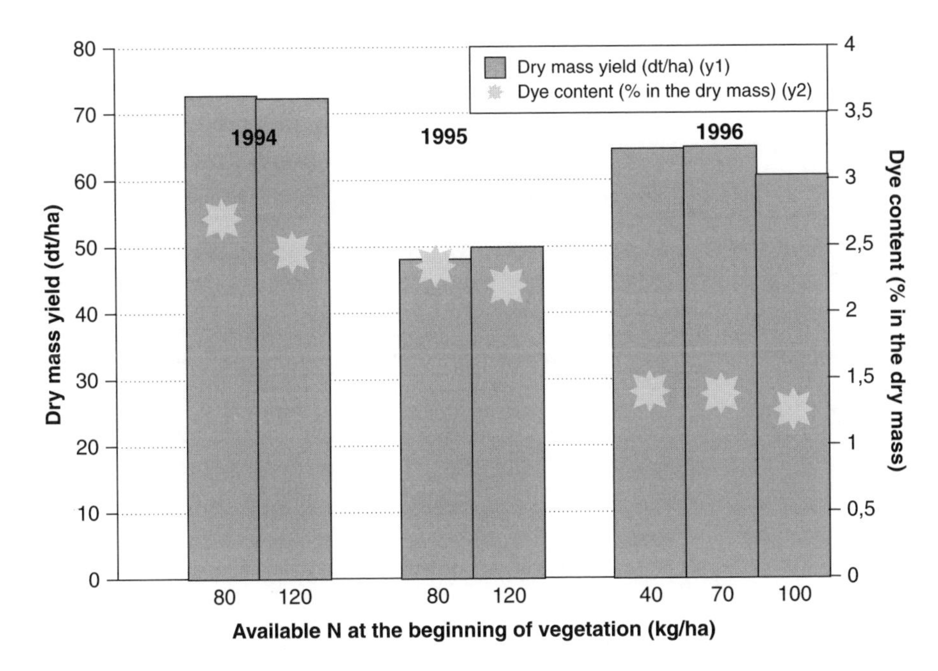

Figure 4.4 *Influence of nitrogen fertilization on the dry mass yield and dye content of weld (Dornburg, 1994–1996)*

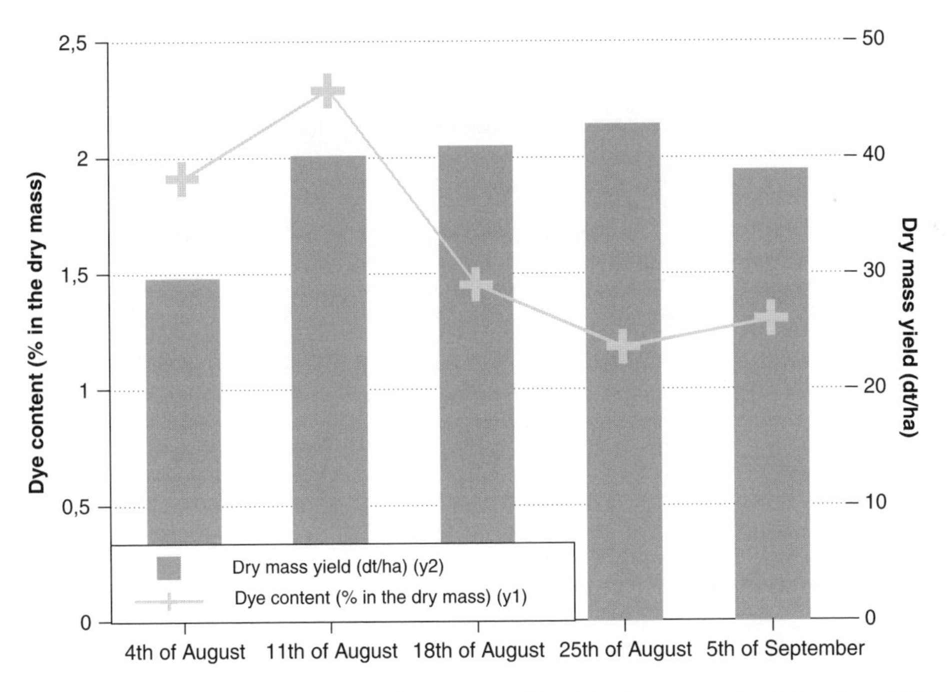

Figure 4.5 *Influence of the harvest time on the dry mass yield and the dye content of weld (Dornburg, 1995)*

Table 4.5 Cultivation instructions for weld and dyer's chamomile

	Weld (*Reseda luteola* L.)	Dyer's chamomile (*Anthemis tinctoria* L.)
Botanical family	Resedaceae	Compositae
Origin	Mediterranean area	South and Middle Europe, West Asia
Cultivation duration	1–2 years	1–2 years
Demands on the climatic conditions	Develops very well everywhere in Europe; in Middle Europe overwintering is possible	No specific demands on the climatic conditions; too high precipitations prevent the flowering and lower the amount of flowers
Demands on the soil	Prefers deep grounded soils, loves sunny areas	Grows on light and heavy soils, especially on dry soils
Crop rotation	No demands on the preceding crop; because of the negative effect on the dye content N-delivering crops should be avoided, such as cereals	No demands on the preceding crop; because of the negative effect on the dye content N-delivering crops should be avoided, such as cereals
Sowing	Drilling with a usual drill for fine seeds	Drilling with a usual drill for fine seeds
Sowing time	In autumn: end of August–middle of September In spring: as soon as possible	In autumn: end of August–middle of September In spring: as soon as possible
Sowing strength	3–5 kg/ha	2 kg/ha
Sowing depth	1–2 cm, as shallow as possible	1–2 cm, as shallow as possible
Germination duration	1–3 weeks	2–3 weeks
Row distance	15–30 cm	20–30 cm
Weed control	Hoeing with a machine is recommended; for the situation of herbicide application see above	
Fertilization	No organic fertilization, N 60 kg of available N/ha, P 50 kg/ha, K 250 kg/ha	No N fertilization, P 15 kg/ha, K 40 kg/ha
Harvest	Optimal harvest time of weld is the end of flowering; harvesting of the whole plant with a green fodder harvester	Repeated flower picking with a chamomile- picking machine
Yields	3.5–5.5 to dry mass/ha (Angelini *et al.*, 2003)	1.5–2.5 to dry mass/ha
Post-harvest treatment	Technical drying of the whole plant at 40–100 °C; following cutting, dyestuff content: 2.0–4.0 % in the dried whole plant (Kaiser, 1993), strongly dependent on the accession	Quick drying of the picked flowers at 40 °C in a thin layer, dyestuff content: 4 – 6 % in the dry mass
Seed production	Harvest with a reap thresher after desiccation of the ripe crop	Harvest with a reap thresher after the ripening of the flowers
Seed yield	0.2–1.0 ton/ha	0.5–1.0 ton/ha

4.5 Production of Dye Extracts

Indigo of natural origin is never pure. It always contains a large amount of inorganic or organic pollutants. They come from the soil particles, which pollute the plant material, and from the extracting agent water. The Ca^{2+}, Mg^{2+}, K^{+}, PO_4^{3-}, . . . ions contained in the water

Table 4.6 *Cultivation instructions for wild marjoram*

Botanical family	Labiatae
Origin	All of Europe, Asia
Duration of vegetation	Perennial, at the cultivation utilization for about 5 years
Demands on the climatic conditions	Easily satisfied, prefers dry and sunny locations
Demands on the soil	Grows on all soils
Crop rotation	No demands on the preceding crops; the following crops should be cereals
Sowing/planting	Sowing with a drill is possible, sowing strength: 4–6 kg/ha at a depth of about 1 cm; better planting is 10–15 plants (grown in a greenhouse) per m^2 at a row distance of 20– 30 cm
Fertilization	N 80–100 kg/ha, P 15–25 kg/ha, K 120–170 kg/ha each year
Weed control	Hoeing with a machine; for the situation of herbicide application see above
Harvest	Cutting of the crop with a green fodder harvester at the time of full flowering of the plants
Yields	4–6 to dry mass/ha from the second year
Post-harvest treatment	Drying of the harvested material at 40 °C following cutting; dyestuff content: 2.0–2.5 % of the dry mass
Seed production	Desiccation of the plants and then harvest with a reap thresher
Seed yield	0.1 0.2 ton of pure seed/ha

and also in the plant material develop water-insoluble compounds during the production process. The indigotin content in the produced indigo has been increased by cleaning procedures (Table 4.7), but not up to 100 %.

Table 4.7 *Indigotin content of raw indigo after different pretreatments, respectively aftertreatments*

Treatment	*Indigotin content (%)*
No treatment	17
Washing of the leaves before the extraction	33
Washing of the leaves, washing of the raw material with acetic acid	58
No pretreatment, washing of the raw material with acetic acid	27

For the extraction of woad a temperature of 70 °C and an extraction time of a half hour are recommended (Stoker *et al*., 1998). The isatan B hydrolyses into indoxyl and sugar. The indican of ai is absolutely stable at this temperature. Its cleavage is difficult. Therefore, 40 °C seems to be the best temperature for the extraction from *Polygonum* leaves. The plant's own β-glucosidase still possesses full activity in this case. At 60 °C the native β-glucosidase is completely inhibited (Minami *et al*., 1996; Maugard *et al*., 2002). The extraction time should be 4 h at 40 °C; at room temperature (25 °C) it should be longer (24 h).

The dyeing compounds of yellow, brown and red dye plants (flavones, flavonoles, anthraquinones, naphtoquinones, etc.) are not very water-soluble. Therefore, other (organic) solvents must be used. Since the solubilities of the single dyestuffs in the solvent are very different, the right choice of solvent, respectively solvent mixture, is of great importance for the success of the extraction (Schweppe, 1992). Because of the acidic character of various aglucones, diluted alkaline hydrous solutions, e.g. 0.1 % sodium carbonate, are suitable for the extraction of weld, madder, etc. The degree of extraction is high and the end-product is water-soluble like the modern technical dyes. This method results in well-reproducible relations between the dye content and the dyeing power (Vetter *et al.*, 1997a, 1997b).

4.6 Relevant Examples for the Application

According to the assessment of Dürbeck and Grieder (2000), the European market for natural indigo comprises 3–5 tons per year. This is only a half per mil of the consumption of the technical product, but it shows that there are many handcrafters, designers, etc., who use natural indigo. The dyeing with natural indigo is only 6 % more expensive than using technical indigo (Schrott and Saling, 2000) with respect to the dyestuff.

The aim of an EU project (QLK5-CT-2000-30962) with partners from England, Germany, Finland, Italy and Spain was to fulfil 5 % of the European indigo market (about 8.000 tons) with the natural product. This hope could not be fulfilled. The demand for the natural indigo has not increased in the last few years, although the technical and the natural indigo are the same chemical compound and the dyeing is the same procedure.

More difficult is the situation for the rest of the natural dyes. Dyeing with chaffed or powdered material, including those in bags, is only suitable for small-scale dyeing, but concentrated dye extracts from weld, madder, etc., can be produced (Wähling and Lange, 1997; Wähling, 1999, 2001; Cerrato *et al.*, 2002) and used without problems on an industrial scale. Most of the natural dyes or dye plants traded on the European market come from southern countries, e.g. India. Recently bioplastics and wood chips for soil covering have frequently been dyed with natural dyes and the leather industry has shown an interest in chromium-free tanned leather dyed with plant pigments.

4.7 Conclusions, Discussion and Summary

Among all of the problems with regard to the reintroduction of natural dyes for dyeing purposes in Europe the question of cultivation on an industrial scale is easy to solve. It has in practice been solved for such important dye plants as woad, ai, weld, madder, golden rod, dyer's chamomile, safflower and wild marjoram through the work of many European research teams. Although most of the available dye plants are still typical wild plant species at the moment, they can be cultivated under European conditions very successfully on a commercial scale with standard equipment currently used by farmers. Their biomass yields per area are high or very high and also their dye content. That presupposes optimal fertilization of the plants, the harvest at the right time and careful processing and storage.

Also, adapting dyeing to modern use and modern machinery is relatively simple. Generally, naturally dyed textiles need not have a poorer fastness, a poorer colour strength and duller shades than synthetic dyes. Weld, like dyer's chamomile, delivers a broad spectrum of yellow shades, which can be varied to yellow-green or olive by changing the mordant combination. Textiles can be dyed with madder in red shades with a yellow tint, red with a blue tint and in brown. The combination of weld and madder leads to different orange shades. However, European dye plants that are missing are those that deliver the basic colours green and black, as well as a bright red. Nevertheless, the industrial use of natural dyes in natural textile production could be much higher than it is today in Europe.

References

Angelini, L. G. (1999) Dyeing plants in Italy, in Gülzower Fachgespräche 'Forum Färberpflanzen', pp. 209–220.

Angelini, L. G., Bertoli, A., Rolanelli, S. and Pistelli, L. (2003) Agronomic potential of *Reseda luteola* L. as new crop for natural dyes in textiles production, *Industrial Crops and Products*, **17**, 199–207.

Angelini, L. G., Pistelli, L., Belloni, P., Bertoli, A. and Panconesi, S. (1997) *Rubia tinctorum* a source of natural dyes: agronomic evaluation quantitative analysis of alizarin and industrial assays, *Industrial Crops and Products*, **6**, 303–311.

Biertümpfel, A., Wurl, G., Vetter, A. and Bochmann, R. (2000) Anbau von Färberpflanzen zur Gewinnung von Farbstoffextrakten für die Applikation auf Textilmaterial, *Berichte über Landwirtschaft Bd.*, **78**(3), 402–420, Landwirtschaftsverlag GmbH, Münster-Hiltrup.

Cardon, D. and du Chatenet, G. (1990) Guide de Teintures Naturelles – Plantes, Liches, Champignons, Mollusques et Insects, Delachaux et Niestle, Paris.

Cerrato, A., de Santis, D. and Moresi, M. (2002) Production of luteolin from *Reseda luteola* and assesment of their dyeing properties, *J. Sci. Food Agr.*, **82**, 1189–1199.

Dürbeck, K.and Grieder, K. (2000) Pflanzliches Indigo in El Salvador – ein Projekt der GTZ, *Drogenreport*, **13**, 63–65.

Kaiser, R. (1993) Qualitiative analysis of flavonoids in yellow dye plant species weld (*Reseda luteola L.) and sawwort (Serratula tinctoria L.), Angew. Botanik*, **67**, 128–131.

Maugard, T., Enaud, E., de La Sayette, A., Choisy, P. and Legoy, M. D. (2002) β-Glucosidase-catalysed hydrolysis of indican from leaves of *Polygonum tinctorium, Biotechnol. Progress*, **18**, 1104–1108.

Minami, Y., Kanafuji, T. and Miura, K. (1996) Purification and characterisation of a β-glucosidase from *Polygonum tinctorium*, which catalyses preferentially the hydrolysis of indican, *Biosci. Biotech. Biochem.*, **60**, 147–149.

Roth, L., Kormann, K. and Schweppe, H. (1992) *Färberpflanzen – Pflanzenfarben*, Ecomed Fachverlag Landsberg a. Lech.

Schrott, W. and Saling, P. (2000) Ökoeffizienz-Analyse – Produkte zum Kundennutzen auf dem Prüfstand, *Melliand Textilberichte*, **3**, 190–194.

Schweppe, H. (1992) *Handbuch der Naturfarbstoffe: Vorkommen, Verwendung, Nachweis*, Ecomed Fachverlag Landsberg a. Lech.

Stoker, K., Cooke, D. and Hill, D. T. (1998) An improved method for the large-scale processing of woad indigo, *S. Agric. Engng. Res.*, **71**, 315–320.

Vetter, A. (1997) Potentielle Pflanzen zur Gewinnung von Naturfarbstoffen – Bedeutung und Markt, in Gülzower Fachgespräche 'Färberpflanzen', pp. 21–38.

Vetter, A., Eggers, U. and Hill, D. T. (1997a) Cultivation and extraction of natural dyes for industrial use in natural textile production, Final report of the EU-project AIR-CT94-0981 (DG12SSMA), Dornburg.

Vetter, A., Wurl, G. and Biertümpfel, A. (1997b) Auswahl geeigneter Färberpflanzen für einen Anbau in Mitteleuropa, *Z. Arzn.Gew.Pfl.*, **2** (4), 186–192.

Voltolina, G. and Valeriani, C. (1996) L'informatore agrario, **L11** (18), Verona.
Wähling, A. (1999) Erste Ergebnisse zur Extraktion von Naturfarbstoffen, in Gülzower Fachgespräche 'Forum Färberpflanzen', pp.142–155.
Wähling, A. (2001) Aktueller Stand der Farbstoffgewinnung aus nachwachsenden Rohstoffen, in Gülzower Fachgespräche'Forum Färberpflanzen', pp. 146–148.
Wähling, A. and Lange, E. (1997) Extraktion von Naturfarbstoffen, in Gülzower Fachgespräche 'Färberpflanzen', pp. 104–109.
Wiesner, J.v. (1927) *Die Rohstoffe des Pflanzenreichs*, Leipzig.
Wurl, G. (1997) Ergebnisse von Anbauversuchen aussichtsreicher Färberpflanzen, in Gülzower Fachgespräche'Färberpflanzen', pp. 47–60.

5

Dyes in South America

Veridiana Vera de Rosso and Adriana Zerlotti Mercadante

5.1 Introduction

Production of natural colorants has became an interesting alternative for the development of certain regions in Central and South America, which survive only from wild collection or subsistence plantations. Some of these regions have received either local government or international organizations support in order to develop sustainable agriculture for production and commercialization of natural colorants. In Brazil, the farming development of *Bixa orellana* in Paraíba State and of *Curcuma longa* in Goiás State, and plantation of *Bixa orellana* in the Cochabamba region of Bolivia, are successful experiences [1–4].

Apart from their colorant properties, some natural colorants are related to important functions and physiological actions. Annatto extracts show antimicrobial activity [5] and antioxidant activity [6–9], as well as protection against lipid peroxidation and chromosome aberration induced by cisplatin [10] and chemopreventive effects [11]. Curcuminoids also show important biological activities, such as anti-inflammatory, anti-HIV and antitumoral [12–15]. Epidemiological studies indicate that lutein, and to a much lesser extent their esters, protect the eyes, skin and the other tissues from destructive oxidation reactions by quenching free radicals and singlet oxygen [16–19], also contributing to prevention of aged-macular degeneration [20]. Few data were found regarding studies of the functional properties of cochineal pigments, such as inhibition of cancer cell proliferation and DNA protection [21].

The acceptable daily intake (ADI) had already been established for some natural colours, and these values have been re-evaluated in recent years. For annatto extracts, ADI is established as 0–12 mg/kg bw (body weight) for bixin and 0–0.6 mg/kg bw for norbixin [22], for turmeric extracts the ADI value is 0–3 mg/kg [23], for lutein the value is 0–2 mg/kg bw [24] and 0–5 mg/kg bw for carminic acid [25].

Handbook of Natural Colorants Edited by Thomas Bechtold and Rita Mussak

Table 5.1 *Overview of the major colorants produced in South America*

Name			*Climate*	*Region*	*Used part*	*Amount*	*Remarks*	*Present use*
English	*Botanical*	*Traditional*[a]						
Annatto	*Bixa orellana* L.	Urucum (Pt); achiote (Sp)	Tropical	Central and South America	Seeds	10 000 to 12 000 tons in Brazil 4 000 tons/year in Peru	2 harvests every year	Food colorant, spice/ condiment, textile colour
Turmeric	*Curcuma longa* L.	Curcuma (Pt)	Tropical	India, China, Central and South America	Rhizomes	1000 to 1500 tons/year in Brazil	9 to 12 months until harvest	Food colorant, spice/ condiment
Marigold	*Tagetes erecta* L.		Tropical	Mexico, Central and South America	Flowers	3000 tons/year in Peru; details not available for other countries	90 to 110 days	Food colorant, colorant for poultry feed
Cochineal	*Dactylopius coccus*	Cochonilha (Pt) Cochinilla or zacatillo (Sp)	Tropical and subtropical	South America	Bodies of female insects	714 tons/year in Peru, 100 to 200 tons/year in Chile, Bolivia and Ecuador	4 harvests each year	Food colorant, cosmetic dye, biological stain

[a]Pt, Portuguese; Sp, Spanish.

Although natural colours play an important role in the economy of some countries from Central and South American, official data are very scarce. In addition, annatto and turmeric are consumed in large quantities as spice in some Latin American countries, and this demand is probably not accounted for in the official statistics. An overview of the major colorants produced in these countries is summarized in Table 5.1.

5.2 Annatto

The tree *Bixa orellana* L. is native to northern South America and was later introduced in Central America, India and Africa. It is one of the four species of the family Bixaceae, order Violales, subclass Archichlamydeae, division Angiosperms/Dicoyledons [26]. The plant was named after Francisco de Orellana, the Conqueror, who explored the Amazon river for the first time in 1541 [27]. The plant thrives from sea level to about 1200 m altitude in a moist climate, especially in deep, loamy and well-drained soil. *Bixa orellana* is a small tree, 2 to 5 m in height, bearing clusters of brownish-crimson capsular fruits, 3 to 4.5 cm, covered by flexible thorns. The fruits are divided inside generally in two cavities containing 15 to 50 small seeds, about the size of grape seeds. The seeds are covered with a thin, highly coloured, orange to red resinous layer from where the natural colorant is obtained. The amount of red pigment in annatto seeds varies from 1.5 to 4 %, depending on the annatto variety. In Brazil, some cultivars were developed or adapted, the principal commercial ones being Peruana Paulista (São Paulo State), Bico de Pato (Bahia State) and Piave Vermelha (Pará State) [28].

There is general agreement [27, 29] that more than 80 % of the carotenoids in the *B. orellana* seed coat consists of bixin (Figure 5.1), which to date has been encountered only in *B. orellana*. As can be seen in Figure 5.2, small amounts of norbixin (Figure 5.1) are also found in annatto seeds; other minor carotenoids, with bixin-related structures and less than 40 carbons, were identified in annatto seeds [30–33].

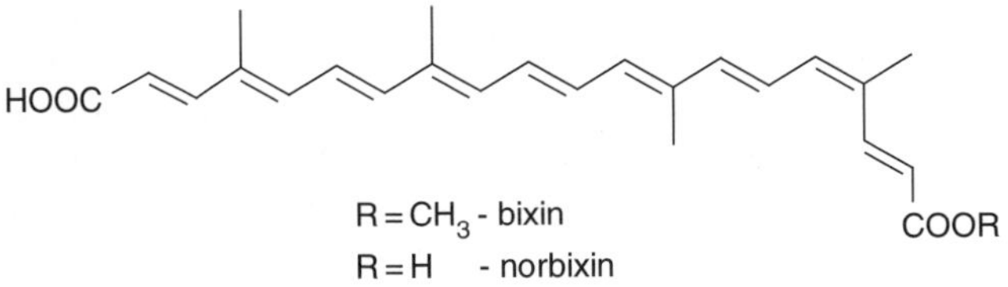

HOOC

COOR

$R = CH_3$ - bixin
R = H - norbixin

Figure 5.1 *Structures of bixin and norbixin*

Commercial annatto preparations are available to impart yellow to red colour to a variety of products. They include oil-soluble colours and annatto powder (which contain bixin as the main pigment), water-dispersible colours (with norbixin as the main colorant) and emulsions, which may contain either a combination of both bixin and norbixin or only norbixin.

Annatto powder or oily annatto extract mixed with cassava or corn flour and added salt or edible oil, known as 'colorifico', is available for domestic use in Brazil [34]. This powder shows a very similar aspect to that of paprika and is mostly used for colouring meat, chicken and fish mainly in the Central, North and Northeast regions of Brazil. The contents of bixin ranged from 154 to 354 mg/100 g, while much less norbixin was found, 2.1–6.6. mg/100 g, in many brands of the condiment 'colorifico' [35].

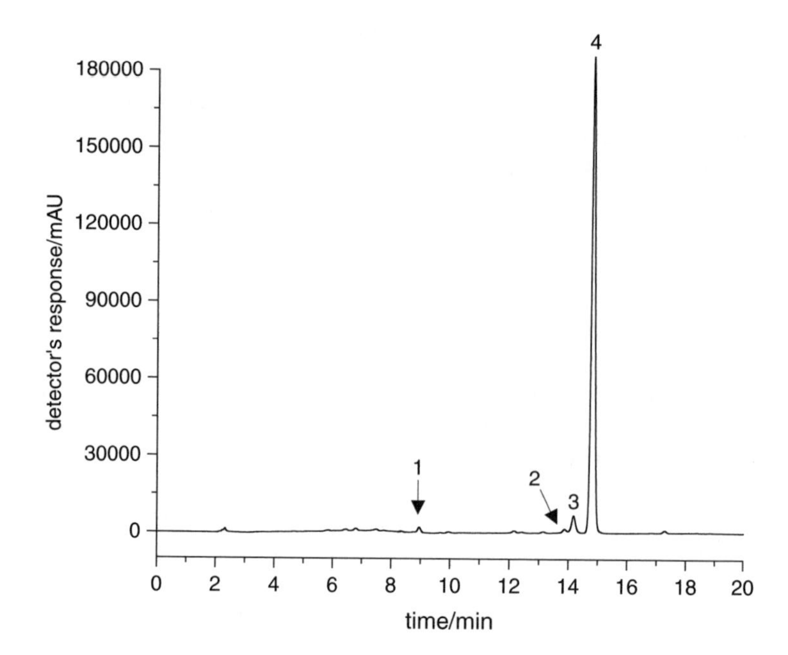

Figure 5.2 *HPLC chromatogram of the methanolic extract from annatto seeds. Conditions: C_{18} synergi fusion colunm (4 μm, 4.6 × 250 mm), linear gradient of acetonitrile/0.5 % formic acid from 65:35 to 85:25 in 20 min at 1 mL/min as mobile phase. Chromatogram processed at 470 nm. Peak identification 1: norbixin, 2: all-trans-bixin, 3: not identified, 4: bixin*

The most commonly employed process for the production of water-soluble annatto extracts involves direct extraction of the seeds with aqueous alkali (usually sodium or potassium hydroxide) as the first step. In general, the seeds are soaked or stirred in dilute aqueous alkali in a stainless steel vessel at temperatures not higher than 70 °C. In this operation, bixin is converted to a water-soluble derivative, the salt of norbixin. The alkaline extract is centrifuged, filtered and then acidified with dilute mineral acid (sulfuric or hydrochloric acid) to precipitate norbixin. The oil-soluble annatto colour is normally produced by direct extraction from the seeds with food-grade vegetable oils. The extraction is carried out by mechanical rasp at temperatures below 70 °C. This extract is a suspension of *cis* and *trans* isomers of bixin in vegetable oil. In the case where a more yellow colour is required, the initial extract is either produced at higher temperatures (over 100 °C) or is subsequently heat-treated. In this case, besides *cis* and *trans* isomers of bixin, the extract contains a yellow pigment, identified as 4,8-dimethyl-tetradecahexaene-dioic acid monomethyl ester [36–38]. Bixin crystals are produced by a similar process employed for spice oleoresins. The seeds are extracted with an organic solvent (hexane, acetone or alcohol), followed by concentration and crystallization by addition of methanol or ethanol, furnishing crude crystals [39]. Bixin extraction by supercritical carbon dioxide was reported in the literature [40–42], but has not yet been commercially adopted.

The traditional major colouring applications have been in hard cheeses, butter, other dairy products and in margarine. Annatto colours are also employed in fish products, salad dressings, confectioneries, bakery products, ice-creams, beverages and snack foods. The type of extract employed is dependent on the food product and the need for the pigment to be oil- or water-soluble. The water-soluble extracts are employed in biphasic or water-based foods and the oil-soluble extracts are used in foods containing a high fat content. The levels of bixin/norbixin allowed to be added in food products usually range between 0.01 and 0.5 % in Europe and the USA [39, 43]. In Brazil, the levels allowed are similar to those from other countries, such as 0.02 % in candies, confectioneries, frozen deserts and pasta, 0.005 % in nonalcoholic soft drinks, 0.002 % in meat and meat products, 0.001 % in sauces, condiments, deserts and bakery products [44].

Nowadays, the major producers of annatto seeds are the countries in South America, mainly Brazil, Peru and Ecuador. It is very difficult to estimate the market precisely for the different annatto preparations because no official statistics that include both export and domestic markets are available in these countries. According to Franco *et al.* [4], the Brazilian production of annatto seeds ranged from 10 000 to 12 000 tons in 1999, 60 % being commercialized as 'colorifico' in the domestic Brazilian market, about 30 % used for colorant production and 10 % of the seeds were exported. On the other side, in Peru, the production of annatto is export-oriented, 80 % of 4482 tons produced in 2000 being exported [45]. Other countries in South America, such as Bolivia and Ecuador, have lower estimated production, each between 500 and 1000 tons (not official data) [1, 43].

In terms of world volume traded, the most important colorant extract is the water-soluble norbixin type, followed by vegetable oil extracts and solvent-extracted ones, both with bixin as the major pigment. The market trend in developed countries has moved towards a progressive increase in imports of extracts, and for more stringent quality requirements (bixin content) for imported seeds [39]. The USA is the largest single market for annatto colours, followed by the European Union and Japan.

In summary, the future of annatto is very promising since there is a general increased demand for natural colours and the acceptable daily intake of annatto colours (ADI) has recently increased [22]. In addition, the Brazilian annatto producers are improving the annatto varieties in terms of bixin contents and plant productivity, and hard efforts have been made on techniques for seed drying and cleaning. The project 'Jatun Sach'a', supported by the United Nations and FAO, focus on income generation activities and environmental protection through implementation of plantations of annatto, among others, as a farming alternative for coca cultivation in the Cochabamba Tropics and the Yungas of La Paz regions in Bolivia [2]. Due to the support of this programme, annatto production in Bolivia is expected to increase from 500 tons in 2003 to 3700 tons in 2009 [1].

5.3 Turmeric

Curcuma longa L., which belongs to the Zingiberaceae family, originates from the Indian sub-continent and possibly neighbouring areas of Southeast Asia, but it is nowadays widely grown throughout the tropics. It is a small perennial herbaceous plant that measures up to 1 m high with a short stem and rhizomes, which show oblong, ovate, pyriform, often short-branched forms, from where the pigments are extracted. Curcuma thrives in a hot, moist tropical climate with rainfall of 1000–2000 mm/year, preferably in a rich well-drained loam soil. Cultivation is possible from sea level up to about 1500 m [46].

The pigments in the colorant extracts obtained from *Curcuma* are collectively known as curcuminoids, the major constituent being curcumin, along with small amounts of demethoxycurcumin and bis-demethoxycurcumin (Figure 5.3). The levels at dry basis of curcuminoid pigments in turmeric cultivated in Brazil ranged from 1.4 to 6.1 g/100 g, while the volatile oil fraction was between 1.0 and 7.6 mL/100 g [47]. These pigments are insoluble in water and soluble in organic solvents, such as ethanol.

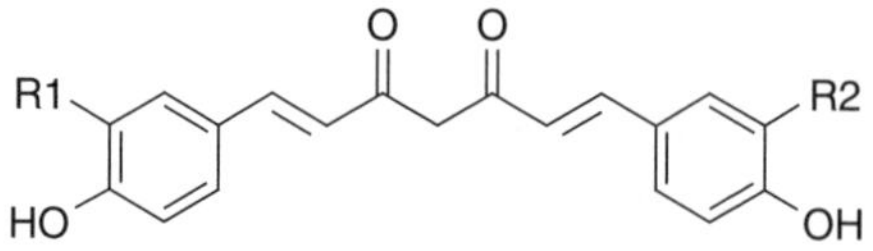

R1 = OCH_3, R2 = OCH_3 - curcumin
R1 = OCH_3, R2 = H - demethoxycurcumin
R1 = R2 = H - bisdemethoxycurcumin

Figure 5.3 *Structures of curcumin and its analogues*

Brazil has favourable conditions for turmeric cultivation; however, the country needs to improve the technology in order to obtain oleoresins of added value and good quality (free of volatiles), although two conventional processes of turmeric are employed in Brazil. One consists of slicing the rhizomes followed by sun drying for 5 days, while in the other the rhizomes are boiled in water during 1 to 4 hours followed by a slow dry for 30 to 35 days [3, 48]. These time-consuming processes furnish products that do not accomplish the necessary uniformity and quality to be exported. The ground turmeric rhizomes are largely commercialized as spice to supply the Brazilian domestic demand, although Brazil imports turmeric oleoresins from India and Peru [3].

Nowadays, food manufacturers employ the turmeric oleoresin or pure curcumin to a greater extent than the powdered spice. The oleoresins may be prepared by a similar procedure widely used for spices, e.g. sequential extraction of the ground material with an organic solvent, such as methanol, ethanol, acetone or dichloromethane, with yields ranging from 8 to 13 %, followed by solvent stripping. The oleoresins contain a mixture of compounds, such as resinous materials, lipids, volatile oils, pigments and others, depending on the process employed. The pure curcumin is obtained by crystallization from the oleoresin, followed by recrystallization to remove volatile oils and other plant extractives. Supercritical fluid extraction was evaluated as an alternative to obtain turmeric oleoresin [49–51]; however, this technique was efficient only when 10 to 90 % of co-solvents, ethanol and/or isopropyl alcohol, were used [51, 52].

Turmeric, its oleoresin and curcumin are universally allowed as food additives; in the USA curcumin, oleoresin, and turmeric are approved as natural colours, whereas curcumin is classified as an additive in Brazil and the European Union [23, 53]. In Brazil, the principal applications of turmeric colours are 0.05 % in pasta, 0.01 % in soft drinks and 0.05 % in salad sauces and jams [44].

The annual world trade for the turmeric spices is estimated between 15 000 and 20 000 tons, while the demand for turmeric oleoresin and pure curcumin is much lower, around

100 to 150 ton/year [46]. India is the major exporter of both turmeric spice and oleoresin, supplying at least 60 % of the two products, followed by China, whilst in the third rank there are other countries in Asia, the Caribbean and South America. However, none of these countries are significant oleoresin suppliers [46]. Unfortunately, the Allepey type of turmeric material commercialized by the Caribbean and South American countries, Jamaica, Haiti and Peru, presents poor colour tone as compared to those exported by India and China [46].

Although the future market trends are expected to increase, the prospects for new producers of turmeric oleoresins are not good in the near future due to the established dominant position of India. On the other hand, the South American market is growing at a rate of 10 %/year, and imports increased by around 40 % by Argentina, Venezuela, Brazil and Colombia between 1997 and 1999. Therefore, the Caribbean and South American countries should make efforts to improve the quality of the turmeric raw material, as well as to improve the processes to obtain different turmeric oleoresins [3, 46].

5.4 Marigold

The common term marigold comprises a diversity of plants with golden flowers, including different *Tagetes* species native to the Americas, which were introduced worldwide either for industrial use or decorative purposes. *Tagetes erecta*, known as Aztec marigold, native to Mexico and Central America, is an annual herb that prefers a warm and low-humidity climate. Their propagation is achieved by seeds and flowering commences 90 to 110 days after field plantation; harvesting of fully developed flowers is regularly carried out throughout the season [46].

The major colouring component of marigold (78–80 %) is all-*trans*-lutein (3*R*,3′*R*,6′*R*-β,ε-carotene-3,3′-diol), an asymmetric carotenoid (xanthophyll). Since lutein bears one hydroxyl group at each ionone ring, it can be esterified with fatty acids in plant cells, resulting in mono- and diacylated derivatives [54]. Gregory *et al.* [55] and Rivas [56] showed that freshly prepared marigold orange flower extract contains no free lutein (Figure 5.4), but instead several lutein diesters with dipalmitate (Figure 5.4) and myristate-palmitate as the major esters. Small amounts of 9-, 9′-, 13- and 13′-*cis* isomers of lutein, along with all-*trans*-zeaxanthin were also found in marigold extracts [57, 58].

There are three forms of commercial marigold colorants. The most used (about 60 %) and easiest prepared is the marigold meal, obtained by sun- or shaded-dried flowers

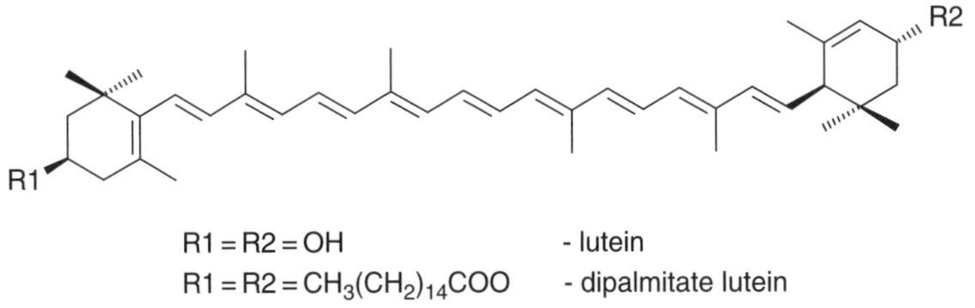

Figure 5.4 *Structure of lutein in both free and esterified forms*

reduced to a powder before packing, and mainly added to poultry feed in order to enhance the yellow yolk colour. The concentrated marigold extract, obtained with organic solvent (hexane is the most used) containing high concentrations of lutein esters, is commercialized mixed with vegetable oil for food application or mixed with soy or corn meal to be added to poultry feed. In order to obtain free lutein, marigold oleoresin, which contains lutein esters, is saponified according to a patented process (US Patents 5,382,714 and 5,648,564). After saponification, lutein is purified, crystallized and commercialized, as free natural lutein, in two vehicles, vegetable oil and dry form (sucrose and starch).

The principal use of marigold is as additive, either in the form of the dried flower meal or as a solvent extract, to poultry feed in order to enhance the yellow colour of the flesh and the egg yolks. When the marigold extracts are used as food colorant, typical applications are salad dressings, ice-cream, dairy products, bakery products, jams and confectionery. The extract containing free lutein is mainly commercialized in dietary supplement products [53]. Lutein, either synthetic or from marigold extract, is allowed to be added in nonalcoholic soft drinks at 0.01–0.05 % levels, at 0.01 % in pasta and at 0.05 % in sauces [44].

Marigold flowers are principally produced in Mexico, Peru, Ecuador, Argentina and Venezuela. According to data available from the Ministerio de Agricultura do Peru [45], Peru produced a total of 197 946 tons of fresh marigold flowers in 2000, with 11 746 tons of marigold meal exported in that year. Apart from Peru, statistics on production and international trade of both marigold meal and extract are scanty and unreliable in the Americas. Mexico is the main producer of extract and imports substantial flower amounts in order to supplement its domestic production for the extraction industry. Other major importers are North America and Western Europe, both of marigold meal and extract.

5.5 Cochineal and Carmine

Cochineal is the name used to describe both the colour and its raw material source, the dried pregnant females of *Dactylopius* species of insects, especially *D. coccus* Costa. The main hosts of these scale insects are the aerial parts of prickly pear *Opuntia* and torch thistle *Nopalea* species of cacti. Other less important host plants in the tropical Americas include *Schinus molle*, *Acacia macarantha* and *Caesalpinia spinos* [46]. The female insects live in close juxtaposition to each other on the cactus phylloclade, with mouths inserted into the surface of the cactus and their bodies covered by a wax layer. The western and southern regions of South America have climatic conditions ideally suited to the growth of members of the subfamily Opuntioideae of the Cactaceae.

The insects and their host cacti are adapted to arid areas of the tropics and subtropics. When cultivated, a host plant is inoculated with a brood of the insects, which are readily collected since they mass together. The females mature over a period of 90–110 days and immediately prior to egg laying the insects are harvested by hand, followed by brushing. Fully developed females are roughly 6 mm long, 4.5 mm in width and 4 mm in height, with an average weight of 45 mg. The process begins with killing and drying female insects, which reduces the weight by 70 %. Cochineal is commercially available in two forms, depending on the drying process employed. Silver cochineal is obtained by simple sun-drying of insects and is considered of inferior quality. Black cochineal is produced by accelerated drying or by solvent extraction. Between 80 000 and 100 000 insects are required to produce 1 kg of dried cochineal [46, 59].

The principal pigment in cochineal is carminic acid, a hydroxyanthraquinone linked by a *C*-glycoside bond to a glucose unity (Figure 5.5), being very soluble in water. The carminic acid content in good quality raw cochineal can reach 18 to 20 % of insect dry weight [60]. Treatment of carminic acid with an aluminium salt produces carmine, an aluminium lake, which is insoluble in water, while carmine precipitated by addition of a calcium salt is soluble in water [46, 61]. In recent years, acid-stable carmine, a new dye, was launched on the market in response to needs in the food industry for a brighter red colour in acidic foods. This pigment was identified as 4-aminocarminic acid and is soluble in water [62].

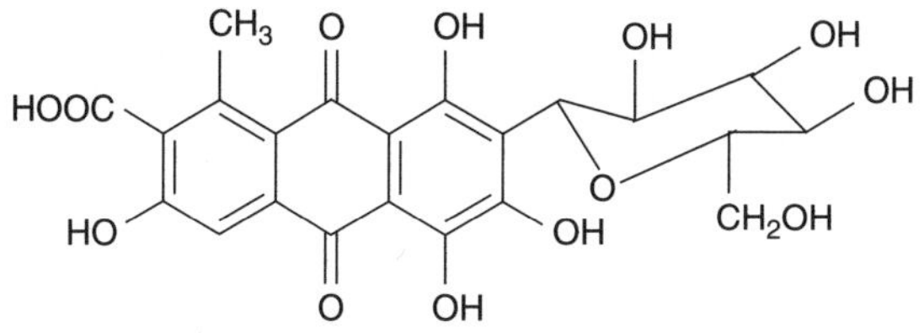

Figure 5.5 *Structure of carminic acid*

Colour changes dramatically according to the pH, depending on the colorant type. Carminic acid goes from magenta-red above pH 12 to violet at 11.5, from dark to lighter violet at 11.0–9, from red to lighter red at 7.7–7.0, orange-red at 6.5, orange at 5.5 and then to light orange at pH below 4.5. Carmine colour changes resemble those of carminic acid, except that below pH 7.0 red shades remain instead of orange ones. Aminocarminic acid is violet at high pH and remains with various red intensities at pH ranging from 9 to 2 [61].

Current major usages of cochineal and its derivates are in the food, drug and cosmetics industries, and as biological stains. Carmine is an important colorant for cosmetics employed near to the eye and carmine aluminium lakes are permitted and widely used in food applications (soft and alcoholic drinks, bakery and dairy products, confectionery and pickles) at dosage levels ranging from 0.1 to 0.5 % [46, 53, 61]. In Brazil [44], carminic acid and its analogues are employed in extruded cereal products (0.02 %), pasta (0.05 %), nonalcoholic beverages (0.01 %), sauces, jams and fillings (0.05 %).

Peru remains the world's largest producer and exporter of cochineal, accounting for at least 90 % of the export market; their production has increased 6 times in the last 23 years, reaching 714 tons in 2004 [63]. The other significant suppliers are the Canary Islands, Chile, Bolivia and Ecuador, with exports fluctuating between 100 and 200 tons per annum [63]. Prior to 1980, all Peruvian exports were in the form of raw cochineal, but in the subsequent period extraction industries were developed in Peru and, nowadays, over 50 % of the annual crop is processed to carmine prior to export, followed by small quantities of carminic acid [64]. The major buyer of carmine is Western Europe, followed by the USA. Japan predominantly imports cochineal for local processing [46].

The consumption of cochineal and its derivative colours substantially increased in the major markets, most probably due to actual trends towards natural colours in foods. Prospects for significant further expansion in these traditional markets are uncertain, but a modest growth is likely to occur globally [46].

Acknowledgements

The authors thank the financial support from the Brazilian Funding Agencies (FAPESP and CNPq) and also from DSM Nutritional Products (Basel, Switzerland).

References

1. Los Tiempos, Cultivos de achiote crecen, pero no cubre la demanda, www.lostiempos.com/noticias/31-08-04/31_08_04_eco6.php, 2004.
2. Servicio de Noticias de las Naciones Unidas, Bolívia, FAO entrega resultados positivos de programa contra deforestación, http://www.un.org/spanish/News/printnews.asp?newsID=6627, 2006.
3. G. Marinozzi, W. Canto-Leite, S. B. Roma and E. Oliveira, Fortalecimento organizacional e institucional para construção coletiva da competitividade nos sistemas agro-alimentares localizados: o caso do cluster da curcuma em Mara Rosa-GO, in *Estabelecimento de Tecnologia para o Fortalecimento de Agronegócio do Açafrão (Curcuma longa L.) em Mara Rosa-GO*, 2001.
4. C. F. O. Franco, F. C. P. Silva, J. Cazé-Filho, M. Barreiro-Neto, A. B. São José, T. N. H. Rebouças and I. S. C. Fontinélli, *Urucuzeiro: Agronegócio de Corantes Naturais*, EMEPA-PB, João Pessoa, 2002.
5. O. N. Irobi, M. Mooyoung and W. A. Anderson, Antimicrobial activity of annatto (*Bixa orellana) extract, Int. J. Pharmacogn.*, **34**, 87–90 (1996).
6. S. Kiokias and M. H. Gordon, Antioxidant properties of annatto carotenoids, *Food Chem.*, **83**, 523–529 (2003).
7. C. R. Cardarelli, M. T. Benassi and A. Z. Mercadante, Characterization of different annatto extracts based on antioxidant and colour properties, *LWT-Food Sci. Technol.*, **41**, 1689–1693 (2008).
8. M. A. Montenegro, A. de O. Rios, A. Z. Mercadante, M. A. Nazareno and C. D. Borsarelli, Model studies on the photosensitized isomerization of bixin, *J. Agric. Food Chem.*, **52**, 367–373 (2004).
9. A. de O. Rios, A. Z. Mercadante and C. D. Borsarelli, Triplet state energy of the carotenoid bixin determined by photoacustic calorimetry, *Dyes Pigm.*, **74**, 561–565 (2007).
10. C. Rodrigues-Silva, L. M. G. Antunes and M. L. Bianchi, Antioxidant action of bixin against cisplatin-induced chromosome aberrations and lipid peroxidation in rats, *Pharmacol. Res.*, **43**, 561–566 (2001).
11. A. R. Agner, A. P. Bazo, L. R. Ribeiro and D. M. F. Salvadori, DNA damage and aberrant crypt foci as putative biomarkers to evaluate the chemopreventive effect of annatto (*Bixa orellana* L.) in rat colon carcinogenesis, *Mut. Res.*, **582**, 146–154 (2005).
12. B. B. Aggarwal, A. Kumar and A. C. Bharti, Anticancer potential of curcumin: preclinical and clinical studies, *Anticancer Res.*, **23**, 363–398 (2003).
13. T. Choudhuri, S. Pal, M. L. Agwarwal, T. Das and G. Sa, Curcumin induces apoptosis in human breast cancer cells through p53-dependent Bax induction, *FEBS Lett.*, **512**, 334–340 (2002).
14. A. Mazumber, K. Raghavan, J. Weinstein, K. W. Kohn and Y. Pommer, Inhibition on human immunodeficiency virus type I integrase by curcumin, *Biochem. Pharmacol.*, **49**, 1165–1170 (1995).
15. H. C. Huang, T. R. Jan and S. F. Yeh, Inhibitory effect of curcumin, an anti-inflamatory agent, on vascular muscle cell proliferation, *Eur. J. Pharmacol.*, **221**, 381–384 (1992).

16. J. A. Mares-Perlman, A. E. Millen, T. L. Ficek and S. E. Hankinson, The body of evidence to support a protective role for lutein and zeaxanthin in delaying chronic disease. Overview, *J. Nutr.*, **132**, 518S–524S (2002).
17. S. Astner, A. Wu, J. Chen, N. Philips, F. Rius-Diaz, C. Parrado, M. C. Mihn, D. A. Goukassian, M. A. Pathak and S. Gonzalez, Dietary lutein/zeaxanthin partially reduces photoaging and photocarcinogenesis in chronically UVB-irradiated Skh-1 hairless mice, *Skin Pharmac. Physiol.*, **20**, 283–291 (2007).
18. L. Brown, E. B. Rimm, J. M. Seddon and E. L. Giovanucci, A prospective study of carotenoid intake and risk of cataract extraction in US men, *Am. J. Clin. Nutr.*, **70**, 517 (1999).
19. M. D. Snodderly, Evidence for protection against age-related macular degeneration by carotenoids and antioxidants vitamins, *Am. J. Clin. Nutr.*, **62**, 1448S–1461S (1995).
20. J. M. Stringham and B. R. Hammond, Dietary lutein and zeaxanthin: possible effects on visual function, *Nutr. Rev.*, **63**, 59–64 (2005).
21. T. S. Ko, H. Y. Maeng, M. K. Park, I. S. Park, B. S. Kim, Cu^{+2}-anthraquinone complexes – formation, interaction with DNA, and biological activity, *B. Korean Chem. Soc.*, **15**, 364–368 (1994).
22. FAO/WHO, Summary and conclusions of the Sixty-Seventh Meeting of the Joint FAO/WHO Expert Committe on Food Additives (JECFA), Rome, 2006.
23. JECFA, Summary of evaluations performed by the Joint FAO/WHO Expert Commitee on Food Additives – Curcumin, http://www.inchem.org/documents/jecfa/jeceval/jec_460.htm, 2004.
24. JECFA, Summary of evaluations performed by the Joint FAO/WHO Expert Commitee on Food Additives – Lutein from Tagetes erecta, http://www.inchem.org/documents/jecfa/jeceval/jec_1289.htm, 2004.
25. J. B. Greig, Cochineal extract, carmine, and carminic acid in WHO Food Additives Series 46, http://www.inchem.org/documents/jecfa/jecmono/v46je03.htm, 2003.
26. W. C. Evans, Trease and Evans' Pharmacognosy, 15th edition, W. B. Saunders, Edinburgh, 2002.
27. H. D. Preston and M. D. Rickard, Extraction and chemistry of annatto, *Food Chem.*, **8**, 47–56 (1980).
28. T. N. H. Rebouças and A. R. São José, A cultura do urucum: práticas de cultivo e comercialização, DFZ/UESB/SBCN, Vitória da Conquista, 1996.
29. G. J. Lauro, A primer on natural colors, *Cereal Foods World*, **36**, 949–953 (1991).
30. A. Z. Mercadante, A. Steck, D. B. Rodriguez-Amaya, H. Pfander and G. Britton, Isolation of methyl 9′Z-apo-6′-lycopenoate from Bixa orellana, Phytochemistry, **41**, 1201–1203 (1996).
31. A. Z. Mercadante, A. Steck and H. Pfander, Isolation and identification of new apocarotenoids from annatto (*Bixa orellana* L.) seeds. *J. Agric. Food Chem.*, **45**, 1050–1054 (1997).
32. A. Z. Mercadante, A. Steck and H. Pfander, Isolation and structure elucidation of minor carotenoids from annatto (*Bixa orellana* L.) seeds. *Phytochemistry*, **46**, 1379–1383 (1997).
33. A. Z. Mercadante, A. Steck and H. Pfander, Three minor carotenoids from annatto (*Bixa orellana* L.) seeds, *Phytochemistry*, **52**, 135–139 (1999).
34. A. Z. Mercadante, Composition of carotenoids from annatto, in J. M. Ames and T. F. Hofmann (eds), *Chemistry and Physiology of Selected Food Colorants*, American Chemical Society, Washington, 2001.
35. L. Tocchini and A. Z. Mercadante, Extraction and determination of bixin and norbixin, by HPLC, in annatto spice ('colorifico'), *Ciênc. Tecnol. Aliment.*, **21**, 310–313 (2001).
36. A. de O. Rios, C. D. Borsarelli, A. Z. Mercadante, Thermal degradation kinetics of bixin in an aqueous model system, *J. Agric. Food Chem.*, **53**, 2307–2311 (2005).
37. M. J. Scotter, Characterization of the coloured thermal degradation products of bixin from annatto and a revised mechanism for their formation, *Food Chem.*, **53**, 177–185 (1995).
38. G. G. Mckeown, Composition of oil-soluble annatto food colors. 3. Structure of the yellow pigment formed by the thermal degradation of bixin, *J. Assoc. Off. Agric. Chem.*, **48**, 835–857 (1965).
39. FAO/WHO, Major colorants and dyestuffs entering international trade, in Natural Colourants and Dyestuffs, Non-Wood Forest Products Series, Rome, 1995.
40. G. F. Silva, F. M. C. Gamarra, A. L. Oliveira and F. A. Cabral, Extraction of bixin from annatto seeds using supercritical carbon dioxide, Braz. J. Chem. Engng, in press (2007).

41. S. G. Andersen, M. G. Nair, A. Chandra and E. Morrison, Supercritical fluid carbon dioxide extraction of annatto seeds and quantification of *trans*-bixin by pressure liquid chromatography, *Phytochem. Anal.*, **8**, 247–249 (1997).
42. A. J. Degnan, J. H. Vonelbe and R. W. Hartel, Extraction of annatto seed pigment by supercritical carbon-dioxide, *J. Food Sci.*, **56**, 1655–1659 (1991).
43. L. W. Levy and D. M. Rivadeneira, Annatto, in G. L. Lauro and F. J. Francis (eds), Natural Food Colorants: Science and Technology, Marcel Dekker, New York, 2000.
44. Ministério da Saúde do Brasil, Legislação Brasileira para Alimentos, http://anvisa.gov.br/legis/index.html, 2007.
45. *Ministerio de Agricultura do Peru, Peru: Estadística Agraria 2000*, DGIA-MINAG, Lima, 2002.
46. FAO/WHO, Major colourants and dyestuffs mainly produced in horticultural systems, in Natural Colourants and Dyestuffs, Non-Wood Forest Products Series, Rome, 1995.
47. C. R. A. Souza and M. B. A. Gloria, Chemical analysis of turmeric from Minas Gerais, Brazil and comparison of methods for flavour free oleoresin, *Braz. Arch. Biol. Technol.*, **41**, 218–224 (1998).
48. M. L. A. Bambirra, R. G. Junqueira and M. B. A. Glória, Influence of post harvest processing conditions on yield and quality of ground turmeric (*Curcuma longa* L.), *Braz. Arch. Biol. Technol.*, **45**, 423–429 (2002).
49. G. Began, M. Goto, A. Kodama and T. Hirose, Response surfaces of total oil yield of turmeric (*Curcuma longa*) in supercritical carbon dioxide, *Food Res. Int.*, **33**, 341–345 (2000).
50. A. L. C. Méndez, N. T. Machado, M. E. Araújo, J. G. Maia and M. A. A. Meireles, Supercritical CO_2 extraction of curcumins and essential oil from the rhizomes of turmeric (*Curcuma longa* L.), *Ind. Eng. Chem. Res.*, **39**, 4729–4733 (2000).
51. M. Æ. M. Braga, P. F. Leal, J. E. Carvalho and M. A. A. Meireles, Comparison of yield, composition, and antioxidant activity of turmeric (*Curcuma longa* L.) extracts obtained using various techniques, *J. Agric. Food Chem.*, **51**, 6604–6611 (2003).
52. M. E. M. Braga and M. A. A. Meireles, Accelerated solvent extraction and fractioned exctraction to obtain the *Curcuma longa* volatile oil and oleoresin, *J. Food Process Engng*, **30**, 501–521 (2007).
53. C. Socaciu, Natural pigments as food colorants, in C. Socaciu (ed.), *Food Colorants – Chemical and Functional Properties*, CRC Press, Boca Raton, Florida, 2007.
54. F. W. Quackenbush and S. L. Miller, Composition and analysis of the carotenoids in marigold petals. *J. Assoc. Off. Anal.Chem.*, **55**, 617–621 (1972).
55. G. K. Gregory, T.-S. Chen and T. Philip, Quantitative analysis of lutein esters in marigold flowers (*Tagetes erecta*) by high performance liquid chromatography, *J. Food Sci.*, **51**, 1093–1094 (1986).
56. J. D. L. Rivas, Reversed-phase high-performance liquid chromatographic separation of lutein and lutein fatty acid esters from marigold flower petal powder, *J. Chromatogr.*, **464**, 442–447 (1989).
57. W. L. Hadden, R. H. Watkins, L. W. Levy, E. Regalado, D. M. Rivadeneira, R. B. van Breemen and S. J. Schwartz, Carotenoid composition of marigold (*Tagetes erecta*) flower extract used as nutritional supplement, *J. Agric. Food Chem.*, **47**, 4189–4194 (1999).
58. F. Delgado-Vargas and O. Paredes-López, Effects of enzymatic treatments of marigold flowers on lutein isomeric profiles. *J. Agric. Food Chem.*, **45**, 1097–1102 (1997).
59. A. G. Lloyd, Extraction and chemistry of cochineal, *Food Chem.*, **5**, 91–107 (1980).
60. L. Salas-Guerra, La tuna: derivados, precios y aproveichamiento en el Perú, monografia in la Facultad Ciencias Administrativas y Gestión de Recursos Humanos de la Universidad San Martín de Porres, Lima, Peru, www.monografias.com, 2007.
61. R. W. Dapson, A method for determining identity and relative purity of carmine, carminic acid and aminocarminic acid, *Biotechnic Histochem.*, **80**, 201–205 (2005).
62. N. Sugimoto, Y. Kawasaki, K. Sato, H. Aoki, T. Ichi, T. Koda, T. Yamazaki and T. Maitani, Structure of acid-stable carmine, *J. Food Hyg. Soc. Japan*, **43**, 18–23 (2002).
63. C. Aldama-Aguillera, C. Llanderal-Cazáres, M. Soto-Hernández and L. Castillo-Márquez, Cochineal (*Dactylopius coccus* Costa) production in prickly pear plants in the open and in microtunnel greenhouses, *Agrociencia*, **39**, 161–171 (2005).
64. C. A. Barriga-Ruiz, Cochineal production in Peru, in FAO Expert Consultation Meeting on *Non-wood Forest Products in Latin America*, Santiago, 1994.

6

Natural Dyes in Eastern Asia (Vietnam and Neighbouring Countries)

Hoang Thi Linh

6.1 Introduction

The tropical climate in Vietnam and neighbouring countries is very suitable for plants to grow. Apart from offering substantial vegetables and fruits for the food industry, some plants can be used for dyeing fabric. In this chapter we will introduce some plant species that have been used for dyeing textile products. Representative examples are listed in Table 6.1 and important plants are described in more detail below.

6.2 Annatto (Botanical Name *Bixa orellana* L., Family Bixaceae)

This species grows up to 5–10 m in height in bush. After planting, a short time later it will flower, which then becomes a cluster of fruit. The fruit ripens in 1 or 2 months and are collected in the dry season. Total time from planting to harvesting *Bixa orellana* is around 18 months. In each nut, there are several tens of seeds and a single plant can provide several hundreds of kilograms of seeds. The seed, after separation from the nutshell, will be dried and packed.

This species grows in basalt soil in tropical climates worldwide, such as Peru, Guatemala, Ecuador, Kenya, Venezuela, Brazil, India, Spain, Philippines, Thailand, Laos, Cambodia and Vietnam. The nut's colour is relatively attractive (from orange to yellow) and it has been long used worldwide. In Vietnam, a kind of *annatto* powder can be

Handbook of Natural Colorants Edited by Thomas Bechtold and Rita Mussak

Table 6.1 *Sources for natural colorants that are of relevance in the Vietnamese climate*

Name			*Region*	*Climate*	*Plant part*	*Substrate*	*Colour of dyeing*
English	*Botanical*	*Traditional*					
1. *Lipsstick tree*	*Bixa orellana L*	Annatto	Tropical	Asia, South America	Seeds		orange, yellow
2. *Tea*	*Camellia sinensis (L.) Kuntze*		Tropical	Asia	Leaves	Silk, Cotton	yellow, brown, green, black
3. *Umbrella tree*	*Terminalia catappa L*		Tropical	Asia	Leaves (Fresh and dried)		yellow, brown
4. *Makua*	*Diospyros L mollis Griff*	mackloeur	Tropical	South East Asia	Fruits	Silk	black
5. *Indigo*	*Indigofera L*	Indigo	Tropical	Asia, America, Africa	Leaves	Cotton, Silk	blue
6. *Henna*	*Lawsonia Spinosa L*		Tropical	Asia	Leaves	Hair, Finger-nail	orange, brown-red
7. *Nacre*	*Khaya senegalensis*			Asia, Africa	Leaves, Barks	Silk	yellow, brown
8. *Sappan wood*	*Caesalpinia sappan L*		Tropical	Asia	Wood	Silk, Leather, Furniture	red
9. *Pagoda tree*	*Sophora Japonica L*		Tropical	Asia	flower buds		
10. *Turmeric*	*Curcuma Longa L*		Tropical	Asia	Bulbs	Silk, Leather, Wood	scarlet, red, violet
11. *Sapodilla*	*Manilkara zapota L*		Tropical	South America, Vietnam	Leaves		brown
12. *Betel*	*Piper betle L*		Tropical	Indonexia, South-East Asia	Leaves		brown
13. *Eucalyptus*	*Eucalyptus*			Australia, Asia, Africa, America, Europe	Leaves		light green
14. Barbados pride	Caesalpinia Pulcherrima L			Mexico, Asia	Barks	silk, cotton	yellow
15. Brow-tuber	Dioscorea Cirrhosa Lour			Vietnam	Bulbs	cotton	brown, black

made out of the seeds, which is used in many dishes. Apart from that, *bixa orellana* seeds are used to dye silk, cotton, visco, lyocell, wool and polyester. The dyeing method is very simple: boil the seeds in water and then use extraction to dye using techniques depending on the material. Dyed fabric has a bright colour from orange to yellow. Extracted colorants include bixin, norbixin and ingredients of the carotenoid family.

6.3 Tea (Botanical Name *Camellia sinensis* (L.) Kuntze, Family Theaceae)

The *tea* family has 82 species of plants. In Vietnam, there are already 56 species with various uses. *Camellia sinensis* is a quite popular one and can grow more than 10 m high, but is usually trimmed to below 2 m to provide more branches and limbs, which then can offer more leaves and buds. The leaves are oval, sharp at the terminal and stem, thick and serrated at the margins along the stem. The flowers are white and blossom from October to February and fruit in April. The leaves can be collected all year round.

Camellia sinensis (L.) Kuntze originated from North India and South China and was then imported to Myanmar, Thailand and Vietnam. The species prefers an acid mountainous soil and wet climate. By processing it differently, people can make different kinds of tea (olong tea, black tea, tea buds, Chinese tea, etc.). However, many leaves are collected to use immediately as green tea or are used for dyeing purposes. These leaves dye fabric in brown, yellow-green and even black shades, after mordanting with iron sulfate. Since the tea yield in Vietnam is high, being planted in many regions, and it is an export product, it is possible to dye fabric with those leaves not used for either small production or industrial purposes. In addition, the leaves can be used for medical applications and for deodorizing.

Colouring ingredients in the leaves include polyphenols and flavonols. To prepare the dye the leaves are placed in boiling water and the extract is used directly for dyeing silk or cotton fabric. The dyeings show good light fastness and wash fastness.

6.4 Umbrella Tree (Botanical Name *Terminalia catappa* L., Family Combretaceae)

This large wood tree grows from 10 to 25 m high and has horizontal whorls of branches with single and unequal leaves. The leaves are upside down and oval, 20–30 cm long and 10–13 cm broad. Young leaves appear in February, the plant flowers in March and July and the fruit ripens in August and September. Leaves fall mostly in winter. Dry, yellow and green leaves can be collected for dyeing.

The species grows in many countries such as Cambodia, Laos, India, China, Malaysia, Thailand, Madagascar and other Pacific islands. In Vietnam, mainly used for shade, it appears everywhere from mountainous areas, remote islands to streets in cities. The wood, fruit, skin, seed and leaf are commonly used for general and medical purposes. Increasing attention is directed to the use of leaves as a source for natural dyes with good colour fastness. To extract the dye, the leaves are boiled in water and the extract is used to dye, for example, silk and cotton. Colours range from green to yellow after mordanting with alum, and iron sulfate can also be used.

6.5 *Diospyros mollis* – Mackloeur (Botanical Name *Diospyros mollis* L. Griff, Family Ebenaceae)

The plant is 10–20 m tall with branches, with leaves that are single, alternate, with an elliptical surface 5.5–13 cm long and 2.5–7 cm broad. The tree has yellow flowers blooming from January to July and fruiting from September to April. The fruit is round, shiny green when young and then becomes green-yellow or pink-yellow and finally black when ripe.

The species is widely planted in Laos, Cambodia, Myanmar and Vietnam in abundant sunlit land and is used for timber, to produce art objects, and for dyeing precious fabric, e.g. silk black, by extracting colorant from its leaves and seeds. In the dyeing process, after soaking the fabric in the extract solution, it is important to dry it out in the air to oxidize the dye into black. This process is repeated many times to achieve a dark black colour.

Colouring ingredients in the seed include hydroquinone, tannin and others. Fruit and seeds also contain an antibiotic substance used to treat parasitic worms.

6.6 Indigo (Botanical Name *Indigofera* L., Family Fabaceae)

In tropical countries there are many different varieties of indigo plant (Vietnam has 25 types), but only some of them can be used for dyeing including the following.

6.6.1 *Indigofera tinctoria* L.

This species grows in bush and is around 1 m in height, with straight branches and 7–10 cm long egg-shaped leaves, double alternate, 1.5–2.5 cm long, 0.6–1.5 cm broad, thin and narrow at the stem. The flowers grow in clusters out of spaces between the leaves, blooming all year round, and the petals are butterfly-wing shaped and yellow red. The fruit body is 2.5–4 cm long, 0.3 cm broad and contains 5–12 small seeds. The seeds are sown and planted from February to May and harvested from August to October (autumn in Vietnam). After harvesting, if roots are left, the plant will continue to grow and a second harvest is possible after several months.

The species is intertropical, lives in mountainous areas from north to south in Vietnam, and in China, Thailand, Malaysia, Cambodia, Laos, Philippine, Myanmar, Indonesia, India, Pakistan, Sri Lanka, America and Africa.

After being picked, fresh leaves are soaked in water. After a few days the extract can be used for dyeing. During this period the extracted indican is hydrolysed to indoxyl. After soaking the fabric in the dye-bath the material is dried in the air to oxidize the reduced indigo. This process is repeated several times to achieve a dark blue colour.

The extract can also be processed to indigo by adding lime and drying the precipitated insoluble indigo. As an estimate, from 100 kg of plant material 3 kg of indigo powder can be obtained, which contains 300–400 g of indigo.

6.6.2 *Indigofera galegoides* DC.

This plant grows in bushes, 1–3 m in height, with branches 10–15 cm long, which have 21 to 25 symmetric leaflets; the leaves are egg-shaped, bigger than those of *Indigofera tinctoria* L., 2–3.5 cm long and 0.5–1.5 cm broad. The flowers are pink and the nuts contain 3 mm square seeds. The harvesting and dyeing processes are similar to those of *Indigofera tinctoria* L.

6.6.3 *Strobilanthes cusia* (*Baphicacanthus*)

This is a broad-leaved plant, with a stem up to 0.4–2 m tall, ramified largely, leaves oval and symmetric and flowers a deep blue purple. It prefers humid valley or milpa near water sources, can be planted by a branch in spring and leaves can be harvested 6 months after. The extraction and indigo production is similar to the other indigo plants. The natural indigo powder contains up to 60–70 % of indigo.

6.7 Henna (kok khan, or khao youak in Laos) (Botanical Name *Lawsonia spinosa* L., Family Lythraceae)

The plant is 3–4 m tall, with board small leaves and fragrant flowers. The plant is propogated by planting branches or seeds. Before sowing seeds are soaked in warm water at around 60 °C. Leaves can be collected twice a year when the plant is 2–3 years old, which can continue for around 20–30 years. Each hectare of plant can give 1.500–2.000 kg of fresh leaves, which are dried and used as ground powder.

The species lives in tropical countries such as Egypt, Iran, etc. In Vietnam, it can grow in hilly areas or midland deltas such as Phu Tho, Ha Tay, etc.

Henna leaves are used as medicine as they have antibiotic ingredients. Traditionally, people used it to dye nails red. In some countries, it is used for dyeing hair and making cosmetics. The leaves contain lawsone, yellow orange crystals that change into red when in contact with air.

6.8 Nacre (Botanical Name *Khaya senegalensis*, Family Meliaceae)

The tree grows to a height of up to 30 m. Branches contain many subbranches with unequal leaves. The young bark is grey, smooth, scaly and becomes dark grey as it grows older. The wood is light pink with twisted fibre and is known as iron-wood, as it is strong and can be used for ship building.

Originally from Africa, nacre was imported to be planted along roads for shade. Leaves and bark are used for dyeing silk in colours ranging from light yellow to brown.

6.9 Sappan Wood (Botanical Name *Caesalpinia sappan* L., Family Fabaceae)

Sappan (East Indian red wood) is a small thorny spreading tree, which is fast growing and reaches up to 10 m in height. The wood reaches 15–30 cm in diameter. Within a year's time the plant reaches a height of 3–5 m and begins to bloom in April, continuing till December.

Sappan is cultivated as a horticultural plant. It grows well in all kinds of soil and lush growth is obtained in red soil of the countries of Sri Lanka, India, Myanmar, Thailand, Cambodia, Laos, South of China and Vietnam.

The important part of this plant is the heartwood, which contains water-soluble dyes such as brazilin, protosappanins, sappan chalcone and haematoxylum. The dye is extracted by boiling chipped wood pieces in water. Brazilin is a red dye used to colour leather, silk, cotton, wool, fibres of different kinds, batik, calico printing, furniture, floors and feathers. Alum or iron sulfate can be used to increase the binding of the dye to the substrate.

6.10 *Sophora japonica* Flowers (Botanical Name *Sophora japonica* L., Family Leguminosae)

The plant has a straight trunk, up to 2–4 m in height, flowering in clusters 10–20 cm long from April to September and fruiting from February to April. Normally, flower buds are picked up before blooming and then dried out.

The plant grows mainly in the Bac Bo delta of Vietnam and in tropical and subtropical countries such as Cambodia, Philippines, Malaysia and India. The flower bud contains rutin, which forms a yellow dye in alkali solution. The substance is often used for dyeing paper or for painting and printing Dong Ho pictures.

6.11 Turmeric (Botanical Name *Curcuma longa* L., Family Zingiberaceae)

This species grows up to 0.6–1 m high in bush, having a big root stem and a cylinder bulb, with leaves that are oval, pointed at two ends, 45 cm long and 18 cm broad. The plant flowers in August. Turmeric bulbs are often collected in the autumn. The bulbs can be sliced or ground and dried for storage.

This plant grows wildly or can be planted and is abundant in tropical countries such as China, India, Indonesia, Laos, Cambodia and Vietnam. It is planted for its bulbs, which are used as spice, food and medicine. The bulbs can also be used to dye wood, silk and leather products. Its main colorant curcumin (brown-red crystals) does not dissolve in water but it does in alcohol, ether and apolar solvents. In acid solution the dye changes colour to scarlet; in alkali solution the colour is first red and then violet.

6.12 Sapodilla (Botanical Name *Manilkara zapota* L. or *Achras zapota*, Family Sapotaceae)

The tree is high and big, whose leaves and branches grow in layers. The trees are propogated by planting branches or seeds and fruiting begins after 3 or 4 years, although leaves can be collected all year round. Sapodilla is grown mainly for its fruit. Also gum can be extracted to produce chewing gum. Bark and seeds are used for medical purposes. The bark includes saponin, tanin and alcaloids. The leaves can be used to dye fabric brown.

The species originated in South America (Mexico) and was introduced to Vietnam, growing both in the north and the south. As it prefers hot weather it is planted in areas that have a high temperature, much sunlight, no hoarfrost, high humidity and rains regularly. After growing for 3 or 4 years the plants can endure any heat and thus can be planted in every soil type.

6.13 Betel (Botanical Name *Piper betle* L., Family Piperaceae)

Betel is a climbing plant, with oval leaves (10–13 cm long, 4–9 cm broad) that grow from the root to the top. Betel can be harvested all year round, particularly in spring and autumn when its leaves grow faster than in winter.

Betel originated in Indonesia, in South-East Asia countries such as Malaysia, Philippines, Thailand, Laos, Cambodia and Vietnam. Solution extracted from boiling leaves is used for dyeing natural fabric to a high-light-fastness brown colour. The leaves contain some attars as chavibetol, chavicol (*p*-allylphenol) and tannin.

6.14 Eucalyptus (Botanical Name *Eucalyptus*, Family Myrtaceae)

The plant grows to more than 10 m in height; some species can be up to 80 m. Eucalyptus is mainly planted for its timber, rubber and its leaves, which can be extracted for oil to make medicine. There are 450 species of eucalyptus grown in Australia, Asia, Africa, America and Europe. Vietnam has 25 species, mostly white ones. They are planted for wood pulp, but when its leaves are boiled in water, the solution extracted can be used for dyeing fabric to a light green colour.

6.15 Caesalpinia Yellow (Botanical Name *Caesalpinia pulcherrima* L., Family Fabaceae)

The trunk is medium sized with a brown lumpy bark that is easy to peel off. The leaves are bipinnately compound and the leaflets are oval (10–20 cm long, 6–10 cm broad). The flat fruit is spear-shaped and contains a flat seed which is oval and brown.

The species lives in Mexico, India, Sri Lanka, China, Japan, Thailand, Myanmar, Malaysia islands, Philippines, Nepal, Cambodia, Laos and Vietnam. The plant is suitable for many kinds of lowland and midland topography. In Vietnam *Caesalpinia pulcherrima* L. is planted everywhere for ornament and shade. In addition, people use its leaves, bark, flowers and seeds for medicine as well as boiling its bark to extract a colorant for dyeing silk and cotton to yellow.

6.16 Brow-tuber (Botanical Name *Dioscorea cirrhosa* Lour, Family Dioscoreaceae)

The plant grows with a round trunk into branches. The leaves are single, oval and alternate at the end. Bulbs grow from the root, each plant having 1 or 2 grey-brown bulbs with brown-red flesh.

In Vietnam, the plants grow in mountainous or forest areas or in the countryside. The leaves can be used for leather tanning or dyeing. Traditionally the solution extracted from the bulb is used to dye cotton in a light to dark brown colour. When the dyed fabric is soaked with black mud or treated with iron-salt mordant black colour with good high-fastness is obtained.

Part III

Colorant Production and Properties

7

Indigo – Agricultural Aspects

Philip John and Luciana Gabriella Angelini

7.1 Introduction

In this chapter we have adopted a perspective that is European for *Isatis* spp., Japanese and European for *Persicaria tinctoria* (*Polygonum tinctorium*)[1] and tropical for *Indigofera* spp., these being the principal zones in which these crops are currently cultivated and are likely to be cultivated in the future.

An extensive body of knowledge on the cultivation of *Isatis* and *Persicaria* (*Polygonum*) in the UK, Spain, Germany, Finland and Italy was accumulated during the Spindigo Project (EU, 2006), which expanded on work that had been carried out in earlier collaborative European Commission-funded projects based in Bristol, UK, and in Thuringia, Germany. The present chapter draws heavily on this knowledge, which relates particularly to times of seed sowing, the control of weeds, pests and diseases, crop requirements for fertilizer and irrigation, seed production and harvest times in relation to final indigo yields. The Project also included trials on new strains and accessions of *Isatis* and *Persicaria* (*Polygonum*). Some of the extensive work carried out in Italy (Angelini *et al.*, 2004, 2005, 2007; Campeol *et al.*, 2006) and in Spain (Sales *et al.*, 2006) has been published, but, except for a summary (EU, 2006), a large proportion of the work documented in the detailed Final Report submitted to the EU in 2004 remains unpublished. A particularly detailed picture of the requirements of the *Isatis* and *Persicaria* (*Polygonum*) crops is available for Central Germany with the work

[1] *Polygonum tinctorium* is now *Persicaria tinctoria* (Mabberley, 2008). While we adopt the new name here, we also include the older name parenthetically, because of its wide use in the literature.

Handbook of Natural Colorants Edited by Thomas Bechtold and Rita Mussak

of Armin Vetter's group and for Central Italy with the work of the group of one of us (L. G. Angelini) at Pisa. The latter collaborated with growers in the region to produce crops on a commercial basis. In addition, other partners of the Spindigo Project provided data on the performance of *Isatis* crops in South Eastern Spain and in Finland. Practical advice to growers was made available in 'Agronomic Blueprints', written specifically for each country and produced in English, Finnish, Italian and German languages. Details can be obtained from the authors' websites (www.spindigo.net).

Indigo crops, like any other, should be grown according to the principles of Good Agricultural Practice (GAP). Although GAP has developed in the context of food crop production (http://www.fao.org/prods/gap/home/principles_en.htm), the GAP concept in reducing environmental impact and maintaining a sustainable production system has much to commend it for nonfood crops. Specifically, GAP recommends addressing environmental, economic and social sustainability by on-farm production and post-production processes, recommendations that are particularly applicable to indigo-yielding crops and the on-farm extraction of indigo.

Assessing the potential and actual impact of nonfood crops on the environment has recently acquired new urgency, with the predicted large-scale development of crops grown for bioethanol and biodiesel. European indigo crops are in fact one of the few nonfood crops for which an Environmental Impact Assessment (EIA) is available. An EIA for the sustainable production of indigo (Maule *et al.*, 2004) was produced by the Spindigo project (EU, 2006). According to EU Directives 85/337/EEC and 97/11/EC, which determine the assessment required for new developments, an EIA is not mandatory for the introduction of novel crops. However, there is provision within the Directives for an EIA being made on a case-by-case basis. The EIA for the sustainable production of indigo (Maule *et al.*, 2004) includes a brief comparison of the various options for obtaining natural indigo and then deals in detail with *Isatis tinctoria* and *Persicaria tinctoria* (*Polygonum tinctorium*) in terms of novel crops. In the case of *Isatis*, spring-sown oilseed rape (*Brassica napus* L.) was chosen as a comparison crop. The potential environmental impact of *Isatis* cultivation – and the corresponding mitigation measures – were assessed in relation to soil, watercourses, wildlife and in its potential to create allergy problems. Similarly, the potential environmental impact of on-farm processing was assessed and mitigation measures identified.

7.2 *Isatis*

7.2.1 Introduction

The genus *Isatis* contains about 30 annual, biennial and perennial species (Ball and Akeroyd, 1991), most of which are believed to produce indigo. All have yellow flowers and seeds contained within winged, indehiscent fruit called a silicula (or silicle), which is a short, broad type of silique. The cultivated species, *Isatis tinctoria* L., is an out-breeding (Spataro and Negri, 2007) biennial probably native to south-west Asia and south-eastern Europe, but because it has been cultivated since prehistoric times in Europe it is now found widely naturalized throughout the continent (Spataro *et al.*, 2007). *I. tinctoria* is a very variable species, and there is no agreement as to the limits of the species itself and the taxonomic status of the subspecies that have been identified. Underlining this intra-species phenotypic diversity, Gilbert *et al.* (2002) identified a high degree of genetic diversity within a selection of 28 *Isatis* landraces using amplified fragment length polymorphism (AFLP). More recently, Spataro *et al.* (2007) have extended this work using selective amplified microsatellite polymorphic locus (SAMPL) molecular markers. They concluded that the clustering of genetic variation within the wild and cultivated populations they sampled was consistent with the crop originating in central Asia and spreading west (Spataro *et al.*, 2007).

Isatis tinctoria was brought by European settlers to North America. In the western USA it has become an invasive weed which is avoided by browsing animals (Farah *et al.*, 1988, Monaco *et al.*, 2005). Other factors that contribute to its invasiveness are believed to include: allelopathic effects on the germination of native species, a high seed output, a two-layered rooting pattern of laterals in the upper 300 mm of soil and a deeply penetrating taproot (Monaco *et al.*, 2005). It may also be relevant that the species is genetically diverse, as noted above, and highly variable in its phenology and morphology.

Isatis indigotica Fortune ex Lindley, known as Chinese woad, resembles *I. tinctoria* in general features but differs as follows: rosette leaves are glaucous, waxy and glabrous; the flowering stems are shorter, flowers are paler and the siliculas have a narrow waist rather than being oval.

7.2.2 Agronomy

Crops of *Isatis* for indigo production are, or have recently been, grown regularly on a commercial scale in at least three areas of Europe: Bleu de Pastel de Lectoure, Toulouse France (http://www.bleu-de-lectoure.com); Woad-inc™, Norfolk, UK (http://www.woad-inc.co.uk); and La Campana, Montefiore sull'Aso, Italy (http://www.lacampana.it).

Isatis is a member of the Brassicaceae (Cruciferae) family. Consequently other brassica crops, particularly oilseed rape (*Brassica napus*), where considerable experience and knowledge have accumulated, can usefully be referred to for matters such as crop rotations and the control of weeds, pests and diseases. *Isatis* can be included in a 4–5-year rotation with other crops or included in a 3-year rotation with wheat–sugarbeet–soya or other grain legume. It can also be included in rotation with wheat or any winter cereal. The sequence of crops before and after *Isatis* affects the productivity of each of the crops. Thus, for example, *Isatis* following soya or other grain legumes yields more than following corn or sugarbeet. When *Isatis* is grown for seed, rather than for the indigo itself, it can be sown at the end of summer or at the beginning of September, with seed being harvested in the following June. In this case it can conveniently follow on from a cereal crop.

Altogether *Isatis* is considered a hardy crop: it is frost-resistant, especially during the rosette stage; to germinate it requires temperatures greater than 2–4 °C, and then, for vegetative growth, a minimum temperature of around 5 °C. Temperatures of around 20–25 °C seem to be optimal for leaf development and expansion. As the temperature rises above 25 °C vegetative development slows and then stops at around 30 °C (L. G. Angelini, unpublished).

7.2.2.1 Seed Sowing

Under natural conditions it is likely that seed of *Isatis* spp. would germinate in late summer, pass the winter in a rosette form and flower in the following spring, with seed production through the summer. However, for the production of indigo, leaves from the rosette form of the plant are harvested. Thus to obtain the maximum yield it is necessary to maximize the rosette stage by sowing as early as possible without allowing the plants to flower prematurely in the first year. The actual sowing dates will depend on the region (see Table 7.1) and on the weather. The plant is able to re-grow after leaves are harvested, as long as the crop has access to sufficient water and mineral nutrients. In this way more than one harvest is possible from a single crop. The number of successive harvests from a crop depends on the sowing date, climate and the prevailing weather: in Southern Spain up to 7 harvests have been recorded, in Italy 4, in Germany 3 and in the UK and Finland 2 (Table 7.1).

Table 7.1 *Sowing and harvesting dates for Isatis tinctoria grown as a crop in different regions of Europe. Unpublished data are from the Spindigo Project (EU, 2006)*

Country (site)	Sowing dates	Number and dates of successive harvests	Trial years	Reference
Spain (Valencia)	Dec–Jan	4–7 harvests May–Aug	2002, 2003	(Sales *et al.*, 2006)
Italy (Pisa)	Feb–April, March (transplants)	Up to 4 harvests Jul–Aug	2001–2003	(Angelini *et al.*, 2005)
Germany (Dornburg)		Up to 3 harvests June–Sept	2001–2003	A. Vetter (unpublished)
UK (Long Ashton)	Feb–March	2 harvests July–Sept		(Hill, 1992)
Finland (Jokioinen)	May	2 harvests July–Sept	2001–2003	A. Vuorema and M. Keskitalo (unpublished)

Isatis prefers an open-structured, well-drained, deep soil rich in humus and minerals. If a soil pan is present, the soil should be initially ploughed to a depth of 30–35 cm in either autumn or spring to permit deep penetration of the taproots. It is important to harrow the soil to a fine tilth. Seeds are sown either by broadcasting followed by a chain harrow and light rolling or by a seed-drill, directly or air assisted so that seeds are only just below the surface, 0.5 cm for the naked seed and 1–1.5 cm for the silicula. Rolling provides a level soil surface, helping the eventual harvest of the leaves, which are cut a short distance from the soil surface. The thousand-seed weight is around 6 g, when still in the silicula, but this is reduced to 1.8–2.0 g for the naked seed. The percentage germination is around 80–90 % for a good seed batch. In calculating the sowing rate, the aim is for a planting density of 30–35 000 plants/ha. Thus a seed bed that has been well prepared with soil in good condition requires 80–100 g of naked seeds/ha, while less optimal conditions with a correspondingly lower germination and seedling establishment rates would require 3–4 kg of naked seeds/ha (15 kg of siliculas/ha). The seed rows need to be about 30 cm apart, to give a final density of 30–40 plants/m^2. Emergence can be expected 2 to 3 weeks after sowing and complete cover in a further 6 to 8 weeks, depending on conditions prevailing. For example, in the dry springs characteristic of Finland, lack of water can hinder germination; conversely a very wet spring, which is not unknown in the UK, will seriously reduce germination and crop establishment.

7.2.2.2 *Developmental Stages*

For any crop, a description of developmental changes, such as emergence, vegetative growth, flowering and fruit development, and their regulation by climate and seasonal changes – the phenology of the crop – is important for its correct management (Mendham and Salisbury, 1995). From a climatological point of view, these phenomena lay the foundation for the interpretation of changes due to bioclimatic factors. From an agronomical point of view, understanding the consequences of a particular microclimate enables the response of the plant to be foreseen. Finally, from an economic point of view, definitions and coding of developmental stages allow us to optimize pollination, predict a likely pest outbreak and apply efficiently fertilizer or herbicide. However, up to the present, no specific description of the developmental stages was available for *Isatis*. One of us (L. G. Angelini) has now provided a description for *Isatis* based on observations made in Central Italy over several years and with a variety of populations. This width of observational experience helps to ensure that the description has a wide generality, given that the duration of inflorescence development, flowering and fruit development can vary with cultivar and with annual climatic variation. The developmental stages for *Isatis* are listed in Table 7.2 and illustrated in Figure 7.1.

Table 7.2 *Development stages of Isatis tinctoria. These are illustrated in Figure 7.1. From Angelini (unpublished)*

Code	Definition
A	Emergence
B1	Two true leaves
B2	Four true leaves
B3	Eight true leaves
C1	Rosette of more than 20 cm diameter
C2	Rosette of more than 30 cm diameter
D1	Appearance of young leaves on resumption of growth
E1	Formation of flower buds; the young leaves curl and enclose flower buds; elongation of the flower stem becomes apparent
E2	Flower buds just visible
E3	Formation of the secondary branches from the main stem, at the level of the nodes of the apical leaves of the flower stem; the main inflorescence is well defined and the flower stem elongation slows compared with the previous stage
E4	First flowers open
F1	Many flowers open, and first petal falls; formation of the first basal siliculas
F2	Complete formation of the siliculas on the whole inflorescence
F3	Siliculas twist
F4	Seeds mature

Development stage A – emergence

The emergence stage coincides with the above ground appearance of the cotyledon. It is important that this phase of development is as rapid as possible to guarantee the maximum competitiveness with weeds, as can be seen in the photograph, *Isatis* seedlings manage to emerge from the soil even if it does not have an optimum structure.

Development stage B1 – two true leaves

This is the stage that follows emergence, in which the first two true leaves are visible. This stage is reached about three weeks after a spring sowing.

Development stage B2 – four true leaves

This stage is easy to distinguish; with four true leaves the plantlets are somewhat more robust, but the root system is not yet sufficiently developed. It is from this stage on that the indigo precursors begin to be actively synthesised.

Figure 7.1 *Developmental stages of Isatis tinctoria. From Angelini (unpublished)*

Development stage B3 – eight green leaves

In this phase the plant reaches a size such that it can start to compete with any weeds. The taproot starts to elongate. To reach this stage takes about 40 days after sowing.

Development stage C1 – rosette of more than 20 cm diameter

The plant assumes the typical characteristics of the vegetative stage, showing morphological features that differ between populations for colour, leaf size, leaf margins and form of the rosette. Plants in cultivation can now compete effectively with any weeds. This phase is reached about 70 days after sowing.

Development stage C2 – rosette of more than 30 cm diameter

In this phase the plant reaches its final size with rosettes exceeding 30 cm diameter (from 30 a 50 cm). To reach this stage takes an average of 80 days after sowing.

This phase closes the vegetative part of the *Isatis tinctoria* life cycle. After this stage is reached the plants are harvested. At the end of the harvest season, plant development slows and continues to produce new leaves only when favourable conditions return. When temperatures drop below 7–8 °C the plants enter into a winter resting phase until they are in their second year.

Figure 7.1 *(Continued)*

Development stage D1 – appearance of young leaves on resumption of vegetative growth

In winter, bud growth stops and the plant is in a resting phase. With day-time temperatures around 10 °C and an absence of night frosts, vegetative growth returns with the emergence from the base of the rosette of new growth, distinguishable by its clear coloration and a greater consistency to the new leaves in comparison to the old. The production of new leaves varies as much due to the genetic characteristics as to the number of harvests taken in the prior vegetative phase. A plant that has been cut repeatedly produces a lower number of leaves, prefering to direct its nutritional reserves towards its reproductive organs.

Development stage E1 – Formation of flower buds

With this stage all production of new leaves stops. There appears a swelling of those leaves that surround the future flower bud. This is recognizable more than from the swelling, because the nearest leaves curl and wrap themselves around the distal part of the stem. The plant has a height of 30 cm and has at the same time the formation and differentiation of the flowering stalk which reaches a height of more than 100 cm at its complete development in the prior vegetative phase. A plant that has been cut repeatedly produces a lower number of leaves, prefering to direct its nutritional reserves towards its reproductive organs.

Figure 7.1 *(Continued)*

Development stage E2 – flower buds just visible

The main flower bud is visible because the leaf cover becomes less. The flower stalk visibly starts to elongate.

Development stage E3 – formation of the secondary racemes

At the nodes of the apical leaves of the flowering stem, secondary racemes form under the main raceme. The main inflorescence at this stage is well defined and the flowering stem continues to elongate although it lengthens less than in the previous stage.

Development stage E4 – first flowers open

In this stage the first flowers begin to open in a sequential manner in the various inflorescences.

Figure 7.1 *(Continued)*

Development stage F1 – many flowers open and formation of the first basal siliculas

In this stage some of the flowers at the base of the inflorescence lose their petals and the outlines of the future siliculas are apparent. The swelling of the siliculas confirms that they have been fertilised.

Development stage F2 – complete formation of the siliculas on the whole inflorescence

The petals fall from the flowers becoming replaced by the siliculas. The siliculas appear green and of variable size moving from the apex to the base of the flower stalk, depending on when they were formed (over as much as 10 days). The basal siliculas, relatively more developed, show an enlargement due to the formation of the seed. The completion of this stage happens about 400 days after sowing.

Figure 7.1 *(Continued)*

7.2.2.3 Weeds, Pests and Diseases

In its early stages of growth, *Isatis* is a poor competitor and thus effective weed control, especially in the establishment phase of the crop, is vital. Once a complete crop canopy is achieved, weeds are suppressed and further control becomes less important. There are several methods of weed control available to the grower. For the small-scale grower, manual weeding, hoeing or other physical methods may be appropriate, but for the large-scale grower herbicides are likely to be the only cost-effective option. The availability and use of herbicides is subject to national and international legislation, which is subject to constant review. Within the European Union, herbicide use is governed by the Plant Protection Products Directive (91/414/EEC), which harmonizes arrangements for authorization of herbicide use within the European Union. Member States can authorize the marketing and use of plant protection products only after an active substance is listed in Annex I of the Directive. Within the European Union further restrictions on herbicide use are likely with the full implementation of the EU Water Framework Directive. Thus it is important for growers to consult authoritative up-to-date guides. For example, in the UK, there are the UK Pesticide Guides issued annually (Whitehead, 2008), also available as http://www.ukpesticideguide.co.uk, published by the British Crop Protection Council, and guidance can be obtained from the Pesticide Safety Directorate, an agency of the Health and Safety Executive (http://www.pesticides.gov.uk/home.asp). As a general guide to the selection of herbicides for specific weed problems, oil-seed rape offers a useful model crop, but Specific Off-Label Approvals (SOLA) would be required for any applications on crops of *Isatis*.

Trials carried out as part of the Spindigo Project (EU, 2006) showed the following to be effective in the UK: Stomp (pendimethalin), Butisan (metazachlor), Starane 180 (fluroxypyr) and Lentagran WP (pyridate); some kill (up to 20 %) of the *Isatis* crop was observed, but with no loss in the final yield (K. G. Gilbert *et al.*, unpublished). In Germany, of a variety of herbicides tested, the best results were obtained with a mixture of Lentagran WP + Butisan + Starane 180; again the herbicide application depressed growth of the *Isatis* crop, but it had no influence on the final yield of leaves (A. Vetter *et al.*, unpublished).

Trails carried out in Italy in 2003 (Habyarimana and Laureti, 2005) showed the effectiveness of oxyfluorfen (active compound 24 %) and metazachlor (active compound 43 %) when applied a day after sowing at rates of 3 kg/ha and 2 kg/ha respectively. Herbicide treatment raised the weight of leaves harvested, especially from the second and third of the three successive harvests, to levels comparable to a crop where the weeds had been eliminated by mechanical hoeing. The herbicide treatments increased indigo yields by about 25 % compared with the untreated plots (Habyarimana and Laureti, 2005).

Isatis appears to share many of the major insect pests that affect established brassica crops. Pest species that were noted to have caused problems in Spindigo Project trials in 2002 were striped flea beetle (*Phyllotreta undulata*) (Finland), cabbage stem flea beetle (*Psylliodes chrysocephala*) (Italy) and large white butterfly (*Pieris brassicae*) (Italy). These insect pests were controlled using conventional insecticides. Birds, particularly pigeons in the UK, can cause considerable feeding damage to seedlings of crops in the brassica family. This was noted in *Isatis* trials in Norfolk, UK, in 2002. However, damage was greatly reduced with subsequent planting away from woodland refuges. Damage can be controlled by the use of bird scarers such as acoustic systems and decoys. Diseases that have been noted to affect *Isatis* crops are powdery mildew (*Erysiphe cruciferarum*) in the UK and white blister (*Albugo candida*) in Italy.

7.2.2.4 Fertilizers and Irrigation

Weedy *I. tinctoria* growing as an invasive in the western rangelands of the USA exhibits low plasticity in growth and physiology in response to variable soil nitrogen, suggesting a low-N requirement (Monaco *et al.*, 2005), but nevertheless cultivated crops require high levels of N (and P and K) for maximum leaf yields. For 1 ton of fresh leaf material, the crop requires 3.5 kg N, 0.45 kg P and 0.7 kg K. As a rule, reasonably good harvests should produce in a season about 9 t/ha dry weight, which is equivalent to 90 t/ha fresh leaf weight. On this basis, recommended fertilizer applications would be 60–80 kg/ha of P_2O_5 and 80–100 kg/ha of K_2O pre-sowing, allowing for a contribution to these two elements from amounts already present in the soil. The fertilizer rate for N should be about 150 kg/ha in total, with 50 kg/ha applied pre-sowing and the remainder as 50 kg/ha after each harvest, except the last. Nitrogen can be supplied as ammonium nitrate (NH_4NO_3), ammonium sulfate ($(NH_4)_2SO_4$) or urea (CH_4N_2O) as granules by spreader. When organic systems of cultivation are employed (e.g. complying with the European Reg. 2092/91 CEE) then the fertility of the soil needs to be conserved with the application of appropriate organic fertilizers within the rotation being employed. Alongside the application of manures derived from the production farm itself, including the spent *Isatis* leaves, suitable sources include chicken manures, dried blood or treated animal residues.

Isatis does not grow well in waterlogged soils; the soil must be free-draining. *Isatis* has a deep taproot system that enables it to withstand a period of low rainfall once the plants are established, giving multiple harvests with high yields under both rain-fed and well-watered conditions (Campeol *et al.*, 2006). In 2002, in trials carried out as part of the Spindigo Project, in Finland and Germany no irrigation was used, while in Central Italy at one site sprinkler irrigation of 30 mm was provided four times during the growing season and at another site a gun irrigator provided 15–20 mm 11 times during the season. In Valencia, south-eastern Spain, flood irrigation was used every 10 days, from sowing (December to March) until final harvest in mid-October and amounted to as much as 5000 m^3/ha, which is equivalent to a layer 500 mm deep.

In general terms, it can be concluded on the basis of data collected over two years in Central Italy that *Isatis* cultivation needs irrigation only when there is a marked drought. Irrigation is not required in years when there is a good distribution of rain during the growing season, but when this is inadequate, irrigation would be required at a rate of about 1300 mm/ha. The water can most effectively be distributed in 3 or 4 operations after each harvest. Such a rate provides less than the seasonal crop requirement, but allows for a good harvest and a reasonable saving of water in line with the tenets of a sustainable agriculture.

7.2.2.5 Harvesting and Yields

To maximize the indigo yield from a crop, the rosette of leaves is harvested repeatedly, the plants rapidly regenerating foliage from basal meristems (Figure 7.1). The re-growth is nourished initially by the persistent root system, with the number of harvests that can be taken determined by the growing conditions and length of the summer season (Table 7.1).

The machinery employed depends to some extent on local availability. For example, in Italy a modified spinach harvester with a cutter bar width of 180 cm has been used to harvest a hectare of *Isatis* (Figure 7.2). In Norfolk, UK, a modified bean harvester has been used successfully for a number of years.

Figure 7.2 *Machine used to harvest Isatis leaves in Italy. From Angelini (unpublished)*

The final yield of indigo is a result of two major factors: the indigo content of the leaves and the yield of leaves. The first of these two factors depends on genetic and environmental factors and is considered in detail in Chapter 8. Table 7.3 shows how the sowing date affects the yield of leaves and of indigo for crops in Italy; a late sowing (for this climatic zone) towards the end of May leads to significantly lower yields. Figure 7.3 shows a pattern of increasing the indigo precursor content and indigo potential as the season advances, with

Table 7.3 *Effect of sowing date on yields in Isatis tinctoria. Field experiments were carried out in Central Italy in 2001. From Angelini (unpublished)*

Sowing date	*Plant fresh yield (t/ha)*	*Leaves fresh yield (t/ha)*	*Plant dry yield (t/ha)*	*Leaves dry yield (t/ha)*	*Indigo (k/ha)*
30 March	103	44	16	6.7	273
26 April	98	41	14	6.5	298
22 May	57	27	8.5	4.2	193
Significance	*	*	*	*	*
LSD (P = 0.05)	11.9	5	1.7	0.8	41.7

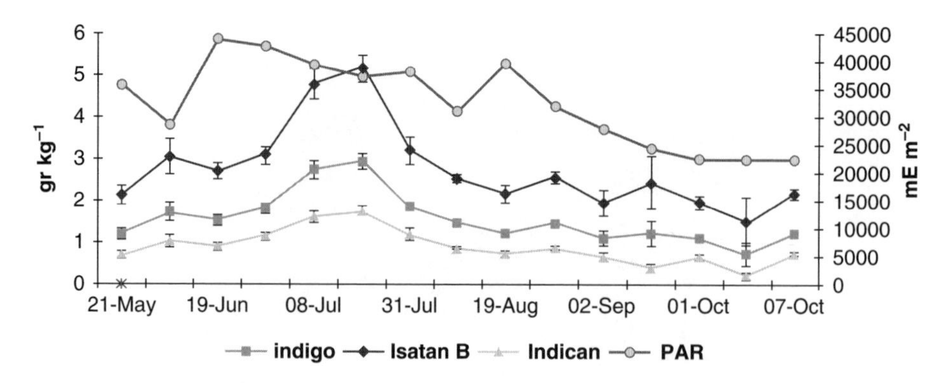

Figure 7.3 Changes in the photosynthetically active radiation (PAR) and in the levels of indigo precursors and of the potential indigo yield during the growing season of Isatis tinctoria cultivated in Pisa, Centeral Italty, 2002. Fram Angeline (unpublished)

a peak around July when temperatures are highest and light intensity is at a maximum. These results with field-grown *Isatis* crops echo earlier experiments (Stoker *et al.*, 1998, Gilbert *et al.*, 2000), in which it was shown that the indigo yield of pot-grown *Isatis* is increased by exposure to higher light intensities prior to harvest. The second factor depends crucially on how many harvests it is economic to take in a season, which in turn depends, among other factors, on the rate of foliage re-growth between harvests. Figure 7.4

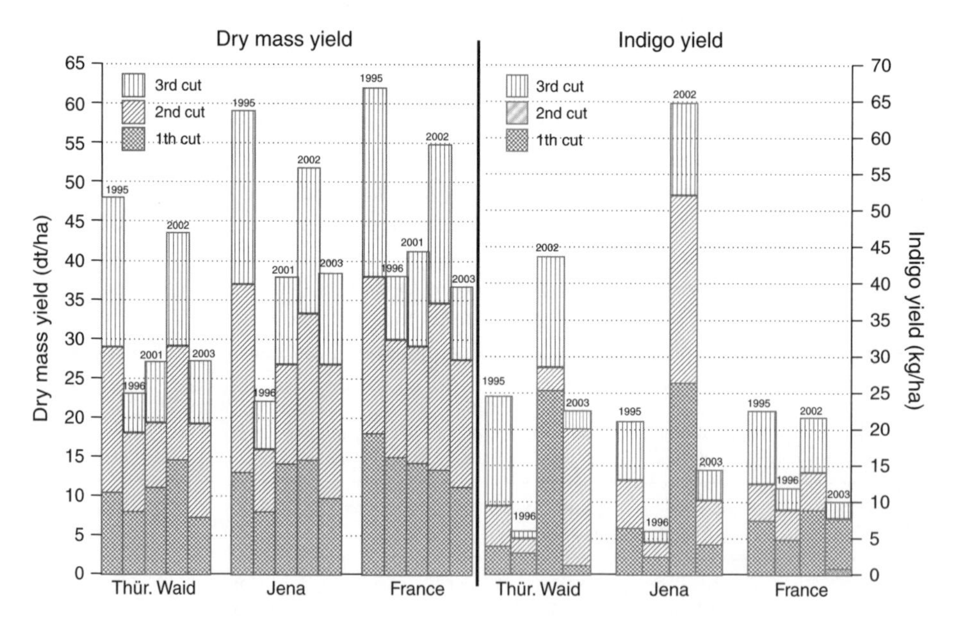

Figure 7.4 Total leaf dry matter and indigo yields from three successive harvests per year taken in 1995, 1996, 2001, 2002 and 2003 for three accessions of Isatis cultivated in Central Germany. From A. Vetter (unpublished). Reproduced by permission of Prof Vetter

illustrates how yields of foliage and of indigo depend on the prevailing conditions and how effectively the season is used by the crop.

In general terms, a year in which the weather has been conducive to a high production of foliage does not necessarily correspond with an equally high yield of indigo. This is exemplified in the data illustrated in Figure 7.4, where 1995 provided high yields of foliage for all three accessions, but 2002 provided for two of the accessions the highest yields of indigo. A high yield of leaves is always going to be necessary for a high yield of indigo, but it is not the only factor. It is important to note that when one is considering environmental factors that are conducive to foliage production and to indigo yield, morphological and phenological variability within *Isatis tinctoria*, which were noted above, come into play, so that a vintage year for one genetic line will not necessarily be a vintage year for another line. Heavy yields of indigo depend on a number of interacting factors and, as illustrated by the data of Figure 7.4, the accession (*Jena*) which in 1996 produced 6 kg of indigo/ha, in 2002 produced 65 kg of indigo/ha. The low yields in 1996 in effect signal a crop failure for that year. In the experience of researchers in the UK and Germany over the decade 1994 to 2004, unsuitable weather will lead to a crop failure of this type one year in four (A. Vetter and K. G. Gilbert, unpublished).

Among agronomic factors that affect yield, row distance can be important. Plants grown in narrow rows (15 cm as opposed to 30 cm apart) are more upright, which makes the mechanical harvest more effective. This in turn leads to a higher mass of harvested leaves, but, on the other hand, exposure of the leaves to sunlight will be lower, a factor that might be important in higher latitudes or cloudier climates. The advantage of narrow rows is illustrated by the data in Figure 7.5, which show that in a trial in Germany, indigo yields per

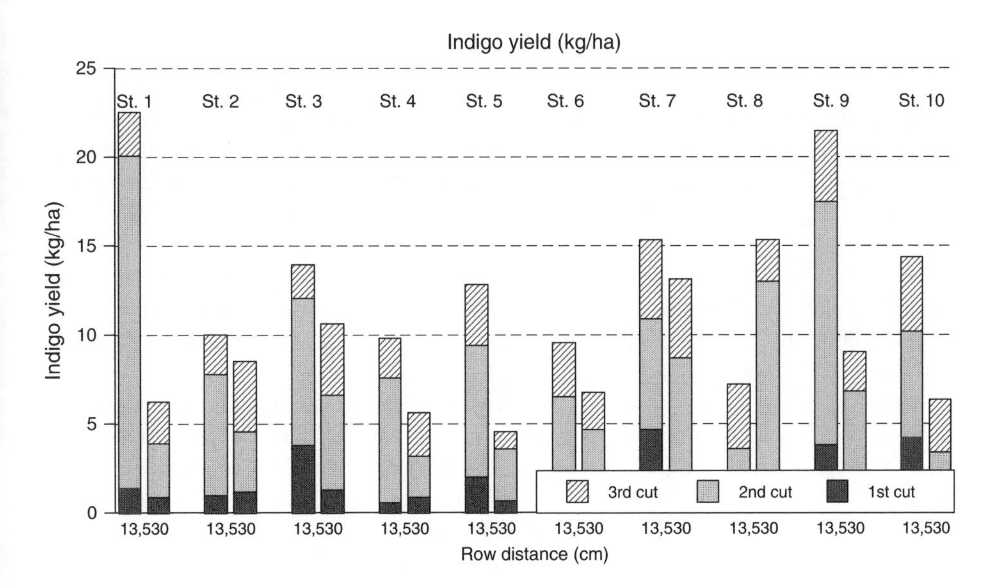

Figure 7.5 *Influence of row distance on indigo yield of 10 different Isatis accessions (St 1–10) cultivated in 2003 in Central Germany. From Vetter (unpublished). Reproduced by permission of Prof Vetter*

hectare were higher for the crops grown in closely spaced rows. This was consistent for 9 out of 10 different accessions, with some accessions showing greater advantage of the narrow row spacing than others, probably due to interaccession variation in rosette morphology, especially in how upright the leaves are. In this trial the second cut provided the highest indigo yield of the three cuts taken, and it is with this harvest that the greatest difference due to row distance becomes evident.

7.2.2.6 Seed production

Isatis can grow as a biennial, monocarpic perennial or as a winter annual, but for seed production in cultivation it is most conveniently grown as a biennial. When plants grown for indigo production are left in the ground in the autumn after leaf harvesting, they will flower the next spring and seed can be harvested, thereby avoiding the need to sow a special crop. The effect of sowing date on seed yield and quality is shown in Table 7.4.

Table 7.4 *Seed yield and seed quality of Isatis tinctoria from different sowing times. Field experiments were carried out in Central Italy. From Angelini (unpublished)*

Sowing dates in 2001	*Harvest maturity date in 2002*	*Cycle length (days)*	*Siliculas (g/m2)*	*Stems (g/m2)*	*HI*[a]	*Naked seeds (g/m2)*	*TSW*[b] *(g)*	*Germination (%)*
23 March	27 May	430	612	782	0.44	209	1.64	91
6 April	27 May	396	478	730	0.40	163	1.61	85
23 May	27 May	369	434	634	0.41	148	1.61	92
27 June	27 June	365	410	519	0.44	140	1.65	77
14 Sept	31 May	259	524	313	0.63	179	1.64	87

[a] Harvest index.
[b] Thousand seed weight.

Each plant produces hundreds of siliculas each containing a single seed. In Italy the yield under experimental conditions is about 0.6 ton/ha of naked seed, while in Finland naked seed yields of 100–250 kg/ha have been recorded (A. Vuorema and M. Keskitalo, unpublished). In Italy three farm-scale productions of *Isatis* seed gave yields of 1.4, 2.4 and 2.4 ton/ha of siliculas with 261, 344 and 408 kg/ha of naked seed (L. G. Angelini, unpublished). Different populations of *Isatis* will have different requirements for vernalization and day-length for the induction of flowers. Different accessions grown under similar conditions have been shown to differ by one month in the timing of flowering (L. G. Angelini, unpublished).

Plants flower and ripen seeds over a period of time, and synchrony of development is not strong in the crop strains used at present. This means that yields from combined harvested seed crops are lower than they could be, and the scattering of early produced seed prior to harvesting can create a weed problem in subsequent crops. However, over a period of successive seed production crops, synchrony would be expected to increase simply on account of selection.

The usual agricultural machinery, e.g. a combine harvester, can be used for harvesting the ripe woad seed. Ripeness is indicated when the husks are blue-black and the stems and leaves are green turning brown. Full ripeness leads to a loss of seed yield. The dry matter content is about 65 % and drying in a drying room is needed. The dried material consists of about 80 % of seeds contained within a silicula, together with stems and leaves. A clover huller can be used to husk the seeds. In seed production, the multiplication factor is high at 1:>60, and loss of seeds at harvest is about 25 % from a reap thresher, a figure comparable with that of other oil seed crops (L. G. Angelini, unpublished).

7.2.2.7 *Isatis indigotica compared with Isatis tinctoria*

I. indigotica has been cultivated for a long time in China (Cardon, 2007) but was introduced to the West only after its description in the 1840s by Robert Fortune and by a French team investigating the Chinese silk industry (Cardon, 2007). Far more experimental work has been done on *Isatis tinctoria* than on *Isatis indigotica*. In terms of cultivation the main differences between *I. indigotica* and *I. tinctoria* are shown in Table 7.5. In Italy, *I. indigotica* shows greater potential indigo yields (Angelini *et al.*, 2007), while in Spain (Sales *et al.*, 2006), potential indigo yields were similar, and *I. tinctoria* gives higher leaf yields. In order to take advantage of the higher potential indigo yield in *I. indigotica*, three agronomic challenges need to be met. First, *I. indigotica* has to be sown later than *I. tinctoria*, because *I. indigotica* is more sensitive to vernalizing triggers, presumably attributable to its adaptation to a warmer climate compared with that to which *I. tinctoria* is adapted. This late start means that *I. indigotica* avoids much of the earlier part of the growing season that is available to *I. tinctoria* (Sales *et al.*, 2006; Angelini *et al.*, 2007). Low temperatures (<5 °C) quite possibly induce flowering when detected at any stage in *I. indigotica* while *I. tinctoria* needs to be in the rosette stage to be induced (A. Vetter and L. G. Angelini, unpublished). Secondly, *I. indigotica* appears to be more susceptible to the

Table 7.5 *Comparison of Isatis indigotica with Isatis tinctoria in cultivation in south-eastern Spain and in Central Italy in 2003*

Characteristic	*I. indigotica*	*I. tinctoria*
Spain (Sales *et al.*, 2006)		
Vernalisation requirement for bolting	More ready to flower prematurely allowing only 4 successive harvests	Resistant to premature flowering allowing up to 7 harvests
Leaf yield (4 harvests)	9.1 ton FW/ha	33 ton FW/ha
Indigo potential/g leaf	0.6 g/kg FW	0.5 g/kg FW
Indigo potential yield/ha	5 kg/ha	17 kg/ha
Italy (Angelini *et al.*, 2007)		
Vernalization requirement for bolting	Few cool nights lead to premature flowering when sown in early spring	Resistant to premature flowering when sown in early spring, allowing up to 4 harvests
Leaf yield (2 harvests)	14 ton FW/ha	19 ton FW/ha
Indigo potential /g leaf	2.21 g/ kg FW	0.98 g/kg FW

pests of *Isatis*, the cabbage webworm (*Hellula undalis*) and nematodes (*Meloidogyne* spp.) (Sales *et al.*, 2006), the flea beetle (*Psylliodes chrysocephala*) and to mildew (*Erysiphe cruciferarum*). Thirdly, *I. indigotica* is less capable of re-growth after successive harvests; for example, in south-eastern Spain under conditions where up to seven harvests of *I. tinctoria* were possible, only four harvests of *I. indigotica* were obtained (Sales *et al.*, 2006).

7.3 *Persicaria* (*Polygonum*)

7.3.1 Introduction

*Persicaria tinctoria (*Ait.) Spach (syn. *Polygonum tinctorium* Ait.) (Mabberley, 2008) is a subtropical, annual member of the Polygonaceae family, which has been grown for centuries in Japan and China as a source of indigo. A native of Vietnam and southern China, *Persicaria* (*Polygonum*) is adapted to growing in wet places, and therefore would be expected to require a warm humid tropical or subtropical climate (Balfour-Paul, 1998; Cardon, 2007). In fact, it can be successfully cultivated in France (Cardon, 2007), Central Europe and Italy (Angelini *et al.*, 2004), but outside the traditional areas of cultivation in Asia, *Persicaria* (*Polygonum*) as a crop is much less well known than *Isatis*.

A detailed account of *Persicaria* (*Polygonum*) cultivation by traditional methods in the Tokushima prefecture, Shikoku Island of Japan, has been provided by Ricketts (2006). Seed is sown into seed beds treated with fertilizer and pesticide, after danger of frost has passed (early March in Tokushima). The bed is covered with a layer of sand and a seedling sheet to retain moisture in the soil and protect the delicate seedlings against a late frost. In mid-April bunches of 5 or 6 seedlings are transplanted 30 cm apart in rows 80 cm apart into fields liberally manured with cow manure and composted indigo stems, and treated with chemical fertilizer, pesticides to eliminate cutworms, lime and soil bacteria. Seedlings are transplanted so that both roots and stem are buried in the finely tilled soil, with just the leaves above soil level to allow the stem to root at the nodes. Fertilizer is added twice and the rows tilled and hoed so that soil builds up as a raised bed around the rows of plants. The rains that regularly fall in June allow the plants to grow enough for the first harvest in late June or early July. Plants are cut a few cm from the ground to allow for re-growth, which provides leaves for a further two harvests.

In Europe, research conducted as part of the Spindigo Project (EU, 2006) found that *Persicaria* (*Polygonum*) is well adapted to cultivation in Central Germany and Central Italy. The growing seasons in the UK and Finland were too short for reliable production, and south-eastern Spain was too hot. However, we note that a limited trial of a *Persicaria* (*Polygonum*) crop grown in Wales, UK, in 1989 has been described (Hill, 1992). Seedlings were planted out in early June from an April sowing, and leaves were harvested for extraction in late August.

From three years of extensive work in Germany (EU, 2006), it was concluded that *Persicaria* (*Polygonum*) was the most suitable indigo-yielding crop plant for Central European conditions. *Persicaria* (*Polygonum*) has similar leaf yields to *Isatis* (2–5 ton/ha), but it has 3–5 times the indigo content (mean of 1.4 % dry weight). In 2002 and 2003, with three cuts per year, the yields of *Isatis* indigo varied from 10 to 65 kg/ha, depending on a combination of weather and variety, and from *Persicaria* (*Polygonum*) with two cuts per year, the indigo yield varied from 50 to 168 kg/ha (A. Vetter, unpublished).

Extensive trials in Central Italy in 2001 and 2002 (Angelini *et al.*, 2004) of three accessions showed that *Persicaria* (*Polygonum*) requires high air temperature with a minimum of 10 °C, typical of warm season crops such as maize. It was found that the *Persicaria* (*Polygonum*) crop requires 2017 growing degree days; three harvests are possible from July to November; and the potential indigo yield was up to 326 kg/ha. The variation in indican content and indigo potential of the crop and the variation of photosynthetically active radiation through the growing season are shown in Figure 7.6. Of three accessions examined, a white-flowered line gave a lower leaf biomass, and thus lower potential indigo yields, than two other pink-flowered lines, when all three lines showed similar indican contents (Angelini *et al.*, 2004).

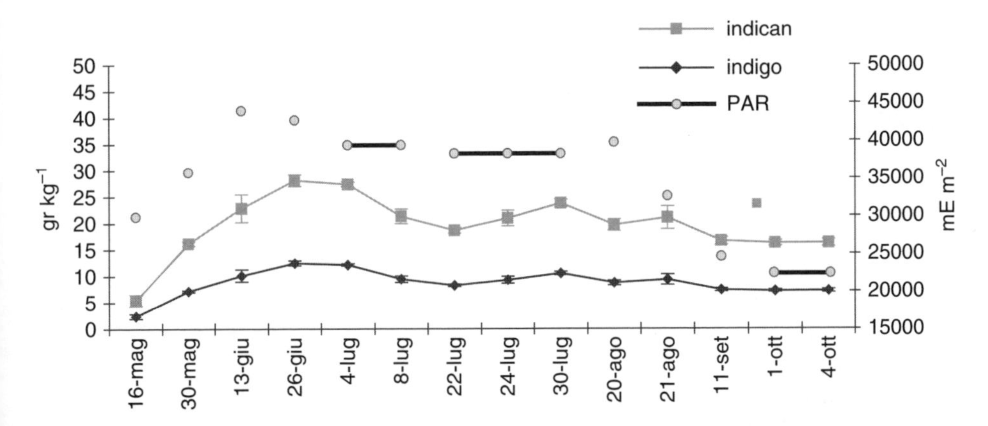

Figure 7.6 *Changes in the photosynthetically active radiation (PAR) and in the levels of indican and of the potential indigo yield during the growing season of Persicaria tinctoria (Polygonum tinctorium) cultivated in Pisa, Central Italy, 2002. From Angelini (unpublished)*

Different yields from different harvests were rationalized by Angelini *et al.* (2004) by showing that the variation in indican content (12–25 g/kg FW) could be related to the total daily values of photosynthetically active radiation for the preceding period (Figure 7.7). Thus, provided sufficient soil water is available, maximum yields are obtained when there is most solar radiation, in June and July (Figure 7.6).

7.3.2 Agronomy

7.3.2.1 Developmental Stages

The rationale and utility of describing a defined set of developmental stages for a crop have already been identified for *Isatis* above. As for *Isatis*, so for *Persicaria* (*Polygonum*), a series of developmental stages have been defined by one of us (L. G. Anelini, unpublished), based on several years of observations of the crop grown in Central Italy. These are shown in Table 7.6 and illustrated in Figure 7.8.

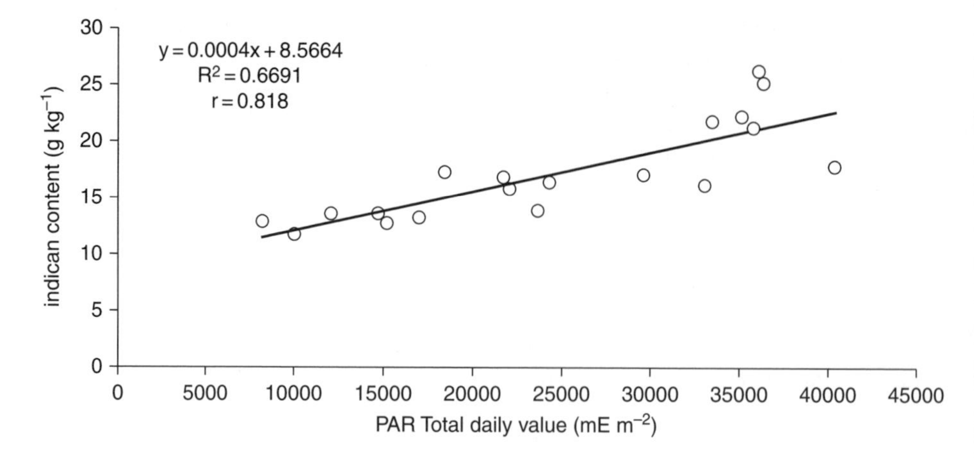

Figure 7.7 Relationship between the indican content of Persicaria tinctoria (Polygonum tinctorium) leaves and the total daily photosynthetically active radiation (PAR). Plants were cultivated in Italy, and data are derived from two years of experiments. From Angelini et al. 2004. Reprinted with permission from Journal of Agricultural & Food Chemistry, **52**, Copyright 2004 American Chemical Society

Table 7.6 Developmental stages of Persicaria tinctoria (Polygonum tinctorium). These are illustrated in Figure 7.8. From Angelini (unpublished)

A1	Emergence
B1	Two true leaves
B2	Four true leaves
C	Branching of main stem; start of secondary stem branching
H20	Height of stems 20 cm
H40	Height of stems 40 cm
D	Differentiation of the raceme: the future raceme of 1–2 cm is seen between the enfolding leaves
F1	Flowering begins: 50 % of the flowers of the racemes present between the first 3 nodes are open
F2	Full flowering: all the flowers of the racemes present between the first 3 nodes are open
M1	Start of seed formation: 50 % of the flowers of the racemes present between the first 3 nodes forming seeds (inside the perianth the achene being formed is visible)
M2	End of seed formation: all the flowers of the racemes present between the first 3 nodes have formed seeds (inside the perianth the already formed achene is visible)
H1	Start of seed ripening: 50 % of the achenes acquire a dark colour
H2	Complete ripening: seeds disperse easily from the raceme

7.3.2.2 *Sowing, Harvesting and Yield*

In Europe, sowing times for *Persicaria* (*Polygonum*) need to take into account two factors. On the one hand, germination requires a warm, moist soil, with temperatures above 10 °C, suboptimal conditions leading to a loss of the ability to germinate. *Persicaria* (*Polygonum*) is frost-sensitive, and all danger of frost must have passed before emergence. On the other hand, the earlier seed is sown, the longer the vegetative phase and the greater the yield of indigo. This means that in Central Europe, sowing should be in April at the earliest, with later sowings up to the middle of May; dye yield is always lower from later sowing dates

Development stage A1 – emergence

Emergence arrives about 15 days after sowing, the seedlings considered germinated when the plumule and radical are clearly visible.

Development stage B1 – two true leaves

About 20 days after sowing 2 true leaves are visible and completely developed.

Development stage B2 – four true leaves

Plant development continues with the appearance of the first four true leaves, about a month after sowing.

Figure 7.8 *Developmental stages of Persicaria tinctoria (Polygonum tinctorium). From Angelini (unpublished)*

Development stage C – Branching of main stem

The stem develops with branching from the base, starting at about 6 weeks after sowing.

Development stage H20 – height of stems 20 cm

Vegetative stage in which the brances reach a height of 20 cm. From here on vegetative growth occurs with lengthening and branching of the stems, average length of 20 cm. The plants arrive to this stage after about 3 months from sowing. It is in this stage that Polygonum needs optimum conditions of water and nutrients to maximise its vegetative development and thus its indigo yield.

Development stage H40 – height of stems 40 cm

Vegetative stage in which further lengthening and branching of the stems leads to a height of 40 cm about 3.5 months after sowing.

Figure 7.8 *(Continued)*

Development stage D – differentiation of the raceme

Stage in which the leaves begin to curl, indicating that the raceme is starting to differentiate. In the following week we pass into the raceme differentiation stage when the apical leaves are seen to assume a more curled appearance and then the flower buds develop, about 115 days after sowing.

Development stage F1 – flowering begins

The start of flowering with 50% of the flowers present between the first 3 nodes open. Influenced by the daylength (actually hours of darkness), this happens about 125 days after sowing.

Developmental stage F2 – full flowering

The full flowering is reached about a month after the earlier stage, and it corresponds to the complete closure of all the flowers present between the first 3 nodes, about 155 days after sowing.

Figure 7.8 *(Continued)*

Developmental stage M1 – start of seed formation

The start of seed formation is characterised by about 50 % of the flowers of the racemes present between the first 3 nodes with seeds being formed (inside the perianth the forming achene is visible) 180 days after sowing.

Developmental stage M2 – end of seed formation

At the completion of seed formation, all the flowers of the racemes present between the first 3 nodes have formed seeds, and visible inside the perianth is the already formed achene, 220 days after sowing.

Developmental stage H1– start of seed ripening

This stage arrives after 1 or 2 weeks, according to the population, and the achenes acquire a dark colour, 230 days after sowing.

Figure 7.8 *(Continued)*

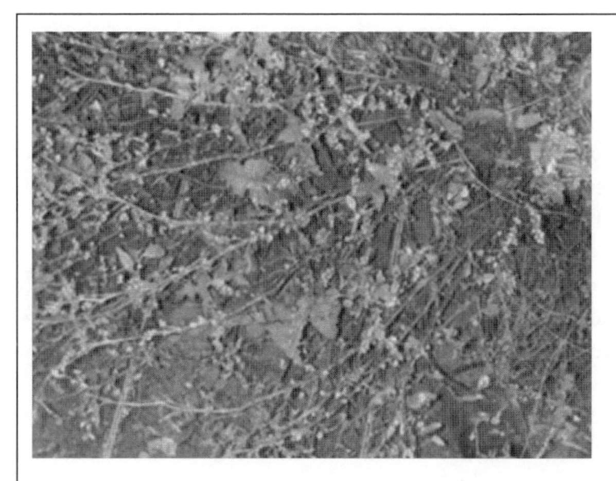

Developmental stage H2 – complete ripening

If the seeds do not reach full ripeness in the field, because of unfavourable meteorological factors, the plants are cut and put to dry in a glasshouse until there is a complete dehiscence of the seeds.

Figure 7.8 *(Continued)*

(A. Vetter, unpublished). In Central Italy sowing dates would be earlier and again later sowing reduces yield; for the 2001 crop, delaying sowing from late March or late April to late May reduced the number of successive harvests from two to one, with a consequent reduction in indigo yield (Table 7.7).

Table 7.7 *Effect of sowing date on yields in Persicaria tinctoria (Polygonum tinctorium). Field experiments were carried out in Central Italy in 2001. From Angelini (unpublished)*

Sowing date	*Plant fresh yield (ton/ha)*	*Leaves fresh yield (ton/ha)*	*Plant dry yield (ton/ha)*	*Leaves dry yield (ton/ha)*	*Indigo (kg/ha)*
11 March	113	45	18	7.2	312
11 April	117	45	17	7.4	357
10 May	82	38	11	5.6	292
11 June	58	25	6	3.5	155
11 July	49	25	5	2.5	137
22 August	7	4	1	0.8	18
Significance	*	*	*	*	*
LSD ($P = 0.05$)	13.2	6.6	1.6	0.87	55.8

A feature of the growth pattern relevant to the indigo yield is that, in the later stages of development, weight increase is represented by the growth of stalks as flowers start to develop. The stalks are free of indigo precursor, which is confined to the leaves. For this reason harvesting is best done before the end of the vegetative growth period. A harvest with a relatively low proportion of stalks needs less space for extraction, transport and needs less water for extraction. A conclusion reached from many trials is that the best time for the first cut of the crop is reached when the ratio between leaves and stalks is about 1:1, when the crop rows are closed by the developing *Persicaria* (*Polygonum*) plants. The time when this takes place under Central European conditions is from the beginning to the middle of August, and about a month earlier in Italy (Figure 7.6). If the harvest is too early, re-growth of the plants, perhaps because of their smaller roots, is slower and the total yield lower. Trials indicate that more than two cuts under Central European conditions are not useful. Even if the amount of indigo obtained from a third cut is significant, the additional costs for the third cut are unlikely to be covered by the extra indigo obtained (A. Vetter, unpublished). The date for the first harvest will depend on prevailing weather conditions. Under Central European conditions it is rarely before the end of July, and generally at the beginning of August. *Persicaria* (*Polygonum*) plants need at least 6 weeks to re-grow fully, and so the stand can be harvested after a second cut around the middle of September. The first harvest should not be made after an extended period of dryness, but one should either wait for rain or irrigate in order to promote re-growth. Under the conditions of Central Italy (Angelini *et al.*, 2004), three successive harvests are possible.

7.3.2.3 Weeds, Pests and Diseases

General considerations regarding herbicides have already been described with regard to *Isatis*. As a crop *Persicaria* (*Polygonum*) is relatively fast-growing, and therefore a single application of a herbicide before the emergence of the seeds or in an early stage of plant development can be enough to control weeds. In later stages the plant is able to suppress germinating weeds. Thus, especially with a later sowing date, a field already prepared for the cultivation of *Persicaria* (*Polygonum*) at the end of March or beginning of April will have weeds that have emerged before sowing. These weeds can be effectively treated with Roundup (glyphosate) shortly before drilling the *Persicaria* (*Polygonum*) crop.

Vetter and colleagues working in Dornburg, Germany (unpublished) have investigated over a number of years the effectiveness of herbicides for the *Persicaria* (*Polygonum*) crop. They found that the pre-emergence herbicide, Patoran FL (metobromuron), applied soon after emergence, was useful with a reasonably good tolerance by *Persicaria* (*Polygonum*), although heavy rain following an application moved the herbicide to lower soil levels causing it to damage the crop. In later trials, Afalon (linuron) performed well, with good tolerance by *Persicaria* (*Polygonum*) and useful levels of control. However, as far as the European Union is concerned, regulation has overtaken this work, and we note that the EC no longer permits use of metobromuron and EC approval for linuron expires in December 2008 (Whitehead, 2008). In Italy, weed development has been monitored during *Persicaria* (*Polygonum*) trials, with *Polygonum aviculare*, *Amaranthus retroflexus* and *Chenopodium album* being noted as serious competitors, especially after the first harvest, and before the *Persicaria* (*Polygonum*) plants had re-grown (L. G. Angelini, unpublished). Weed control has been accomplished with glyphosate as a pre-sowing and Fusilade (fluazifop-P-butyl) as a post-emergence herbicide.

Regarding pests, diseases and other problems, *Persicaria* (*Polygonum*) seems to be relatively trouble-free, although plants may be affected by strong winds, which injure leaves and cause indigo losses. In Italy, in some years a massive presence of *Ostrinia nubilaris* (Lepidoptera) has been noted from June onwards. The attack was treated with Confidor (imidacloprid) (L. G. Angelini, unpublished).

7.3.2.4 *Fertilizer and Irrigation Requirement*

An important requirement for maximizing yield in *Persicaria* (*Polygonum*) is to ensure an adequate supply of water to meet the demands for re-growth after the first and second harvests. An adequate supply of water is also necessary to ensure that the plants have sufficient nitrogen, the availability of which is related to both leaf growth and indican synthesis. The effect of water supply in Central Italy can be seen when yields are compared between 2001 (a relatively dry year) and 2002 (a relatively wet year), which were respectively 82 and 120 tons fresh weight of plant/ha (Angelini *et al.*, 2004). The beneficial effects of irrigation on indigo yields in *Persicaria* (*Polygonum*) grown in Central Italy have been fully documented (Campeol *et al.*, 2006).

Persicaria (*Polygonum*) as an organic crop has been trialled in a field trial at two sites in Lower Austria in 1999 (Hartl and Vogl, 2003). Seed was sown in mid-May in rows 45 cm apart and with 20 cm spacing (10 plants/m^2). Irrigation was supplied, weeds were controlled mechanically and successive harvests were carried out at the end of July and August and then in September and October. Leaf dry matter yield was comparable to that obtained from conventional farming, and indican contents varied between different accessions to the indigo equivalent of 0.22–0.64 % dry matter, which were lower values than those obtained with conventional farming.

7.3.2.5 *Seed Production*

Persicaria (*Polygonum*) is obligately monocarpic (dies after flowering) and flowering requires short days (<14 hours/day), which means that flowering starts around mid-August in Central Europe, where seed production occurs only with a long warm autumn without frosts. Seed yields under such conditions have been about 500 kg/ha (A. Vetter, unpublished). As an example of the pattern of *Persicaria* (*Polygonum*) seed production in Northern Italy (Cesena), a crop was sown in April, flowering began in early July, seed filling started in mid- August, the seeds started to mature in mid-September and seeds were completely mature by November (L. G. Angelini, unpublished). However, in this region, climatic conditions for seed filling and maturity are not always favourable, with the possibility in some years of low temperatures and frequent rainfalls. In Central and Southern Italy, autumn weather is more reliable for seed production, but seed production is possible only where water is available in nonlimiting quantities during the spring and summer. With favourable conditions in Northern Italy, seed production can amount to 400 kg/ha.

7.4 *Indigofera*

Indigofera is a large genus of some 700 species, the main cultivated species being *Indigofera tinctoria* L. (=*I. sumatrana* Gaertn.), *I. suffruticosa* Miller (=*I. anil* L.), *I. arrecta* Hochst. Ex A. Rich., *I. micheliana* Rose and *I. coerulea* Roxb. *Indigofera* spp. are annual or perennial herbs or small shrubs growing from 1 to 4 m high with pinnate leaves and papilionaceous flowers characteristic of the Fabaceae (Leguminosae) family. The different species are adapted to a range of tropical and subtropical climates (Table 7.8).

Table 7.8 Indigofera species cultivated for indigo

Indigofera species	Common name	Region where cultivated
I. tinctoria L.	Indian indigo	South Asia, widely in tropics
I. suffruticosa Miller	Platanillo	Lowland Central and South America, southern Japan
I. micheliana Rose	Jiquilite	Central America especially El Salvador, higher altitudes
I. coerulea Roxb.	Indigotier sauvage	Drier tropics
I. arrecta Hochst. Ex. A.Rich.	Natal indigo, Java indigo	Africa, India, Java

In recent years a revival of commercial *Indigofera* cultivation has occurred, notably in India (see, for example, www.thecoloursofnature.com), Bangladesh (Cardon, 2007), Mexico and El Salvador (Cardon, 2007).

A detailed description of *Indigofera* cultivation as it was carried out in early 19th century Bengal and in North India is given by Darrac and van Schendel (2006). At that time the average annual production from Bengal was substantial at around 3 million kg of indigo cake (Darrac and van Schendel, 2006), and their descriptions are sufficiently detailed to be of interest to anyone wishing to revive the cultivation of *Indigofera* in these regions. Darrac and van Schendel described the kind of land that was most suitable, how it was ploughed, the source and varieties of seed used and when it was sown. They also provided an account of how the crop was weeded, harvested and transported to the indigo extraction factories. A brief description of current cultivation practice is given by Cardon (2007) on which the following is based. Seeds are sown at a rate of 20–30 kg/ha and field spacing is about 50–60 cm. Branches are harvested after 4–5 months by cutting about 10–20 cm above soil level. A second cut is made after 2–4 months. Up to three harvests per year are taken, with the crop remaining productive for 2 or 3 years. *I. suffruticosa*, mainly grown in Central and South America, is usually propagated by 30 cm-long cuttings and harvested as described for *I. tinctoria*.

The indigo precursor in *Indigofera* is indican and is confined to the leaves. Yield therefore depends on the mass of leaves per unit area cultivated and on the leaf content of indican. Research carried out by the British in colonial India in the first decade of the 20th century and summarized by Perkin and Everest (1918) established that the leaf normally constituted about 40 % of the harvested material and that the indican content (measured as indigo yielding capacity) varied during the year from 0.2 % to 0.76 % of the leaf dry weight. Comparison of young and mature leaves gave figures of 0.71 % and 0.35 % respectively. Although exceptional samples of *I. tinctoria* might yield indigo to a value of 3.5 %, higher indican contents (1.8 %) were routinely observed with leaves from *I. arrecta* compared with the generally grown *I. tinctoria*, and this species was adopted once problems of seed germination had been solved (see also Kumar, 2004).

More recent data have been reported by Sandoval-Salas *et al.* (2006) who worked with *I. suffruticosa* growing in Veracruz, Mexico, at an altitude of 800 m. The indigo yield in terms of dry weight was about 1 % of leaf dry weight up to 150 days after sowing; yield then fell as the plants aged. The biomass of leaves increased up to 150 days after sowing, and then entered a stationary phase. Thus the indigo yield per hectare peaked at 150 days after sowing at about 10 kg/ha, falling to 6 kg/ha with later harvests.

There have been no published determinations by high-performance liquid chromatography (HPLC) and other modern techniques of the indican content of *Indigofera* leaves, and therefore it is difficult to compare the indigo-yielding potential of *Indigofera* with that of other major sources of indigo, such as *Isatis* and *Persicaria* (*Polygonum*). In the absence of this information, it is also difficult to estimate the efficiency of the methods used to extract indigo from *Indigofera* (Cardon, 2007). Clearly the kind of research that has been carried out on other indigo crops (Angelini *et al.*, 2004, 2005, 2007, Sales *et al.*, 2006) urgently needs to be applied to *Indigofera*.

Acknowledgements

The authors are grateful to partners of the Spindigo Project and Dr R. J. Froud-Williams for useful discussions.

References

Angelini, L. G., Tozzi, S. and Di Nasso, N. Nassi O. (2004) Environmental factors affecting productivity, indican content, and indigo yield in *Polygonum tinctorium* Ait., a subtropical crop grown under temperate conditions, *Journal of Agricultural and Food Chemistry*, **52**, 7541–7547.

Angelini, L. G., John, P., Tozzi, S. and Vandenburg, H. (2005). Extraction of indigo from *Isatis tinctoria* L. and *Polygonum tinctorium* Ait. as a basis for large-scale production, in M. J. Pascual-Villalobos, F. S. Nakayama, C. A. Bailey, E. Correal and W. W. Schloman Jr. (eds), *Proceedings of 2005 Annual Meeting of the Association for the Advancement of Industrial Crops: International Conference on Industrial Crops and Rural Development*, Murcia, Spain, 17–21 September 2005, pp. 521–534, IMIDA (Istituto Marciano de Investigation y Desarollo Agrario y Alimentario) Selegrafica, S. L. Murcia, ISBN 84-689-3363-5.

Angelini, L. G., Tozzi, S. and Di Nasso, N. Nassi O. (2007) Differences in leaf yield and indigo precursors production in woad (*Isatis tinctoria* L.) and Chinese woad (*Isatis indigotica* Fort.) genotypes, *Field Crops Research*, **101**, 285–295.

Balfour-Paul, J. (1998) *Indigo*, British Museum Press, London.

Ball, P. and Akeroyd, J. (1991) Flora Europea, in S. Tutin, N. Burges, A. Chater, J. Edmondson, V. Heywood, D. Moore, D. Valentine, S. Walters and D. Webb (eds), *Flora Europea*, Cambridge University Press, Cambridge.

Campeol, E. Angelini, L. G., Tozzi, S. and Bertolacci, M. (2006) Seasonal variation of indigo precursors in *Isatis tinctoria* L. and *Polygonum tinctorium* Ait. as affected by water deficit, *Environmental and Experimental Botany*, **58**, 223–233.

Cardon, D. (2007) *Natural Dyes: Sources, Tradition, Technology and Science*, Archetype Publications, London.

Darrac, P.-P. and van Schendel, W. (2006) *Global Blue: Indigo and Espionage in Colonial Bengal*, Mohiuddin Ahmad, The University Press Limited, Dhaka.

EU (1998–2006) *Spindigo. Sustainable Production of Plant-Derived Indigo*, Sustainable agriculture, fisheries and forestry. Research Results 1998–2006, Fifth Framework Programme Key Action 5, Project Synopses, page 120; http://ec.europa.eu/research/agriculture/pdf/research_results.pdf, April 2008 (2006).

Farah, K., Tanaka, A. and West, N. (1988) Autecology and population biology of dyer's woad (*Isatis tinctoria*), Weed Science, **36**, 186–193.

Gilbert, K. G., Hill, D. J., Crespo, C., Mas, A., Lewis, M., Rudolph, B. and Cooke, D. T. (2000) Qualitative analysis of indigo precursors from woad by HPLC and HPLC-MS, *Phytochemical Analysis*, **11**, 18–20.

Gilbert, K. G., Garton, S., Karam, M. A., Arnold, G. M., Karp, A., Edwards, K. J., Cooke, D. T. and Barker, J. H. A. (2002) A high degree of genetic diversity is revealed in Isatis spp. (dyer's woad) by amplified fragment length polymorphism (AFLP), *Theoretical and Applied Genetics*, **104**, 1150–1156.

Habyarimana, E. and Laureti, D. (2005) Diserbo di pre-emergenza e produzione d'indaco, in *Isatis tintoria* L., *in XXXVI Convegno della Società italiana di agronomia- Ricerca ed innovazione per le produzioni vegetali e la gestione delle risorse agro-ambientali*, Foggia.

Hartl, A. and Vogl, C. R. (2003) The potential use of organically grown dye plants in the organic textile industry: experiences and results on cultivation and yields of dyer's chamomile (*Anthemis tinctoria* L.), dyer's knotweed (*Polygonum tinctorium* Ait.), and weld (*Reseda luteola* L.), *Journal of Sustainable Agriculture*, **23**, 17–40.

Hill, D. J. (1992) Production of natural indigo in the United Kingdom, *Beiträge zur Waidtagung*, **4/5**, 23–26.

Kumar, P. (2004) Facing competition: the history of indigo experiments in colonial India, 1897–1920, Georgia Institute of Technology, Georgia.

Mabberley, D. (2008) *Mabberley's Plant-Book: A Portable Dictionary of Plants, Their Classification and Uses*, 3rd edition, Cambridge University Press, Cambridge.

Maule, H., Hill, D., Gilbert, K. and Cooke, D. (2004) *Environmental Impact Assessment for the Sustainable Production of Indigo* (unpublished).

Mendham, N. and Salisbury, P. (1995) Physiology: crop development, growth and yield, in D. Kimber and D. McGregor (eds), *Brassica Oilseeds: Production and Utilization*, CAB International, Wallingford, Oxon.

Monaco, T., Johnson, D. and Creech, J. (2005) Morphological and physiological responses of the invasive weed *Isatis tinctoria* to contrasting light, soil-nitrogen and water, *Weed Research*, **45**, 460–466.

Perkin, A. and Everest, A. (1918) *The Natural Organic Colouring Matters*, Longmans Green and Co, London.

Ricketts III, R. (2006) *Polygonum tinctorium:* contemporary indigo farming and processing in Japan, in L. Meijer, N. Guyard, L. Skaltsounis and G. Eisenbrand (eds), *Indirubin, The Red Shade of Indigo*, Life in Progress Editions, Roscoff, France.

Sales, E., Kanhonou, R., Baixauli, C., Giner, A., Cooke, D., Gilbert, K., Arrillaga, I., Segura, J. and Ros, R. (2006) Sowing date, transplanting, plant density and nitrogen fertilization affect indigo production from *Isatis species* in a Mediterranean region of Spain, *Industrial Crops and Products*, **23**, 29–39.

Sandoval-Salas, F., Gschaedler-Mathis, A., Vilarem, G. and Mendez-Carreto, C. (2006) Effect of harvest time on dye production in *Indigofera suffruticosa* Mill., *Agrociencia*, **40**, 585–591.

Spataro, G. and Negri, V. (2007) Assessment of the reproductive system of *Isatis tinctoria* L., *Euphytica*, **159**, 229–331.

Spataro, G., Taviani, P. and Negri, V. (2007) Genetic variation and population structure in a Eurasian collection of *Isatis tinctoria* L., *Genetic Resources and Crop Evolution*, **54**, 573–584.

Stoker, K. G., Cooke, D. T. and Hill, D. J. (1998) Influence of light on natural indigo production from woad (*Isatis tinctoria*), *Plant Growth Regulation*, **25**, 181–185.

Whitehead, R. (2008) *The UK Pesticide Guide*, CABI and BCPC, Wallingford, UK.

8

Indigo – Extraction

Philip John

8.1 Introduction

Indigo is a dye with a long history (Balfour-Paul, 1998), but one that has retained a contemporary importance because of the global popularity of blue denim. Until the commercialization of chemically synthesized indigo at the end of the 19th century, indigo was extracted from plants. The current revival of interest in plant-derived indigo is driven by the consumer's concern for the environmental impact and sustainability of manufactured products and the consumer's increasing awareness of the provenance of these products (www.ritegroup.org).

Unlike other naturally sourced dyes, indigo does not occur in the plants themselves, but is made during the extraction process from precursors accumulated by the plants. The precursors are compounds containing the indoxyl group. During extraction, the indoxyl is released and is spontaneously oxidized by atmospheric oxygen to indigo.

The main sources of indigo are *Isatis tinctoria* (woad) which is cultivated in temperate climates, *Indigofera* spp. in tropical climates and *Polygonum tinctorium* (dyers' knotweed) in subtropical and temperate climates (Cardon, 2007). *Isatis indigotica* (Chinese woad) is cultivated for indigo in China, and a variety of species acquire a local importance around the world as a source of indigo (Cardon, 2007).

Before the process of extraction itself is dealt with, there are two topics that require some consideration. The first is the quantitative determination of indigo; the second is the nature of the precursors in the plants. Both topics are important when the efficiency of extraction procedures is considered.

Handbook of Natural Colorants Edited by Thomas Bechtold and Rita Mussak

8.2 Methods of Determining Indigo

In order to quantify the overall yield and the purity of the indigo won from a crop, a simple and reliable method of determining indigo is needed. However, no standard, generally accepted method appears to be available.

The insoluble indigo has been determined spectrophotometrically after converting it to derivatives that are soluble in water by various methods (Table 8.1): chemical reduction to the soluble *leuco* indigo (Bechtold *et al.*, 2002; Fanjul-Bolado *et al.*, 2004, 2005), derivitization by boiling in $K_2Cr_2O_7$ (Campos *et al.*, 2001) and the traditional method of sulfonation to the water-soluble indigo carmine using concentrated sulfuric acid. Indigo has been determined electrochemically as the soluble, reduced *leuco* indigo (Govaert *et al.*, 1999; Westbroek *et al.*, 2003; Fanjul-Bolado *et al.*, 2004, 2005). It has also been quantified by standard dyeing experiments (Bechtold *et al.*, 2002). The disadvantage of all these methods is that they require careful control to obtain repeatable results, and for the evaluation of indigo yields during production, they have the additional disadvantage of being time-consuming. With regard to the method involving sulfonation, the spectrophotometric determination of indigo carmine can be complicated by the formation of a monomer–dimer equilibrium in dilute aqueous solution (Shen *et al.*, 1991), so that a voltammetric determination of indigo carmine (Fanjul-Bolado *et al.*, 2004) provides an interesting alternative.

Table 8.1 *Methods for determining indigo spectrophotometrically*

Method	Reference
Spectrophotometry in ethyl acetate at 600 nm	Stoker *et al.* (1998a)
Spectrophotometry in *N,N*-dimethyl formamide at 615 nm	Kokubun *et al.* (1998)
Spectrophotometry in *N,N*-dimethyl formamide at 600 nm	Woo *et al.* (2000)
Derivitization by boiling in $K_2Cr_2O_7$, then absorbance at 441 nm	Campos *et al.* (2001)
Reduction to *leuco* form and then spectrophotometry	Bechtold *et al.* (2002)
Spectrophotometry in *N*-methyl-2-pyrrolidone at 614 nm	García-Macías and John (2004)

A number of authors have chosen the simpler technique of dissolving the indigo in an organic solvent and determining the absorbance at wavelengths greater than 600 nm of the dissolved blue pigment (Table 8.1), for example in ethyl acetate (Stoker *et al.*, 1998a) and *N,N*-dimethyl formamide (Kokubun *et al.*, 1998; Woo *et al.*, 2000). Although such procedures can apparently be useful, caution is required, as the behaviour of indigo in these solvents may be complicated by the formation of aggregates. Indeed it is well established (Zollinger, 2003; Christie, 2007) that single molecules of indigo, such as are found in the gas phase, are red (absorption maximum 540 nm), while in the solid state and in polar solvents indigo becomes blue, due to intermolecular hydrogen bonding leading to aggregation and to solvent interaction (Christie, 2007). The actual absorption maximum depends on the extent of the interaction of the indigo molecules with other molecules. Thus amorphous solid indigo has an absorption maximum at 650 nm and crystalline indigo at 675 nm (Christie, 2007); in solution the absorption maximum depends on the solvent polarity (Table 8.2). This means that when indigo solutions are quantified by absorption in the blue region, the measured absorption can potentially be influenced by the degree of aggregation into dimeric and other polymeric forms, and to interaction with impurities present. To take but one example, if as seems likely (Novotná *et al.*, 2003) synthetic indigo, which is routinely used to standardize assays, is more highly aggregated in solution (lower specific absorption) than is natural indigo (higher specific absorption), then the indigo content of samples of natural indigo may be overestimated.

Table 8.2 *Absorbance maxima for indigo in solvents of various polarities*

Solvent	*Dielectric constant* (ϵ)	*Absorbance maxima*
Benzene	2.2	595[a]
Carbon tetrachloride	2.2	594[a]
Chloroform	4.9	605[a]
Trichlorethylene	8.2	613[a]
Ethanol	24.6	608[a]
Dimethylsulfoxide	46.7	620[a]
N-methylpyrrolidone	31.5	610[b]

Data from [a] Jacquemin (2006); [b] Green and Daniels (1993).

The association of indigo molecules by hydrogen bonding in solution has long been recognized (Holt and Sadler, 1958). Additional, more recent evidence for indigo aggregation in solution comes from a variety of sources. Bond *et al.* (1997) concluded from the pattern of electrochemical reduction of indigo dissolved in *N,N*-dimethyl formamide that there was intermolecular hydrogen bonding to form aggregates. Further evidence came from the EPR spectrum of an electrochemically generated indigo radical anion, which was 'consistent with a range of species, rather than only a monomer (Bond *et al.*, 1997). A similar conclusion was reached by Miliani *et al.* (1998) from concentration and temperature effects on the absorption spectrum of indigo dissolved in *sym*-dichloroethane. Miliani *et al.* (1998) suggested that the aggregation took the form of hydrogen-bonded dimer formation. In a study of light-induced indigo degradation in dichloromethane solution, Novotná *et al.* (2003) found that solutions of natural indigo faded more rapidly than solutions of synthetic indigo, and suggested that this was attributable to the impurities in natural indigo reducing the size of the indigo aggregates in solution.

This aggregation of indigo in solution means that spectrophotometric determinations of indigo in organic solvents are likely to be susceptible to anomalies. Such a case is illustrated by the data of García-Macías and John (2004), which showed that when samples of indigo were reduced and then re-oxidized, the absorption of the indigo taken up in ethyl acetate *increased*. This increase was attributed (García-Macías and John, 2004) to the indigo particles being finer on re-oxidation of the dissolved *leuco*-indigo (Padden *et al.*, 1998, 1999), and thus having a higher specific absorbance than the relatively large particles in the unreduced sample. The effect was also seen when indigo formed from indoxyl acetate hydrolysis (relatively fine particles) gave a *higher* absorbance than that expected from the reaction stoichiometry when calibrations were made with the (relatively large) particles of the commercial synthetic indigo product. Particle size analysis (García-Macías and John, 2004) showed that the particles of indigo made from indoxyl acetate were indeed significantly smaller than those obtained by sonication of the synthetic indigo used to calibrate the assay.

Green and Daniels (1993) introduced *N*-methyl-2-pyrrolidone (NMP) as a solvent for extracting indigo from textile materials. This solvent was subsequently adopted (García-Macías and John, 2004) as the basis for a spectrophotometric assay of indigo that did not show the anomalies of ethyl acetate. When NMP was used as solvent, the re-oxidized *leuco* indigo had a lower absorption than did the original oxidized sample, consistent with the known relative instability of the reduced compared with the oxidized form of indigo (Russell and Kaupp, 1969) and with the loss of yield due to the formation of indirubin and other non-indigo by-products (Perkin and Everest, 1918; Russell and Kaupp, 1969).

Indigo fades easily when dissolved in NMP, even in the absence of light (Green and Daniels, 1993). Presumably this instability in solution is attributable to a disruption of the intermolecular hydrogen bonds (Süsse *et al.*, 1988) that stabilize the aggregates of the microcrystalline indigo molecules, with the central double bond linking the two indoxyl groups being the most vulnerable to attack (Green and Daniels, 1993). Fading of indigo dissolved in NMP can be prevented by including in the NMP solution butylated hydroxytoluene (BHT) as an antioxidant (García-Macías and John, 2004). Because natural indigo is made after extracting the crop leaves in water, determinations often involve aqueous samples. Routinely García-Macías and John (2004) used an NMP solution containing 8 % water, which provided an absorbance near the maximum obtained with pure NMP.

Even though NMP did not show the anomalies apparent with ethyl acetate, quantification of indigo dissolved in NMP, or in any other solvent, requires further refinement before it can be accepted as reliable. It is particularly important to evaluate the influence on the observed absorption of impurities on the observed absorption, and of the aggregation state of the indigo, because it might differ between synthetic and natural indigo samples.

8.3 Precursors in the Plants and Indigo Formation

Indican (Figure 8.1), indoxyl-β–D-glucoside (systematic name 1-*O*- (1 *H*-indol-3-yl)- β–D-glucoside), was identified as the sole precursor of indigo in *Indigofera* and in *Polygonum* in 1900 (Gilbert and Cooke, 2001). Largely through the work of Minami and co-workers (Minami *et al.*, 2000) in *Polygonum*, we know that indican is mainly found in the leaves, while the roots, stems, flower buds, flowers and cotyledons contain less than 3 % of the concentration in the leaves. Within the leaf cells, the indican is localized in vacuoles (Minami *et al.*, 2000).

In *Isatis*, indican is a minor source of indigo, but the main indigo precursor in this plant is a different indoxyl derivative, which was identified by Epstein *et al.* (1967) as indoxyl-5-ketogluconate and named isatan B (Figure 8.1). Its ready hydrolysis by mild base suggested that it was an ester, and the product of alkaline hydrolysis was identified by its chemical reactivity and chromatographic behaviour. This identification was accepted as correct for nearly 30 years by researchers investigating precursor biosynthesis (Strobel and Gröger, 1989; Maier *et al.*, 1990; Xia and Zenk, 1992; Stoker *et al.*, 1998b), processing of indigo leaves (Kokubun *et al.*, 1998; Stoker *et al.*, 1998a; Gilbert *et al.*, 2000) and the indigo potential of *Isatis* leaves (Gilbert *et al.*, 2000; Maugard *et al.*, 2001; Campeol *et al.*, 2006; Angelini *et al.*, 2007).

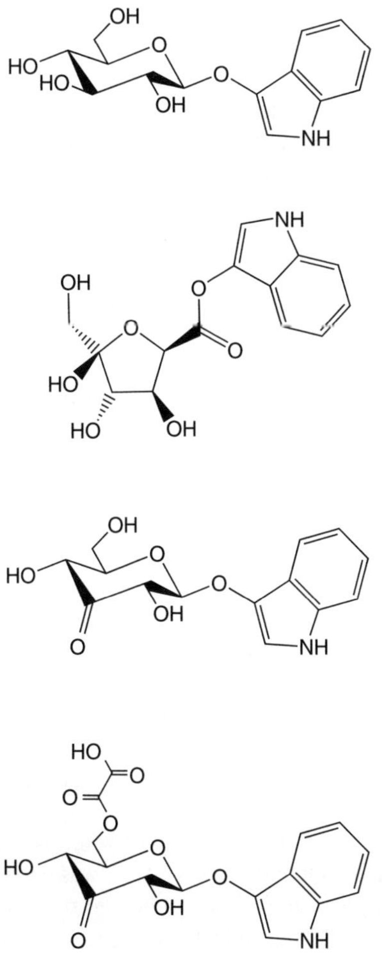

Figure 8.1 Indigo precursors: 1, indican; 2, isatan B of Epstein et al. (1967); 3, isatan B of Oberthür et al. (2004a); 4, isatan A of Oberthür et al. (2004a)

Maugard *et al.* (2001) provided evidence that *Isatis* leaves contain, in addition to isatan B (which was characterized using NMR) as the major precursor, and indican as a minor precursor, a compound, isatan C, which was proposed to be an ester of dioxindole; while alkaline hydrolysis of isatan B yielded mainly indigo, isatan C yielded mainly isoindirubin. Isatan C has not been fully characterized, but whatever its actual identity, its importance to indigo production is likely to be small, given its low concentration relative to indican and isatan B, and the low level of conversion to indigo (Maugard *et al.*, 2001).

In 2004, Oberthür *et al.* (2004b) revised the identification of the major precursor of indigo in *Isatis*. This group, working on *Isatis* leaf extracts for their anti-inflammatory properties (Hamburger, 2002), identified the structure of isatan B as 1-*H*-indol-3-yl β-D-ribohex-3-ulopyranoside, and also identified its malonyl derivative 1-*H*-indol-3-yl 6′-*O*-(carboxyacetyl)-

β-D-ribohex-3′-ulopyranoside, which was called isatan A (Figure 8.1). Isatan B was present on a leaf dry weight basis at about 1–5 %, isatan A at about 5–20 % and indican at about 0.2–1 % (Oberthür *et al.*, 2004a). The isatans were extracted with methanol from frozen leaves, separated by column chromatography and characterized from ESI-mass spectra and 1H-, 13C- and HMQC-NMR spectra (Oberthür *et al.*, 2004b). NMR evidence was also obtained for the presence of H_2O adducts, and it was suggested that these hydrates were the native form of the compounds in the plant vacuole (Oberthür *et al.*, 2004b). It was also suggested that minor indigogenic spots on TLC plates that were sensitive to HCl hydrolysis might be acyl derivatives of indican.

Using their corrected identification of isatans in *Isatis*, Oberthür *et al.* (2004a) have described changes in precursor content with different *Isatis* accessions and post-harvest treatments. They observed that their isatan A, the major precursor in *Isatis*, could not be reliably measured by HPLC, which had been used by others (Gilbert *et al.*, 2000; Maugard *et al.*, 2001) to measure isatan. Instead, Oberthür *et al.* (2004a) used a densitometer to estimate the indigo formed after precursor spots on TLC plates had been treated with HCl (for indican) or NaOH (for the isatans). They found, as others had previously (Kokubun *et al.*, 1998; Gilbert and Cooke, 2001), that the isatan content of *Isatis* is enormously variable: they found that different harvest dates gave up to eightfold differences in isatan A and B contents for *Isatis tinctoria* and tenfold differences for *I. indigotica*. The total amounts of isatans A and B were 2.5–8.8 % for *I. tinctoria* and 6.8–23.6 % for *I. indigotica*. By comparison, indican contents were lower at 0.2–0.9 % for *I. tinctoria* and 0.1–0.% for *I. indigotica*. The isatan/indican ratio was about 2–3:1 for *I. tinctoria*, as observed previously (Kokubun *et al.*, 1998). Oberthür *et al.* (2004a) also observed, like others previously (Kokubun *et al.*, 1998), that isatans disappeared when the leaves were dried, but Oberthür *et al.* (2004a) found unaccountably that drying the leaves caused the indican content to increase fivefold.

The work of Oberthür *et al.* (2004a, 2004b) used reference compounds of known chromatographic and NMR purity. Quantitative analysis was carried out with calibration curves based on reference samples run on each plate and the *Isatis* leaves were shock-frozen in liquid N_2 and freeze dried. Thus, this work represents an important step forward in the qualitative and quantitative analysis of the indigo precursors in *Isatis* spp. However, much potentially valuable quantitative work (e.g. the analyses of Campeol *et al.* (2006) had been done assuming isatan B to be the indoxyl 5-ketogluconate of Epstein *et al.* (1967) and separating indican and isatan B by HPLC essentially as described by Gilbert *et al.* (2004). Confidence in the HPLC-based method derived in part from the demonstration (Gilbert *et al.*, 2004) that indigo quantified spectrophotometrically after precursor extraction and conversion, correlated with the peak areas after HPLC separation of the precursors.

With regard to the content of indigo precursors in *Isatis tinctoria*, it is instructive to compare the results obtained using HPLC, on the one hand, and HPTLC, on the other hand (different structures of the major isatan precursor being assumed). Such a comparison shows that the contents of indican were comparable: 2–9 g/kg dry weight (Oberthür *et al.*, 2004a) and 2–6 g/kg dry weight (Campeol *et al.*, 2006). However, the contents of total isatans were lower at 2–20 g/kg dry weight (Campeol *et al.*, 2006) for the HPLC-based analyses compared with 25–88 g/kg dry weight (Oberthür *et al.*, 2004a) for the HPTLC-based analyses. The relatively low value for isatan (2 g/kg dry weight) obtained by Gilbert *et al.* (2004) may be related to the fact that they boiled the samples of *Isatis* leaf, a procedure known (Kokubun *et al.*, 1998) to severely reduce the indigogenic potential of *Isatis* leaves. Isatans are recognized to be unstable compounds, both within the leaf and after extraction, and it is possible that a higher proportion of the isatan may have been retained by the gentler extraction procedure adopted by Oberthür *et al.* (2004a). Although it is notable that the HPTLC method adopted by Oberthür *et al.* (2004a) involves drying the HPTLC plates, a treatment to which the purified isatans appear to be stable, unlike the isatans *in situ* in the leaves (Kokubun *et al.*, 1998; Oberthür *et al.*, 2004a). Table 8.3 summarizes recent determinations of precursor content in the leaves of indigo-yielding crops.

For thc production of indigo from *Isatis*, the critical feature of the revised identity for isatan B provided by Oberthür *et al.* (2004b) is that the indoxyl is now linked to its sugar group by an ether link, rather than by the ester link in the 5-ketogluconate derivative of

Table 8.3 *Indigo precursors determined in leaves of (a) Isatis tinctoria, (b) Isatis indigotica and (c) Polygonum tinctorium. The indigo equivalent was calculated assuming a complete conversion of two moles of indoxyl to one mole of indigo*

Precursor	*Methodology*	*Precursor content*		*Indigo equivalent*		*Reference*
		g/kg FW	*g/kg DW*	*g/kg FW*	*g/kg DW*	
(a) *Isatis tinctoria*						
Indican	HPTLC-densitometry		2–9		0.9–4.0	Oberthür et *al.* (2004a)
Indican	HPLC-ELSD		2–6		0.9–2.7	Campeol et *al.* (2006)
Indican	HPLC-ELSD	0.4[c]		0.2		Angelini et *al.* (2007)
Indican	Spectrophotometry Rhodanine derivative	4–6		2–2.7		Kokubun et *al.* (1998)
Isatan B[a]	Spectrophotometry Rhodanine derivative	10–18		4–8		Kokubun et *al.* (1998)
Isatan B[a]	HPLC-ELSD		2–20		0.9–8.5	Campeol et *al.* (2006)
Isatan B[b]	HPTLC-densitometry		4–23		1.7–9.8	Oberthür et *al.* (2004a)
Isatan A	HPTLC-densitometry		10–76		3.5–26	Oberthür et *al.* (2004)
Isatan B[a]	HPLC-ELSD	1.9[c]		0.8		(Angelini et *al.* (2007)
(b) *Isatis indigotica*						
Indican	HPTLC-densitometry		1–5		0.4–2.2	Oberthür et *al.* (2004a)
Indican	HPLC-ELSD	0.4[c]		0.2		Angelini et *al.* (2007)
Isatan B[b]	HPTLC-densitometry		6–49		2.5–21	Oberthür et *al.* (2004a)
Isatan A	HPTLC-densitometry		19–218		6.6–75	Oberthür et *al.* (2004a)
Isatan B[a]	HPLC-ELSD	4.9[c]		2.0		Angelini et *al.* (2007)
(c) *Polygonum tinctorium*						
Indican	HPLC-ELSD	10		4.4		Gilbert et *al.* (2004)
Indican	HPLC-fluorescence	6–15		2.7–6.7		Minami et *al.* (2000)
Indican	HPLC	3–15		1.3–6.7		Maugard et *al.* (2002)
Indican	HPLC-ELSD	5–28		2.2–12.4		Angelini et *al.* (2003)
Indican	HPLC-ELSD	12–25		5.3–11.1		Angelini et *al.* (2004)
Indican	HPLC-ELSD		20–200		8.9–89	Campeol et *al.* (2006)

[a]Measured as indoxyl-5-ketogluconate.
[b]Measured as indoxyl-ribohex-3-ulopyranoside.
[c]Mean values, two harvests in 2003.

Epstein *et al.* (1967). An ester link, but not necessarily an ether link, would be expected to be easily broken by alkaline hydrolysis. However, it is clear from the work of Oberthür *et al.* (2004b) that both isatan A and isatan B are sensitive to alkaline hydrolysis. Presumably, the malonyl group of isatan A is easily split off by esterase (under acid conditions) or by alkali hydrolysis, and the double bond of the ribohex-3-ulopyranose has an electron-withdrawing effect leading to *enol-keto* tautomerization, making it into an acidic group easily split from the indoxyl by alkali. Oberthür *et al.* (2004b) found that the minor isatan, their isatan B, like indican, can be hydrolysed by β-glucosidase, but the major isatan, their isatan A, cannot. All three precursors released some indoxyl when HCl was applied to spots on TLC plates, but only their isatans A and B were sensitive to hydrolysis by NaOH under these conditions. Thus the revised isatan structure (Oberthür *et al.*, 2004), as did the previous isatan B structure (Epstein *et al.*, 1967), provides for the release of indoxyl from the major isatan by alkali, but not by β-glucosidase.

The reactions leading to the formation of indigo from indoxyl are as shown in Figure 8.2. The initial step is the release of indoxyl from the precursor, whether it is an isatan or indican. In the case of the former, hydrolysis can be catalysed by alkali (Oberthür *et al.*, 2004b) or by a glycosidase (but not a β-glucosidase, see above) of plant or of microbial origin (Minami *et al.*, 1996, 1997, 1999; Oberthür *et al.*, 2004b). The fact that indigo is formed in the acidic environment of the woad ball and that precursors are not detected

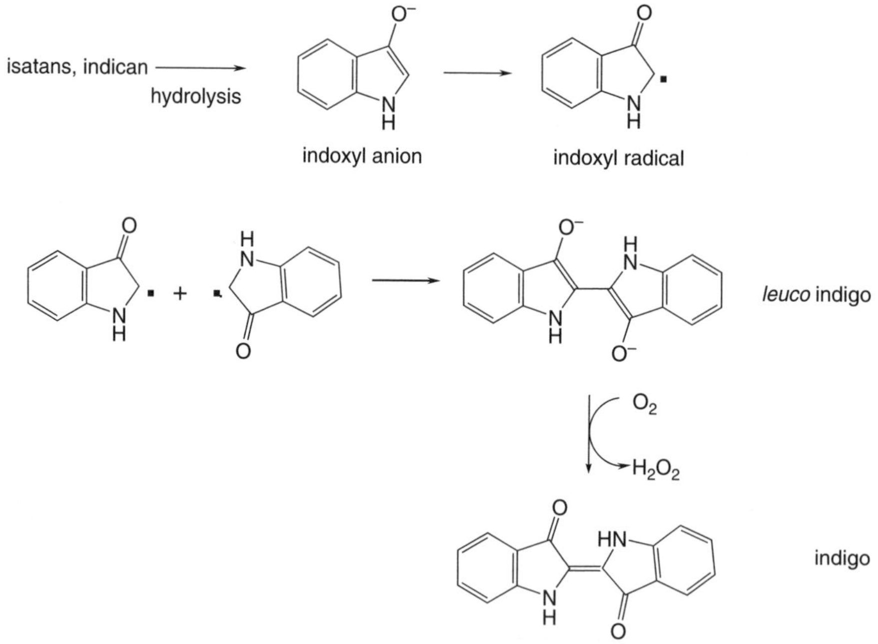

isatans, indican → (hydrolysis) indoxyl anion → indoxyl radical

indoxyl radical + indoxyl radical → *leuco* indigo

leuco indigo → (O_2 → H_2O_2) → indigo

Figure 8.2 The reactions leading to the formation of indigo from indoxyl generated by the hydrolysis of the indigo precursors

(Kokubun *et al.*, 1998) means that the precursors, both indican and the isatans, can be hydrolysed completely and the free indoxyl converted to indigo without the need for alkali. However, invariably, alkali is added to accelerate the conversion of indoxyl to indigo (Perkin and Everest, 1918; Cotson and Holt, 1958). Given the importance in indigo formation of this conversion, little attention appears to have been made to its chemistry. One of the few published studies of these reactions under the aqueous conditions of natural indigo manufacture was made in the context of cytochemical staining. Cotson and Holt (1958) studied the kinetics of indigo formation from indoxyl that had been released by alkaline hydrolysis of indoxyl acetate. The reaction was first-order with respect to indoxyl; raising the pH from pH 6 to pH 8 increased the rate of indoxyl formation tenfold and raising the pH to pH 8.5 further stimulated the conversion. The percentage yield of indigo measured after conversion at pH 11 (NaOH) was unchanged by varying the initial concentration of indoxyl, but was reduced to 85 % by the inclusion of 1 % gelatine, or by oxidation at acid pH. Conversion of indoxyl to indigo is always less than 100 % and Cotson and Holt (1958) identified anthranilic acid as a major by-product. They suggested that anthranilic acid was formed by the oxidation of isatin by the hydrogen peroxide formed during the oxidation of indoxyl to indigo (Figure 8.3).

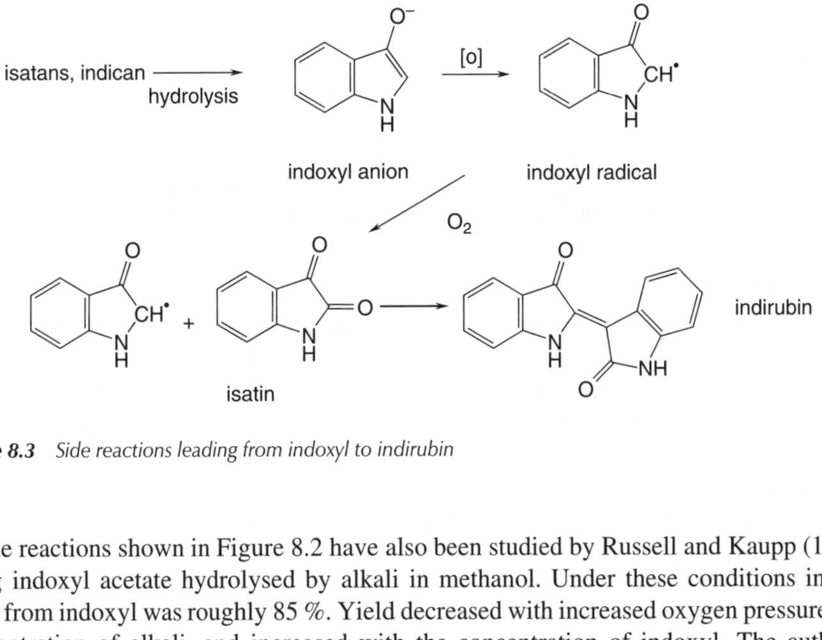

Figure 8.3 *Side reactions leading from indoxyl to indirubin*

The reactions shown in Figure 8.2 have also been studied by Russell and Kaupp (1969) using indoxyl acetate hydrolysed by alkali in methanol. Under these conditions indigo yield from indoxyl was roughly 85 %. Yield decreased with increased oxygen pressure and concentration of alkali, and increased with the concentration of indoxyl. The authors' explanation (Russell and Kaupp, 1969) of these observations is that the reactions of indoxyl, or of their derivatives, with each other to form *leuco* indigo compete with those side-reactions with alkali and oxygen that lead to indoxyl destruction. As shown previously (Cotson and Holt, 1958), the products of indoxyl destruction are probably isatin and its derivative, anthranilic acid. Two other conclusions from Russell and Kaupp (1969) relevant to the chemistry of indigo formation were, firstly, the involvement of the cherry-red (Bruin *et al.*, 1963) *leuco* indigo free radical and, secondly, the identification of the rate-determining stage as the oxidation of indoxyl itself, with the oxidation of *leuco* indigo being relatively fast.

The indoxyl that is oxidized to isatin readily combines with indoxyl radical to form indirubin (Figure 8.3). This red dye is also produced during the production of synthetic and biotech indigo (Berry *et al.*, 2002), but in these cases, it is removed from the synthetic product during manufacture and is virtually absent from the finished product. It can also be removed from biotech indigo by the introduction of the gene for the enzyme isatin hydrolase (Berry *et al.*, 2002). Indirubin formation lowers both the yield and purity of natural indigo, and thus it would be beneficial to be able to define conditions that would favour the formation of indigo over indirubin, but the factors that determine indirubin formation are not well understood (Gillam *et al.*, 2000; Berry *et al.*, 2002).

8.4 Extraction Procedures

8.4.1 Traditional Process Using Crushed Leaf Material

The principle employed in these procedures is to allow indigo to form within the leaf material by crushing the leaves, and then to put the crushed leaf material through a process called couching which reduces the mass and fibrous nature of the leaf material, thus concentrating the indigo and allowing it to become available for dyeing. The whole procedure is time- and labour-consuming, and the final product is an undefined, crude mixture of decomposed plant material and indigo. However, this can be used directly in a fermenting vat, where bacteria (Padden *et al.*, 1998, 1999) reduce the indigo to the *leuco* form, without the addition of a reductant (Nicholson and John, 2005).

8.4.1.1 Isatis

Until it was abandoned in the 19th century, the procedure used for the preparation of *Isatis* for the woad vat had remained virtually unchanged since prehistoric times, and varied little between the European centres of woad cultivation. One of the earliest records comes from Irish Law Texts of the 7th and 8th centuries (Kelly, 1997). The classic account of the medieval process is given by Hurry (1930) who witnessed the closure of the last of the traditional woad mills in Europe, when the Parson Drove Woad Mill shut down around 1910. An essentially similar indigo production procedure was followed in Southern Italy until about 1830 (Guarino *et al.*, 2000). Briefly, freshly harvested *Isatis* leaves were crushed to a pulpy paste and kneaded into balls (woad balls) about the size of a grapefruit. These were left to dry on specially constructed frames. When dry they could be stored indefinitely. In preparation for use in the dye vat, the balls were crushed to a rough powder. This was spread in heaps and watered. The ensuing aerobic microbial activity heated the material, and over a period of some nine weeks the temperature of the mass was controlled to about 50 °C by careful management of watering and turning the material. Finally, the so-called couched material was then dried, stored in barrels and was ready for use in the woad vat. Kokobun *et al.* (1998) demonstrated that all of the indican and isatans had been converted to indigo in the dried woad balls. Presumably what happens is that enzymes released by disruption of cell structure in the leaves catalyse hydrolysis of the precursors in the mildly acidic medium of the chopped leaf paste. It is known that indican and the minor isatan of Oberthür *et al.* (2004a, 2004b), isatan B, are susceptible to hydrolysis by β-glucosidase. Possibly an esterase releases the malonyl group from the major isatan, isatan A, rendering it also sensitive to β-glucosidase. Enzymes released by bacterial activity, which is dependent on sugars and other substrates released from the disrupted leaf cells, could also contribute to the release of the indoxyl groups. Respiration of bacteria and the disrupted plant tissues must consume oxygen from the interior of the wet woad balls, and it is only when the woad balls dry that enough oxygen penetrates the woad ball for the conversion of the indoxyl to indigo. Indirubin is also measurable in the woad balls (Kokubun *et al.*, 1998). The factors that determine the relative conversion to indirubin and to indigo are not well understood.

In the absence of any systematic study of the chemistry and microbiology of the couching process, it is assumed that during this process there is an active breakdown of cellulose, hemicelluloses and other structural materials so as to allow the indigo to be liberated for dyeing. An analysis of the bacterial inhabitants of the woad vat (Lawson *et al.*, in press) identified thermophilic aerobes such as *Geobacillus pallidus* and *Ureibacillus thermosphaericus*, which are characteristic of hot composts, large populations of which were probably carried over into the woad vat with the couched material.

In the procedure followed by Kokubun *et al.* (1998), the final yield of indigo was 14 % of the potential measured in the leaves. This value compares favourably with recoveries from cleaner and more modern processes (see below), in which indigo precursors are leached from the leaves. This is surprising when the conditions for indoxyl conversion to indigo within the woad ball appear to be unfavourable: firstly, the released cell sap (pH around 4.5) provides an acidic environment which will stabilize the indoxyl, but delay its conversion to indigo (Cotson and Holt, 1958), and at the same time not allow the intermediate *leuco* indigo to ionize; secondly, the range and concentration of plant-derived compounds within the mass of the woad ball provide side-reactions for the released indoxyl; thirdly, until the woad ball dries the oxygen tension within the plant mass is low, delaying the formation of the stable indigo end-product, and extending the life of the relatively unstable indoxyl and *leuco* indigo intermediates. There appears to be ample scope to improve the efficiency of conversion in the traditional method: the conversion of the precursors to free indoxyl could be accelerated so as to favour indigo formation, possibly by raising the pH; at the same time, given the reactivity of indoxyl to form side-products, the admission of oxygen by drying the woad balls could be speeded up. To determine the optimum time before drying is initiated, it might well be useful to monitor the formation of indoxyl and its conversion to indigo. The crushing of the leaves and the formation of the woad balls results in a considerable amount of juice that would normally be allowed to run off. This juice turns blue on standing, and thus represents a loss of indigo. Again, improvements in methodology could help reduce this loss of indigo.

8.4.1.2 Polygonum

Ricketts (2006) has described the historical context in which the production of sukumo, the couched leaves of *P. tinctorium*, has survived in Japan. This survival contrasts with the commercial defeat of the similar woad process by indigo powder imported from the tropics. The Japanese tradition is retained today by a small number of producers in Tokushima prefecture, whose skills have been recognized by the Japanese Ministry of Culture as a 'Living Treasure' or 'Important Intangible Cultural Asset'. The accolade is double edged: while providing recognition of their skills and heritage, and a degree of commercial protection, at the same time the producers are obliged to remain faithful to traditional methods, thus limiting mechanization and other technical innovations.

Unlike *Isatis*, *Polygonum* leaves can be dried before use. The explanation for this lies in the fact that *Polygonum* contains only indican as a precursor, and while in *Isatis* indican is a minor precursor compared to the isatan, on drying, isatan is almost completely lost while indican is unaffected (Kokubun *et al.*, 1998, Oberthür *et al.*, 2004a, 2004b). The harvested *Polygonum* plants are chopped using a cutting machine and electric fans blow the leaves clear of the heavier stem material. The leaves are dried for a day in the sun, then overnight in tobacco driers and set aside in straw bags until the autumn. It is not known whether these dried leaves contain indican or whether it has been converted to indigo at this stage. The next stage is equivalent to the couching process in *Isatis* processing. The dried *Polygonum* leaves are spread on the floor of a building dedicated to this stage of processing. The floor of the building is built up of successive layers of large stones, pebbles, sand, rice husks and clay. Both its structure and its sloping away from the centre act to draw water away from the composted leaves. Judicious sprinkling of water and a regular weekly (or more frequent) turning of the composting material facilitate the microbial activity that breaks down the leafy material. This causes the piles to shrink, allowing more material to be added. The most important factor is the moisture content: too much will allow anaerobic pockets to develop and too little will prevent optimum microbial activity. Other critical factors are the turning of the piles to provide an even decomposition and an appropriate temperature of around 70 °C. During the composting, any clumps that have formed are split up in a sieving machine. During the final period straw mats are placed on top and around the piles of decomposing leaves to insulate them against the lowest of the winter temperatures. Finally, the straw mats are removed, the piles are allowed to cool, excess moisture steams off and the sukumo is packed into straw bags. These will gradually lose further moisture and can be kept indefinitely. The couching lasts 100 days and results in an enormous reduction in the volume and a corresponding concentration of indigo. Cardon (2007) provides the following figures: 300 kg of leaves (presumably in the dried state) yield 195 kg of sukumo containing 3–8 % indigo. This represents an indigo recovery of 20–50 g indigo/kg leaf dry weight, a value which speaks for a high recovery in the processing of the leaves, since the indican content of dried *Polygonum* leaves has been measured (Angelini *et al.*, 2004) to be such (80 g indican/kg dry weight) as to yield a maximum of 37 g indigo/ kg dry weight of leaf.

In conclusion, the traditional processes of producing indigo from *Isatis* and *Polygonum* appear to be irretrievably obsolete, and the product is such that any commercial dyer after a century of experience with the pure synthetic product would need a great deal of convincing to use it. Nevertheless, the traditional processes could repay a revisit. Mechanical crushing of the leaves, a controlled drying of the crushed material, more efficient couching, and one has a reasonably efficient conversion to indigo in a process that has virtues of a lower energy input and almost no chemical inputs compared with the processes that involve steeping in water, which are described below.

8.4.2 Steeping in Water

In principle, the production of indigo by steeping indigogenic leaves in water does not impose a substantial problem: the precursors are water-soluble and on hydrolysis, they easily release free indoxyl, which on aeration readily forms indigo, an insoluble and readily sedimentable product. The challenge in achieving the highest possible yield and purity is, and always has been (Perkin and Everest, 1918), to maximize the precursor extraction, maximize conversion of indoxyl to indigo and minimize contamination of the indigo.

8.4.2.1 Indigofera

The traditional method of producing indigo from *Indigofera* has been by water extraction. There were three stages to the operation, as described for the colonial period in Bengal (Perkin and Everest, 1918; Kumar, 2004; Darrac and van Schendel, 2006). In the first stage, freshly harvested bundles of plants were soaked tightly packed in the steeping vat. An anaerobic fermentation ensued, and after 10–15 h the resulting yellow-green solution was run off into a second tank, the beating vat. In this second stage, mechanical paddles introduced air by creating a spray of water (the shower bath method), by pumping in air or simply by manually beating the solution with paddles. The oxidative conditions led to the formation of a blue precipitate of indigo, which settled to the bottom of the tank. This was run off as sludge into a cauldron where, in the third stage, it was boiled to help purify the indigo, which was filtered, washed and finally dried in the form of cakes.

Although the chemical basis of the extraction process has not yet been analysed by modern methods, it was apparent to the original investigators (Perkin and Everest, 1918) that the indican is attacked by β-glucosidases from the leaves themselves and, to a lesser extent, from bacteria to generate free indoxyl in the steeping vat. Interestingly, it was noted that during the fermentation stage, indican itself does not pass from the leaf into the surrounding liquid (Perkin and Everest, 1918). It was recognized that the indican is hydrolysed within the decaying leaf tissue and that the resulting indoxyl is leached from the leaves. The tight packing of the fermentation vat rapidly leads to the exhaustion of oxygen in the solution, thus preventing the released indoxyl from converting prematurely to indigo, which would be trapped in the plant tissues. It was believed that indoxyl passes into the beating vat (Perkin and Everest, 1918), but it may be that this indoxyl is partially or completely converted to *leuco* indigo (Russell and Kaupp, 1969) under the predominantly anaerobic conditions of the fermenting vat. Darrac and van Schendel (2006) distinguished between a 'good fermentation' and an 'excessive fermentation'. The former was marked by a greenish tinge to the water, which carried a light bluish foam. The latter was marked by frequent bubbles rising to the surface, a putrid odour and patches of thick foam. The dyeing quality of the final product (which was presumably determined largely by its purity) was greater with a 'good' than with an 'excessive' fermentation. Darrac in the early 19th century recognized that the fermentation stage needed to be long enough to extract the indigogenic factors (indoxyl), but not be so long as to contaminate the steep water with plant-derived materials. That remains the key to an effective extraction.

Faced with what turned out to be overwhelming competition from the newly marketed synthetic product, the predominantly English and Dutch indigo producers around the turn of the 19th century employed a number of investigators to examine the efficiency of the extraction process. The aim was to enhance product yield and purity (Perkin and Everest, 1918; Kumar, 2004). One of the investigators' recommendations was that the oxidation stage be started immediately after completion of the steeping stage. It was noted that delays at this stage led to losses by unwanted side reactions due to the reactivity of the free indoxyl towards a range of compounds leached from the *Indigofera* leaves in the steeping stage. The instability of indoxyl relative to indican and indigo itself was recognized in this early work (Perkin and Everest, 1918), and prolonging the fermentation stage led to what was referred to as the 'decay' of indoxyl to products that did not yield indigo. It was also noted from experiments in which known amounts of indican were hydrolysed and the resulting indoxyl oxidized to indigo that the conversion of indoxyl to indigo was not stoichiometric and that the extent of the losses depended on the conditions employed. The technical advisors also recommended the addition of a 'small quantity' of an alkali, such as ammonia, sodium carbonate or calcium hydroxide (as slaked lime), after the fermentation stage to improve the yield of indigo. It was known from early practice (Darrac and van Schendel, 2006) that a modest alkalinity increased the rate of indoxyl oxidation to indigo, but an excess of alkali decreased the yield indigo. The acceleration by high pH of indoxyl conversion to indigo has been described quantitatively by Cotson and Holt (1958). However, when calcium hydroxide was used as alkali, there was a copious production of solid calcium and magnesium carbonates from the dissolved carbon dioxide that had come over from the fermentation vat. These carbonate precipitates helped carry down the newly formed indigo, but contributed to impurities in the indigo sludge. A third improvement introduced at this time was to wash the indigo precipitate with dilute HCl, in order to help remove solid impurities that interfered with the dyeing and printing processes. The HCl was of course especially useful when lime had been added, as it dissolves the resulting calcium carbonate particles present in the indigo. This is discussed further in the section on indigo purity below.

As recognized a century ago (Perkin and Everest, 1918), the reliability of estimates of indigo yield from *Indigofera* depends on the reliability of the indigo assays, as discussed at the beginning of this chapter. Bearing this caveat in mind, in the early years of the 20th century the efficiency of the extraction was estimated to be 60–70 % or less, with losses of indigo attributed to failure to sediment completely in the oxidation tanks, a retention of indoxyl in the leaves and conversion of indoxyl into compounds other than indigo by side-reactions with other compounds released from the leaves (Perkin, 1900; Perkin and Everest, 1918).

From data produced by Rawson (see Kumar, 2004) it can be calculated that in 1900 the yield of indigo from *Indigofera tinctoria* was 7 kg pure indigo from 3700 kg of leaves, a yield of just 0.2 %, based on leaf fresh weight. However, it was not only the poor yield that made the natural product uncompetitive with the recently introduced synthetic indigo; natural indigo was less pure. Interestingly, it was demonstrated on a laboratory scale that natural indigo could be readily separated from the largely insoluble impurities after reducing it to the soluble *leuco* indigo with alkaline hydrosulphite (dithionite), but this procedure was not practical on a larger industrial scale (Perkin and Everest, 1918). Despite improvements in yield that arose from the advice of the botanists and chemists, the *Indigofera* indigo produced in the tropics lost its once lucrative market to the synthetic product, predominantly because the synthetic product was reliably over 90 % pure when the natural product varied from 20 % to 90 % pure, and a good sample of Bengal indigo was only 61 % indigo (Perkin, 1900).

8.4.2.2 *Isatis*

The early development of water-based extraction methods for *Isatis* has been described by Gilbert and Cooke (2001). A water-based extraction of indigo from *Isatis* was described by Plowright (1901), who provides a sketch published in 1812 which shows an apparatus constructed in Vienna by a J. B. Heinrich. Essentially the apparatus allowed indigo leaves to be soaked in water in a raised vessel; the steep water was then allowed to flow into a second vessel containing quicklime, where it was aerated. The resulting precipitated indigo was drawn off this vessel. Plowright (1901) was aware from earlier investigators that *Isatis*, unlike *Indigofera* and *Polygonum*, released indoxyl from leaves treated with hot or cold water, apparently without the aid of enzymes. More recent developments have been described by Stoker *et al.* (1998a) and by Angelini *et al.* (2005), although the process has not changed in its essentials in 200 years.

Production of indigo from *Isatis* is currently carried out on a commercial scale in at least three areas of Europe: Bleu de Pastel de Lectoure, Toulouse, France (http://www.bleu-de-lectoure.com), Woad-inc ™, Norfolk, UK (http://www.woad-inc.co.uk) and La Campana, Montefiore sull'Aso, Italy (http://www.lacampana.it). These commercial producers extract indigo by processes essentially similar to those described below.

The basic process of indigo extraction from *Isatis* resembles that used for the extraction of indigo from *Indigofera*. Essentially, *Isatis* leaves are steeped in heated water for a certain period of time; the liquid is then removed, made alkaline and oxygenated vigorously. The indigo formed is allowed to sediment, and is filtered, washed and dried.

With respect to the extraction process, the main differences between *Indigofera* and *Isatis* are that where the former is grown and extracted, ambient temperatures are higher, which obviates heating the steep water, and in *Isatis* an isatan, rather than indican, is the principal indigo precursor.

Stoker *et al.* (1998a) carried out laboratory-scale extractions in which leaves were briefly scalded with boiling water to break down the cuticle wax; the leaves were steeped overnight in an acidified solution and the solution was then made alkaline and oxygenated. Under these conditions it was found that the optimum yield (380 mg indigo/kg FW leaves) was obtained at 30 °C, with much less obtained at 60 °C and 90 °C (47 and 18 mg indigo/kg FW leaves, respectively). Stoker *et al.* (1998a) reported that these steeping times were unnecessarily long, and especially at high temperature; 5 min at 100 °C gave a similar yield to an overnight steeping at 30 °C.

On the basis of the findings made on the laboratory scale, a larger, pilot-scale production was carried out. Water heated to 60 °C was poured over the freshly harvested leaves contained in gauze bags (2 L water/kg leaves). This was followed by the addition of sufficient cold water to cover the leaves and bring the temperature to about 40 °C. Sulfuric acid was added to reduce the pH to 3.5. The steep water was circulated for 2 h and the leaves were steeped overnight, during which time the temperature dropped to 25 °C. The steep water was then pumped into a settling tank, the pH adjusted to between pH 9 and 10 with ammonia and aerated for 2–4 h. The indigo was allowed to settle for 24 h, woad

filtered and dried at room temperature. The final yield was about 20 kg pure indigo/ton FW leaves, with an indigo purity of 20–40 %.

For Stoker *et al.* (1998a) the critical step for indigo extraction was the release from the leaves of isatan by the hot water treatment of the leaves and the stabilization of this precursor by addition of acid. It is noteworthy that Stoker *et al.* (1998a) treated the *Isatis* leaves with water at 100 °C, while Kokubun *et al.* (1998) described the almost complete loss of isatan in leaves treated with water at 100 °C for just 10 s. Nevertheless, the yields described by Stoker *et al.* (1998a) were higher than those achieved by later developments (Angelini *et al.*, 2005) and 10 times the yield described by Bleu de Pastel de Lectoure, Toulouse France (http://www.bleu-de-lectoure.com).

The procedure described by Stoker *et al.* (1998a) was important in preparing the way for a much larger scale development of a commercial on-farm extraction procedure developed within the EU-funded Spindigo Project (2001–2004) as described by Angelini *et al.* (2005). The aim of the development was to produce a robust, mobile, on-farm extraction unit that would have a sufficiently high throughput to enable the rapid extraction of extensive crops of *Isatis*. The extraction unit that was developed (see Figure 8.4) was preceded by two prototypes, all constructed by Critical Processes Ltd (UK). The extraction units were tested in various locations in Europe over the years 2002–2004, with a large-scale trial in Pisa, Italy.

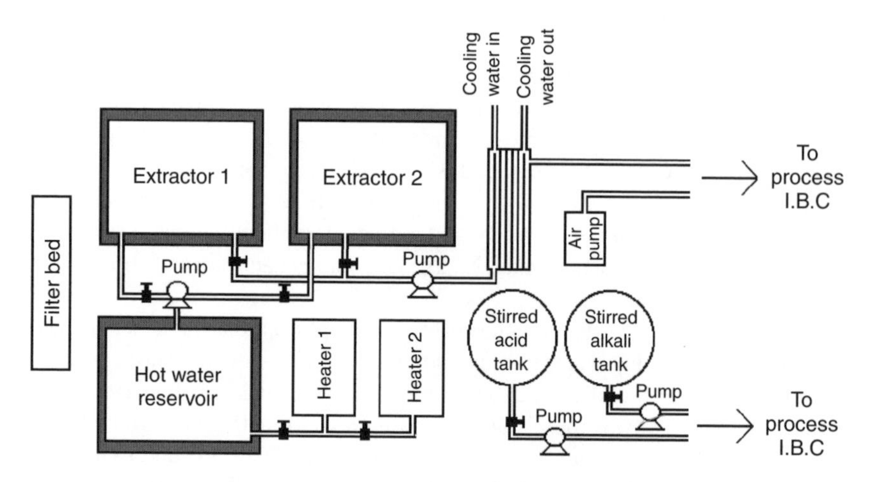

Figure 8.4 *Diagrammatic representation of the high-throughput equipment for the production of indigo from Isatis. Reproduced with permission of Critical Processes Ltd UK*

The extraction unit is based on two 750 litre dip tanks provided with water at 75 °C from the main heating and reservoir system. The primary heater unit holds 3500 litres and the secondary recirculating tank 1500 litres. The heating is via an oil-fired boiler capable of heating 750 litres of water in 8 min. Cooling is carried out in a brazed heat exchanger unit cooled by recirculation of water from a mobile 8 ton reservoir. The heat exchanger flow rate allows the cooling water passing through it, which becomes warmed during the process, to be passed directly to the main heating reservoir of the unit, thus saving considerable energy input. Cooling water passes into the reservoir tank at 50 °C so that the heat input required from the burner to elevate this water to the extraction temperature of 75 °C is reduced. This energy recovery process allows the unit to produce up to 60 tons of water per day at 75 °C. This gives a maximal daily throughput of between 20 and 30 tons of leaf material. The alkali treatment and oxygenation of the extraction water are carried out in 1000 litre standard plastic bulk container units, which allow for easy transport on site and easy transfer handling.

The extraction procedure optimized from preliminary experiments with the prototype extraction units was as follows. The harvested leaves are washed in cold water to remove soil particles, which can contaminate the indigo product (García-Macías and John, 2004). In Italy a washer designed for leafy vegetables was successfully used. The loose leaves are placed in about 3000 litres of water, and then carried on a conveyor belt while being sprayed with clean water before falling into the metal frame steeping baskets, each of which contains about 100 kg of leaf material (Angelini *et al.*, 2005). The baskets are moved by means of a mechanical hoist (Figure 8.5). Water heated to 75 °C is introduced into the tanks by gravity feed to just cover the leaves, giving a water:leaf ratio of 4:1; the tanks are closed with lids to preserve heat and allowed to steep for 15 min. The baskets are then removed, allowed to drain and the leaves removed for eventual composting. The extract is pumped through the heat exchanger, where it cools to 25 °C, and passes through a filter to the standard plastic bulk container unit. Alkali in the form of $Ca(OH)_2$ (slaked lime), NH_3 or KOH is now added to raise the pH to 10 –11, and the extraction liquid is aerated by pumping the liquid via a spray back into the tank with a submersible pump. For alkalization, $Ca(OH)_2$ has an advantage in that the particles of $CaCO_3$ formed help the indigo to settle more rapidly, but NH_3 gives rise to a cleaner indigo product with less $CaCO_3$

Figure 8.5 *The high throughput extraction unit capable of producing indigo from Isatis. Leaves are dipped using the wire basket shown for 15 min in one of the two tanks containing heated water. Equipment constructed by Critical Processes Ltd UK. Reproduced by permission of L.G. Angelini*

contamination. After about 20 min, the liquor is pumped to a settling tank, where it remains undisturbed. When the indigo is seen to have settled (about 1 h), the yellowish supernatant is pumped off and the settled indigo slurry from a number of settling tanks is combined. Addition of citric or hydrochloric acid to a pH of 3–4 dissolves any carbonates in the indigo, which, after settling (at least overnight), is transferred to cloth filters and allowed to dry at room temperature.

Trials carried out in Pisa, Italy in August 2003 (Angelini *et al.*, 2005), using the extraction procedure described here, was able to process 1 ton of leaves per day with two personnel. The yield varied from preparation to preparation in the range of 1.2–2.6 kg of crude indigo product/t of leaves. The crude indigo was assayed as being 7–10 % pure. As discussed above, the yield and purity values are subject to uncertainties related to the indigo assay, but taking a value of 10 % purity, it can be estimated that yields of pure indigo were around 0.12–0.26 kg/ton of leaves. The yield of leaves was about 50 ton/ha, giving the yield from this high throughput procedure of 0.6–13 kg of pure indigo/ha. The potential yield of the leaves grown under these conditions was estimated to be 2 g/kg fresh weight of leaves (Angelini *et al.*, 2005). Thus the efficiency of the process appears to be relatively low at around 10 % of the potential yield.

There are a number of possible sources of loss in the extraction process: failure to extract all of the precursors, loss of indoxyl due to side reactions and failure to recover all the indigo formed. García-Macías and John (2004) carried out model, laboratory-scale experiments to determine indigo yield from alkaline hydrolysis of indoxyl acetate under conditions similar to those of the farm-scale trials. They found that inclusion of an extract of *Isatis* leaf and a sample of soil during the hydrolysis, in amounts comparable to those present in farm-scale extractions, reduced the indigo yields: from 57 % conversion of the indoxyl acetate to 46 % after inclusion of the leaf extract and to 41 % after inclusion of the soil. Later similar experiments by A. Vuorema and P. John (unpublished) showed that indoxyl derived from hydrolysis of either indoxyl phosphate at pH 4.8 or of indoxyl acetate at pH 11 gave a maximum conversion to indigo of 60 %, and when chopped leaves of *Isatis*, or of a non-indigogenic member of the Brassicaceae family, were added to the hydrolysis at levels comparable to that present in the large-scale extraction, the yield of indigo was reduced by 40 %. This latter loss of indigo yield is consistent with the finding (Cotson and Holt, 1958) that inclusion of biological material in the form of 1 % gelatine decreased by 15 % the yield of indigo from indoxyl oxidation.

Thus about 40 % of the loss of potential yield can be ascribed to a simple failure to convert the available indoxyl to indigo, and further similar levels of loss can be ascribed to side reactions due to contaminants that divert the indoxyl to products other than indigo. The loss of yield in this way was recognized a century ago when Perkin and Everest (1918) were concerned that the overall manufacturing process for indigo from *Indigofera* was yielding only (!) 62–72 % of the theoretical yield (Perkin and Everest, 1918).

It is assumed (see, for example, García-Macías and John, 2004, and Angelini *et al.*, 2005) that indoxyl is extracted in the form of the predominant precursor, isatan, which then releases indoxyl on addition of alkali. However, the extent to which indoxyl, rather than isatan, is extracted from the steeped leaves in a water-based extraction will depend on the activity of the hydrolysing enzymes, which are inhibited partially or wholly by the high temperatures (70 °C) used in the water-based extraction. Kokubun *et al.* (1998) showed that in the traditional woad ball, isatan was absent and all the indoxyl was represented by indigo or indirubin. Under these conditions isatan had been converted to indigo via indoxyl without the addition of alkali to the naturally acid conditions of the crushed leaves. In this case, the period of exposure of the precursors to hydrolyzing enzymes is very much extended compared with the period of exposure in a water-based extraction.

It is possible to monitor continuously the release of indoxyl and its conversion to indigo by virtue of the fluorescence of indoxyl and *leuco* indigo. The high fluorescence yields of indoxyl and *leuco* indigo were first used as the basis for a sensitive assay of the nerve gas, sarin (Gehauf and Goldenson, 1957); later, the fluorescence of indoxyl was used by Woo *et al.* (2000) to follow the bacterial oxidation of indole. García-Macías and John (2004), studying natural indigo production from *Isatis*, showed the pattern of fluorescence to increase when indoxyl acetate was hydrolysed by KOH addition. In the absence of oxygen, the fluorescence remained relatively stable, and could be shown to be proportional to the indoxyl acetate added, while in the presence of oxygen, fluorescence was transitory, but even then was appreciable for minutes. Inclusion of isatin, which combines with indoxyl to form indirubin (Figure 8.3) led to a lower fluorescence and it was concluded that at least a proportion of the fluorescence observed is attributable to free indoxyl (Gehauf and Goldenson, 1957), although one would expect a contribution from *leuco* indigo, which would also be present (Russell and Kaupp, 1969). When oxygen was admitted to an anoxic system, there was a rapid loss of fluorescence (Figure 8.6) and the solution turned blue, consistent with the oxidative formation of indigo from free indoxyl and *leuco* indigo. Addition of oxygen after isatin had been included showed no loss of fluorescence (Figure 8.6), consistent with the formation of indirubin from free indoxyl and isatin (Figure 8.6). Unfortunately, the fluorescence method could not be used (García-Macías and John, 2004) to monitor intermediate formation with extracts of *Isatis* because the *Isatis* extracts interfered with the fluorescence yield.

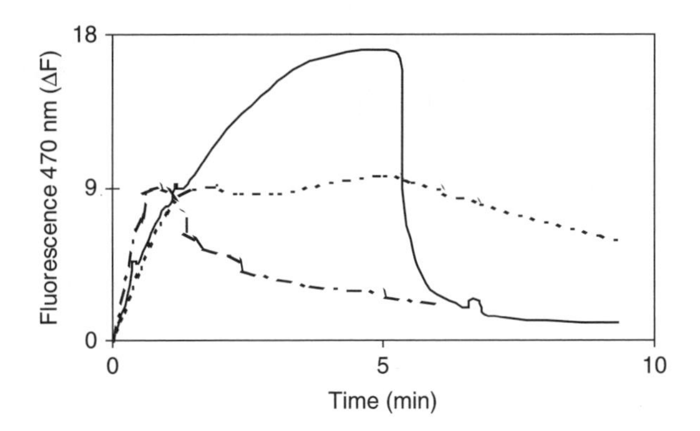

Figure 8.6 *Determination of fluorescent intermediates during indigo formation from 1 mM indoxyl acetate. Continuous line, flushed with nitrogen; broken line, not flushed with nitrogen; dotted line, flushed with nitrogen, with addition of isatin (30 μL of 10 mM in methanol). Oxygen was added at 5 min. Fluorimetric cuvettes, previously flushed with N_2 to remove O_2 and sealed were maintained at 30 °C and at zero time KOH to release the indoxyl. The resulting fluorescence was measured with excitation and emission wavelengths at 365 and 470 nm respectively. From García-Macías and John (2004). Reprinted with permission from Journal of Agricultural & Food Chemistry,* ***52****, Copyright 2004 American Chemical Society*

The indigo precursors and indoxyl appear in the steep water much diluted. In principle it ought to be possible to capture and concentrate these compounds from the steep water after adsorbing them on appropriate resins. García-Macías and John (2004) described the use of two exemplar resins: the nonionic Amberlite XAD16 and the weakly basic Amberlyst A-21. Preliminary experiments in which 1 litre volumes of 1 mM indoxyl acetate or standard *Isatis* extracts were passed though small resin columns, containing 1 g of resin, showed that 30–70 % of the indigo-forming potential could be captured by the resins. However, after about 1 h the beads became red, and while the Amberlite XAD16 beads stayed red indefinitely, the Amberlyst A-21 beads rapidly turned blue. The chemical changes that indoxyl underwent on the beads were not examined. When the columns were eluted with acetone, methanol and ethanol, indigo remained on the resins; the red eluates had indistinct absorption maxima, which indicated that compounds other than indirubin were present. One possibility is that the red *leuco* indigo radical was stabilized by adsorption to the resin beads. The beads, as judged by their yellow colour, were able to retain indigo reduced by dithionite to the yellow *leuco* form. From experiments in which the indigo precursor from woad extract was adsorbed on continuously stirred Amberlite XAD16, García-Macías and John (2004) obtained a Langmuir isotherm, from which they obtained adsorbent loadings at equilibrium for each extract. The loadings for four of the extracts at equilibrium varied from 42 to 45 % of the total precursor content. This work showed that resin adsorbents in the form of a removable trap are potentially useful for extracting indigo. They would be particularly useful if the *leuco* form could be held on the resin trap for direct transmission to the dyer, but more work would be needed to devise a practical and economic means of releasing the dye.

There is a further consideration with the use of resins: the rate at which *Isatis* steep water could flow through the resin bed. According to the manufacturers (Rohm and Haas, personal communication) the maximum flow rate for XAD-16 is 16 bed volumes/hour. This translates into 16 L/h for a trap of 1 litre notional volume. If the high throughput steep tank of 400 litre liquid volume (Figures 8.4 and 8.5) is taken as an example of field levels of extraction, then the volume of the tank would require 400/16 = 25 h to flow through the resin trap, when the high-throughput method described above extracted 400 litres in about 15 min with a further period for the alkalization and settling. Thus resin traps may well be too slow to be practical.

8.4.2.3 *Polygonum*

Bechtold *et al.* (2002) described a simple, low-technology process for the production of indigo from *Polygonum* based on a hot water extraction of indican. Batches of plant material (15–20 kg) were soaked in successive baths of water: 40 litres at 50–70 °C for 4 h and then 20 litres at 20 °C for 2 h. The resulting combined extracts were then stored for a period of days with occasional stirring. To improve precipitation of the indigo formed, $CaCl_2{\cdot}2H_2O$ was added at 1.5 g/litre. The sedimented indigo was filtered and dried. Yields of indigo and purity depended on the storage time during the oxidation stage: yields fell from about 50 g of pure indigo/100 kg fresh weight of plant material (0.05 % yield) with a maximum purity of 12 % when harvested leaves had been stored for 1 day, to less than 10 g of pure indigo/100 kg with a purity of around 1 % when harvested leaves had been stored for 2–9 days.

The emphasis in this work (Bechtold *et al.*, 2002) was on simplicity and sustainability. Thus although the yields and purity were low, so were the inputs, with, for example, no addition of alkali to speed up the conversion of the released indoxyl to indigo (Cotson and Holt, 1958). The most significant inputs in the production process itself were the consumption of water and of energy for heating water and drying the product.

Some authors have augmented with commercial β-glucosidase the native plant and bacterial activities to speed up the release of indoxyl from *Polygonum* indican: Maugard *et al.* (2002) noted that the commercially available β-glucosidases, Novozym 199 (a cellobiase) and Novarom G, were able to hydrolyse the indican completely in an aqueous extract from *Polygonum*, raising the yield of indigo from 0.5 to 5.9 g/kg fresh weight of leaves. The pH optimum for both commercial enzymes was at pH 3.0. The authors point out the advantage of carrying out the reaction at this relatively low pH is that the indoxyl is stabilized. At pH 5.0, the optimum of the endogenous β-glucosidase, the hydrolysis of indican is accompanied by a simultaneous oxidation to indigo, but at pH 3.0 the released indoxyl appears to be unable to convert to indigo at an appreciable rate. They also noted (Maugard *et al.*, 2002) that it may be advantageous to control the pH of the oxidation step. Thus, after oxidation at pH 5.0 the indigo/indirubin ratio was 97/3 (32:1) and after oxidation at pH 11 (the usual pH after alkali addition) the ratio was 88/12 (7:1).

Angelini *et al.* (2003) pointed out that the extracts used by Maugard *et al.* (2002) contained the native β-glucosidase activity, and so the specific contribution of the added commercial enzymes could not be easily determined. Consequently, Angelini *et al.* (2003) used extracts isolated from *Polygonum* at 100 °C, a temperature that inactivates the native β-glucosidase. This enabled them to examine the specific effect of Novarom G and a β-glucosidase from sweet almond. The relative stability of indican in extracts made at 100 °C compared with extracts made at 25 °C showed that the native β-glucosidase rapidly hydrolyses indican released from the damaged leaf cells. This is reminiscent of the observation made a century earlier (Perkin and Everest, 1918) with *Indigofera* (see above) that the native β-glucosidase was responsible for hydrolysing the indican within the *Indigofera* leaves, and that indoxyl, rather than indican, was leached from the steeped leaves. Angelini *et al.* (2003) also noted that the yield of indigo, either from indican in *Polygonum* extracts or from a standard purified indican, was only 10–24 % of the yield expected from the complete conversion of indoxyl to indigo. Presumably under the conditions employed (pH 5.0, 25 °C) a proportion of the indoxyl will be lost to indirubin formation and to other side reactions.

Angelini *et al.* (2005) described the use for indigo extraction from *Polygonum* of the high-throughput extraction system designed for *Isatis* (Figures 8.4 and 8.5). Unlike *Isatis* leaves, those of *Polygonum* do not retain soil particles, and therefore they were not washed prior to extraction, which was carried out essentially as described for *Isatis* (see above). The leaves were added at a ratio of 1:3 or 1:4 to water heated to a temperature of 50 °C and steeped for 30 min. The heating could be avoided, but for water at ambient temperature of 25 °C, a longer steeping of 72 h was necessary. The leaves were removed, the steep water cooled to 25 °C, $Ca(OH)_2$ was added and the extract was aerated to accelerate the formation of indigo, which was allowed to settle overnight and then washed in acid.

Angelini *et al.* (2005) reported that a series of extractions yielded 3.2–7.8 g of crude indigo/kg fresh weight of leaves after extraction at 25 °C and 1.8–10 g of crude indigo/g fresh weight of leaves after extraction at 50 °C. The crude indigo had a purity of 5–13 %.

In Italy about 40 tons of leaves per ha can be expected, which would yield (at 10 g of 13 % pure indigo per kg fresh weight of leaves) 52 kg of pure indigo per ha.

8.5 Purity of Natural Indigo

From the earliest modern research into the production of natural indigo (Perkin and Everest, 1918) its purity has been of concern. The demise of *Indigofera* indigo in the face of competition from synthetic indigo can be attributed as much to the synthetic indigo's higher and reliable purity as to its lower price (Balfour-Paul, 1998). The purity of the best-quality *Indigofera* indigo of the 19th century was about 60 % (Perkin, 1900). Later work identified the impurities as soil, clay, calcium salts and various derivatives of indoxyl (Perkin and Everest, 1918).

Indigo sublimes at about 300 °C and decomposes at 390 °C. Thus if the impure indigo extracted from the crop is heated to 300 °C, or just above, a vapour phase of indigo is obtained, which can be readily separated from impurities remaining in the solid phase. When the temperature of the indigo vapour phase is brought below 300 °C, indigo desublimes and reappears in the solid state. C. Nicolella (unpublished) has exploited this rationale to obtain a solid phase more concentrated in indigo than the original material. A block diagram of the purification process based on sequential indigo sublimation and desublimation is given in Figure 8.7. The process is constituted by two consecutive phases: sublimation and desublimation.

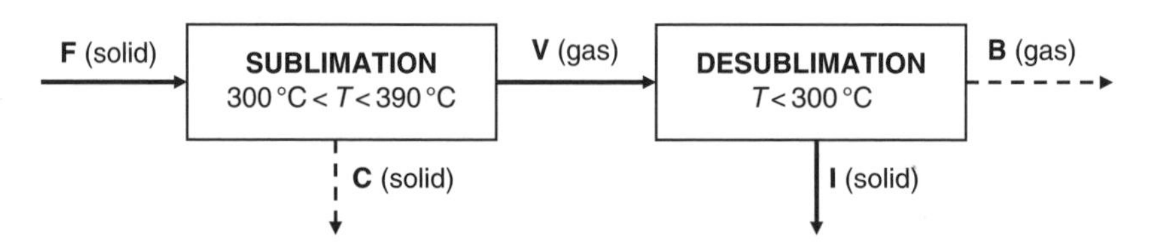

Figure 8.7 Block diagram for an indigo purification process based on sequential sublimation–desublimation. F, feed; C, carbonaceous residue; V, gas phase containing sublimated indigo; I, purified indigo; B, by-products. Solid lines represent streams containing indigo; dashed lines represent streams not containing indigo. From C. Nicolella (unpublished). Reproduced by permission of Dr C. Nicolella

In the sublimation phase, a feed (F) containing indigo and other components (i.e. organic and inorganic components) is heated to a temperature in the range 300–390 °C. Indigo sublimes and is carried away in the gas stream (V). Heat-sensitive materials (i.e. organic components) may decompose at the indigo sublimation temperature into volatile molecules (carried away in the gas stream together with indigo vapour) and produce a carbonaceous residue (C) (which remains together with any inorganic component of the feed in the solid state). Close control of temperature in the sublimation phase is crucial to optimize the ratio between the rate of indigo sublimation and the rate of by-product formation. A degree of overheating will, however, be required to avoid immediate desublimation.

In the desublimation phase, the gas stream (V) containing indigo and volatile molecules (i.e. tar vapours and light hydrocarbon gases) produced by the eventual thermal decomposition of organic components in the feed is cooled down below 300 °C. Indigo desublimes and reappears in the solid state. If the temperature of the desublimation phase is higher than the temperature of tar condensation (generally below 200 °C), a condensed phase (I) containing pure indigo will be easily separated from the gas stream (B) containing any by-products. Control of temperature in the desublimation phase is therefore crucial to avoid product contamination.

Some preliminary tests using synthetic indigo were performed in the Spindigo Project (C. Nicolella, unpublished) to assess the applicability of an indigo purification process based on sequential sublimation and desublimation. Two experimental techniques were used towards this aim: thermogravimetric analysis and a fixed-bed reactor. Thermogravimetric analysis provided weight loss curves during indigo sublimation in an inert atmosphere of pure nitrogen under both isothermal (310 and 320 °C) and constant heating rate conditions (10 °C/min). As expected, the main weight loss step occurred between 300 and 390 °C, but there was some degradation of the indigo evidenced by visual inspection. A fixed-bed reactor was used for sublimation runs to determine whether sublimed indigo could be recovered by condensation. The tubular reactor was preheated to 320–360 °C fluxed by a nitrogen stream to produce a stream of indigo vapour continuously condensed for recovery in a water trap maintained at 0 °C. A significant outcome of this preliminary trial was that to ensure a near full recovery of indigo it would be necessary to heat parts of the reactor not directly heated by the furnace to avoid condensation before the water trap.

The only recent research work to have focused on the question of indigo purity is that of García-Macías and John (2004) who worked with *Isatis*. They used a variety of approaches to show that impurities were trapped within the indigo aggregates that were formed in the oxygenation stage of indigo production. The impurities came from three major sources: dust and soil particles on the surface of the *Isatis* leaves, particulate and colloidal materials derived from the leaf itself and calcium salts formed when calcium hydroxide is the alkalizing agent.

Generating indoxyl by the alkaline hydrolysis of indoxyl acetate, inclusion of soil and *Isatis* extract reduced the yield and purity of the indigo formed (García-Macías and John, 2004). Moreover, when soil was included with the indoxyl acetate at levels equivalent to normal field contamination, the stable fluorescence seen on addition of KOH anaerobically (see Figure 8.6) was decreased by 20 % (García-Macías and John, 2004). This soil-dependent decrease in fluorescence may be due to soil-related side reactions that lead to indoxyl loss or to adsorption of indoxyl to the soil particles quenching the fluorescence.

An obvious point perhaps, but one that needs to be made explicit when considering the purity of the indigo produced from plants, is that the greater the indigo precursor content of the leaves, the greater the purity. Contamination from the plant material and residual soil is relatively constant for particular extraction conditions, while the precursor content can vary greatly: older leaves and leaves harvested early in the year contain relatively low precursor levels. Thus we observed (P. John, K. Seymour and P. García-Macías, unpublished) that in extractions from 300–600 g batches of leaves grown in the UK, a harvest early in the season (April) gave low yields of indigo and purities of 32–36 % for unwashed and 50–56 % for washed leaves. The highest purities were obtained when leaves were harvested in the period July to September. Harvesting in this period and selecting smaller, cleaner leaves, the purities of eight extractions varied from 57 % to 91 % (mean of

74 % ± 10 SEM) with an average yield of 0.5 (± 0.08 SEM) g per kg fresh weight of leaves. To obtain the highest purity, it was necessary to sediment the indigo in an acid medium containing citric acid or to wash the sedimented indigo with acid subsequently. In a series of three experiments, sedimentation in acid compared to that in alkali gave rise to improvements in purity from 17 % to 56 %, 43 % to 68 % and 15 % to 60 %. Similarly, washing the raw indigo in dilute HCl after grinding the dried product provided dramatic increases in purity. For example, the purities of three large-scale preparations of indigo were raised by a wash in 2M HCl from 21 % to 71 %, from 0.6 % to 63 % and from 19 % to 72 %.

From laboratory-scale experiments in which indoxyl was generated by the alkaline hydrolysis of indoxyl acetate, García-Macías and John (2004) found that the purity of the resulting indigo fell from 78 % to 48 % when KOH was substituted by $Ca(OH)_2$, to 39% on inclusion of *Isatis* leaf extract and to 15 % on inclusion of soil. The calcium hydroxide, leaf extract and soil were each added in amounts equivalent to those present with farm-scale indigo production. In the absence of any contaminating additions, the indigo was deposited as a fine powder with no visible aggregation (Figure 8.8(a)). In similar experiments, when indigo was made from unwashed *Isatis* leaves instead of from indoxyl acetate, the indigo was seen as clumps with a diameter of roughly 50 μm (Figure 8.8(b)). Similar clumps were formed from the hydrolysis of indoxyl acetate when calcium nitrate was included (García-Macías and John, 2004). Scanning electron microscopy (SEM) and energy dispersive X-ray analysis have been used to show that under these conditions calcium is concentrated into certain areas, while the rest of the indigo remains calcium-free (García-Macías and John, 2004).

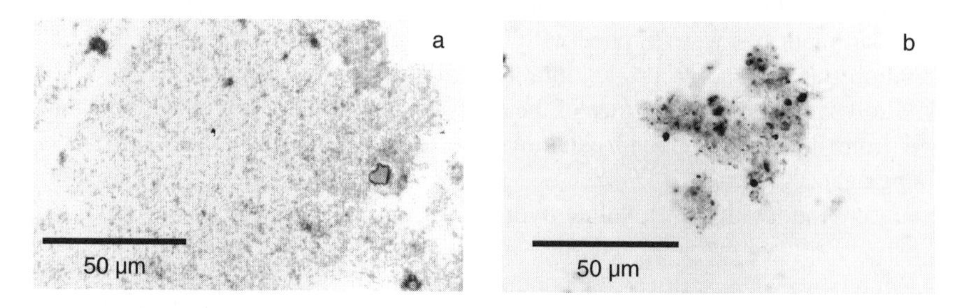

Figure 8.8 *Impurities in indigo observed under bright field microscopy. Indigo was made from alkaline hydrolysis of: (a) 1 mM indoxyl acetate; (b) unwashed woad leaves. From García-Macías and John (2004). Reprinted with permission from Journal of Agricultural & Food Chemistry, Copyright 2004 American Chemical Society*

On the basis of this work, García-Macías and John (2004) proposed a model of natural indigo formation (Figure 8.9) in which the indoxyl released from the leaves is adsorbed on to solid particles in the steep water. On subsequent alkalization and oxygenation the indoxyl converts to indigo, which coats the particulate impurities. This means that the impurities are protected from being easily washed from the indigo product. This model is supported by the following observations (García-Macías and John, 2004): fluorescence studies showed free indoxyl is stable for a measurable time after release, indoxyl forms indigo at the surface of synthetic polymeric adsorbents such as Amberlyst A-21, indigo forms sedimentable particles nucleated by a dispersion of calcium hydroxide and particles derived from soil and *Isatis* extract, both of which reduce indigo yield and purity, can be

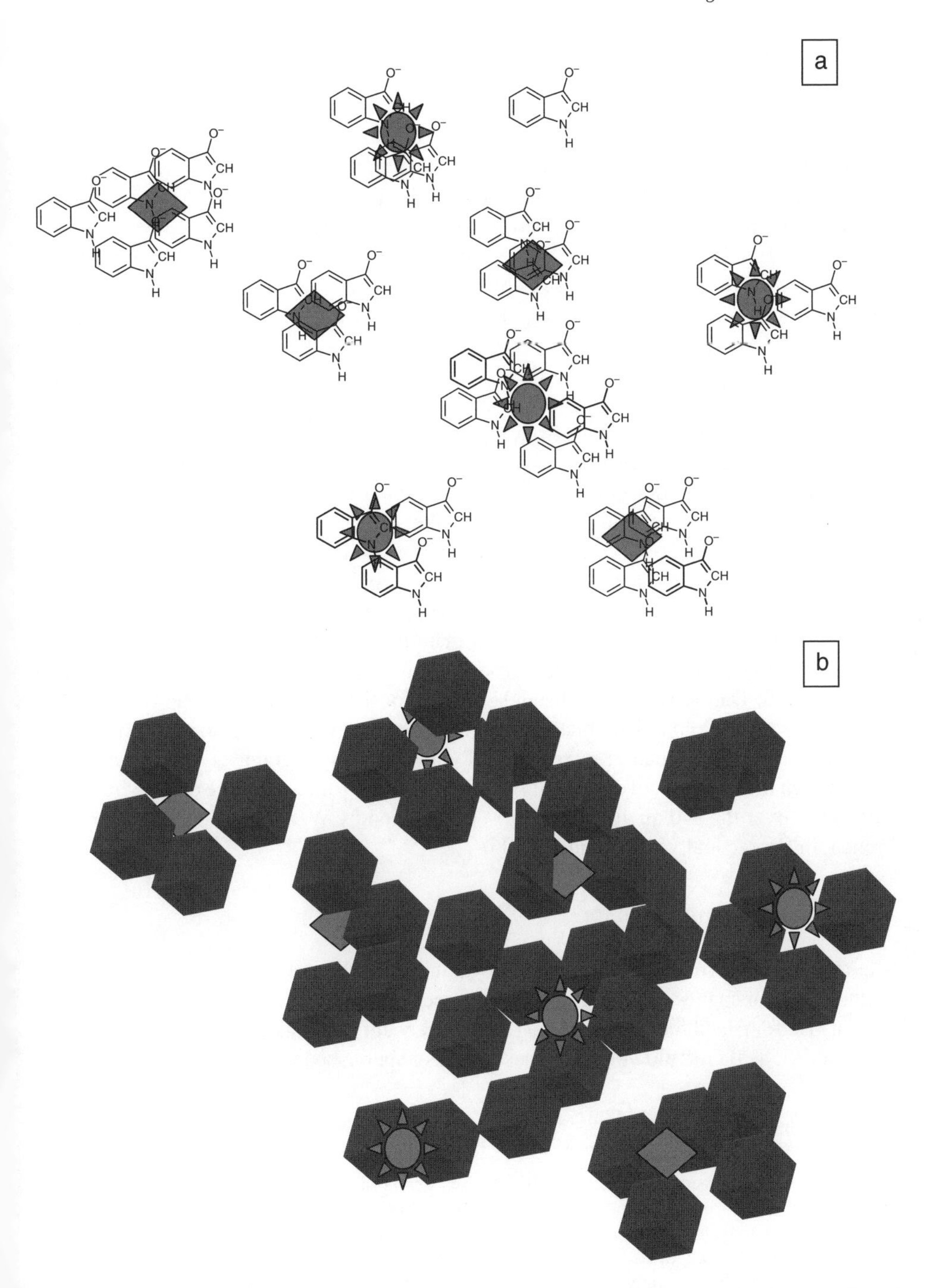

Figure 8.9 *Scheme for the interaction of natural indigo with particulate impurities: (a) indoxyl groups released from the indigo precursors adsorb to particles of soil and plant origin; (b) on oxygenation, solid indigo is formed predominantly around the particulate impurities*

observed within the indigo product. The model is also consistent with what is currently understood of the molecular basis of the interactions between xenobiotic substances and soil particles, both mineral and organic (Schulten and Leinweber, 2000). In the proposed model (Figure 8.8) the impurities are effectively 'protected' by a layer of indigo from simple washes that might otherwise remove them from the indigo product. The nucleation effect of the soil particles and colloidal plant materials might suggest that a purer product could be obtained by using indigo itself to 'seed' the nucleation process. However, it has been found (P. John and P. García-Macías, unpublished) that neither synthetic nor natural indigo is effective in this way, presumably because the affinity of the negatively charged indoxyl groups for uncharged indigo is lower than their affinity for charged mineral and colloidal organic particles.

This model (Figure 8.9) explains the difficulty in releasing impurities from the indigo once it is formed, and implies that the production of natural indigo of high purity is via the exclusion of particulate impurities from the process; i.e. make a clean product, rather than clean up a dirty product. García-Macías and John (2004) concluded that indigo can be extracted from *Isatis* with a purity of 90 % if three conditions are met: the leaves contain a sufficiently high yield of indigo precursors, the leaves are rinsed free of soil before extraction and the indigo is sedimented in an acid medium.

The recent work with *Isatis* indigo (García-Macías and John, 2004) simply extends what was understood a century ago. In the large-scale manufacture of *Indigofera* indigo during the 19th century it was recognized that plant material was a major impurity that needed to be removed (by boiling) for the final product to be acceptable to the dyers (Perkin and Everest, 1918; Darrac and van Schendel, 2006). Consistent with the model in which the indigo coats the colloidal plant material (Figure 8.9), the ever astute Darrac had noted that 180 years ago 'Vegetal matter mixes so well with the [indigo] granules that it escapes the eyes of even the best trained experts. It is however very detrimental to the dyers who are the first to discover the bad quality of the mix' (translation by W. van Schendel from original French, in Darrac and van Schendel, 2006).

Acknowledgements

The author is grateful to Chris Cooksey, Matthias Hamburger, Peter Leinweber and all partners and colleagues of the EU Spindigo Project (2001–2004) for invaluable discussions, to Jenny Balfour-Paul for alerting me to Darrac's account of Bengal indigo production and to Cristiano Nicolella for allowing me to describe his unpublished work on indigo sublimation.

References

Angelini, L. G., Campeol, E., Tozzi, S., Gilbert, K., Cooke, D. and John, P. (2003) A new HPLC-ELSD method to quantify indican in *Polygonum tinctorium* L. and to evaluate ß-glucosidase hydrolysis of indican for indigo production, *Biotechnology Progress*, **19**, 1792–1797.

Angelini, L. G., Tozzi, S. and Di Nasso, N. Nassi O. (2004) Environmental factors affecting productivity, indican content, and indigo yield in *Polygonum tinctorium* Ait., a subtropical crop grown under temperate conditions, *Journal of Agricultural and Food Chemistry*, **52**, 7541–7547.

Angelini, L. G., John, P., Tozzi, S. and Vandenburg, H. (2005) Extraction of indigo from *Isatis tinctoria* L. and *Polygonum tinctorium* Ait. as a basis for large-scale production, in M. J. Pascual-Villalobos,

F. S. Nakayama, C. A. Bailey, E. Correal and W. W. Schloman Jr. (eds), *Proceedings of 2005 Annual Meeting of the Association for the Advancement of Industrial Crops: International Conference on Industrial Crops and Rural Development*, Murcia, Spain, 17–21 September 2005, pp. 521–534, IMIDA (Istituto Marciano de Investigation y Desarollo Agrario y Alimentario) Selegrafica, S.L. Murcia, ISBN 84-689-3363-5.

Angelini, L. G., Tozzi, S. and Di Nasso, N. Nassi O. (2007) Differences in leaf yield and indigo precursors production in woad (*Isatis tinctoria* L.) and Chinese woad (*Isatis indigotica* Fort.) genotypes, *Field Crops Research*, **101**, 285–295.

Balfour-Paul, J. *Indigo*, British Museum Press, London, 1998.

Bechtold, T., Turcanu, A., Geissler, S. and Ganglberger, E. (2002) Process balance and product quality in the production of natural indigo from *Polygonum tinctorium* Ait. applying low-technology methods, Bioresource Technology, **81**, 171–177

Berry, A., Dodge, T., Pepsin, M. and Weyler, W. (2002) Application of metabolic engineering to improve both the production and use of biotech indigo, *Journal of Industrial Microbiology and Biotechnology*, **28**, 127–133.

Bond, A., Marken, F., Hill, E.,. Compton, R. and Hügel, H. (1997) The electrochemical reduction of indigo dissolved in organic solvents and as a solid mechanically attached to a basal plane pyrolytic graphite electrode immersed in aqueous electrolyte solution, *Journal of the Chemical Society Faraday Transactions*, **2**, 1735–1742.

Bruin, F., Heineken, F. W. and Bruin, M. (1963) Electron spin resonance spectra of the basic indigoid dye radicals, *Journal of Organic Chemistry*, **28**, 562–564.

Campeol, E., Angelini, L. G., Tozzi, S. and Bertolacci, M. (2006) Seasonal variation of indigo precursors in *Isatis tinctoria* L. and *Polygonum tinctorium* Ait. as affected by water deficit, *Environmental and Experimental Botany*, **58**, 223–233.

Campos, R., Kandelbauer, A., Robra, K., Cavaco-Paolo, A. and Gübitz, G. (2001) Indigo degradation with purified laccases from *Trametes hirsuta* and *Sclerotium rolfsii*, *Journal of Biotechnology*, **89**, 131–139.

Cardon, D. (2007) *Natural Dyes: Sources, Tradition, Technology and Science*, Archetype Publications, London.

Christie, R. (2007) Why is indigo blue?, *Biotechnic and Histochemistry*, **82**, 51–56.

Cotson, S. and Holt, S. (1958) Studies in enzyme cytochemistry. IV. Kinetics of aerial oxidation of indoxyl and some of its halogen derivatives, *Journal of the Royal Society of London, Series B, Biological Sciences*, **148**, 506–519.

Darrac, P.-P. and van Schendel, W. (2006) *Global Blue: Indigo and Espionage in Colonial Bengal*, Mohiuddin Ahmad, The University Press Limited, Dhaka.

Epstein, E., Nabors, M. W. and Stowe, B. B. (1967) Origin of indigo in woad, *Nature*, **216**, 547–549.

Fanjul-Bolado, P., Gonzáles-García, M. B. and Costa-García, A. (2004) Voltammetric determination of alkaline phosphatase and horseradish peroxidase activity using 3-indoxyl phosphate as substrate. Application to enzyme immunoassay, *Talanta*, **64**, 452–457.

Fanjul-Bolado, P., Gonzáles-García, M. B. and Costa-García, A. (2005) Detection of leucoindigo in alkaline phosphatase and peroxidase based assays using 3-indoxyl phosphate as substrate, *Analytica Chimica Acta*, **534**, 231–238.

García-Macías, P. and John, P. (2004) Formation of natural indigo derived from woad (*Isatis tinctoria* L.) in relation to product purity, *Journal of Agricultural and Food Chemistry*, **52**, 7891–7896.

Gehauf, B. and Goldenson, J. (1957) Detection and estimation of nerve gases by fluorescence reaction, *Analytical Chemistry*, **29**, 276–278.

Gilbert, K. G. and Cooke, D. T. (2001) Dyes from plants: past usage, present understanding and potential, *Plant Growth Regulation*, **34**, 57–69.

Gilbert, K. G., Hill, D. J., Crespo, C., Mas, A., Lewis, M., Rudolph, B. and Cooke, D.T. (2000) Qualitative analysis of indigo precursors from woad by HPLC and HPLC-MS, *Phytochemical Analysis*, **11**, 18–20.

Gilbert, K. G., Maule, H. G., Rudolph, B., Lewis, M., Vandenburg, H., Sales, E., Tozzi, S. and Cooke, D. T. (2004) Quantitative analysis of indigo and indigo precursors in leaves of *Isatis* spp. and *Polygonum tinctorium*, *Biotechnology Progress*, **20**, 1289–1292.

Gillam, E., Notely, L., Cai, H., De Voss, J. and Guengerich, F. (2000) Oxidation of indole by cytochrome P450 enzymes, *Biochemistry*, **39**, 13817–13824.

Govaert, F., Temmerman, E. and Kiekens, P. (1999) Development of voltammetric sensors for the determination of sodium dithionite and indanthrene/indigo dyes in alkaline solutions, *Analytica Chimica Acta*, **385**, 307–314.

Green, L. and Daniels, V. (1993) The use of *N*-methyl-2-pyrrolidone (NM2P) as a solvent for the analysis of indigoid dyes, *Dyes in History and Archaeology*, **11**, 10–18.

Guarino, C., Casoria, P. and Menale, B. (2000) Cultivation and use of *Isatis tinctoria* L. (Brassicaceae) in Southern Italy, *Economic Botany*, **54**, 395–400.

Hamburger, M. (2002) *Isatis tinctoria* – from the rediscovery of an ancient medicinal plant towards a novel anti-inflammatory phytopharmaceutical, *Phytochemistry Reviews*, **1**, 333–344.

Holt, S. and Sadler, P. (1958) Studies in enzyme cytochemistry. III. Relationships between solubility, molecular association and structure in indigoid dyes, *Proceedings of the Royal Society of London, Series B, Biological Sciences*, **148**, 495–505.

Hurry, J. (1930) *The Woad Plant and Its Dye*, Oxford University Press, London.

Jacquemin, D., Preat, J., Wathelet, V. and Perpète, E. (2006) Substitution and chemical environment effects on the absorption spectrum of indigo, *The Journal of Chemical Physics*, **124**, Article number 074104.

Kelly, F. (1997) *Early Irish Farming*, Early Irish Law Series, Dublin Institute for Advanced Studies, Dublin, Ireland.

Kokubun, T., Edmunds, J. and John, P. (1998) Indoxyl derivatives in woad in relation to medieval indigo production, *Phytochemistry*, **49**, 79–87.

Kumar, P. (2004) *Facing Competition: The History of Indigo Experiments in Colonial India, 1897–1920*, Georgia Institute of Technology, Georgia, USA.

Lawson, P. A., Nicholson, S. K., John, P. and Collins, M. D. (in press) Bacteria of the medieval woad (*Isatis tinctoria* L.) vat, *Dyes in History and Archaeology*.

Maier, W., Schumann, B. and Gröger, D. (1990) Biosynthesis of indoxyl derivatives in *Isatis tinctoria* and *Polygonum tinctorium*, *Phytochemistry*, **29**, 817–819.

Maugard, T., Enaud, E., Choisy, P. and Legoy, M. (2001) Identification of an indigo precursor from leaves of *Isatis tinctoria* (woad), *Phytochemistry*, **58**, 897–904.

Maugard, T., Enaud, E., de la Sayette, A., Choisy, P. and Legoy, M. (2002) ß-Glucosidase-catalyzed hydrolysis of indican from leaves of *Polygonum tinctorium*, *Biotechnology Progress*, **18**, 1104–1108.

Miliani, C., Romani, A. and Favaro, G. (1998) A spectrophotometric and fluorimetric study of some anthroquinoid and indigoid colorants used in artistic paintings, *Spectrochimica Acta Part A*, **54**, 581–588.

Minami, Y., Kanafuji, T. and Miura, K. (1996) Purification and characterization of a ß-glucosidase from *Polygonum tinctorium*, which catalyzes preferentially the hydrolysis of indican, *Bioscience Biotechnology Biochemistry*, **60**, 147–149.

Minami, Y., Takao, H., Kanafuji, T., Miura, K., Kondo, M., Hara-Nishimura, I., Nishimura, M. and Matsubara, H. (1997) ß-Glucosidase in the indigo plant: intracellular localization and tissue specific expression in leaves, *Plant and Cell Physiology*, **38**, 1069–1074.

Minami, Y., Shigeta, Y., Tokumoto, U., Tanaka, Y., Yonekura-Sakakibara, K., Oh-Oka, H. and Matsubara, H. (1999) Cloning, sequencing, characterization and expression, of a ß-glucosidase cDNA from indigo plant, *Plant Science*, **142**, 219–226.

Minami, Y., Nishimura, O., Hara-Nishimura, I., Nishimura, M. and Matsubara, H. (2000) Tissue and intracellular localization of indican and the purification and characterization of indican synthase from indigo plants, *Plant and Cell Physiology*, **41**, 218–225.

Nicholson, S. and John, P. (2005) The mechanism of bacterial indigo reduction, *Applied Microbiology and Biotechnology*, **68**, 117–123.

Novotná, P., Boon, J., van der Horst, J. and Pacáková, V. (2003) Photodegradation of indigo in dichloromethane solution, *Color Technology*, **119**, 121–127.

Oberthür, C., Graf, H. and Hamburger, M. (2004a) The content of indigo precursors in *Isatis tinctoria* leaves – a comparative study of selected accessions and post-harvest treatments, *Phytochemistry*, **65**, 3261–3268.

Oberthür, C., Schneider, B., Graf, H. and Hamburger, M., (2004b) The elusive indigo precursors in woad (*Isatis tinctoria* L.) – identification of the major indigo precursor, isatan A, and a structure revision of isatan B, *Chemistry and Biodiversity*, **1**, 174–182.

Padden, A., Dillon, V., John, P., Edmonds, J., Collins, M. and Alvarez, N. (1998) *Clostridium* used in medieval dyeing, Nature, **396**, 225.

Padden, A., Dillon, V., Edmonds, J., Collins, M., Alvarez, N. and John, P. (1999) An indigo-reducing moderate thermophile from a woad vat, *Clostridium isatidis* sp. nov., *International Journal of Systematic Bacteriology*, **49**, 1025–1031.

Perkin, F. (1900) The present condition of the indigo industry, *Nature*, **63**, 7–9.

Perkin, A. and Everest, A. (1918) *The Natural Organic Colouring Matters*, Longmans Green and Co, London.

Plowright, C. (1901) On woad as a prehistoric pigment, *Journal of the Royal Horticultural Society*, **26**, 33–40.

Ricketts III, R. (2006) *Polygonum tinctorium*: contemporary indigo farming and processing in Japan, in L. Meijer, N. Guyard, L. Skaltsounis and G. Eisenbrand, *Indirubin, The Red Shade of Indigo*, Life in Progress Editions, Roscoff, France.

Russell, G. and Kaupp, G. (1969) Oxidation of carbanions IV. Oxidation of indoxyl to indigo in basic solution, *Journal of the American Chemical Society*, **91**, 3851–3859.

Schulten, H.-R. and Leinweber, P. (2000) New insights into organic-mineral particles: composition, properties and models of molecular structure, *Biology of Fertile Soils*, **30**, 399–432.

Shen, B., Olbrich-Stock, M., Posdorfer, J. and Schindler, R. (1991) An optical and spectroelectrochemical investigation of indigo carmine, *Zeitschrift für Physikalische Chemie*, **173**, 251–255.

Stoker, K. G., Cooke, D. T. and Hill, D. J. (1998a) An improved method for the large-scale processing of woad (*Isatis tinctoria*) for possible commercial production of woad indigo, *Journal of Agricultural Engineering Research*, **71**, 315–320.

Stoker, K. G., Cooke, D. T. and Hill, D. J. (1998b) Influence of light on natural indigo production from woad (I*satis tinctoria*), *Plant Growth Regulation*, **25**, 181–185.

Strobel, J. and Gröger, D. (1989) Indigo precursors of *Isatis* species, *Biochemie und Physiologie der Pflanzen*, **184**, 321–327.

Süsse, P., Steins, M. and Kupcik, V. (1988) Indigo: crystal structure refinement based on synchrotron data, *Zeitschrift für Kristallographie*, **184**, 269–273.

Westbroek, P., DeClerck, K., Kiekens, P., Gasana, E. and Temmerman, E. (2003) Improving quality and reproducibility of the indigo dye process by measuring and controlling indigo and sodium dithionite concentrations, *Textile Research Journal*, **73**, 1079–1084.

Woo, H.-J., Sanseverino, J., Cox, C., Robinson, K. and Sayler, G. (2000) The measurement of toluene dioxygenase activity in biofilm culture of *Pseudomonas putida* F1, *Journal of Microbiological Methods*, **40**, 181–191.

Xia, Z.-Q. and Zenk, M. (**1992**) Biosynthesis of indigo precursors in higher plants, *Phytochemistry*, **31**, 2695–2697.

Zollinger, H. (2003) *Color Chemistry*, Wiley–VCH, Zurich.

9

Anthocyanins: Nature's Glamorous Palette

Maria J. Melo, Fernando Pina and Claude Andary

9.1 Chemical Basis

9.1.1 Chemical Structures

The colours of flowers and fruits have attracted the attention of man throughout history: the flowers as a fundamental source of artistic inspiration and beauty and the fruits as an important part of the human diet. Anthocyanins are the ubiquitous water-soluble pigments that are found in most flowers and fruits and are responsible for their impressive red and blue colours [1]. Several references on the use of these compounds as colour pigments in decoration, namely to substitute inorganic pigments when not available, as described by Vitruvius, or in medieval illuminations to give special lighting effects, have been reported and were summarized earlier in Chapter 1.

The name anthocyanins derives from the Greek *Anthos* (flower) and *Kyanos* (blue) and is used to designate the 2-phenylbenzopyrylium derivatives bearing an hydroxyl in position 7 and 4′ and a glycoside in position 3 (monoglycoside) or 3 and 5 (diglycoside) (Figure 9.1). This biosynthetic pathway begins with tetrahydroxychalcone (open ring) or naringenin (closed ring) and is included in the more general flavonoid biosynthesis [1d]. The term anthocyanidin was coined by Richard Willstätter [2] and describes the anthocyanin's aglycone, the 2-phenylbenzopyrylium free from sugar substitution and bearing hydroxyl groups at the positions occupied by the glycosides (Figure 9.1); it is important to remember that anthocyanidins do not exist in Nature and lead to unstable structures in aqueous solution. On the other hand, deoxyanthocyanidins are anthocyanidins lacking the hydroxyl substituent in position 3. The other 2-phenylbenzopyrylium derivatives are natural

Handbook of Natural Colorants Edited by Thomas Bechtold and Rita Mussak

flavylium compounds [3, 4] as well as synthetic flavylium compounds obtained by the creativity of synthetic organic chemists since the beginning of the last century [5]. From the point of view of biological evolution, deoxyanthocyanidins seem to be the ancestors of anthocyanins because they have been found in primitive plants such as mosses and ferns. An interesting question is the establishment of the frontier between synthetic flavylium compounds and natural flavylium compounds. We have recently found that dracoflavylium (7,4′-dihydroxy-5-methoxyflavylium) and 7,4′-dihydroxyflavylium, the last one having been accepted by the scientific community as a synthetic flavylium compound, are present in Dragon's blood resins, obtained from trees such as *Daemonorops draco* and *Draceana draco* [4]. It is worthy of note that the history of anthocyanins and synthetic flavylium compounds have been crossed over since the beginning: synthetic flavylium compounds were synthesized and characterized by Büllow (1901) before the discovery of anthocyanins by the Nobel Prize in Chemistry, Richard Willstätter (1915), and of natural flavylium compounds (1927) [2–5].

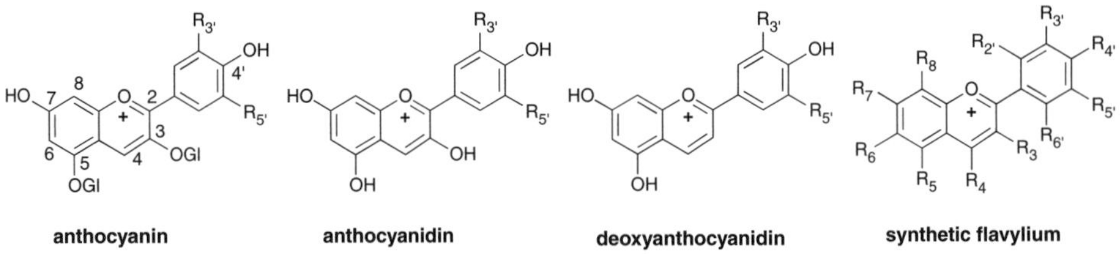

anthocyanin | anthocyanidin | deoxyanthocyanidin | synthetic flavylium

Figure 9.1 *Molecular structures for anthocyanins, anthocyanidins, deoxyanthocyanins and synthetic flavylium chromophores; OGl=glycoside*

Anthocyanins are flavonoids. There are eighteen basic structures [1, 6], six of which are the most commonly found (Table 9.1). They exist in plants in the form of glycosides, the most common sugar moieties being glucose, galactose, rhamnose and arabinose. Cyanidin derivatives are the most abundant anthocyanins in plants [1a]. Glycosylation confers solubility and a certain degree of stability on these compounds [7] and is sometimes accompanied by acylation. According to the degree of hydroxylation and methylation, the dominant anthocyanin colour varies from orange (pelargonidin) to violet (delphinidin) (Table 9.1).

Table 9.1 *Chemical structures of principal anthocyanidins*

Name	$R_{3'}$	$R_{5'}$	λ_{max}[a] *(nm)*	*Colour*
Pelargonidin	H	H	520	Orange
Cyanidin	OH	H	535	Red orange
Delphinidin	OH	OH	545	Violet
Peonidin	OCH_3	H	532	Red
Petunidin	OCH_3	OH	543	Violet
Malvidin	OCH_3	OCH_3	542	Violet

[a] Absorbance maxima in the visible spectrum in methanol with 0.01 % HCl.

9.1.2 Equilibria in Solution

For the applications of the natural 2-phenylbenzopyrylium derivatives as food colorants it is important to understand their chemical behaviour in solution, and in this context the studies on synthetic flavylium compounds were particularly useful [8, 9]. In the 1970s, it was firmly established by Brouillard and Dubois (anthocyanins) [10] and McClelland (synthetic flavylium salts) [8] that both families of compounds undergo multiple structural transformations in aqueous solution, following the same basic mechanisms (see Scheme 9.1). In particular, the work of McClelland and co-workers was important to indicate the role of the *trans*-chalcone into the network of chemical reactions involving anthocyanins, showing that a unique general scheme, Scheme 9.1, can be used to account for the thermodynamic as well as the kinetics of all 2-phenylbenzopyrylium derivatives, natural or synthetic [8].

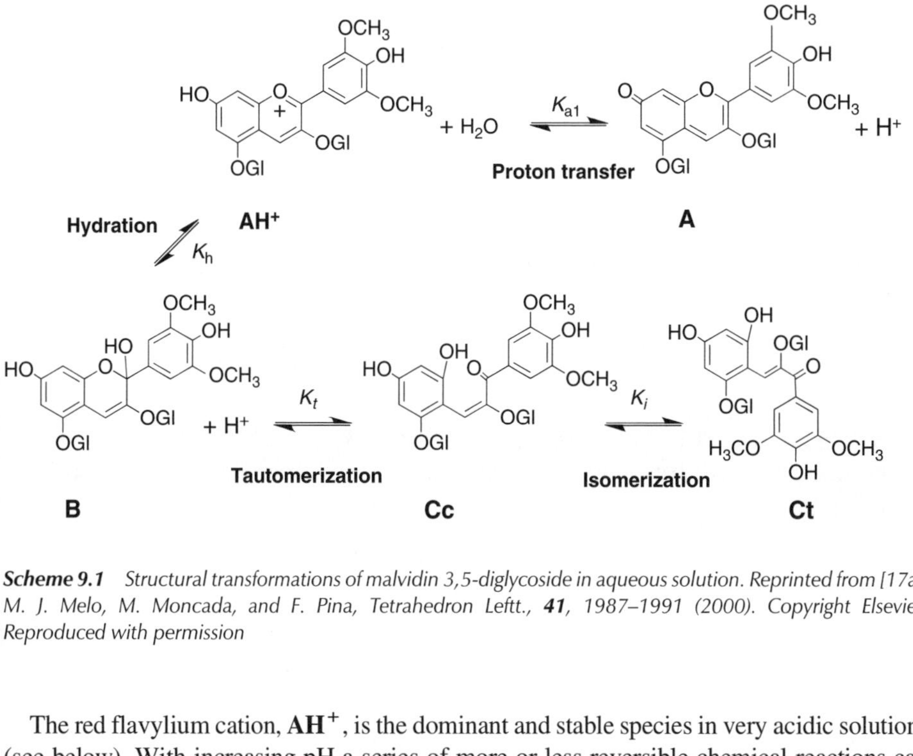

*Scheme 9.1 Structural transformations of malvidin 3,5-diglycoside in aqueous solution. Reprinted from [17a], M. J. Melo, M. Moncada, and F. Pina, Tetrahedron Leftt., **41**, 1987–1991 (2000). Copyright Elsevier. Reproduced with permission*

The red flavylium cation, $\mathbf{AH^+}$, is the dominant and stable species in very acidic solutions (see below). With increasing pH a series of more or less reversible chemical reactions can occur: (a) proton transfer leading to the blue quinoidal (or quinoid) base, **A**, (b) hydration of the flavylium cation giving rise to the colourless hemiketal, **B**, (c) tautomerization reaction responsible for ring opening, to give the pale yellow *cis*-chalcone form, **Cc**, and, finally, (d) *cis–trans* isomerization to form the pale yellow *trans*-chalcone, **Ct**. When the flavylium solutions are made less acidic ($4<pH<6$) the blue colour of the base **A** immediately appears but may fade over time. This is a result of the kinetic processes reported in Scheme 9.1. The proton transfer from $\mathbf{AH^+}$ species to water to give the base **A** is very fast. However, the $\mathbf{AH^+}$

species can also be attacked by water at position 2, leading to the colourless hemiketal, **B**. This reaction is slower than the previous one, but if **B** is thermodynamically more stable than **A** the blue colour tends to disappear over time. An important point in this kinetic process, clarified by Brouillard and Dubois, was that **A** is transformed into **B** through $\mathbf{AH^+}$. In other words, at less acidic pH values $\mathbf{AH^+}$ equilibrates with **A** (to an extent that depends on pH), but only the remaining $\mathbf{AH^+}$ makes the system shift towards **B**. On the other hand, **B** equilibrates with **Cc** in the subsecond scale of time, and **Cc** produces **Ct** is a time scale of a few hours [11]. In the case of anthocyanins, at higher pH values, both **Cc** and **Ct** are minor species and **B** is the major species, but in some natural flavylium systems we have observed that **A** can be the major species [4, 12c], and in many synthetic flavylium compounds the chalcones are practically the only stable species at these pH values. In these cases photochromic systems can be obtained taking profit from the photoinduced *cis–trans* isomerization [12].

The network of chemical reactions reported in Scheme 9.1 can be accounted for by the following set of chemical equlibria:

$$\mathbf{AH^+} + H_2O \rightleftharpoons \mathbf{A} + H_3O^+ \qquad K_a \tag{9.1}$$

$$\mathbf{AH^+} + 2H_2O \rightleftharpoons \mathbf{B} + H_3O^+ \qquad K_h \tag{9.2}$$

$$\mathbf{B} \rightleftharpoons \mathbf{Cc} \qquad K_t \tag{9.3}$$

$$\mathbf{Cc} \rightleftharpoons \mathbf{Ct} \qquad K_i \tag{9.4}$$

Equations (9.1) to (9.4) can be substituted for by a single acid–base equilibrium:

$$\mathbf{AH^+} + H_2O \rightleftharpoons \mathbf{CB} + H_3O^+ \qquad K'_a = \frac{[\mathbf{CB}][H_3O^+]}{[\mathbf{AH^+}]} \tag{9.5}$$

where $[\mathbf{CB}] = [\mathbf{A}] + [\mathbf{B}] + [\mathbf{Cc}] + [\mathbf{Ct}]$ and $K'_a = K_a + K_h + K_hK_t + K_hK_tK_i$.

The mole fraction distribution of the species in solution is determined by the solution pH according to

$$\chi_{AH^+} = \frac{[H^+]}{[H^+] + K'_a} \tag{9.6}$$

$$\chi_A = \frac{K_a}{[H^+] + K'_a} \tag{9.7}$$

$$\chi_B = \frac{K_h}{[H^+] + K'_a} \tag{9.8}$$

$$\chi_{Cc} = \frac{K_hK_{'}}{[H^+] + K'_a} \tag{9.9}$$

$$\chi_{Ct} = \frac{K_hK_tK_i}{[H^+] + K'_a} \tag{9.10}$$

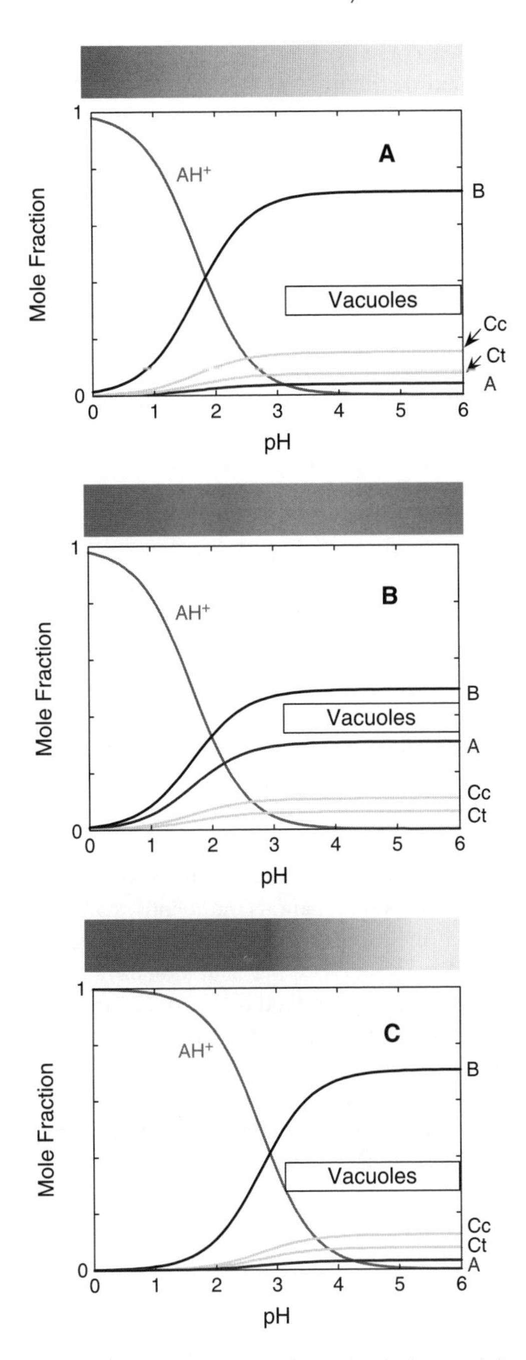

Figure 9.2 *Mole fraction distribution of the several species of malvidin 3,5-diglycoside:* ***A****-free ($pK'_a = 1.7$);* ***B****-simulation in the case of co-pigmentation with A for a copigment concentration of 0.1 M and an association constant, $K'_{cp} = 100\ M^{-1}$; ($pK'_a = 1.7 = 1.68$)* ***C****-the smae as in B but when exclusive copigmentation with* ***AH^+*** *occurs, $K_{cp} = 100\ M^{-1}$ ($pK'_a = 1.7 = 2.74$). The band at the top simulates the colour that is obtained at different pH values. Reprinted with permission from Reference [17a], © 2000, Elsevier Science (See Colour Plate 4)*

Equations (9.1) to (9.10) account for the thermodynamic equilibrium of the network permitting the mole fraction distribution to be obtained of all the species provided that the equilibrium constants have been obtained experimentally (Figure 9.2(a)). This can be done by a routine procedure that is summarized in Reference [11a].

9.1.3 Colour and Colour Stability

With regards to colour, and within the pH expected in the vacuoles, the contributing species are the flavylium cation, $\mathbf{AH^+}$, and the quinoidal base, **A**. A useful way to illustrate the pH domain of the red and blue colours is to represent the pH dependence of the mole fraction distribution of the species (Figure 9.2(a)) when the system has reached the thermodynamic state. The key parameter is pK'_a, defined as the pH value for which the mole fraction distribution of the flavylium cation is reduced by one half (Equation (9.5)).

Inspection of Figure 9.2(a) shows that the appearance of the red colour occurs only at very acidic pH values and that the blue colour is not thermodynamically stable. This is the dilemma that we have to face when projecting the use of anthocyanins as food colorants or for dyeing purposes: on the one hand, the red colour implies a very acidic medium, not easily available, but on the other hand, at moderately acidic and neutral pH values the blue fades over time. This is also the same problem that Nature had to solve. Anthocyanins are located inside vacuoles, where the pH can assume a large range of values [13]. However, as can be visualized in the band at the top of Figure 9.2(a), the colouring power of malvidin 3,5-diglycoside is very low, even for the most favourable pH values that can appear in the vacuoles of certain plants. There is no doubt that malvidin 3,5-diglycoside, for example at pH *ca.* 3.1 (a common pH value of fruits), does not exhibit any substantial colouring power, neither red nor blue.

The question is that for the formation of a blue colour some interaction between the blue quinoidal base and other compounds is required (Figure 9.2(b)). This phenomenon is called co-pigmentation and can involve other natural compounds, such as flavonoids and proteins (AVIs-anthocyanic vacuolar inclusions) and metals [14]. Several examples of intermolecular, intramolecular or self-association co-pigmentation have been reported [1, 14, 15], in particular the fascinating structures described by Goto, Kondo and collaborators [14]. As an example, the mole fraction distribution of the several malvidin 3,5-diglycoside forms, obtained upon complexation, can be simulated for a simplified model system consisting of a single 1:1 equilibrium with the association constant K'_{cp} involving the species A and the co-pigment. The results are shown in Figure 9.2(b), and indicate that a blue colour can now be obtained at moderately acidic pH values.

In many cases co-pigmentation is also necessary for obtaining red colours. A red flower or fruit in which the pH of the vacuoles is larger than 3.0 and the colorant is malvidin 3,5-diglycoside ($K'_a = 1.7$) or cyanidin 3,5-diglycoside ($K'_a = 1.85$), for example, immediately suggests the existence of co-pigmentation with a flavylium cation, because at this pH value the concentration of AH^+ is very low. This is the case for the red buds of *Ageratum*, described sixty years ago by Robinson [16b], which contain acylated cyanidin diglycoside at pH *ca.* 6. In this and other similar cases, a red adduct between AH^+ and the co-pigment should be formed in order to shift the pK'_a to high pH values. The resulting mole fraction distribution can also be easily simulated as shown in Figure 9.2(c), where enlargement of the pH domain of the red species is evident.

An alternative to obtain the red colour was reported for the case of raspberry (*Rubus idaeus*) [17a]. This fruit presents a very intense red colour, and its major anthocyanins are cyanidin and 3-glycosides; 3-glycoside derivatives exhibit higher pK'_a (≈ 3) values compared with 3,5-diglycosides. In this case, it was demonstrated that the red colour could be obtained without co-pigmentation and that the strategy was to make use of a large concentration of the colorant, *ca.* 2.4 mg of anthocyanin (expressed in cyanidin 3-glycoside) per gram of fresh fruit. In a certain way, Nature uses this variable to compensate for the fact of not using the total colouring power of the anthocyanins.

Colour in Nature, with anthocyanins, depends also on the substitution pattern of the benzopyrilium ring (Table 9.1), namely on the number and position of the hydroxy or methoxy substituents in ring A or B [17b]. For example, by comparing pelargonidin (four groups of OH) with cyanidin (five groups of OH) and delphinidin (six groups of OH), a shift to the visible from 520, 535 and finally 545 nm, respectively, is observed. On the other hand, position 3 is particularly important when colour is concerned, for example by comparing pelargonidin with luteolinidin (same number of OH groups, but luteolinidin as no OH in position 3) a difference of 25 nm is found [17b].

9.1.4 Anthocyanins as Antioxidants

Many major human degenerative diseases may be triggered by free radical processes and reactive oxygen species have been implicated in various human diseases including the processes of ageing, cancer, multiple sclerosis, Parkinson's disease, autoimmune disease, senile dementia, inflammation and arthritis, and atherosclerosis. Oxidative stress in human cells leads to the formation of radicals, namely phospholipids, oxidation of cellular membranes, resulting in cellular death, damaging of proteins with lost of important functions, as in lipoproteins (atherosclerosis), damaging of nucleic acids, giving rise to cancer and genetic diseases, etc. Therefore, lipid oxidation may be related to atherosclerosis and coronary arterial disease, and possibly is associated with arthritis, cancer and atherogenesis [1b, 18]. There are numerous vital processes in our organism that generate free radicals, which must be neutralized. These radicals, as, for example, superoxide and hydrogen peroxide, can be deactivated by specific enzymes (catalase and superoxide-dismutase). For other free radicals that are more dangerous, as, for example, singlet oxygen and peroxyl, the organism does not possess specific enzymes, and the survival of the cell depends on the antioxidant properties of vitamins A, E and C. Neutralization of peroxyl radicals is carried out by a series of coupled electron and proton transfer reactions involving vitamins E and C, gluthatione and NADPH, as generally depicted in Scheme 9.2. Within this framework, several studies indicated that improving the dietary intake of nutrients with antioxidant properties could ameliorate, if not prevent, the above-described diseases. The role of nutrients with antioxidant properties, such as vitamin E, vitamin C, carotenoids and flavonoids, has been described and epidemiological evidence points to a reduced risk of certain degenerative diseases by consumption of beverages rich in polyphenols, in particular green tea, red wines or berries such as raspberry [1b]. It was also proved that the hemiketal form of common anthocyanins, such as cyanidin-3-OGl and malvidin-3-OG,l has the redox potential necessary to reduce ascorbic acid, acting as an electron donor, eventually by a mechanism involving the simultaneous transference of an electron and a proton [19a]. Therefore, these anthocyanins could participate in the antioxidant cycle [19b] (Scheme 9.2), contributing to the neutralization of an oxidized lipid to a less-reactive species, a hydroperoxide, by regenerating ascorbic acid. Furthermore, these compounds have also been shown to be capable of inhibiting oxidation of lipoproteins and platelet aggregation *in vitro* [18d, 19c].

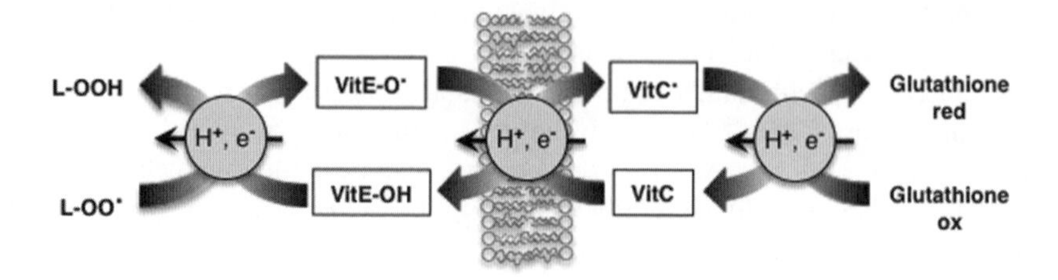

Scheme 9.2 *Anthocyanins as antioxidants: possible role based on the redox potentials of the hemiketal species*

9.2 Natural Sources for Anthocyanins

9.2.1 Plant Sources, Content, Influencing Parameters

The red, violet or blue colours of numerous fruit, flowers and vegetables are due to the presence of anthocyanins, as seen in Table 9.2. Research into the distribution of anthocyanins in plants was briefly reported at the end of the 1930s, and was resumed with the work of Harborne in 1967 [21a] and Harborne and Grayer in 1988 [21c]. Anthocyanins are abundantly present in fruit and flowers and present in lesser quantities in other plant parts such as leaves, stems, roots and wood. They are usually localized in the vacuoles of cells in the epidermis and the hypodermis, e.g. in the skins of berries, or they may be found in the entire fruit, as in the case of strawberries and blackcurrants.

Table 9.2 *Main anthocyanins in different natural sources (according to Reference [20])*

Anthocyanins	*Plant sources*
Pelargonidin	Strawberry, banana, red radish, potato
Cyanidin	Apple, blackberry, elderberry, peach, pear, fig, cherry, onion, gooseberry, red cabbage, rhubarb
Cyanidin and Delphinidin	Blackcurrant, blood orange, gooseberry, red cabbage, purple carrot
Delphinidin	Passion fruit, eggplant, green bean, pomegranate
Cyanidin and Peonidin	Cranberry, plum
Petunidin and Malvidin	Bilberry, red grape

Table 9.3 Anthocyanin contents in ripe fruits (according to Reference [22])

Family	Species	Common name	Anthocyanin content[a]	References
Caprifoliaceae	*Sambucus nigrum*	Elderberry	448	[23]
Empetraceae	*Empetrum nigrum*	Crowberry	300–460	[24]
Ericaceae	*Vaccinium myrtillus*	Bilberry	460	[25]
Ericaceae	*V. corymbosum*	Blueberry	160–503	[26]
Ericaceae	*V. macrocarpon*	Cranberry	46–172	[27]
Rosaceae	*Fragaria* sp.	Strawberry	28–70	[28]
Rosaceae	*Rubus fructicosus*	Blackberry	82–180	[29]
Rosaceae	*R. occidentalis*	Black raspberry	214–428	[29]
Rosaceae	*R. idaeus*	Red raspberry	23–59	[29]
Saxifragaceae	*Ribes nigrum*	Blackcurrant	250	[30]
Vitaceae	*Vitis vinifera*	Red grape	8–388	[31]

[a] All values in mg/100 g fresh weight.

There is a great diversity of anthocyanin concentrations depending on the species (Table 9.3). The cellular distribution of anthocyanins is not homogeneous and substantial differences in concentration (and therefore in colour) are observed throughout the ripening of red fruits.

The nature and accumulation of anthocyanins are principally regulated by genetic factors, but also by external factors such as light and temperature. Observations in the field and laboratory experiments have demonstrated the importance of light in anthocyanin accumulation in fruit and flowers. Light acts directly via red, blue and UV radiation, which trigger different photoreceptors [32]. It also acts by inducing the synthesis of several enzymes involved in phenolic metabolism, e.g. PAL (polyphenol-ammonia-lyase) and CHS (chalcone synthase) [22, 33].

Temperature also acts on the expression of phenolic metabolism and very often in interaction with light [22]. For instance, a decrease in temperature associated with sufficient light frequently induces the accumulation of anthocyanins in numerous fruit, e.g. apples, grapes. Here, again, the process involves PAL, inhibitors of which can be produced at high temperatures. Growth substances also affect phenolic metabolism. For example, auxins such as 2,4-dinitrophenol, zeatine, etc., at concentrations greater than 10^{-8} M have a negative effect on the production of anthocyanins in carrot cell cultures. Conversely, at the same concentrations cytokines (plant hormones) stimulate anthocyanin accumulation [34]. Lastly, it is possible using genetic engineering techniques to modify plant colours by introducing genes coding for new activities, by activation or deactivation of structural genes [35].

The different colours of vegetables are due not only to anthocyanins but also to other flavonoids, and thereby colour plays an important role in the interaction between the plant and its environment. For example, insects and birds are attracted by certain colours (e.g. the colibri is attracted by hibiscus flowers) and are essential for pollination and dissemination of seeds. Thus these polyphenolic compounds participate in the co-evolution between plants and animals [36].

9.3 Applications

9.3.1 Food Colorants

The use of anthocyanins as food additives dates back to antiquity. The Romans used highly coloured berries to intensify the colour of wine. Anthocyanins are also used in numerous food preparations such as jellies, jams and fruit juices, as the reference E163 N (where N signifies natural origin) in the list of food additives authorized in the European Union. The permitted dose of anthocyanin additives is 200 ppm. These colorants do not pose any particular problems in terms of food safety. On the contrary, as already described, they are antioxidants and have many health benefits for humans (see Section 9.1.4). Cyanidin, for example, possesses an antioxidant activity fourfold that of α-tocopherol and twice that of catechin [37]. Much research has been undertaken to identify compounds that may be exploited both as colorants and as antioxidants.

From the second half of the 19th century, most natural colorants were replaced by synthetic colorants, which are less expensive and more stable. From the end of the 19th century to the beginning of the 21st century the number of synthetic colorants rose from 90 to 5000, engendering a multimillion dollar industry. However, various food crises coupled to an increasing awareness in relation to matters of ecology and the environment has led agrifood industrialists to seek other colorants in diverse plant sources. Moreover, increasing strict legislation regarding the use of certain synthetic colorants in foods and cosmetics has largely reduced the palette of colours at the disposal of the industry [6]. No particular colour has been spared but red colorants were most affected. Thus, the banning of Red 2 (Amaranth) by the American Food and Drug Administration (FDA), RN orange in Great Britain, red Ecarlate and Ponceau GR in France, the toxic and even cancer risks of tartrazine or of erythrosine (if it is converted into fluorescein) has led to the increasing use of natural water-soluble red colorants [38]. Other synthetic red colorants are in the process of being banned by the FDA (Red 40 or Allura Red AC and Red 3) and by UK agrifood industrialists (Tartrazine (E102), Carmoisine (E122 or Azorubine) and Cochineal A Red (E124 S)).

There are several sources of anthocyanins, but their lack of availability, the cost of certain plant sources and the low stability of certain extracts has hitherto limited the commercial exploitation of natural sources to a small number of species, such as:

Red fruit: concentrated extracts of grapes, blackcurrant, whortleberry (roselle), *Hibiscus sabdariffa*, elderberry, etc.
Vegetables: red cabbage, black carrot, radish, sweet potato.

The principal difficulty in terms of commercializing anthocyanins as colorants is the low stability of their extracts vis-à-vis to pH, light, temperature, etc. Several studies have shown that anthocyanins that contain several glycosylated groups acylated by aromatic acids, such as *p*-coumaric acid, caffeic acid, ferulic acid or sinapic acid, or by aliphatic acids (*p*-hydroxybenzoic, malonic, acetic, succinic, oxalic or malic) are more stable than the nonacylated equivalent [39a, 40].

For the capture of colour, *in vitro*, intramolecular co-pigmentation (Figure 9.3) appears to be more efficient than intermolecular [15a, 40] co-pigmentation, which requires the presence of other compounds such as planar flavonoids, e.g. quercetin, *C*-glucosyl flavone, etc., which essentially protect the chromophore from water addition reactions [39a, 41].

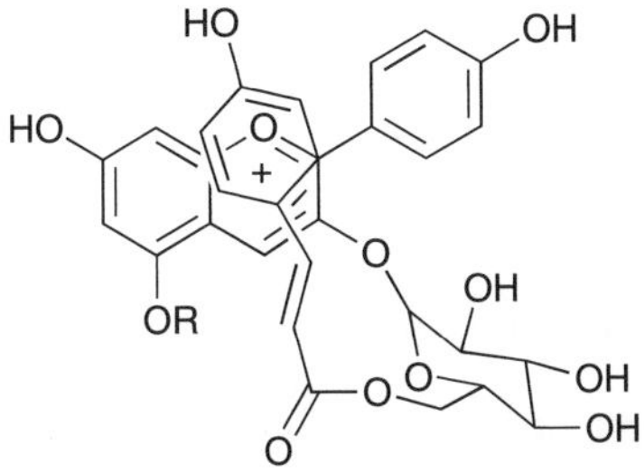

Figure 9.3 *Example of intramolecular co-pigmentation with an acylated anthocyanin: pelargonidin glycoside acylated by p-coumaric acid [15e]. Adapted from Reference [39b]*

A study carried out on several anthocyanin extracts (grape pomace, elderberry, purple carrot, red cabbage, red radish, blackcurrant and chokeberry) showed that extracts containing diacylated anthocyanins (red radish and red cabbage) were significantly more stable to light and heat than extracts that either did not contain these compounds (elderberry, blackcurrant, chokeberry and grape pomace) or contained only monoacylated anthocyanins (purple carrot) [41]. This is explained by the fact that the aromatic acyl groups in the diacylated anthocyanins can simultaneously stack on both faces of the chromophore (sandwich-type complex), thereby offering optimal protection [39a, 42]; the monoacylated anthocyanin present in purple carrot has only one face of its pyrylium ring protected against water addition [41].

9.3.2 Other Uses

Apart from their use in food and nutritional supplements, anthocyanins have been widely used as dye substances, which has been observed in other continents. In Japan, for example, 'Awobana' paper [43] is tinted blue with an extract of the flowers of *Commelina communis*, var. *hortensis* (tsuyukusa), which contain a blue anthocyanin–metal–flavonol complex that is particularly stable. The most common sources of anthocyanins used in areas other than food colouring are fruit (berries essentially) that are very rich in colouring compounds. Examples are berries of the Ericaceae family – bilberry (*Vaccinum myrtillus*), blueberry (*V. corymbosum*), cranberry (*V. macrocarpon*) – or grape pomace (Vitaceae), a highly abundant by-product of wine production. Other sources that are worth citing are hollyhock (*Alcea rosea* var. *nigra*, Malvaceae) and sorghum (*Sorghum bicolor*, Poaceae). Varieties of the latter species are particularly rich in anthocyanins in the seeds and above all in leaf stalks, for instance in the '*colorans*' variety (the remains of which have been found in archaeological sites dating back a thousand years [43]), and are still widely used as dye plants in Africa (Chad, the Sudan and north-east Nigeria).

The dyeing quality of sorghum was verified by comparing it with other anthocyanin-containing species measuring, with professional equipment, the colour characteristics in addition to light and water stability of dyed wool and cotton [44]. The relative stability of the colour was explained by the nature of the anthocyanins contained in the extract (apigeninidin, luteolinidin and fisetinidin), which belong to the group of 3-deoxyanthocaynidins and are more stable than the classic anthocyanins [21d], which are analogous to the flavones, luteolin and apigenin that do not have a hydroxyl group in position 3 either [45].

In the case of dyeing solid materials (textiles, skin, wood, etc.), it should be remembered that the stability of the colour will also depend on the mordant used to fix the colour. However, this aspect will not be treated in this chapter.

9.4 Examples of Commercial Products and Processing

The world market of food colorants is currently estimated at 940 million dollars [46], two-thirds of which is represented by natural colorants. The United States, Europe and Asia represent 73 % of the market in equal parts. This market is supported by the increasing number of bans on synthetic colorants for toxicological and ecological reasons. The market for natural colorants is therefore expanding with an increase of 4–6 % per annum as compared to 2–3 % for artificial colorants. In Europe, it is the Germans who most consume natural colorants (24 % of sales), followed by the French (17 %) and the English (16 %).

There are few companies that supply a wide variety of colorants and the best industrial example remains grape production, which represents the principal source of colorants based on anthocyanins. Grapes are the most widespread harvested fruit crop, with 65 million tons produced annually, 80 % of which is used for the production of wine. Anthocyanins are extracted from grape pomace, which is composed of grape skins and stalks after pressing and which still contains almost 20 % in anthocyanin dry residue. Grape pomace is therefore the least costly starting material for this type of colorant [47, 48].

In France one distillery company alone (Groupe GRAP'SUD, Cruviers-Lascours, France) treats 20 % of the worldwide quantity of grape pomace (approximately 100 000 tons) for the production of anthocyanin colorants (E163). To be more specific, this company manufactures and commercializes 1000 tons of colorant E163 annually.

Another valid reason for valorizing this relatively inexpensive by-product is that it reduces the pollution load of distillery waste products, though a disadvantage is the seasonal nature of the starting material, which is irregular both in terms of quality and of quantity. There is therefore an increasing interest on the part of industrialists in the agrifood sector in the development of research aimed at a greater understanding of the phytochemistry of new sources of anthocyanins that are naturally acylated and highly concentrated in colorants. Typical examples are red radish and sweet potato.

The extraction of anthocyanins, which are highly polar, is a delicate technological process due to their instability and the necessity to use water as the extracting solvent. Several trials have been carried out on grape anthocyanins by comparing different solvents (water, methanol, ethanol) and different acids (organic and mineral acids). Acidified methanol is more effective than ethanol and water [49] and it has been shown that methanolic 5 % citric acid and 10 % HCl are the most effective [50]. However, in an industrial context, ethanol acidified with hydrochloric acid would be the preferred solvent due to the toxicity of methanol.

Aqueous solutions containing SO_2 at variable concentrations have also been used to extract anthocyanins. SO_2 is an antioxidant that enhances the solubilization of anthocyanins and protects them from oxidation. The optimal extraction temperature would appear to be between 30 and 35 °C in order to avoid degradation of the anthocyanins [51].

Other extraction trials were carried out using ultrasonication of purple carrot fragments in nonacidified water (data not shown) using an extraction time of 20 minutes at ambient temperature. These extracts were twice as concentrated as extracts obtained in water acidified with HCl (pH 1) at 80 °C without ultrasonication.

It is necessary to improve the purification of anthocyanins in the enriched extract in order to improve their stability. For this purpose the extract may be passed through adsorbent resins such as styrene co-polymers, phenol-formaldehyde supports or polymethacrylate resins [20]. Generally, the colorant in the extract is 50 times less concentrated than the accompanying substances, sugars in particular, and the most widely used procedure to simultaneously filter and concentrate at low temperature is cross-flow membrane ultrafiltration.

In conclusion, the potential colour linked to anthocyanins can be very significant, depending on the plant sources. The colour obtained varies from red to blue-violet, which may be employed in diverse manners, such as addition to solid or liquid foodstuffs to enhance the appearance or recognition of the product or to impart beneficial health properties. They may also be used to colour various solid materials such as textile, paper, skin, wood, walls, etc. However, the low stability of anthocyanin colorants continues to attract much research into methods to circumvent this disadvantage and therefore encourage substantially more widespread use of these products.

References

1. (a) J. B. Harborne, in W. Goodwin (ed.), *Functions of Flavonoids in Plants*, Vol. 1, Academic Press, London, 1976. (b) E. Haslam, *Practical Polyphenolics: From Structure to Molecular Recognition and Physiological Action*, Cambridge University Press, Cambridge, 1998. (c) O. M. Andersen and M. Jordheim (eds), *Flavonoids: Chemistry, Biochemistry and Applications*, Taylor & Francis Group, 2006. (d) B. Winkel-Shirley, Flavonoid biosynthesis. A colorful model for genetics, biochemistry, cell biology, and biotechnology, *Plant Physiol.*, **126**, 485–493 (2001). (e) F. C. Stintzing and R. Carle, Functional properties of anthocyanins and betalains in plants, food, and human nutrition, *Trends Food Sci. Technol.*, **15**, 19–38 (2004). (f) D. Strack and V. Wray, Anthocyanins, in J. B. Harborne (ed.), *The Flavonoids, Advances in Research*, Chapman and Hall, London, 1994.
2. (a) R. Willstätter and A. E. Everest, Untersuchungen über die Anthocyane. I. Über den Farbstoff der Kornblume, *Justus liebigs Ann. Chem.*, **401**, 189–232 (1913). (b) R. Willstätter and H. Mallison, X. Über Variationen der Blütenfarben, *Justus liebigs Ann. Chem.*, **408**, 147–162 (1915).
3. (a) E. Chapman, A. G. Perkin and R. Robinson, The colouring matters of Carajura, *J. Chem. Soc.*, 3015–3041 (1927). (b) D. D. Pratt, R. Robinson and A. Robertson, A synthesis of pyrylium salts of antho-cyanidin type. Part XII, *J. Chem. Soc.*, 1975–1983 (1927).
4. (a) M. J. Melo, M. Sousa, A. J. Parola, J. Seixas de Melo, F. Catarino, J. Marçalo and F. Pina, Identification of 7,4′-dihydroxy-5-methoxyflavylium in 'Dragon's blood': to be or not to be an anthocyanin, *Chem. Eur. J.*, **13**, 1417–1422 (2007). (b) M. Sousa, M. J. Melo, A. J. Parola, J. Sérgio Seixas de Melo, F. Catarino, F. Pina, Frances Cook, Monique S. J. Simmonds and João A. Lopes, On natural flavylium chromophores as species markers for Dragon's blood resins from Dracaena and Daemonorops trees, *J. Chromatogr. A.*, **1209**, 153–161 (2008).
5. C. Büllow and H. Wagner, Derivatives of 1.4-benzopyranol, the mother substance of a new class of pigments. II, *Ber.*, **34**, 1782–1804 (1901).
6. C. Malien-Aubert and M. J. Amiot-Carlin, Pigments phénoliques – structure, stabilité, marché des colorants naturels, in P. Sarni-Manchado and V. Cheynier (eds), *Les Polyphénols en Agroalimentaire*, TEC & DOC, London, Paris, New York, 2006.
7. P. Markakis, Stability of anthocyanins in foods, in P. Markakis (ed.), *Anthocyanins as Food Colors*, Academic Press, New York, 1982.
8. (a) R. A. McClelland and S. Gedge, Hydration of the flavylium ion, *J. Am. Chem. Soc.*, **102**, 5838–5848 (1980). (b) R. A. McClelland and G. H. McGall, Hydration of the flavylium ion. 2. The 4′-hydroxyflavylium ion, *J. Org. Chem.*, **47**, 3730–3736 (1982).

9. H. Santos, D. L. Turner, J. C. Lima, P. Figueiredo, F. Pina and A. L. Maçanita, Elucidation of the multiple equilibria of malvin in aqueous solution by one- and two-dimensional NMR, *Phytochemistry*, **33**, 1227–1232 (1993).
10. J. R. Brouillard and J. E. Dubois, Mechanism of structural transformations of anthocyanins in acidic media, *J. Am. Chem. Soc.*, **99**, 1359–1364 (1977).
11. (a) F. Pina, Thermodynamics and kinetics of flavylium salts – Malvin revisited, *J. Chem. Soc. Faraday Trans.*, **94**, 2109–2116 (1998). (b) F. Pina, M. Maestri and V. Balzani, Photochromic systems based on synthetic flavylium compounds and their potential use as molecular-level memory devices, in H. S. Nalwa (ed.), *Handbook of Photochemistry and Photobiology: Supramolecular Photochemistry*, Vol. **3**, American Scientific Publishers, 2003. (c) F. Pina, M. J. Melo, A. J. Parola, M. Maestri and V. Balzani, pH-Controlled photochromism of hydroxy-flavylium ions, *Chem. Eur. J.*, **4**, 2001–2007 (1998).
12. (a) F. Pina, M. Maestri and V. Balzani, Photochromic flavylium compounds as multistate/multifunction molecular-level systems, *Chem. Commun.*, 107–114 (1999). (b) F. Pina, J. C. Lima, A. J. Parola and C. A. M. Afonso, *Angew. Chem.*, **116**, 1551–1553 (2004); *Angew. Chem. Int. Ed.*, **43**, 1525–1527 (2004). (c) M. J. Melo, S. Moura, A. Roque, M. Maestri and F. Pina, Photochemistry of Luteolinidin. Write–lock–read–unlock–erase with a natural compound, *J. Photochem. Photobiology*, **135**, 33–39 (2000). (d) F. Pina, M. J. Melo, M. Maestri, R. Ballardini and V. Balzani, Artificial chemical systems capable of mimicking some elementary properties of neurons, *J. Am. Chem. Soc.*, **119**, 5556–5561 (1997).
13. R. N. Steward, K. H. Norris and S. Asen, Microspectrophotometric measurement of pH and pH effect on color of petal epidermal-cells, *Phytochemistry*, **14**, 937–942 (1975).
14. (a) T. Goto and T. Kondo, Structure and molecular stacking of anthocyanins – flower color variation, *Angewandte Chemie Int. Ed. Engl.*, **30**, 17–33 (1991). (b) T. Kondo, K. Yoshida, A. Nakagawa, T. Kawai, H. Tamura and T. Goto, Structural basis of a blue-colour development in flower petals from *Commelina communis*, *Nature*, **358**, 515–518 (1992). (c) K. Yoshida, T. Kondo, Y. Okazaki and K. Katou, Cause of blue petal colour, *Nature*, **373**, 291 (1995).
15. (a) S. Asen, R. N. Steward and K. H. Norris, Co-pigmentation of anthocyanins in plant tissues and its effect on color, *Phytochemistry*, **11**, 1139–1144 (1972). (b)Y. Cai, T. H. Lilley and H. Haslam, Polyphenol anthocyanin copigmentation, *J. Chem. Soc. Chem. Commun.*, 380–383 (1990). (c) O. Dangles, N. Saito and R. Brouillard, Kinetic and thermodynamic control of flavylium hydration in the pelargonidin cinnamic acid complexation – origin of the extraordinary flower color diversity of *Pharbitis-nil*, *J. Am. Chem. Soc.*, **115**, 3125–3132 (1993). (d) T.Hoshino,An approximate estimate of self-association and the self-stacking conformation of malvin quinoidal bases studies by 1H-NMR, *Phytochemistry*, **30**, 2049–2055 (1991). (e) K. Yoshida, T. Kondo and T. Goto, Intramolecular stacking conformation of gentiodelphin, a diacylated anthocyanin from *Gentiana makinoi*, *Tetrahedron*, **48**, 4313–4326 (1992).
16. (a) K. R. Markham, K. S Gould, C. S. Winefield, K. A. Mitchell, S. J. Bloor and M. R. Boase, Anthocyanic vacuolar inclusions – their nature and significance in flower coloration, *Phytochemistry*, **55**, 327–336 (2000). (b) G. M. Robinson, Notes on variable colors of flower petals, *J. Am. Chem. Soc.*, **61**, 1606–1607 (1939).
17. (a) M. J. Melo, M. Moncada and F. Pina, On the red colour of Raspberry (*Rubus idaeus*), *Tetrahedron Lett.*, **41**, 1987–1991 (2000). (b) M. J. Melo, Missal blue: anthocyanins in nature and art, in J. Kirby (ed.), *Dyes in History and Archaeology*, Vol. **21**, Archetype Publications, in press.
18. (a) J. T. Kumpulainen and J. T. Salonen (eds), *Natural Antioxidants and Food Quality in Atherosclerosis and Cancer Prevention*, The Royal Society of Chemistry, Cambridge, 1996. (b) K. Kondo, A. Matsumoto, H. Kurata, H. Tanahashi, H. Koda, T. Amachi and H. Itakura, Inhibition of oxidation of low-density lipoprotein with red wine, *Lancet*, **344**, 1152–1152 (1994). (c) P. Bridle and C. F. Timberlake, Anthocyanins as natural food colours – selected aspects, *Food Chem.*, **58**, 103–109 (1997). (d) M. T. Satué-Garcia, M. Heinonen and E. N. Frankel, Anthocyanins as antioxidants on human low-density lipoprotein, *J. Agric. Food Chem.*, **45**, 3362–3367 (1997).

19. (a) F. Pina, M. C. Moncada, M. J. Melo, M. M. Correia dos Santos and M. Fernanda Mesquita, Some aspects of the chemistry of anthocyanins, in M. J. Halpern (ed.), *Lipid Metabolism and Related Pathology*, FCT, 2000. (b) M. Jordheim, K. Aaby, T. Fossen, G. Skrede and O. M. Andersen, Molar absorptivities and reducing capacity of pyranoanthocyanins and other anthocyanins, *J. Agric Food Chem.*, **55**, 10591–10598 (2007). (c) A. Ghiselli, M. Nardini, A. Baldi and C. Scaccini, Antioxidant activity of different phenolic fractions separated from an Italian red wine, *J. Agric. Food Chem.*, **46**, 361–367 (1998).
20. A. J. Shrikhande, Extraction and intensification of anthocyanins from grape pomace and other material. US Patent 4,452,822 (1984).
21. (a) J. B. Harborne, *Comparative Biochemistry of the Flavonoids*, Academic Press, New York, 1967. (b) J. B. Harborne and P.M. Dey, *Methods in Plant Biochemistry*, Academic Press, London, New York, 1989. (c) J. B. Harborne and R. J. Grayer, The Anthocyanins, *Chapman & Hall*, London, 1988. (d) J. B. Harborne, Comparative biochemistry of flavonoids – II. 3-Deoxyanthocyanins and their systematic distribution in ferns and gesnerads, *Phytochemistry*, **5**, 589–600 (1966).
22. J. J. Macheix, A. Fleuriet and J. Billot, *Fruit Phenolics*, CRC Press, Boca Raton, Florida, 1990.
23. K. Bronnum-Hansen and S. H. Hansen, High performance liquid chromatography separation of anthocyanins of *Sambucus nigra* L., *J. Chromatogr.*, **262**, 385 (1983).
24. J. Kärppä, Analysis of anthocyanins during ripening of crowberry, *Empetrum nigrum* Coll., *Lebensm. Wiss. Technol.*, **17**, 175 (1984).
25. R. Linko, J. Kärppä, H. Kallio and S. Ahtonen, Anthocyanin contents of crowberry and crowberry juice, *Phytochemistry*, **16**, 343 (1983).
26. W. E. Ballinger, G. J. Galletta and E. P. Maness, Anthocyanins of fruits of *Vaccinium*, sub-genera *Cyanococcus* and *Polycodium*, *J. Am. Soc. Hortic. Sci.*, **104**, 554 (1979).
27. G. M. Sapers, J. G. Phillips, H. M. Rudolf and A. M. Di Vito, Cranberry quality: selection procedures for breeding programs, *J. Am. Soc. Hortic. Sci.*, **108**, 241 (1983).
28. R. E. Woodward, Physical and chemical changes in developing strawberry fruits, *J. Sci. Food Agric.*, **23**, 465 (1972).
29. L. C. Torre and B. H. Barrit, Quantitative evaluation of *Rubus* fruit anthocyanin pigment, *J. Food Sci.*, **42**, 488 (1977).
30. B. H. Koeppen and K. Herrmann, Flavonoid glycosides and hydroxycinnamic esters of black-currants, *Z. Lebensm. Unters. Forsch.*, **164**, 203 (1988).
31. D. D. Crippen and J. C. Morrisson, The effect of sun exposure on the phenolic content of Cabernet Sauvignon berries during development, *Am. J. Enol. Vitic.*, **37**, 243 (1986).
32. E. Schäfer, T. Kunkel and H. Frohnmeyer, Signal transduction in the protocontrol of chalcone synthase, *Plant Cell. Environ.*, **20**, 722–727 (1997).
33. K. Hahlbrock, D. Scheel, E. Logemann, T. Nurnberger, M. Parniske, S. Reinold, W. R. Sacks and E. Schmelzer, Oligopeptide elicitor-mediated defense gene activation in cultured parsley cells, *Proc. Natl Acad. Sci.*, **92**, 4150–4157 (1995).
34. Y. Ozeki and A. Komamine, Induction of anthocyanin synthesis in relation to embryogenesis in a carrot suspension culture: correlation of metabolic differenciation with morphological differ-enciation, *Physiol. Plant.*, **53**, 570–577 (1981).
35. J. Mol, E. Grotewold and R. Koes, How genes paint flowers and seeds, *Trends Plant Sci.*, **3**, 212–217 (1998).
36. J. J. Macheix, A. Fleuriet and P. Sarni-Manchado, Composés phénoliques dans la plante – structure, biosynthèse, répartition et roles, in P. Sarni-Manchado and V. Cheynier (eds), *Les Polyphénols en Agroalimentaire*, TEC & DOC, London, Paris, New York, 2006.
37. N. J. Miller, The relative antioxidant activities of plant-derived polyphenolic flavonoids, in J. T. Kumpulainen and J. T. Salonen (eds), *Natural Antioxidants and Food Quality in Atherosclerosis and Cancer Prevention*, The Royal Society of Chemistry, Cambridge, 1996.
38. B. K. Wallin and B. J. Smith, Grape anthocyanins as food colourings: sources compared, *IFFA*, **May–June**, 10–15 (1977).
39. (a) O. Dangles, N. Saito and R. Brouillard, Anthocyanin intramolecular copigment effect. *Phytochemistry*, **34**, 119–124 (1993). (b) O. Dangle, Propriétés chimiques des polyphénols, in

P. Sarni-Manchado and V. Cheynier (eds), *Les Polyphénols en Agroalimentaire*, TEC & DOC, London, Paris, New York, 2006.
40. N. Saito, M. Oda, T. Matsui, Y. Osajima, K. Toki and T. Honda, Acylated anthocyanins of *Clitoria ternatea* flowers and their acyl moieties, *J. Nat. Prod.*, **59**, 139–144 (1996).
41. C. Malien-Aubert, O. Dangles and M. J. Amiot, Color stability of commercial anthocyanin-based extracts in relation to the phenolic composition. Protective effect by intra- and intermolecular copigmentation, *J. Agric. Food Chem.*, **49**, 170–176 (2001).
42. R. Brouillard, Chemical structure of anthocyanins, in P. Markakis (ed.), *Anthocyanins as Food Colors*, Academic Press, New York, 1982.
43. (a) D. Cardon, *Le Monde des Teintures Naturelles*, Belin, Paris, 2003. (b) D. Cardon, *Natural dyes. Sources, Tradition, Technology and Science*, Archetype Publications, 2007.
44. P. Guinot, A. Rogé, A. Gargadennec, M. Garcia, D. Dupont, E. Lecoeur, L. Candelier and C. Andary, Dyeing plants screening: an approach to combine past heritage and present development, *Color. Technol.*, **122**, 93–101 (2006).
45. G. J. Smith, S. J. Thomsen, K. R. Markham, C. Andary and D. Cardon, The photostabilities of natural occuring 5-hydroxyflavones, flavonols, their glycosides and their aluminium complex, *J. Photochem. Photobiol. A: Chemistry*, **136**, 87–91 (2000).
46. A. Downham and P. Collins, Colouring our food in the last and next millennium, *Int. J. Food Sci. Technol.*, **35**, 5–22 (2000).
47. F. J. Francis, Food colorants: anthocyanins, *Critic. Rev. Food Sci. Nutr.*, **28**, 273–312 (1989).
48. C. F. Timberlake and B. S. Henry, Plant pigments as natural food colours, *Endeavour New Series*, **10**, 31–36 (1986).
49. Y. D. Hang, Recovery of food ingredients from grape pomace, *Process Biochem.*, **February**, 2–4 (1988).
50. R. P. Métivier, F. J. Francis and F. M. Clydesdale, Solvent extraction of anthocyanins from wine pomace, *J. Food Sci.*, **45**, 1099–1100 (1980).
51. J. E. Cacace and G. Mazza, Extraction of anthocyanins and other phenolics from blackcurrant with sulfured water, *J. Agric. Food Chem.*, **50**, 5939–5946 (2002).

10

Natural Colorants – Quinoid, Naphthoquinoid and Anthraquinoid Dyes

Thomas Bechtold

10.1 Introduction

In this part of the book natural dyes that contain a quinoid group are overviewed. The classification of the dyes has been made according to their basic chemical structure into benzoquinone dyes, naphthoquinone dyes and anthraquinone dyes [1]. Representative examples for such dyes are carthamine (CI Natural Red 26), which contains benzoquinone groups, juglon and lawson, which are naphthoquinone dyes, and alizarin (1, 2-dihydroxy-9,10- anthraquinone).

A considerable number of these dyes have been used since historical times and are still in use for selected applications at present. Examples of such crops are green walnut shell (juglon), henna (lawson) and madder (alizarin).

Due to the high number of plant sources and the great variety of substances that are extracted as a complex mixture, representatives that are still of importance have been selected and will be discussed as examples for this class of natural dyes. The article concentrates on plant dyes, insect dyes like lac or cochenille, which both belong to the group of anthraquinoid dyes, will not be included.

10.2 Benzoquinone Dyes

The π-electron system that forms the chromophore in benzoquinone is quite small; thus such dyes contain further unsaturated groups conjugated to the benzoquinone ring. Carthamin (CI 75140, Natural Red 26) (Figure 10.1) is the red dye in safflower

Handbook of Natural Colorants Edited by Thomas Bechtold and Rita Mussak

(*Carthamus tinctorius* L.). The structure of carthamin is rather complex and sensitive to hydrolysis [2]. The red-orange product cartharmon is formed by oxidation of the yellow cartharmin. The oxidation can also be induced chemically by use of oxidizing chemicals, e.g. $KMnO_4$, H_2O_2, or enzymes [3].

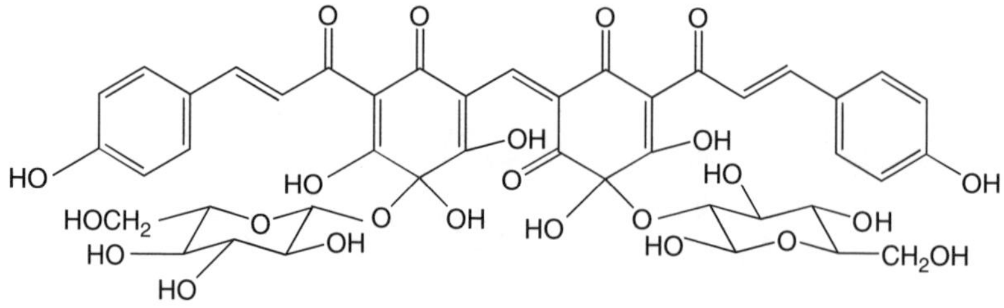

Figure 10.1 *Structure of carthamin*

The decomposition rate of carthamin is dependent on pH in solution, temperature, light and also the presence of metal ions, e.g. Fe(II) ions [4]. The stability of the red colour can be improved by addition of sugar alcohols e.g. glycerol and polyethylene glycol [5].

Safflower (*Carthamus tinctorius* L.) is a typical subtropical plant which has been cultivated in Egypt for more than 4000 years. The plant has been cultivated in India, China, Northern Africa, North and South America and can also be grown in Europe.

In the dry flower the content of red safflorcarmin is approximately 0.4 %. For dyeing purposes the cold water soluble yellow dye (safflor yellow) is removed by cold water extraction. Then the red safflorcamin (carthamon) is extracted in diluted sodium carbonate solution. After neutralization the extract can be used for dyeing wool, silk and cotton [6, 7]. Fastness to light could be improved by addition of nickel hydroxy-aryl-sulfonates while UV absorbers did not suppress photofading to the same extent [6].

Carthamin has a wide range of uses in traditional Chinese medicine and can be used as a colorant for decorative cosmetics, paints and food [8, 9]. Besides the extraction from plants, production in cell-culture conditions has also been studied [10].

Benzoquinone dyes can also be extracted from mushrooms and lichens. For extraction, the use of diluted ammonia solution has been recommended. However, the difficulty to crop mushrooms and lichens in conventional farming limits their wider application in high-volume markets.

10.3 Naphthoquinone Dyes

The two most important representatives for natural colorants with naphthoquinone structure are lawson (2-hydroxy-1,4-naphthoquinone) and juglon (5-hydroxy-1,4, naphthoquinone). The chemical formulas are given in Figure 10.2. Shikonine and alkanine, which are both substituted dihydroxy-1,4-naphthoquinones, can be extracted from Ratanjot (*Onosma echioides*), e.g. by means of supercritical carbon dioxide [11].

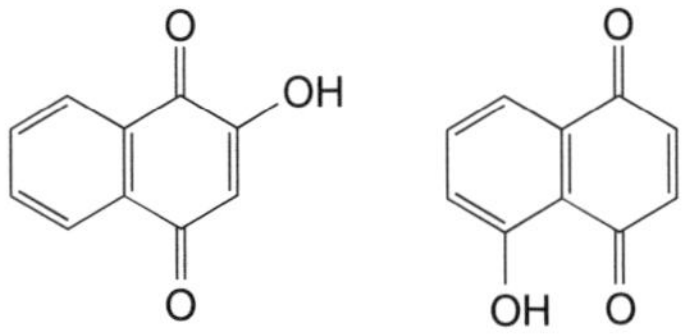

Figure 10.2 Chemical structure of lawson (2-hydroxy-1,4-naphthoquinone) and juglon (5-hydroxy-1, 4-napthoquinone)

10.3.1 Lawson (2-hydroxy-1,4-naphthoquinone, CI Natural Orange 6)

10.3.1.1 Properties and Use

Lawson is extracted from henna (Egyptian privet, *Lawsonia inermis* L.). The plant is cultivated in oriental countries, India, Northern and Eastern Africa and also Australia. An excellent summary about the practice of henna including historical aspects is given by Cartwright-Jones [12]. In the plant the average content of lawson is near 1 % [1]. Besides the lawson, the henna leaves also contain the flavonoid colorant luteolin [13]. The naphthoquinone is supposed to be present in plants in the glycosidic form named Hennosid A, B and C, from which the lawson is released by hydrolysis and oxidation of the aglycon. The production of lawson by hairy root cultures also has been investigated and a lawson content up to 0.13 % of dry plant weight has been analysed [14].

The quantitative analysis of lawson in extracts obtained from henna leaves can be performed by high-performance liquid chromatography (HPLC) on a reverse-phase C_{18} column [15]. HPLC also can be used to determine the content of glycosides in henna leaf extracts [16]. In another publication to determine the content of lawson in henna leaves the chloroform extract was analysed by high-performance thin layer chromatography (HPTLC) and thin layer chromatography (TLC) [14–18]. Variation in lawson content from 0.004 to 0.608 % was reported, indicating high variability of lawson content in commercial plant products. Thin layer chromatography followed by Fourier transform infrared (FTIR) respectively ultraviolet–visible (UV–VIS) spectrophotometry has also been proposed to analyse naphthoquinoid-containing extracts [11].

Henna leaves are usually extracted with hot water, but the use of aqueous alkaline solutions has also been proposed. Due to the dissociation of the phenolate, higher solubility of the extracted lawson can be expected in alkaline aqueous medium. Due to the phenolic group in position 2, lawson dissociates in water and forms the corresponding phenolate [19]. For preparation of henna paste the rushed plant leaves are mixed with a mild acidic ingredient such as lemon juice [20].

As expected, the 1,4-naphthoquinone group can be reduced electrochemically at the cathode to yield the 1,2,4-trihydroxy-naphthoquinone, which is supposed to be the aglycone of Hennosid. The electrochemical reduction of lawson also permits identification of the naphthoquinone [18, 20]. Lawson is able to form 1:2 complexes with iron(II) ions, and 1:1 and 1:2 complexes with Mn(II) ions [22–24].

The traditional use of lawson containing extracts from henna leaves is dyeing of hair and nails as well as traditional paintings on skin [25]. The use of natural colorants for hair dyeing will be reviewed in a separate part of the book. Lawson can also be used for dyeing textile materials, e.g. wool and silk, where orange to brown colours are obtained [2, 26–29]. The extracted dyestuff can be applied as direct dye or in combination with an alum mordant.

The highest uptake of both henna extracts and lawson was found for wool to occur at pH 3. The affinity of the rather small molecule is mainly attributed to ion–dipole forces between the polar groups of the dye and protonated amino groups of the wool [15]. After-chroming of lawson dyeing causes a shift in shade, which is attributed to a chromium complex formation.

Agarwal *et al.* [29] studied the influence of different mordants (alum, chromium, tin and iron salts) in different mordanting procedures on the final result of the dyeing. Shades of the henna dyeings varied from chestnut (without mordant) to moss green (iron salt). Also cotton substrates have been dyed with henna extracts using different mordants [30–32]. Thermodynamic studies of the dyeing behaviour of lawson on wool, silk and polyester indicated aggregation of lawson molecules in solution and also in the dyed substrate [33]. Lawson showed the highest affinity to polyester, followed by polyamide, which is explained by the chemical nature of lawson, which is similar to a disperse dye [33–39]. For dyeing polyester fibres the use of water-solvent mixtures has also been reported in the literature [40]. The application of henna extracts has also been investigated for dyeing of cellulose-triacetate [41].

The simple application procedure of henna permits the use of such dyes in combination with other synthetic dyes, e.g. reactive dyes [20].

The use of plant extracts for dyeing of leather has found growing interest in recent years, particularly because of the growing environmental concerns of conventional leather processing [42]. Use of henna extracts as retanning agents in leather production has been compared with wattle retanned leather, with promising results for a possible application of henna [43]. Another application of henna pigments is in paints [44].

Henna extracts have even been studied as a corrosion prevention agent for metals, e.g. steel, nickel, zinc and aluminium [45–49]. The inhibitive action of the extract is discussed in view of adsorption of lawson molecules on the metal surface and the formation of complexes between metal cations and lawson.

Besides the use as a colorant henna has also found various medical applications, particularly in traditional medicine [45, 50–53]. Further activity of henna in creams to treat infestination of humans with head lice has been described in the literature [54].

Due to the wide application of henna in hair colourants and body paints numerous studies about potential health hazards have been performed. As a general result the studies demonstrate that henna is a natural product with low health risk potential [55, 56]. From an assessment of the genotoxicity of 2-hydroxy-1,4-naphthoquinone, the natural dye ingredient of henna, Kirkland and Marzin suggested from their study that henna and 2-hydroxy-1,4-naphthoquinone pose no genotoxic risk to the consumer [57].

10.3.1.2 Agricultural Aspects

In 2003–2004 the production of henna in India, which is the biggest producer, is estimated at 10.500 tons, with 70 % of the crop of henna being used in India and 30 % being exported, mainly to Middle Eastern clients.

A review about the cultural aspects of henna, including application and farming, has been published by Cartwright-Jones [12]. Henna grows best in tropical savannah and tropical arid zones. The highest dye content is produced by plants growing at temperatures between 35 and 45 °C. Lowest temperatures should not fall below 5 °C. Intensive cultivation of henna is found in regions with 400 to 500 mm of rain per year. In regions with high precipitation and heavy, damp soil the risk of root rot, aphid and insect attack increases.

Prathiba *et al.* [58] have published a field study that was conducted for six years (1998–2004) at Hyderabad to evaluate the influence of planting materials, microsite improvement and fertilizers on productivity, lawson content and economic returns of henna (*Lawsonia inermis* L.). The leaf yield and quality were influenced by different treatments. During the rainy season the yield increase in cuttings over seedlings was 100, 50.7, 37.7, 63.8, 29.6 and 61.3 % during 1998, 1999, 2000, 2001, 2002 and 2003, respectively, whereas the increase in the post-rainy season yields were 360, 26.7, 100, 49.3, 41.7 and 60.7 % during 1998, 1999, 2000, 2001, 2002 and 2003, respectively. The increase in yields was more in high rainfall years (1998, 2001, 2003) compared with the low rainfall years (1999, 2000, 2002).

A significant increase in leaf yields was observed with microsite improvement during low rainfall years, when compared to no microsite improvement. Higher lawson content was reported for treatments having seedlings, microsite improvement and fertilizer application. Seedlings were found to be more drought tolerant than the cuttings [58].

In another study a field experiment was presented which was conducted to find out the effect of conjunctive use of organics, i.e. farmyard manure and inorganic sources of nutrients and row spacing of henna (*Lawsonia inermis* L.) during 2003–2004 to 2005–2006 at Pali (Rajasthan). The application of farmyard manure of 5 ton/ha led to a significant increase in dry leaf yield of 11.4 % compared to crops with no farmyard manure treatment. For fertilizer application at 80 kg/ha N and 40 kg/ha P_2O_5, 19.3 % higher dry leaf yield was recorded. Significant improvement in dry leaf yield was also obtained by increasing the width of row spacing from 30 to 45 cm or 60 cm. The highest crop determined as dry leaf yield of 1302.6 kg/ha was obtained under 80 kg/ha N and 40 kg/ha P_2O_5 with 45 cm row spacing. With no fertilizer and 30 cm row spacing the dry leaf yield reached 945.2 kg/ha. The results indicated further that at higher levels of nutrient supply, the plant density as plants/unit area should be increased to exploit fully the higher fertility potentials [59].

A field experiment was conducted from 1997 to 2001 at the Regional Research Station, Central Arid Zone Research Institute, Pali-Marwar, to study the effect of five crop spacings and four nitrogen levels on henna (*Lawsonia inermis* L.) leaf production. A maximum dry leaf production of 692.5 kg/ha was recorded under 45 cm $\times$ 30 cm, followed by 652.9 kg/ha under 60 cm $\times$ 30 cm, as compared with 556.8 kg/ha under 30 cm $\times$ 30 cm spacings. N fertilization at 60 kg/ha resulted in a significantly higher dry leaf yield of 603.8 kg/ha compared with the control (543.0 kg/ha of dry leaf yield) [60].

A field experiment was conducted during 1994–1995 and 1995–1996 at Udaipur, India, to study the effect of a plant growth regulator (cycocel, kinetin) and thiourea on vegetative growth, essential oil content, yield and quality of henna. The plant height, number of inflorescences, maximum yield of foliage, henna powder per plant, maximum weight of flowers, essential oil content in flowers and yield of essential oil per plant were recorded. The colour intensity of henna powder was shown to be maximal with a combination of plant growth regulator and thiourea treatment [61].

A suppressive effect was obtained by the henna plant against *Meloidogyne incognita* development. Henna reduced tomato root gall numbers, number of the egg-laying females and rate of the nematode reproduction when tomato and henna were grown together [62].

An analysis of the economics of henna production in Pali district of Rajasthan has been published by Chand *et al.* [63]. This region is the main henna-growing belt in Rajasthan. The analysis showed that in the initial year of establishing this crop the cost distribution was as following: 55 % of the costs for labour, 30 % for seedlings and 15 % for ploughing.

Mechanization was expected to increase the profitability [63].

10.3.2 Juglone (5-hydroxy-1,4-naphthoquinone, CI Natural Brown 7)

Juglone is another important representative for a natural dye with the naphthoquinone structure. The dyestuff can be extracted from different parts of nut trees (Juglandaceae). Dependent on the variety of tree, different parts of the plant are used for dyestuff extraction. In the plant the juglone is mainly present as a glycoside, e.g. 5-hydroxy-naphthohydroquinone-4-β-D-glycoside [1]. In a detailed analysis of bark extracts from *Juglans mandshurica* a number of components could be identified, among them dihydrokaempferol, juglone, kaempferol, naringenin and quercetin [64]. In Table 10.1 relevant sources for juglone-containing plant material are shown.

The potential of walnut to serve as a source for natural dyes can be estimated from the volume of walnut harvested in Austria per year. In 2003 approximately 20 300 tons of walnut were collected in Austria alone [65].

Table 10.1 *Sources for juglone-containing plant material*

Source	*Part of plant*	*Cultivated*
Walnut tree	Leaves harvested in early summer	Southwest, Central, Eastern Asia
Juglans regia L.	Shells of green walnut	Middle and Southern Europe
		North Africa,
		North America
Black walnut	Shell of fruit	North America
Juglans nigra L.	Bark, bark of roots	Europe
Butternut tree	Shell of fruit	North America
Juglans cinerea L.	Bark of branches	Canada
	Bark of roots	Europe

Wool dyed with extracts from walnut also showed improved insect resistance for the carpet beetle, but the mordant used can also contribute to the increase in insect resistance [66]. Extracts of walnut leaves and dried shells of green walnut can be used for various purposes: production of liqueur, colour in sunscreen oil and dyeing of textiles [67]. Another application of walnut extracts is in human hair dyeing.

Dyeing of textile substrates with aqueous walnut extracts yields brown shades. A wide range of substrates can be dyed, among them wool, human hair, silk, nylon, polyester and cellulose fibres including leather [39, 67–74]. Studies of the dyeing behaviour showed that the sorption follows the partition mechanism, as expected for a small uncharged molecule [75]. Higher affinity was found for hydrophobic fibres. The application of juglon containing extracts for surface dyeing of wood has also been proposed [76].

The cytotoxicity of juglone and plumbagin (5-hydroxy-3-methyl-1,4-naphthoquinone) has been studied by Inbaraj and Chignell [77]. Herbal preparations derived from black walnut have been applied topically for the treatment of acne, inflammatory diseases, ringworm and fungal, bacterial or viral infections [74, 75]. The cytotoxicity of the two napthoquinones was attributed due to two different mechanisms: redox cycling, which generates semiquinone radicals, and oxidation of glutathione due to H_2O_2 formation. Probably the antifungal, antiviral and antibacterial properties of these quinones are the result of the redox cycling mechanism. Inbaraj and Chignell suggest that care should be taken when using preparations containing juglone and plumbagin to avòid the risk of damage to the skin.

10.4 Anthraquinone Dyes

Anthraquinone dyes represent the biggest group of quinone dyes. Anthraquinone dyes can be extracted from a high number of plants. Here representative compounds are presented as examples of different groups of dyes, following a classification according to Schweppe [1].

10.4.1 Main Components Emodin and Chrysophanol – *Rheum* Species and *Rumex* Species

Rhubarb has been in use for more than 400 years because of its pharmaceutical value. Relevant plant varieties are *Rheum palmatum* L., *Rheum rhabarberum* L., *Rheum rhaponticum*, which grow in China, Tibet, Siberia and also Europe.

The dye (CI Natural Yellow 23) is extracted from the roots. The plant contains 3–5 % anthracene derivatives. In the plant the anthraquinoid components are present in reduced form as anthraglycosides, which hydrolyse and oxidize to the respective anthraquinone [1]. The extracted dye contains a complex mixture of anthraquinones; relevant compounds are

Emodin, Chrysophanol, Aloeemodin, Rhéin and Physcion. The structures of the molecules are shown in Figure 10.3.

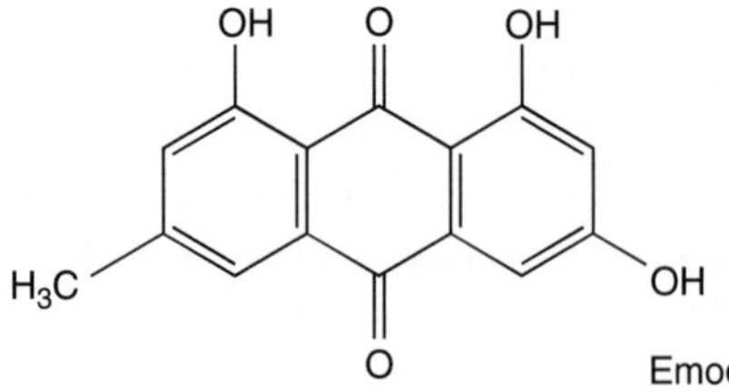

Emodin

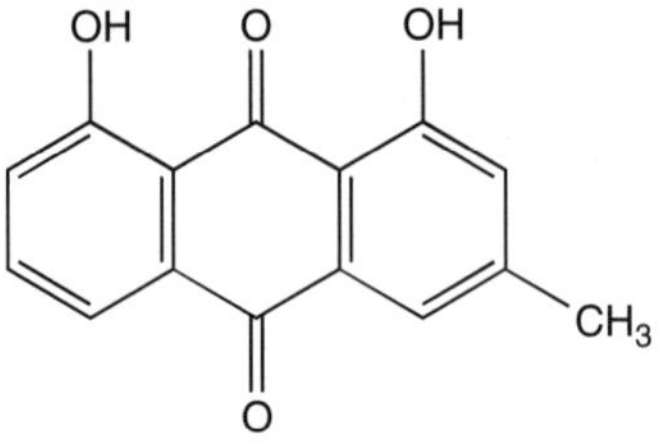

Chysophanol

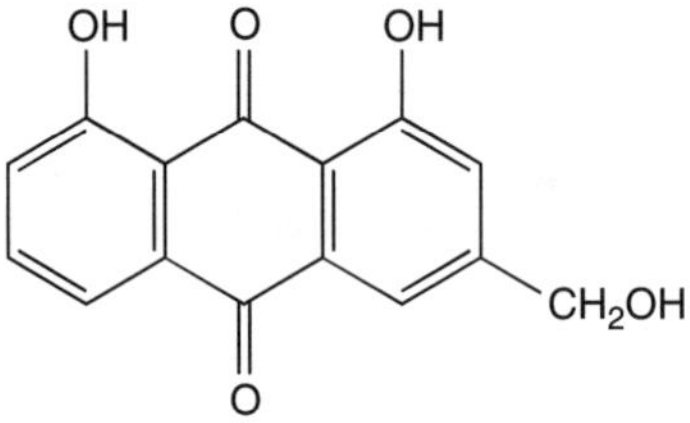

Aloeemodin

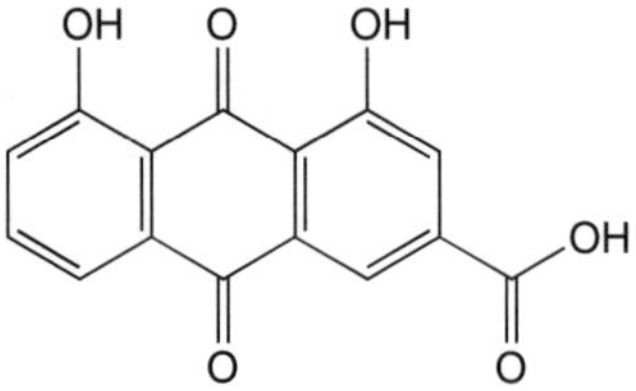

Rhéin

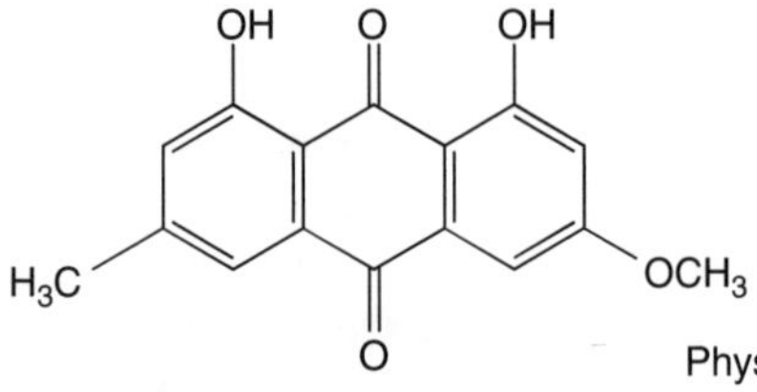

Physcion

Figure 10.3 *Representative examples for the anthraquinoid dye structure: Emodin, Chrysophanol, Aloeemodin, Rhéin and Physcion*

Rheum extracts can be used for textile dyeing. On wool, yellow to orange colour can be obtained with alum mordant; with chromium mordant a red shade is obtained. The formation of complexes due to the presence of metal salt from mordanting considerably improves light fastness of dyeings [79]. Man-made fibres, e.g. cellulose triacetate, can also be dyed [41]. Furthermore, rhubarb extracts have been proposed for hair dyeing.

Other *Rumex* species (e.g. *Rumex crispus* L., curled dock; *Rumex acetosa* L., common sorrel; and *Rumex hymenosepalus Torr.*, Tanner's dock) also contain similar colorants to those of the *Rheum* species, but their content in tannins is higher. *Rumex maritimus* extracts have been studied for their antimicrobial activity against various pathogenic bacteria. Among the studied bacteria, antimicrobial activity was only detected against *Klebsiella pneumoniae* [80].

10.4.2 Main Components Alizarin and/or Pseudopurpurin/Purpurin

10.4.2.1 Plant Sources

The majority of plants that can be used as a source for the extraction of such natural colorants belong to the Rubiaceae family. Relevant plant sources are summarized in Table 10.2. Formulas of the chemical compounds are shown in Figure 10.4.

Table 10.2 *Relevant plant sources for extraction of alizarin, pseudopurpurin and purpurin*

Plant	*Botanical name*	*Colour index (CI)*	*Relevant components in colorant*
Madder	*Rubia tinctorum* L.	Natural Red 8	Alizarin
Wild madder	*Rubia peregrina* L.	Natural Red 8	Pseudopurpurin
Indian madder	*Rubia cordifolia* L.	Natural Red 16	Rubiadin
Japanese madder	*Rubia akane* Nakai	—	Purpurin
	Rubia iberika	—	Nordamnacanthal
			Lucidin
Lady's bedstraw	*Galium verum* L.	Natural Red 14	Alizarin
Hedge bedstraw	*Galium mollugo* L.	Natural Red 14	Pseudopurpurin
Common cleavers	*Galium aparine* L.	Natural Red 14	Purpurin
Sweet woodruff	*Galium odoratum* (L.) *Scop.*	Natural Red 14	Lucidin
Dyer's woodruff	*Asperula tinctoria*	Natural Red 13	Rubiadin
Relbún	*Relbunium hypocarpium* (L.) *Hemsl.*		Pseudopurpurin
	Morinda angustifolia Roxb.		Morindone

For all plants the roots are used for extraction of the natural dye. In the case of madder, the highest concentration of dye is found in the intermediate layer between the blackish bark, the husk, outside and the innermost layer of the root, and the pith. The dyestuff most probably is stored in compartments outside the cells, which possibly are divided by a pectin layer [81].

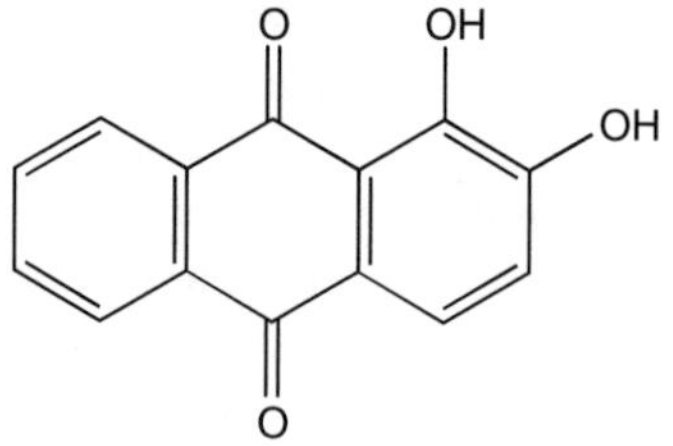

Alizarin

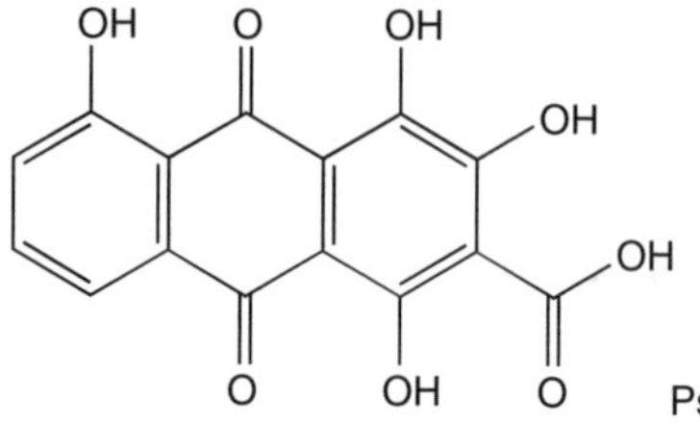

Pseudopurpurin

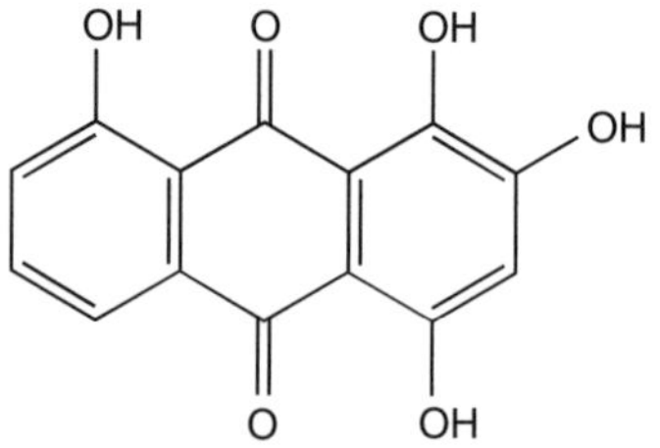

Purpurin

Figure 10.4 *Formulas of alizarin, pseudopurpurin and purpurin*

10.4.2.2 Madder CI Natural Red 8

Properties Madder has been cultivated as a source for red plant dyes in Europe, Asia and Northern and Southern America in huge amounts. The dyestuff is extracted from the dried roots of the plant. The bark of the roots contains a higher amount of dyestuff, the wooden parts less [1]. The roots contain 2–3.5 % of di- and trihydroxyanthraquinone-glycosids. During storage, hydrolysis of the glycosides occurs, which is completed under acidic conditions, established, for example, during the dyeing procedure, when the textile substrate has been pre-mordanted with alum or tartar.

Typical contents of alizarin found in different cultivars vary from 6.1 to 11.8 mg/g of roots. Average values for the amount of anthraquinones in three year old cultivars of *Rubia tinctorum* are: pseudopurpurin 7.4 mg/g, munjistin 6.2 mg/g, alizarin 8.7 mg/g, purpurin 3.5 mg/g and nordamnacanthal 13.4 mg/g.

The composition of the extracted anthraquinones differs between the varieties of *Rubia*. In the European madder (*Rubia tinctorum*) the major component forming the natural dye is alizarin (1,2-dihydroxyanthraquinone). Analysis of the extracts from Indian madder (*Rubia cordifolia*) showed that the major components are purpurin (1,2,4-trihydroxyanthraquinone), which forms approximately 66 % of the colorant, approximately 10 % are munjistin (1,3-dihydroxy-2-carboxyanthraquinone) and another 10 % are nordamnacanthal (1,3-dihydroxy-2-formlyanthraquinone) [82].

In the literature a total of more than 35 anthraquinoids has been reported to be extractable from madder roots [81]. However, part of the compounds is believed to be artefacts formed during extraction or drying, such as:

- Anthraquinones, which contain a 2-methoxymethyl- or a 2-ethoxyethyl group, are formed during extraction with hot methanol or ethanol [81, 83].
- Purpurin (1,2,4-trihydroxyanthraquinone) and purpuroxanthin (1,3-dihydroxy- anthraquinone) are supposed to be formed during the drying of the roots from pseudopurpurin (3-carboxy-1,2,4-trihydroxyanthraquinone) and munjistin (2-carboxy-1,3-dihydroxyanthraquinone), respectively [1].

The extracted colorants can be purified by column chromatography on silica gel. Quantification and determination of purity can be done with HPLC [5, 84].

The high-performance liquid chromatographic method was applied for the determination of nonglycosidic anthraquinones alizarin (1,2-dihydroxy-9,10-anthracenedione), emodin (1,3,8-trihydroxy-6-methyl-9,10-anthracenedione) and anthraquinone (9,10-anthracenedione). The anthraquinones were separated by isocratic elution on reversed-phase material using methanol–acetic acid as the mobile phase.

Due to the heterogeneous extraction conditions, efficiency in dyestuff extraction is a major problem and studies to optimize extraction efficiency have been performed. The use of ultrasound assisted extraction has been studied with regard to time, solvent composition and temperature. In extraction with ultrasound assistance the solvent methanol/water with the ratio of 1:2 at a temperature of 36 °C has been found to be favourable [85].

For dyeing purposes the glycoside ruberythric acid has to be hydrolysed to alizarin. Comparison of different extraction conditions, a strong acid, e.g. 1 M sulfuric acid, a strong base, e.g. 1 M sodium hydroxide or enzymatic hydrolysis with hydrolases, showed that ruberythric acid can be hydrolysed to alizarin without formation of the mutagenic lucidin [81]. While hydrolysis in strong bases yields a high number of anthraquinones, the acid hydrolysis and conversion of glycosides by endogenous enzymes present in madder both yield nonmutagenic products. In the case of acid hydrolysis it is supposed that lucidin disappears in the extracts because of a covalent reaction with the undissolved plant material. In the case of enzyme hydrolysis lucidin is detected in the absence of air oxygen. When air oxygen is present, lucidin primeveroside hydrolyses to lucidin, which then is oxidized to nordamnacanthal. Studies performed by Derksen indicated that the oxidation of lucidin to nordamnacanthal occurs in the presence of enzymes extracted form the madder roots, while lucidin oxidation is not observed after an inactivation of these enzymes [81].

In the presence of metal salt mordant the basic chemical principle in the dyeing process is the formation of coloured metal complexes. In traditional dyeing the complexes formed contained both aluminium and calcium. Other suited mordants can contain iron(II), tin(II) or copper. However, due to restrictions of the heavy metal content in dyeings on clothing, nowadays only aluminium, calcium and iron are recommended as a mordant for textile dyeing [67].

Detailed studies about the complex formation have been published in the literature [81]. For alizarin both hydroxyl groups are involved in the complex formation. The pK_a values of the hydroxyl groups are 12.0 for the 1-hydroxyl group and 8.2 for the 2-hydroxyl group. The calcium ion reacts with the 2-hydroxyl group and aluminium forms a complex with involvement of the 1-hydroxyl group. The stoichiometry between alizarin, aluminium and calcium has been reported to be 2:1:1 [86].

The stability constants for Ca(II) and Zn(II) complexes with alizarin have been determined by potentiometric titration and spectrophotometry. Stability constants of the binary systems have been evaluated by Bilgic *et al.* according to the method suggested by Irving–Rossotti [87].

Colour and shade of the complexes are dependent on the metal ion involved. Tin(II) mordants form orange complexes, alum mordants form red complexes, the mixed calcium–iron–alizarine complex exhibits a violet colour. Also, light fastness depends on the complex formed; e.g. a study performed with purpurin (1,2,4-trihydroxyanthraquinone) and munjistin (1,3-dihydroxy-2-carboxyanthraquinone) on polyamide as substrate showed higher light stability for copper and ferrous mordants, while resistance to fading was lower in the case of stannous chloride or alum mordant dyeings.

Usage The use of madder extracts for textile dyeing purposes has a very long tradition, as the red shades obtained are brilliant and can be obtained with high colour depth. Numerous papers dealing with applicatory aspects have been published in the literature [88–92]: dyeing on silk [93–97], wool [1, 94, 98–110], cotton and flax [94, 99, 100, 103, 111–117], cori (coconut fibre) [115] and man-made fibres, cellulose-triacetate [39, 41, 80, 119–122].

A major problem of natural dyes is their relatively low affinity to the plant material, which leads to considerable concentrations of not-exhausted dyestuff in the used dye-bath. Thus aspects of bath reuse and recycling have been studied to lower the amount of plant material required to dye a certain amount of textile material [123]. An overview of the potential of natural colorants for textile dyeing and the selection of plants, including production costs, has been published for the European/Austrian climate regions [124–126].

Dyeing properties of *Galium molugo* extracts on wool and cellulose fibre and fastness properties have also been studied [127] and the surface dyeing of wood from yellow pine (*Pinus sylvestris*) and beech (*Fagus orientalis*) with extracts from walnut (*Juglans regia*) shells, oleander (*Nerium oleander*), saffron (*Crocus sativus*) and madder root (*Rubia tinctorum*) has been reported in the literature [128].

An estimation of the cost structure of madder cultivation showed that on the basis of a three-year cultivation, a crop of 3 ton/ha is required to cover the production costs. The major part of costs results from the cultivation of young plants and the drying of the harvested roots. For an effective weed control a combination of mechanical and chemical methods has been shown to be practicable [129].

Purpurin exhibits high sensitivity to changes in pH as well as temperature, undergoing decomposition at temperatures above 90 °C. The affinity values obtained for munjistin and nordamnacanthal extracted form Indian madder on synthetic hydrophobic fibres, e.g. polyester or polyamide, are much less than values obtained for naphthoquinones like lawson or juglon, which can be explained by the bigger molecule size and the rather low solubility of the anthraquinone in the aqueous dye-bath [82, 119].

Rubia cordifolia extracts have been studied for their antimicrobial activity against various pathogenic bacteria. Antibacterial and antifungal activities of aqueous (anthraquinone gylcosides) and ether extracts (aglycones) from madder roots have been reported in the literature [130–134]. Among the studied bacteria, antimicrobial activity was only detected against *Klebsiella pneumoniae* [135].

Wool dyed with extracts from madder also showed improved insect resistance for the carpet beetle [66].

Madder extracts can also be used for staining of histological samples [135]. The formation of coloured calcium complexes can be used to stain calcium deposits in soft tissues. A diagnostic application of alizarin is its use as a marker for the study of bone growth [136]. Furthermore, extracts also contain compounds of pharmacological interest, which can be used for the treatment of bladder and kidney stones, particularly ones containing oxalate or phosphate. A possible explanation is the prevention of calcium phosphate and calcium oxalate formation by ruberythric acid [137].

Madder extracts can also be used as colorants for food, e.g. confections, boiled fish and soft drinks, because of its advantageous resistance to heat and light [138, 139]. In a 13-week repeated oral dose toxicity study of madder colour, which was performed using F344 rats, the animals were fed a diet containing 0, 0.6, 1.2, 2.5 or 5.0 % madder colour. The results suggested that madder colour exerts mild toxicity, targeting liver, kidneys and possibly red blood cells and white blood cells, some renal changes being evident from 0.6 % madder colour in diet. This is considered to be the lowest-observed adverse effect level (305.8–309.2 mg/kg of body weight/day) [140, 141].

Antioxidant activity of alizarin extracted from madder has been studied on different test systems, showing antioxidant properties of alizarin in an *in vitro* biological assay [142]. Also, the use of extracts in phytopharmaceuticals has been reported [81, 143].

Traditionally, Indian madder (*Rubia cordifolia*) is used in many polyherbal formulations for various ailments and cosmetic preparations because of its inflammatory, antiseptic and galactopurifier activity. In a study with rats, various extracts of roots of *Rubia cordifolia* were screened for possible hepatoprotective activity using thioacetamide-induced hepatotoxicity. The methanolic extract was found to protect the liver of the animals against thioacetamide-induced hepatotoxicity [144]. A preparation produced from Indian madder (*Rubia cordifolia*) cell culture was found to exhibit anti-inflammatory activity [145].

The antimutagenic effect of purpurin has been found in the Ames *Salmonella* bacterial mutagenicity assay. The antigenotoxic effect was observed in *Drosophila melanogaster* against a range of environmental carcinogens. Inhibition of the formation of hepatic DNA adducts in male C57bl6 mice after a single dose of the heterocyclic amine dietary carcinogen Trp-P-2 (30 mg/kg) was observed by short-term dietary supplementation with purpurin. The inhibition of adduct formation was dose-dependent [146, 147].

In another study purpurin (1,2,4-trihydroxy-9,10-anthraquinone) was found to show inhibition of mutagenicity of a number of heterocyclic amines in the Ames mutagenicity test [148].

In the presence of purpurin (1,2,4-trihydroxy-9,10-anthraquinone) in bacterial mutagenicity assays a marked inhibition of mutagenicity induced by food-derived heterocyclic amines was observed. Purpurin showed better inhibition of Trp-P-2-dependent mutagenicity than epigallocatechin gallate or chlorophyllin, both of which are known as well-established antimutagenic components of diet. The inhibition effect of purpurin was dependent upon pH, being better in neutral than acidic conditions [149].

The use of madder in combination with BCG (*Bacille* Calmette-Guérin) for prophylactic treatment of stage T1 superficial tumors has also been described in the literature [150].

Properties – Extraction and Purification The extraction of anthraquinones from plant, root or cell cultures for *Rubia tinctorum* has been studied extensively [81, 151–160]. Dependent on the polarity of the solvent used free aglycones or glycosides can be extracted. The use of solvents with increasing polarity permits at least partial separation. When hot methanol or ethanol are used as solvents lucidin (1,3-dihydroxy-2-hydroxymethyl-anthraquinone) can partly be converted to the corresponding 2-methoxymethyl respectively 2-ethoxymethyl derivatives. As this reaction was found to be highly temperature dependent these solvents are not recommended for hot extraction [81].

For the production of a commercially useful dye extract from madder, the glycoside ruberythric acid has to be hydrolysed to the aglycone alizarin, which is the main dye component. An intrinsic problem is the simultaneous hydrolysis of the glycoside lucidin primeveroside to the unwanted mutagenic aglycone lucidin. For comparison of different extraction methods the madder root was treated with strong acid, strong base or enzymes to convert ruberythric acid into alizarin and the anthraquinone compositions of the suspensions were analysed by HPLC. A cheap and easy method to hydrolyse ruberythric acid in madder root to alizarin without the formation of lucidin turned out to be the stirring of dried madder roots in water at room temperature for 90 min. This gave a suspension containing pseudopurpurin, munjistin, alizarin and nordamnacanthal. Native enzymes are responsible for the hydrolysis, after which lucidin is converted to nordamnacanthal by an endogenous oxidase [78, 161].

Due to the phenolic character of most anthraquinones in madder alkaline extraction conditions have also been recommended, such as sodium carbonate, sodium bicarbonate, sodium hydroxide or potassium hydroxide [81, 152, 153, 159].

The extracted glycosides are subsequently hydrolysed to their corresponding aglycone form. For this purpose diluted mineral acids, sulfuric acid or hydrochloric acid can be used at elevated temperature. Enzymatic hydrolysis also can be used to convert the ruberythric acid (alizarine-2-β-primveroside) into alizarin [158, 162].

The anthraquinones in the extracts can be purified further by precipitation or by chromatographic methods. Precipitation, e.g. by addition of calcium carbonate, has been used in traditional processes for the removal of purpurin by formation of the insoluble calcium-purpurin salt [1].

Analysis of the thermal stability of the coloring material in madder by differential scanning calorimetry, synchronized thermogravimetry/differential thermal analysis and UV–Vis spectrophotometry showed stability in water at 100 °C for at least 2 hours, which is sufficient to be used as dye at that temperature. At 145 °C thermal decomposition was observed. Alcohol extracts were found to be stable for at least 90 days. In the acidic pH range, higher stability is observed, for example, at pH 3 than above pH 7 [163].

An extraction unit optimized for the extraction of madder roots to obtain a higher quality of extract and equilibrium data for the solid/liquid extraction, including different extraction temperatures and solvents, has been evaluated by Hatamipour and Shafikhani [164].

Analysis Crude extracts of anthraquinones can be fractionated further for separation and quantification of the individual components or purification of a desired component by chromatographic methods. An excellent literature overview about chromatographic methods used and application of HPLC (high-pressure liquid chromatography) and HPLC-MS (mass spectrometry) coupling for analysis of *Rubia* extracts is given by Derksen [81, 165–174]. Electrospray mass spectrometry has also been successfully used for direct analysis of natural dyes in dyeings. This method, for example, permits direct identification and quantification of alizarin and purpurin, lucidin, ruberythric acid and also aluminium- and calcium-alizarin lake in materials used in art [175–177].

Pre-purification and concentration of plant extracts can be achieved by use of solid phase extraction. A high number of studies using paper chromatography and TLC (thin layer chromatography) have been described, including preparative TLC [178]. Detection of the anthraquinones can be done due to their fluorescence under UV_{254} light and partly under UV_{366} light. Colour development from a yellow-orange colour to a red-purple colour can also be achieved with hydroxyanthraquinones by spraying a reagent solution containing KOH or NaOH in methanol (Bronträger reaction).

For gas chromatography the hydroxyanthraquinones have to be derivatized to more volatile compounds such as methyl ethers, trimethylsilyl ethers or trifluoroacetyl derivatives. When separation of anthraquinones has to be achieved without derivatization, HPLC is the favoured method. For detection of UV absorbance at 254 nm or 280 nm, however, absorbance in the visible range of the light spectrum can also be used. Depending on the stationary phase and the solvent used a separation of the glycosides or aglycones can be achieved. Mostly the extracts are hydrolysed first to the aglycones and then separated on C_{18} reversed-phase material as the stationary phase [81]. Other separation techniques that have been applied are centrifugal partition chromatography [179] and capillary electrophoresis [180, 181].

In another study hairy root cultures of *Rubia tinctorum* were cultured on solid and in liquid media in a shaking cabinet and in a bioreactor. The methanolic extracts of lyophilized hairy roots were hydrolysed, purified by solid-phase extraction and analysed with HPLC for alizarin and purpurin on an RPC8 column using 45:55 (v/v) acetonitrile:20 mM ammonium formate–formic acid buffer (pH 3.00) as the mobile phase [182].

Thermally assisted hydrolysis and methylation by tetramethylammonium hydroxide followed by gas chromatography/mass spectrometry has also been applied for the analysis of red dyes based on the anthraquinone structure, such as alizarin, quinizarin, purpurin and carminic acid. Thermochemolysis of alizarin and quinizarin afforded 1,2-dimethoxyanthraquinone and 1,4-dimethoxyanthraquinone, respectively. The thermally assisted hydrolysis has been proposed as a rapid technique for the identification of dyes in real samples, e.g. madder lake [183].

An analysis of ethanol extracts of air-dried powdered roots of *Morinda angustifolia* has been performed by separation on a preparative liquid chromatography followed by analysis of the fractions by IR and UV/VIS spectroscopy [98]. An analysis of madder has also been performed by a new procedure based on pyrolysis-gas chromatography/mass spectrometry (Py-GC/MS), which includes the online derivatization of the natural dyes using hexamethyl-disilazane (HMDS) [184]. Jegorov *et al.* have published crystallographic data of mollugin and lucidin [185].

Toxicology and Safety Aspects Due to its use as phytopharmaceuticals and food colorants the safety of madder (*Rubia tinctorum*) extracts has been studied extensively. Lucidin primeveroside present in madder root is the source of the genotoxic lucidin [186]. Lucidin (1,3-dihydroxy-2-hydroxymethyl-anthraquinone) showed mutagenicity in numerous tests.

In an extensive study the roots of *Rubia tinctorum* were extracted using different solvents and extracts were fractionated with chromatography. In tests for mutagenicity using a *Salmonella typhimurium* strain a number of components showed mutagenicity: 1-hydroxy-2-methylanthraquinone, ludidin-ω-methylether, rubiadin, xanthopurpurin, 7-hydroxy-2-methyl-anthraquinone, lucidin-ω-ethylether, lucidin-primveroside and mollugin, a nonanthraquinone compound [81, 151, 187, 188]. From structure mutagenicity studies it was concluded that 1,3-dihydroxyanthraquinones which bear a methyl or hydroxymethyl group on carbon-2 are mutagenic (Figure 10.5). For direct mutagenicity an oxygenated state of the benzylic carbon-2 is required [151].

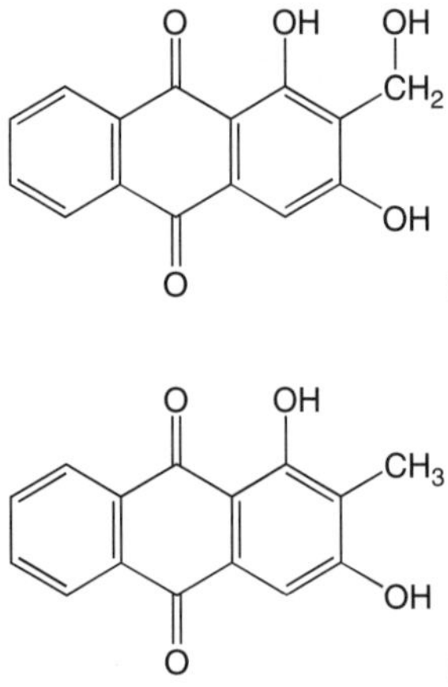

Figure 10.5 *Structures of lucidin and rubiadin*

Mutagenic studies about lucidin showed that a reactive compound is formed from the metabolism of lucidin, which then reacts with DNA and possibly other macromolecules to form covalent adducts. The reaction of lucidin to form methyl or ethyl ethers when heated with alcohols supports the reactive character of lucidin. These ethers were also found to show mutagenicity [151]. The reaction scheme of lucidin to form adducts with adenine and guanidine is shown in Figure 10.6 [189–191].

The proposed electrophilic intermediate, which contains an exomethylenic group, is suggested to react with the bases. Thus other 1,3-dihydroxyanthraquinones that do not possess a hydroxymethyl group in position 2, such as nordamnacanthal (2-formyl-1,3-dihydroxyanthraquinone) and munjistin (2-carboxy-1,3-dihydroxyanthraquinone), are not found to be mutagenic, since the dehydration to the exomethylenic compound is not possible under physiological conditions.

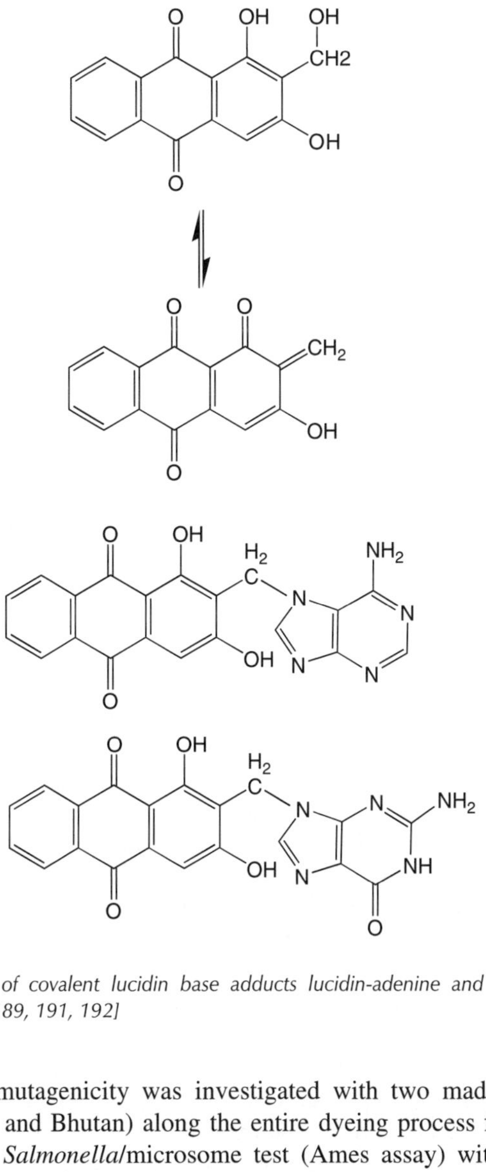

Figure 10.6 *Formation of covalent lucidin base adducts lucidin-adenine and lucidin-guanidine under physiological conditions [189, 191, 192]*

In another study mutagenicity was investigated with two madder root samples of different origin (Iran and Bhutan) along the entire dyeing process from root extracts to the dyed wool. The *Salmonella*/microsome test (Ames assay) with the strains TA98, TA100 and TA1537 was used. Significant mutagenic effects could be detected in madder root extracts and also in the final product, the dyed wool. Madder root from Iran showed considerably higher mutagenic responses than samples from Bhutan. As mutagenic compounds, lucidin, rubiadin and purpuroxanthine could be detected with HPLC [193].

In a long-term study madder roots were added to the feed of rats to study possible carcinogenic potential. A dose-dependent increase in benign and malignant tumour formation was observed in the liver and kidneys of animals fed with addition of madder to their diet [194]. Thus the use of madder roots in long-term medical treatment or food has to be considered with care [81].

In a medium-term multiorgan carcinogenesis bioassay in rats, possible tumour-promoting effects of madder colour extracted from the root of madder was assessed. Male F344 rats were given DMD treatment, consisting of the multicarcinogens, *N*-nitrosodiethylamine (DEN), *N*-methyl-*N*-nitrosourea (MNU) and *N*-bis-(2-hydroxypropyl)-nitrosamine (DHPN), for 4 weeks. The animals were then administered a basal diet containing madder colour at doses between 0 and 5.0 % for the following 28 weeks. As a conclusion of this study, madder colour demonstrated significant tumour-promoting effects in the liver and kidneys in the DMD model [195].

Agricultural Aspects The biosynthesis of anthraquinones in plants can follow two different pathways: the polyketide pathway or the shikimate pathway. Anthraquinones that are synthesized following the polyketide pathway show substituents on both rings A and C, examples being emodin and chrysophanol. Representative plants are Rhamnaceae, Polygonaceae and Leguminosae. Anthraquinones that are synthesised via the shikimate pathway show substitution only in one ring. Typical examples are alizarin, pseudopurprin and lucidin found in Rubiaceae [81, 196–200]. The biosynthesis of these anthraquinones is similar to the naphthoquinone synthesis in plants, e.g. juglone and lawson, which branch at 1,4-dihydroxy-2-naphthoic acid into the biosynthesis of napthoquinones and anthraquinones [83].

The anthraquinones in *Rubia tinctorum* are mainly present as glycosides, which show increased solubility in water and are less reactive towards the cell wall compounds. Other plants, e.g. *Morinda*, store the anthraquinone glycosides in the vacuole.

The madder plant grows to a height of up to 100 cm and blooms in the European climate from June to August. It grows on loamy soil and seeds should be made in April, with a seed strength of 8–10 kg of seed per hectare. When seedlings are cultivated 50 000 to 60 000 plants are required per hectare in rows at a distance between 30 and 40 cm. In autumn, after the foliage has died the plant is covered with soil to protect against frost.

The cultivation requires approximately 80 kg/ha of nitrogen per year. For a three-year cultivation a total of 25 kg/ha of phosphor and 240 kg/ha of potassium is recommended [114].

Studies of the dyestuff content of two- and three-year old madder roots showed that an increase in alizarin content with factor 1.3 occurs between the second and the third year. Also a 10–15 % increase in the harvested amount of dry roots can be obtained in the third year, which is around 200 kg of dry roots per hectare (10.000 m^2) [81].

In another field trial study fertilization experiments were performed and the dyestuff content in the plant was analysed by photometry and test dyeing. However, the results of this study did not show distinct yield effects that correlated with fertilization and thus were not sufficient for recommendations of the nutrient requirement of madder [201].

The roots are harvested after a growth period of 3 years, preferably in September or October (European climate). Dependent on the age of a plantation, different crops can be obtained. Table 10.3 shows average harvests for different cultivation times and techniques (seed and plant cultivation) according to References [81] and [114].

The impurities (sand, soil) are removed mechanically from the harvested roots which are then dried at 60–80 °C and purified again [114]. Higher drying temperatures, e.g. 105 °C,

Table 10.3 *Crop of madder roots as a function of cultivation conditions (cultivation in Germany)*

	Seed		*Plant cultivation*	
	Two-year cultivation	*Three-year cultivation*	*Two-year cultivation*	*Three-year cultivation*
Fresh plant material (ton/ha)	10.0	13.2	14.0	17.6
Dried plant material (ton/ha)	2.5	3.3	3.5	4.4

were found to cause negative effects on the dye quality [201]. If the husk is removed, the (robbed) roots are ground to powder. Different fractions are then separated and collected by sieving. A special procedure includes a hydrolysis step of ground madder in diluted sulfuric acid, which removed glycosides and thus increased the colour strength of the product considerably (the traditional name of the product is 'madder flower') [81].

In field tests in Brandenburg (Germany) madder crops up to 4 ton/ha of dry material (18 ton/ha of fresh plant material) could be obtained. The average dyestuff content in the plant was 3.6 % [202–205]. During purifying and drying the hydrolysis of the glycosides into the anthraquinones proceeds, but does not result in a quality problem as in the later dyeing processes anthraquinoid components and not the glycosides are required.

Studies to investigate plant diseases showed a remarkable risk of attack/infection with ascochyta (fungi) on the leaves, stem and flowers, which lowers the amount of seed significantly. Other fungi (*Fusarium* spp., *Alternaria* spp., *Gotrytis* spp., *Phoma* sp.) were detected on the seed material and precautions against infection are recommended [205].

The revival of agricultural madder production led to the opening of a madder plantation and recovery factory in Steenbergen, Netherlands, in 2006 with a capacity of 70 000 kg of plant dye per annum. From 1 kg of dried madder root approximately 400 g of dye powder can be produced [206].

In an experimental study the agronomic potential and industrial value of madder plants under rain-fed conditions in Southwest Anatolia, Turkey, was investigated. Different propagation materials (seeds, seedlings and root cuttings) and different propagation methods (autumn root transplanting, spring root transplanting, autumn seed sowing, spring seed sowing and spring seedling transplanting) were compared in the study. At the end of the three-year growing period, the fresh root yield varied from 1640.1 kg/ha (in the spring root transplanting) to 4813.2 kg/ha (in the spring seedling transplanting). The spring seedling transplanting was found to show the best performance, producing higher root and dye yields. The dry root content was determined to range from 34.0 % to 37.5 % of the weight of fresh roots. The roots from the seedlings gave the highest dye content (2.20 %). August was found to be the optimum harvest time for obtaining the highest dry matter and dye accumulation. The highest dyestuff content was determined in the cortex [207].

In a comparative study the anthraquinone composition in root stock and root samples of plants originating from 11 different habitats were analysed after the hydrolysis of samples with hydrochloric acid, by HPLC with UV–VIS and MS detection. Six anthraquinone aglycones were identified. The main components were alizarin (9.6–21.8 mg/g), purpurin (3.7–123 mg/g) and lucidin (1.8–5.7 mg/g). The total anthraquinone derivative content varied between 15.6 and 39.4 mg/g. The most favourable ratio of total anthraquinone amount to lucidin amount (11.97) and one of the highest total anthraquinone derivative contents (38.1 mg/g) were found in a sample originating from the Aachen Region (Germany) [208].

Siebenborn *et al.* [209] described *Rubia* accessions from different locations in the wild flora of Turkey and from Western Europe, with regard to morphological characteristics, root yield and dye content. Compared to the genotypes from Turkey (Cesme region) and Western Europe, the plants from the Turkish location of Usak are low growing with small leaves and numerous shoots, and show good values for root yield and dye content [209].

Dormancy breakage of madder seed collected from the Yazd zone in Iran was studied, subjecting the seed to different treatments such as concentrated sulfuric acid for 10–20 minutes, hot water at 70–90 °C, exposure to light for 24 h and gibberellic acid. The highest percentage of germination was observed in the case of 90 % sulfuric acid applied for 15 minutes or hot water treatment at 90 °C. The radicle and plumule length, vigour index, day of first emergence and 50 % emergence, mean germination time and fresh weight were found to be affected by treatments [210].

Cell cultures derived form roots of madder have also been studied for production of anthraquinone dyes. The yield reached about one-third of that of field grown roots [211]. The influence of shear stress on anthraquinones production by *Rubia tinctorum* suspension cultures has also been studied. Suspension cultures of *Rubia tinctorum* were grown both in Erlenmeyer flasks at 100 rpm and in a 1.5 L mechanically stirred tank bioreactor operating at 450 rpm. The biomass obtained in the bioreactor was 29 % lower than that attained in the Erlenmeyer flasks, but the anthraquinone content in the bioreactor was substantially higher than that in the Erlenmeyer flasks. Authors concluded that *Rubia tinctorum* suspension cultures are able to grow in stirred tanks, responding to the hydrodynamic stress with higher concentrations of anthraquinones [212].

In another study an elicitation method was used to increase the synthesis of anthraquinone derivatives occurring in cell suspension cultures of *Rubia tinctorum*. Analysis with high- performance liquid chromatography coupled with diode array detection and mass spectrometry (HPLC-DAD-MS) was used for the detection and quantification of the two glycosides, lucidin primeveroside and ruberithric acid, and eight aglycones, namely pseudopurpurin, lucidin, alizarin, purpurin, alizarin-2-methyl ether, lucidin-ω-ethyl ether, nordamnacanthal and munjistin ethyl ester. Different elicitors, such as fungal polysaccharides and endogenous signal molecules (salicylic and jasmonic acid) were studied. The total yield of anthraquinones increased by a factor of 3–4, while for pseudopurpurin an increase by a factor of 28 was observed [213].

The effect of two elicitor types prepared from fungi on the alizarin content and the ultrastructure of *Rubia tinctorum* cells has been reported in the literature. The number of living cells decreased significantly after a period of 96 h. Phytium elicitor showed the lowest effect while the Botrytis elicitor seemed to be the most effective [214, 215].

Elicitation of genetically transformed madder (*Rubia tinctorum* L.) roots with methyl jasmonate resulted in a five- to eightfold increase in anthraquinone content. Growth of the culture and qualitative pigment composition remained unchanged. It is suggested that methyl jasmonate induces the biosynthesis of one of the key enzymes involved in the first stages of the anthraquinone biosynthesis in madder roots [216, 217].

A positive effect on the anthraquinone production in *Rubia tinctorum* suspension cultures was observed for the addition of proline and aminoindan-2-phosphonic acid. Phenylalanine ammonia liase activity was higher in the presence of proline. Addition of aminoindan-2-phosphonic acid produced phenylalanine ammonia liase inhibition and lower levels of phenolic acid content. The results showed that an increase in the anthraquinone level in *Rubia tinctorum* suspension culture can be achieved by metabolic manipulation and by inhibition of the key enzymes of the metabolic pathways that compete with anthraquinones for a common substrate [218].

Vasconsuelo *et al.* showed in a study on *Rubia tinctorum* cultures that elicitation with chitosan (200 mg/L) significantly stimulates anthraquinone synthesis. Spectrofluorimetric measurements showed that chitosan increases intracellular Ca^{2+} concentration in a medium devoid of calcium. Chitosan induction of anthraquinone synthesis in *Rubia tinctorum* involves the stimulation of phospholipase C, intracellular Ca^{2+} mobilization and phosphoinositide 3-kinase, which mediate mitogen-activated protein kinase (MAPK) activation [176, 219, 220].

The use of permeabilizing agents to enhance the productivity of anthraquinone colorants of madder (*Rubia akane Nakai*) cell cultures brought an increase in total and released concentrations of anthraquinones of about 1.6 times (159 mg/L) and 14 times (71 mg/L), respectively. In the presence of surfactant (Tween 80), chitosan and an ion-exchange sorbent the amount of total anthraquinone production was increased to 220 mg/L, which is 2.2 times the level of control culture [221].

Membrane technologies in general and reverse osmosis in particular have been employed as energy-saving processes for concentration of the plant extracts. Using crossflow reverse osmosis the heat-sensitive alizarin extracted from madder root could be concentrated successfully at 1.0 m/s crossflow velocity, 16 bar transmembrane pressure, at a solution pH of 7 [222].

10.5 Other Sources of Anthraquinoid Dyes

Numerous other sources for anthraquinoid dyes have been used in the past, some of which are mentioned in other chapters of the book. An extensive summary of structures of anthraquinoid dyes and respective plant sources is given by Schweppe [1].

The extraction of anthraquinoid dyes from mushrooms and lichens and the properties of the dyes will be discussed in another chapter of the book. For example, useful amounts of anthraquinoid dyes can be extracted from *Cortinarius sanguineus* and also *Fusarium oxysporum* [223–225].

In cell cultures of *Morinda citrifolia* L. substantial amounts of anthraquinones are formed. An important precursor in the biosynthesis by the shikimate pathway is chorismate (*trans*-3-([1-carboxy-ethenyl]-oxy)-4-hydroxy-1,5-cyclohexadiene-1-carboxylic acid) [226].

References

1. H. Schweppe, *Handbuch der Naturfarbstoffe*, ISBN 3-933203-46-5, Ecomed Verlagsgesellschaft, Landsberg/Lech, Germany, 1993.
2. H. Obara and J. Onodera, Structure of carthamin, *Chemistry Letters*, 201–204 (1979).

3. K. Saito and Y. Fukaya, The assessment of current protocols for preparing edible carthamin dye from dyer's saffron flowers, *Acta Alimentaria*, **26**(2), 141–152 (1997).
4. T. Kanehira, A. Naruse, A. Fukushima and K. Saito, Decomposition of carthamin in aqueous solutions: influence of temperature, pH, light, buffer systems, external gas phases, metal ions, and certain chemicals, *Zeitschrift für Lebensmittel-Untersuchung und -Forschung*, **190**(4), 299–305 (1990).
5. K. Saito, K.-I. Miyamoto and M. Katsukura, Influence of external additives on the preservation of carthamin red colour: an introductory test for utilizing carthamin as a herbal colorant of processed foods, *Zeitschrift für Lebensmittel-Untersuchung und -Forschung*, **196**(3), 259–260 (1993).
6. H. Oda, Improvement of light fastness of natural dyes. Part 2: effect of functional phenyl esters on the photofading of carthamin in polymeric substrate, *Coloration Technology*, **117**(5), 257–261 (2001).
7. H. Oda, Improvement of the light fastness of natural dyes: the action of single oxygen quenchers on the photofading of red carthamin, *Coloration Technology*, **117**(4), 204–208 (2001).
8. T. Blaszczyk, Portrait of an old Chinese medicinal plant – saffron (*Carthamus tinctorius* L.) [Saflor. Portrait einer alten chinesischen Heilpflanze], *Pharmazeutische Zeitung*, **145**(15), 30–33 (2000).
9. K. Saito, The tinctorial stability of carthamin on polysaccharides and related substances, *Acta Botanica Croatica*, **57**, 123–136 (1998).
10. I. Karube, Red pigment production by *Carthamus tinctorius* cells in a two-stage culture system, *Journal of Biotechnology*, **37**(1), 59–65 (1994).
11. N. Bairagi and M. L. Gulrajani, Studies on dyeing with shikonin extracted from Ratanjot by supercritical carbon dioxide, *Indian Journal of Fibre and Textile Research*, **30**, 196–199 (2005).
12. C. Cartwright-Jones, *Developing Guidelines on Henna: A Geographical Approach*, Essay for Masters Degree, Kent State University, Kent, Ohio, USA, www.hennapage.com, August 2006.
13. Z. F. Mahmoud, N. A. Abdel Salam and S. M. Khafagy, Constituents of henna leaves (*Lawsonia inermis* L.) growing in Egypt, *Fitoterapia*, **51**(3), 153–155 (1980).
14. M. S. Karawya, S. M. Abdel Wahhab and A.Y. Zaki, A study of the lawsone content in henna, *Lloydia*, **32**(1–4), 76–78 (1969).
15. B. M. Badri and S. M. Burkinshaw, Dyeing of wool and Nylon 6.6 with henna and lawsone, *Dyes and Pigments*, **22**, 15–25 (1993).
16. T. Kawamura, Y. Hisata, K. Okuda, Y. Noro, Y. Takeda and T. Tanaka, Quality evaluation of plant dye henna with glycosides, *Natural Medicines*, **54**(2), 86–89 (2000).
17. N. S. El-Shaer, J. M. Badr, M. A. Aboul-Ela and Y. M. Gohar, Determination of lawsone in henna powders by high performance thin layer chromatography, *Journal of Separation Science*, **30**(18), 3311–3315 (2007).
18. A. T. Bakkali, M. Jaziri, A. Foriers, Y. Vander Heyden, M. Vanhaelen and J. Homes, Lawsone accumulation in normal and transformed cultures of henna, *Lawsonia inermis, Plant Cell, Tissue and Organ Culture*, **51**(2), 83–87 (1997).
19. S. A. Petrova, M. V. Kolodyazhnyi and O. S. Ksenzhek, Electrochemical properties of sole naturally occurring quinones, *Journal of Electroanalysis of Chemical Interface Electrochemistry*, **277**(1–2), 189–196 (1990).
20. Kh. El-Nagar, S. H. Sanad, A. S. Mohamed and A. Ramadan, Mechanical properties and stability to light exposure for dyed Egyptian cotton fabrics with natural and synthetic dyes, *Polymer – Plastics Technology and Engineering*, **44**(7), 1269–1279 (2005).
21. A. Domenech-Carbo, M. T. Domenech-Carbo, M. C. Sauri-Peris, J. V. Gimeno-Adelantado and F. Bosch-Reig, Electrochemical identification of anthraquinone-based dyes in solid microsamples by square wave voltammetry using graphite/polyester composite electrodes, *Analytical and Bioanalytical Chemistry*, **375**(8), 1169–1175 (2003).
22. M. E. Bodini, P. E. Bravo and V. Arancibia, Voltammetric and spectroscopic study of the iron(II) complexes with the semiquinone of 2-hydroxy-1,4-napthoquinone (lawson) in aprotic medium, *Polyhedron*, **13**(3), 497–503 (1994).
23. P. Garge, S. Padhye and J. P. Tuchagues, Iron(II) complexes of ortho-functionalised *para*-naphthoquinones. 1. Synthesis, characterisation, electronic structure and magnetic properties, *Inorganica Chimica Acta*, **157**(2), 239–249 (1989).

24. V. Arancibia and M. Bodini, Redox chemistry of 2-hydroxy-1,4-naphthoquinone (lawsone) and of its manganese complexes in aprotic media, *Anales de Quimica, Serie B: Quimica Inorganica y Quimica Analitica*, **82**(3), 309–313 (1986).
25. P. L. K. Spurles, 'This is Different, this is the Plaza': space, gender, and tactics in the work of Moroccan tourist sector henna artisans, *Research in Economic Anthropology*, **25**, 99–123 (2006).
26. O. A. Hakeim, S. H. Nassar and K. Haggag, Greener printing of natural colour using microwave fixation, *Indian Journal of Fibre and Textile Research*, **28**(2), 216–220 (2003).
27. M. L. Gulrajani, D. B. Gupta, V. Agarwal and M. Jain, Some studies on natural yellow dyes: III. Quinones: henna dolu, *Indian Textile Journal*, **102**, 6 (1992).
28. N. Gogoi and B. Kalita, Dyeing of silk with natural dyes (Part I), *Colorage*, **46**(1), 23–26 (1999).
29. A. Agarwal, A. Garg and K. C. Gupta, Development of suitable dyeing process for dyeing of wool with natural dye – henna *(Lawsonia inerma), Colourage*, **39**(10), 43–45 (1992).
30. S. Ali, T. Hussain and R. Nawaz, Optimization of alkaline extraction of natural dye from henna leaves and its dyeing on cotton by exhaust method, *Journal of Cleaner Production*, **17**(1), 61–66 (2009).
31. L. Young and M. Mitchell, Colours of Southern India, *Embroidery*, **55**(11–12), 41–43 (2004).
32. J. Paliwal, Effect of mordants on henna, dyed cotton and silk fabrics, *Textile Magazine*, **42**(11), 79–81 (2001).
33. D. B. Gupta and M. L. Gulrajani, Kinetic and thermodynamic studies on 2-hydroxy-1, 4,-naphthoquinone (lawsone), *JSDC*, **110**, 112–115 (1994).
34. M. K Yakubu and A. M. Baba, Extraction, characterisation and application of henna (lele) on Nylon 6,6 fabrics, *Man-Made Textiles in India*, **43**(7), 293–298 (2000).
35. B. M. Badri and S. M. Burkinshaw, Dyeing of wool and Nylon 6.6 with henna and lawsone, *Dyes and Pigments*, **22**(1), 15–25 (1993).
36. S. Houlton, Henna dyes nylon a greener shade of yellow, *New Scientist*, **139**, 1885 (1993).
37. K. C. James, S. P. Spanoudi and T. D. Turner, The absorption of lawsone and henna by bleached wool felt, *Journal of the Society of Cosmetic Chemists*, **37**(5), 359–367 (1986).
38. J. E. Park and K. W. Oh, Characterization of wool dyeing with henna, *Journal of the Korean Fiber Society*, **41**(5), 322–327 (2004).
39. D. Gupta, Mechanism of dyeing synthetic fibres with natural dyes, *Colourage*, **47**(3), 23–26 (2000).
40. K. Singh, V. Kaur, S. Mehra and A. Mahajan, Solvent-assisted dyeing of polyester with henna, *Colourage*, **53**(10), 60–64 (2006).
41. M. D. Teli, R. V. Adivarekar, R. Shah and A. G. Sabale, Dyeing of cellulose triacetate with natural colourants, *Journal of the Textile Association*, **64**(6), 285–294 (2004).
42. J. R. Rao, A. Prakash, E. Thangaraj, K. J. Sreeram, S. Saravanabhavan and B. U. Nair, Natural dyeing of leathers using natural materials, *Journal of the American Leather Chemists Association*, **103**(2), 68–75 (2008).
43. A. E. Musa, B. Madhan, W. Madhulatha, S. Sadulla and G. A. Gasmelseed, Henna extract: Can it be an alternative retanning agent?, *Journal of the American Leather Chemists Association*, **103**(6), 188–193 (2008).
44. Z. H. Z. Abidin, R. M. Taha, R. Puteh and A. K. Arof, Characteristics of paints prepared from *Lawsonia* pigment and dammar from *Dipterocarpus grandifoleus*, *Materials Science Forum*, **517**, 290–293 (2006).
45. R. Petkewich, Henna, *Chemical and Engineering News*, **84**(6), 28 (2006).
46. A. Y. El-Etre, M. Abdallah and Z. E. El-Tantawy, Corrosion inhibition of some metals using *Lawsonia* extract, *Corrosion Science*, **47**(2), 385–395 (2005).
47. H. H. Rehan, Corrosion control by water-soluble extracts from leaves of economic plants, *Materialwissenschaft und Werkstofftechnik*, **34**(2), 232–237 (2003).
48. A. Chetouani and B. Hammouti, Corrosion inhibition of iron in hydrochloric acid solutions by natural henna, *Bulletin of Electrochemistry*, **19**(1), 23–25 (2003).
49. H. Al-Sehaibani, Evaluation of extracts of henna leaves as environmentally friendly corrosion inhibitors for metals, *Materialwissenschaft und Werkstofftechnik*, **31**(12), 1060–1063 (2000).

50. H. S. Muhammad and S. Muhammad, The use of *Lawsonia inermis Linn.* (henna) in the management of burn wound infections, *African Journal of Biotechnology*, **4** (9), 934–937 (2005).
51. B. R. Mikhaeil, F. A. Badria, G. T. Maatooq and M. M. A. Amer, Antioxidant and immunomodulatory constituents of henna leaves, *Zeitschrift für Naturforschung – Section C Journal of Biosciences*, **59**(7–8), 468–476 (2004).
52. T. Dasgupta, A. R. Rao and P. K. Yadava, Modulatory effect of henna leaf (*Lawsonia inermis*) on drug metabolising phase I and phase II enzymes, antioxidant enzymes, lipid peroxidation and chemically induced skin and forestomach papillomagenesis in mice, *Molecular and Cellular Biochemistry*, **245**(1–2), 11–22 (2003).
53. H. Changf and S. E. Suzuka, Lawsone (2-OH-1,4-naphthoquinone) derived from the henna plant increases the oxygen affinity of sickle cell blood, *Biochemical and Biophysical Research Communications*, **107**(2), 602–608 (1982).
54. Z. M. El-Basheir and M. A. Fouad, A preliminary pilot survey on head lice, pediculosis in Sharkia Governorate and treatment of lice with natural plant extracts, *Journal of the Egyptian Society of Parasitology*, **32**(3), 725–736 (2002).
55. D. C. McMillan, S. D. Sarvete Jr, J. E. Oatis and D. J. Jollow, Role of oxidant stress in lawsone-induced hemolytic anemia, *Toxicological Sciences*, **82**(2), 647–655 (2004).
56. D. Marzin and D. Kirkland, 2-Hydroxy-1,4-naphthoquinone, the natural dye of henna, is non-genotoxic in the mouse bone marrow micronucleus test and does not produce oxidative DNA damage in Chinese hamster ovary cells, *Mutation Research – Genetic Toxicology and Environmental Mutagenesis*, **560**(1), 41–47 (2004).
57. D. Kirkland and D. Marzin, An assessment of the genotoxicity of 2-hydroxy-1,4-naphthoquinone, the natural dye ingredient of henna, *Mutation Research – Genetic Toxicology and Environmental Mutagenesis*, **537**(2), 183–199 (2003).
58. G. Pratibha, G. R. Korwar, D. Palanikumar and V. Jois, Effect of planting materials, fertilizers and microsite improvement on yield and quality of henna (*Lawsonia inermis*) in Alfisols of semi-arid regions, *Indian Journal of Agricultural Sciences*, **77**(11), 721–725 (2007).
59. S. S. Rao, P. L. Regar, P. K. Roy and Y. V. Singh, Effect of nutrient management and row spacing on henna (*Lawsonia inermis*) leaf production and quality, *Indian Journal of Agricultural Sciences*, **77**(8), 486–489 (2007).
60. S. S. Rao, P. K. Roy and P. L. Regar, Effect of crop geometry and nitrogen on henna (*Lawsonia inermis*) leaf production in arid fringes, *Indian Journal of Agricultural Sciences*, **73**(5), 283–285 (2003).
61. S. K. Khandelwal, N. K. Gupta and M. P. Sahu, Effect of plant growth regulators on growth, yield and essential oil production of henna (*Lawsonia inermis* L.), *Journal of Horticultural Science and Biotechnology*, **77**(1), 67–72 (2002).
62. A. M. Korayem and H. A. Osman, Nematicidal potential of the henna plant *Lawsonia inermis* against the root knot nematode *Meloidogyne incognita* [Über nematizide Wirkungen der Henna-Pflanze *Lawsonia inermis* gegen den Wurzelnematoden *Meloidogyne incognita*], *Anzeiger für Schädlingskunde Pflanzenschutz Umweltschutz*, **65**(1), 14–16 (1992).
63. K. Chand, B. L. Jangid and B. L. Gajja, Economics of henna in semi-arid Rajasthan, *Annals of Arid Zone*, **41**(2), 175–181 (2002).
64. J.-L. Wang, S.-X. Zhang, T.-J. Li, W.-Q. Zhang, J.-J. Wang and S.-J. Zhang, Chemical constituents from bark of *Juglans mandshurica*, *Chinese Traditional and Herbal Drugs*, **39**(4), 490–493 (2008).
65. T. Bechtold, A. Mahmud-Ali and R. Mussak, Natural dyes from food processing wastes – useage for textile dyeing, Chapter 31 in *Waste Management and Co-product in Food Processing*, pp. 502–533, Woodhead Publishing Ltd, Cambridge, England, March 2007, ISBN 1 84569 025 7.
66. J. H. Park, B. M. Gatewood, and G. N. Ramaswamy, Naturally occurring quinones and flavonoid dyes for wool: insect feeding deterrents, *Journal of Applied Polymer Science*, **98**(1), 322–328 (2005).
67. T. Bechtold, A. Turcanu, E. Ganglberger and S. Geissler, Natural dyes in modern textile dyehouses – How to combine experiences of two centuries to meet the demands of the future?, *Journal of Cleaner Production*, **11**, 499–509 (2003).

68. E. Grover, A. Sharmal, B. Rawat, S. Paul and S. Jahan, Dyeing of silk with natural dyes, *International Dyer*, **190**(10), 9–16 (2005).
69. M. Gahlot, N. Papnai, N. Fatima and S. Singh, Printing with natural dyes, *International Dyer*, **190**(10), 18–20 (2005).
70. A. Onal, N. Camci and A. Sari, Extraction of total dyestuff from walnut leaves (*Juglans regia* L.) and its dyeing conditions for natural fibres, *Asian Journal of Chemistry*, **16**(3–4), 1533–1539 (2004).
71. S. Paul, A. Sharmal and E. Grover, Standardization of dyeing variables for dyeing of cotton with walnut bark, *Man-Made Textiles in India*, **46**(4), 152–155, 158 (2003).
72. S. Paul, E. Grover and A. Sharmal, Process development for dyeing of wool with walnut bark, *Asian Textile Journal*, **12**(2), 72–76 (2003).
73. D. Parac-Osterman, B. Karaman, A. Horvat and M. Pervan, Dyeing wool with natural dyes in the light of ethnological heritage of Lika [Bojadisanje vune prirodnim bojilima u svjetlu etnografske baštine Like], *Tekstil*, **50**(7), 339–344 (2001).
74. E. Bollhalder, Dyeing with plant dyes and natural insect dyestuffs [Farben mit Pflanzenfarben und Insekten naturfarbstoffen], *DWI Reports*, **122**, 28–39 (1999).
75. D. B. Gupta and M. L. Gulrajani, Studies on dyeing with natural dye juglon (5-hydroxy-1, 4-napthoquinone), *Indian Journal of Fibre and Textile Research*, **18**(12), 202–206 (1993).
76. O. Goktas, R. Mammadov, M. E. Duru, E. Baysal, A. M. Colak and E. Ozen, Development of new environment friendly natural colored preservatives for wood surface dyeing derived from different tree and herbaceous plant extracts and determination of their color parameters, *Ekoloji*, **15**(60), 16–23 (2006).
77. J. J. Inbaraj and C. F. Chignell, Cytotoxic action of juglone and plumbagin: a mechanistic study using HaCaT keratinocytes, *Chemical Research in Toxicology*, **17**(1), 55–62 (2004).
78. T. Aburjai and F. M. Natsheh, Plants used in cosmetics, *Phytotherapy Research*, **17**(9), 987–1000 (2003).
79. S. R. Maulik and S. C. Pradhan, Dyeing of wool and silk with hinjal bark, jujube bark and Himalayan rhubarb, *Man-Made Textiles in India*, **48**(10), 396–400 (2005).
80. R. Singh, A. Jain, S. Panwar, D. Gupta and S. K. Khare, Antimicrobial activity of some natural dyes, *Dyes and Pigments*, **66**, 99–102 (2005).
81. G. C. H. Derksen, Red, Redder, Madder – Analysis and Isolation of Anthraquinones from Madder Roots *(*Rubia tinctorum*)*, Dissertation, Wangeningen University, The Netherlands, 2001, ISBN 90-5808-462-0.
82. D. Gupta, S. Kumari and M. Gulrajani, Dyeing studies with hydroxyanthraquinones extracted from Indian madder. Part 1: dyeing of nylon with prupurin, *Coloration Technology*, **117**, 328–332 (2001).
83. R. Wijnsma and R. Verporte, in R. A. Hill and H. C. Krebs (eds), *Fortschritte der Chemie Organischer Naturstoffe, Progress in the Chemistry of Organic Natural Products*, Vol. **49**, Springer-Verlag, Wien, 1986, pp. 79–141.
84. A. Ashnagar, N. G. Naseri and A. S. Zadeh, Isolation and identification of 1,2-dihydroxy-9, 10-anthraquinone (Alizarin) from the roots of maddar plant (*Rubia tinctorum*), *Biosciences Biotechnology Research Asia*, **4**(1), 19–22 (2007).
85. G. Cuoco, C. Mathe, P. Archier, F. Chemat and C. Vieillescazes, A multivariate study of the performance of an ultrasound-assisted madder dyes extraction and characterization by liquid chromatography–photodiode array detection, *Ultrasonics Sonochemistry*, **16**(1), 75–82 (2009).
86. E. G. Kiel, *Metaallcomplexen von Alizarinerood (1,2-dihydroxyanthraquinone)*, Thesis, Technische Hogeschool, Delft, 1961, p. 198.
87. D. Bilgic, S. Karaderi and I. Bapli, The determination of the stability constants binary complexes of alizarin with Ca(II) and Zn(II) by potentiometric and spectrophotometric methods, *Reviews in Analytical Chemistry*, **26**(2), 99–108 (2007).
88. M. Subramanian, S. Kannan and R. Geethamalini, Natural dyes – a glimpse, *Pakistan Textile Journal*, **54**(4), 53 (2005).
89. M. S. S. Kannan and R. Geethamalini, Natural dyes, a brief glimpse, *Textile Asia*, **36**(2–3), 32–39 (2005).

90. A. Onal, I. Kahveci and M. Soylak, Investigation of the effect of (urea + ammonia + calcium oxalate) mordant mixture on the dyeing of wool, feathered-leather and cotton, *Asian Journal of Chemistry*, **16**(1), 445–452 (2004).
91. K. Yoshizumi and P. C. Crews, Characteristics of fading of wool cloth dyed with selected natural dyestuffs on the basis of solar radiant energy, *Dyes and Pigments*, **58**(3), 197–204 (2003).
92. M. D. Teli, R. V. Adivarekar and P. D. Pardeshi, Multiple dips method for application of natural dyes, *Asian Textile Journal*, **11**(12), 77–81 (2002).
93. S. Das, Application of natural dyes on silk, *Colourage*, **39**(9), 52–54 (1992).
94. M. N. Micheal, F. M. Tera and S. A. Aboelanwar, Colour measurements on colorant estimation of natural red dyes on natural fabrics using different mordants, *Colourage*, **50**(1), 31–41 (2003).
95. M. R. Katti, R. Kaur and N. Shrihari, Dyeing of silk with mixture of natural dyes, *Colourage*, **43**(12), 37–40 (1996).
96. K. Patil, Comparative study of fastness properties of silk dyed with synthetic and natural dye, *Journal of the Textile Association*, **64**(3), 137–141 (2003).
97. S. Waheed and A. Alam, Effect of mordants on color shade and color fastness of silk dyed with kikar and madder barks, *Pakistan Journal of Scientific and Industrial Research*, **47**(6), 423–429 (2004).
98. R. Bhuyan and C. N. Saikia, Extraction of natural colourants from roots of *Morinda angustifolia Roxb.* – their identification and studies of dyeing characteristics on woool, *Indian Journal of Chemical Technology*, **10**(3), 131–136 (2003).
99. R. Bochmann and M. Weiser, Pflanzenfarbstoffe in der Textilveredlung, *Melliand Textilberichte*, **84**(3), 198–201 (2003).
100. R. Bochmann and H. Erth, Färben mit Extrakten aus Pflanzen und Hölzern, *Melliand Textilberichte*, **88**(5), 344–347 (2007).
101. A. Aydin and T. Zeki, The sorption behaviors between natural dyes and wool fibre, *International Journal of Chemistry*, **13**(2), 85–91 (2003).
102. M. Montazer, F. A. Taghavi, T. Toliyat and M. B. Moghadam, Optimization of dyeing of wool with madder and liposomes by central composite design, *Journal of Applied Polymer Science*, **106**(3), 1614–1621 (2007).
103. D. De Santis and M. Moresi, Production of alizarin extracts from *Rubia tinctorum* and assessment of their dyeing properties, *Industrial Crops and Products*, **26**(2), 151–162 (2007).
104. C. Clementi, W. Nowik, A. Romani, F. Cibin and G. Favaro, A spectrometric and chromatographic approach to the study of ageing of madder (*Rubia tinctorum* L.) dyestuff on wool, *Analytica Chimica Acta*, **596**(1), 46–54 (2007).
105. M. Parvinzadeh, Effect of proteolytic enzyme on dyeing of wool with madder, *Enzyme and Microbial Technology*, **40**(7), 1719–1722 (2007).
106. L. Wei, X.-L. Hou, Q.-C. Zhou, X.-L. Zhang and X.-F. Hu, The extraction of madder dye and its dyeing properties on wool fabric, *Wool Textile Journal*, **12**, 5–8 (2006).
107. M. A. Khan, M. Khan, P. K. Srivastava and F. Mohammad, Extraction of natural dyes from cutch, ratanjot and madder, and their application on wool, *Colourage*, **53**(1), 61–68 (2006).
108. G. J. Smith, I. J. Miller and V. Daniels, Phototendering of wool sensitized by naturally occurring polyphenolic dyes, *Journal of Photochemistry and Photobiology A: Chemistry*, **169**(2), 147–152 (2005).
109. S. Agarwal and K. C. Gupta, Optimization of dyeing conditions for natural dye-madder roots (*Rubia cordifolia*), *Colourage*, **50**(10), 43–46 (2003).
110. Anon, Wool: pigment printing with natural dye, *Tinctoria*, **98**(1), 37–41 (2001).
111. P. S. Vankar, R. Shanker, D. Mahanta and S. C. Tiwari, Ecofreindly sonicator dyeing of cotton with *Rubia cordifolia* Linn. using biomordant, *Dyes and Pigments*, **76**, 207–212 (2007).
112. H. T. Deo and P. Roshan, Eco-friendly mordant for natural dyeing of denim, *International Dyer*, 49–52 (2003).
113. D. Cristea and G. Vilarem, Improving light fastness of natural dyes on cotton yarn, *Dyes and Pigments*, **70**, 238–245 (2006).
114. C. K. K. Choo and Y. E. Lee, Analysis of dyeings produced by traditional Korean methods using colorants from plant extracts, *Coloration Technology*, **118**, 35–45 (2002).

115. A. K. Samanta, D. Singhee and M. Sethia, Application of single and mixture of selected natural dyes on cotton fabric: a scientific approach, *Colourage*, **50**(10), 29–42 (2003).
116. O. A. Hakeim, Application of tannic acid–metal salt combination in cotton printing with natural dye, *Journal of the Textile Association*, **63**(6), 281–286 (2003).
117. A. K. Samanta, D. Singhee, A. Sengupta and Sk. A. Rahim, Application of selective natural dyes on jute and cotton fabrics by different techniques, *Journal of the Institution of Engineers (India), Part TX: Textile Engineering Division*, **83**(FBR), 22–33 (2003).
118. M. L. Gulrajani, D. Gupta and P. Gupta, Application of natural dyes on bleached coir yarn, *Indian Journal of Fibre and Textile Research*, **28**(12), 466–470 (2003).
119. D. Gupta, S. Kumari and M. Gulrajani, Dyeing studies with hydroxyanthraquinones extracted from Indian madder. Part 2: dyeing of nylon and polyester with nordamncanthal, *Coloration Technology*, **117**, 333–336 (2001).
120. D. Gupta, M. L. Gulrajani and S. Kumari, Light fastness of naturally occurring anthraquinone dyes on nylon, *Coloration Technology*, **120**(5), 205–212 (2004).
121. A. Korkmaz, E. Ozdogan, T. Oktem and N. Seventekin, An alternative approach for polyester dyeing, *Colourage*, **54**(6), 44–48 (2007).
122. T. Wakida, S. Cho, S. Choi, S. Tokino and M. Lee, Effect of low temperature plasma treatment on color of wool and Nylon 6 fabrics dyed with natural dyes, *Textile Research Journal*, **68**(11), 848–853 (1998).
123. M. D. Teli, R. V. Adivarekar and P. D. Pardeshi, Reuse and replenishment of dye bath for economy and ecology in wool dyeing, *Journal of the Textile Association*, **63**(3), 119–124 (2002).
124. B. Rappl, C. Pladerer, M. Meissner, E. Ganglberger and S. Geissler, *Farb&Stoff™ – Von der Idee zum marktfähigen Handelsprodukt: Pflanzenfarben für die Textilindustrie*; Programmlinie 'Fabrik der Zukunft', Bundesministerium für Verkehr, Innovation und Technologie, Wien, 2005.
125. S. Geissler, E. Ganglberger, T. Bechtold, A. Mahmut-Ali, A. Hartl and O. Schütz, *Farb & Stoff – Sustainable Development durch neue Kooperationen und Prozesse*; Wien, Programmlinie 'Fabrik der Zukunft', Bundesministerium für Verkehr, Innovation und Technologie, Wien, 2003.
126. S. Geissler, E. Ganglberger, T. Bechtold, S. Sandberg, O. Schütz, A. Hartl and R. Reiterer, *Potential an nachwachsenden Rohstoffen unter Aspekten der Nachhaltigkeit: Produktion von farbstoffliefernden Pflanzen in Österreich und ihre Nutzung in der Textilindustrie*, Bundesministerium für Verkehr, Innovation und Technologie, Wien, 2001.
127. T. Bechtold, P. Kaulfuss, A. Mahmut-Ali, S. Geissler S. and E. Ganglberger, Naturfarbstoffe in Mitteleuropa – Rohstoffquellen und färberische Qualität, *Melliand Textilberichte*, **85**(4), 268–272 (2004).
128. O. Goktas, R. Mammadov, M. E. Duru, E. Baysal, A. M. Colak and E. Ozen, Development of new environment friendly natural colored preservatives for wood surface dyeing derived from different tree and herbaceous plant extracts and determination of their color parameters, *Ekoloji*, **15**(60), 16–23 (2006).
129. B. Dittmann and L. Adam, Weed control in wild mignosette (*Reseda luteola* L.) and madder (*Rubia tinctorum* L.) [Unkrautbekampfung in farber-resede (*Reseda luteola* L.) und krapp (*Rubia tinctorum* L.)], *Zeitschrift fur Pflanzenkrankheiten und Pflanzenschutz*, **107**(Special Issue 17), 479–484 (2000).
130. I. Formanek and G. Racz, Die antibiotische Wirkung der Krappwuzel (Rubia tinctorum*),* Pharmazie, **30**, 617–618 (1975).
131. G. Rath, M. Ndonzoa and K. Hostettmann, Antifungal anthraquinones from *Morinda lucida.*, *International Journal, Pharmacogology*, **33**, 107–114 (1995).
132. F. Kalyoncu, B. Cetin and H. Saglam, Antimicrobial activity of common madder (*Rubia tinctorum* L.), *Phytotherapy Research*, **20**(6), 490–492 (2006).
133. N. T. Manojlovic, S. Solujic, S. Sukdolak and M. Milosev, Antifungal activity of *Rubia tinctorum*, *Rhamnus frangula* and *Caloplaca cerina*, *Fitoterapia*, **76**(2), 244–246 (2005).
134. S. Mehrabian, A. Majd and I. Majd, Antimicrobial effects of three plants (*Rubia tinctorum*, *Carthamus tinctorius* and *Juglans regia*) on some airborne microorganisms, *Aerobiologia*, **16**(3–4), 455–458 (2000).
135. N. Cucer, N. Guler, H. Demirtas and N. Imamoglu, Staining human lymphocytes and onion root cell nuclei with madder root, *Biotechnic and Histochemistry*, **80**(1), 15–20 (2005).

136. A. D. Dixon and D. A. N. Hoyte, A comparison of autoradiographic and alizarin techniques in the study of bone growth, *Anatomical Records*, **145**, 101–113 (1963).
137. H. Schilcher, Pflanzliche Urologika, *Deutsch Apoth. Ztg.*, **124**, 2429–2436 (1984).
138. N. Ino , T. Tanaka, A. Okumura, Y. Morishita, H. Makita, Y. Kato, M. Nakamura and H. Mori, Acute and subacute toxicity tests of madder root, natural colorant extracted from madder (*Rubia tinctorum*), in (C57BL/6XC3H)F_1 mice. *Toxicology in Industrial Health*, **11**, 440–458 (1995).
139. Y. Kawasaki, Y. Goda, T. Maitani, K. Yoshihira and M. Takeda, Determination of madder color in foods by high performance liquid chromatography, *Shokuhin Eiseigaku Zasshi, Journal of Food Hygiene Society of Japan*, **33**, 563–568 (1992).
140. K. Inoue, M. Shibutani, N. Masutomi, K. Toyoda, H. Takagi, C. Uneyama, A. Nishikawa and M. Hirose, A 13-week subchronic toxicity study of madder color in F344 rats, *Food and Chemical Toxicology*, **46**(1), 241–252 (2008).
141. N. Masutomi, M. Shibutani, K. Toyoda, N. Niho, C. Uneyama and M. Hirose, A 90-day repeated dose toxicity study of madder color in F344 rats: a preliminary study for chronic toxicity and carcinogenicity studies, *Kokuritsu Iyakuhin Shokuhin Eisei Kenkyujo hokoku = Bulletin of National Institute of Health Sciences*, **118**, 55–62 (2000).
142. Z. Jian and S. Xun, Antioxidant activities of baicalin, green tea polyphenols and alizarin *in vitro* and *in vivo*, *Journal of Nutritional and Environmental Medicine*, **7** (2), 79–89 (1997).
143. S. Kultur, Medicinal plants used in Kırklareli Province (Turkey), *Journal of Ethnopharmacology*, **111**(2), 341–364 (2007).
144. M. H. Babita, G. Chhaya and P. Goldee, Hepatoprotective activity of *Rubia cordifolia*, *Pharmacologyonline*, **3**, 73–79 (2007).
145. N. P. Mishchenko, S. A. Fedoreev, V. M. Bryukhanov, Ya. F. Zverev, V. V. Lampatov, O. V. Azarova, Shkryl'Yu.N. and G. K. Chernoded, Chemical composition and pharmacological activity of anthraquinones from *Rubia cordifolia* cell culture, *Pharmaceutical Chemistry Journal*, **41**(11), 605–609 (2007).
146. T. Marczylo, C. Sugiyama and H. Hayatsu, Protection against Trp-*P*-2 DNA adduct formation in C57bl6 mice by purpurin is accompanied by induction of cytochrome P450, *Journal of Agricultural and Food Chemistry*, **51**(11), 3334–3337 (2003).
147. E. Takahashi, T. H. Marczylo, T. Watanabe, S. Nagai, H. Hayatsu and T. Negishi, Preventive effects of anthraquinone food pigments on the DNA damage induced by carcinogens in *Drosophila*, *Mutation Research – Fundamental and Molecular Mechanisms of Mutagenesis*, **480–481**, 139–145 (2001).
148. T. Marczylo, S. Arimoto-Kobayashi and H. Hayatsu, Protection against Trp-*P*-2 mutagenicity by purpurin: mechanism of *in vitro* antimutagenesis, *Mutagenesis*, **15**(3), 223–228 (2000).
149. T. H. Marczylo, T. Hayatsu, S. Arimoto-Kobayashi, M. Tada, K.-I. Fujita, T. Kamataki, K. Nakayama and H. Hayatsu, Protection against the bacterial mutagenicity of heterocyclic amines by purpurin, a natural anthraquinone pigment, *Mutation Research – Genetic Toxicology and Environmental Mutagenesis*, **444**(2), 451–461 (1999).
150. J. L. M. Calvo, A. O. Gamiz, A. R. Diaz, E. S. Sanchez, E. B. Palenciano, J. M. A. Rodriguez and J. C. Fernandez, Profilaxis de los tumores superficiales de vejiga estadio T1 con 27 mg. de BCG semanal durante seis semanas, *Archivos Espanoles de Urologia*, **52**(7), 760–768 (1999).
151. Y. Kawasaki, Y. Goda, K. Yoshihira, The mutagenic constitutents of *Rubia tinctorum*, *Chemical And Pharmaceutical Bulletin* (Tokyo), **40**, 1504–1509 (1992).
152. A. R. Burnett and R. H. Thomason, Naturally occuring quinones. Part XV. Biogenesis of the anthraquinones in *Rubia tinctorum* L. (madder), *Journal of the Chemical Society*, 2437–2441 (1968).
153. J. Kuiper and R. P. Labadie, Polyploid complexes within the genus *Galium*. Part I: anthraquinones of *Galium album*. *Planta Medica*, **42**, 390–399 (1981).
154. N. A. El-Emary and E. Y. Backheet, Three hydroxymethylanthraquinone glycosides from *Rubia tinctorum*, *Phytochemistry*, **49**, 277–279 (1998).
155. L. G. Angelini, L. Pistelli, P. Belloni, A. Bertoli and S. Panconesi, *Rubia tinctorum*, a source of natural dyes: agronomic evaluation, quantitative analysis of alizarin and industrial assays, *Industrial Crop Production*, **6**, 303–311 (1997).

156. Y. Kawasaki, Y. Goda and K. Yoshihira, Anthraquinones from *Rubia tinctorum*, *Shoyakugaku Zasshi*, **42**, 166–167 (1988).
157. Z. A. Tóth, O. Raatikainen, T. Naaranlathi and S. Auriola, Isolation and determination of alizarin in cell cultures of *Rubia tinctorum* and emodin in *Cortinarius sanguineus* using solid-phase extraction and high performance liquid chromatography, *Journal of Chromatography*, **630**, 423–428 (1993).
158. V. V. S. Murti, T. R. Seshadri and S. Sivakumaran, A study of madder, the roots of *Rubia tinctorum* L., *Indian Journal of Chemistry*, **8**, 779–782 (1970).
159. Y. Yasui and N. Takeda, Identification of a mutagenic substance in *Rubia tinctorum* L. (madder) root, as lucidin, *Mutation Research*, **121**, 185–190 (1983).
160. A. G. Ercan, K. M. Taskin, K. Turgut and S. Yuce, *Agrobacterium rhizogenes*-mediated hairy root formation in some *Rubia tinctorum* L. populations grown in Turkey, *Turkish Journal of Botany*, **23**(6), 373–377 (1999).
161. G. C. H. Derksen, M. Naayer, T. A. van Beek, A. Capelle, I. K. Haaksman, H. A. van Doren and A. E. de Groot, Chemical and enzymatic hydrolysis of anthraquinone glycosides from madder roots, *Phytochemical Analysis*, **14**(3), 137–144 (2003).
162. K. Sato, T. Yamazaki, E. Okuyama, K. Yoshihira and K. Shimoura, Anthraquinone production by transformed root cultures of *Rubia tinctorum*; influence of phytohormones and sucrose concentration, *Phytochemistry*, **30**, 1507–1509 (1991).
163. X.-L. Hou, L. Wie, X.-L. Zhang, Q.-C. Zhou, L. Wang, S. H. Jia, X.-F. Hu and S.-Y. Wang, Study on stability of vegetable dyestuff madder, *Wool Textile Journal*, (8), 24–27 (2007).
164. M. S. Hatamipour and H. Shafikhani, Design of extraction unit for madder root, *Iranian Journal of Science and Technology, Transactions B: Technology*, **23**(3), 180–181 (1999).
165. G. C. H. Derksen, G. P. Lelyveld, T. A. van Beek, A. Capelle and A. E. de Groot, Two validated HPLC methods for the quantification of alizarin and other anthraquinones in *Rubia tinctorum* cultivars, *Phytochemical Analysis*, **15**(6), 397–406 (2004).
166. L. Rafaelly, S. Heron, W. Nowik and A. Tchapla, Optimisation of ESI-MS detection for the HPLC of anthraquinone dyes, *Dyes and Pigments*, **77**(1), 191–203 (2008).
167. I. Boldizsar, Z. Szucs, Zs. Fuzfai and I. Molnar-Perl, Identification and quantification of the constituents of madder root by gas chromatography and high-performance liquid chromatography, *Journal of Chromatography A*, **1133**(1–2), 259–274 (2006).
168. G. C. H. Derksen, H. A. G. Niederlander and T. A. Van Beek, Analysis of anthraquinones in *Rubia tinctorum* L. by liquid chromatography coupled with diode-array UV and mass spectrometric detection, *Journal of Chromatography A*, **978**(1–2), 119–127 (2002).
169. N. Bhattacharyya and S. Vairagi, Natural dye – its authenticity and identification, *Colourage*, **49**(4), 45–53 (2002).
170. Z. Bosakova, J. Persl and A. Jegorov, Determination of lucidin in *Rubia tinctorum* aglycones by an HPLC method with isocratic elution, *Journal of Separation Science*, **23**(10), 600–602 (2000).
171. Z. Bosakova, J. Persl and A. Jegorov, Determination of lucidin in *Rubia tinctorum* aglycones by an HPLC method with isocratic elution, *HRC Journal of High Resolution Chromatography*, **23**(10), 600–602 (2000).
172. B. Bozan, M. Kosar, C. Akyurek, K. Ertugrul and K. Husnu Can Baser, Alizarin and purpurin contents of *Rubia tinctorum* L. roots collected from various regions of Turkey, *Acta Pharmaceutica Turcica*, **41**(4), 187–190 (1999).
173. P. Novotna, V. Pacakova, Z. Bosakova and K. Stulik, High-performance liquid chromatographic determination of some anthraquinone and naphthoquinone dyes occurring in historical textiles, *Journal of Chromatography A*, **863**(2), 235–241 (1999).
174. G. C. H Derksen, T. A. Van Beek, A. De Groot and A. Capelle, High-performance liquid chromatographic method for the analysis of anthraquinone glycosides and aglycones in madder root (*Rubia tinctorum* L.), *Journal of Chromatography A*, **816**(2), 277–281 (1998).
175. M. Puchalska, M. Orlinska, M. A. Ackacha, K. Polee-Pawlak and M. Jarosz, Identification of anthraquinone coloring matters in natural red dyes by electrospray mass spectrometry coupled to capillary electrophoresis, *Journal of Mass Spectrometry*, **38**(12), 1252–1258 (2003).

176. A. Vasconsuelo, A. M. Giuletti, G. Picotto, J. Rodriguez-Talou and R. Boland, Involvement of the PLC/PKC pathway in Chitosan-induced anthraquinone production by *Rubia tinctorurn* L. cell cultures, *Plant Science*, **165**(2), 429–436 (2003).
177. M. A. Aakacha, K. Polec-Pawlak and M. Jarosz, Identification of anthraquinone coloring matters in natural red dyestuffs by high performance liquid chromatography with ultraviolet and electrospray mass spectrometric detection, *Journal of Separation Science*, **26**(11), 1028–1034 (2003).
178. M. Hirokado, K. Kimura, K. Suzuki, Y. Sadamasu, Y. Katsuki, K. Yasuda and M. Nishijima, Detection method of madder color, cochineal extract, lac color, carthamus yellow and carthamus red in processed foods by TLC, *Journal of the Food Hygienic Society of Japan*, **40**(6), 488–493 (1999).
179. A. C. J. Hermans-Lokkerbol, R. van der Heijden and R. Verpoorte, Solvent system selection for separation of anthraquinones by means of centrifugal partition chromatography; application to an extract of a *Rubia tinctorum* hairy root culture, *Journal of Liquid Chromatography*, **16**, 1433–1451 (1993).
180. W. C. Weng and S. J. Sheu, Separation of anthraquinones by capillary electrophoresis and high-performance liquid chromatography, *Journal of High Resolution Chromatography*, **23**, 143–148 (2000).
181. X. Sun, X. Yangy and E. Wang, Chromatographic and electrophoretic procedures for analyzing plant pigments of pharmacologically interests, *Analytica Chimica Acta*, **547**(2), 153–157 (2005).
182. P. Banyai, I. N. Kuzovkina, L. Kursinszki and E. Szoke, HPLC analysis of alizarin and purpurin produced by *Rubia tinctorum* L. hairy root cultures, *Chromatographia*, **63**(Suppl. 13), S111–S114 (2006).
183. D. Fabbri, G. Chiavari and H. Ling, Analysis of anthraquinoid and indigoid dyes used in ancient artistic works by thermally assisted hydrolysis and methylation in the presence of tetramethylammonium hydroxide, *Journal of Analytical and Applied Pyrolysis*, **56**(2), 167–178 (2000).
184. M. J. Casas-Catalan and M. T. Domenech-Carbo, Identification of natural dyes used in works of art by pyrolysis-gas chromatography/mass spectrometry combined with in situ trimethylsilylation, *Analytical and Bioanalytical Chemistry*, **382**(2), 259–268 (2005).
185. A. Jegorov, L. Cvak, J. Cejka, B. Kratochvil, P. Sedmera and V. Havlicek, Crystal structures of mollugin and lucidin, *Journal of Chemical Crystallography*, **35**(8), 621–627 (2005).
186. F. Nakanishi, Y. Nagasawa, Y. Kabaya, H. Sekimoto and K. Shimomura, Characterization of lucidin formation in *Rubia tinctorum* L., *Plant Physiology and Biochemistry*, **43**(10–11), 921–928 (2005).
187. J. Westendorf, H. Marquardt, B. Poginsky, M. Dominiak, J. Schmidt and H. Marquardt, Genotoxicity of naturally occurring hydroxyanthraquinones, *Mutation Res.*, **240**, 1–12 (1990).
188. F. Marec, I. Kollarova and A. Jegorov, Mutagenicity of natural anthraquinones from *Rubia tinctorum* in the Drosophila wing spot test, *Planta Medica*, **67**(2), 127–131 (2001).
189. B. Poginsky, J. Westendorf, B. Blömeke, H. Marquardt, A. Hewer, P. L. Grover and D. H. Phillips, Evaluation of DNA-binding activity of hydroxyanthraquinones occurring in *Rubia tinctorum* L., *Carcinogenesis*, **12**, 1265–1271 (1992).
190. B. Poginsky, Vorkommen und Genotoxizität von Anthracen-Derivate in Rubia tinctorum L., Thesis, Universität of Hamburg, 1989, p. 174.
191. Y. Kawasaki, Y. Goda, H. Noguchi and T. Yamady, Identification of adducts formed by reaction of purine bases with a mutagenic anthraquinone, lucidin: mechanism of mutagenicity by anthraquinones occurring in Rubiaceae plants, *Chemical and Pharmaceutical Bulletin (Tokyo)*, **42**, 1971–1973 (1994).
192. J. Westendorf, W. Pfau and A. Schulte, Carcinogenicity and DNA adduct formation observed in ACI rats after long-term treatment with madder root, *Rubia tinctorum* L., *Carcinogenesis*, **19**(12), 2163–2168 (1998).
193. I. Jager, C. Hafner, C. Welsch, K. Schneider, H. Iznaguen and J. Westendorf, The mutagenic potential of madder root in dyeing processes in the textile industry, *Mutation Research – Genetic Toxicology and Environmental Mutagenesis*, **605**(1–2), 22–29 (2006).
194. J. Westendorf, W. Pfau and A. Schulte, Carcinogenicity and DNA adduct formation observed in ACI rats after long term treatment with madder root, *Rubia tinctorum* L., *Carcinogenesis*, **19**, 2163–2168 (1998).

195. M. Yokohira, K. Yamakawa, K. Hosokawa, Y. Matsuda, T. Kuno, K. Saoo and K. Imaida, Promotion potential of madder color in a medium-term multi-organ carcinogenesis bioassay model in F344 rats, *Journal of Food Science*, **73**(3), T26–T32 (2008).
196. E. Leister, Biosynthesis of morindon and alizarine in intact plants and cell suspension cultures of *Morinda citrifolia*, *Phytochemistry*, **12**, 1669–1674 (1973).
197. E. Leister, Mode of incorporation of precursors into alizarin (1,2,-dihydroxy-9,10-anthraquinones), *Phytochemistry*, **12**, 337–345 (1973).
198. A. R. Burnett and R. H. Thomson, Biogenesis of anthraquinones in Rubiaceae, *Chemical Commununications* (London), 1125–1126 (1967).
199. J. W. Fairbairn and F. J. Muthadi, The biosynthesis and metabolism of anthraquinones in *Rumex obtusifolius*, *Phytochemistry*, **11**, 215–219 (1972).
200. D. Eichinger, A. Bacher, M. H. Zenk and W. Eisenreich, Quantitative assessment of metabolic flux by 13C NMR analysis. Biosynthesis of anthraquinones in *Rubia tinctorum*, *Journal of the American Chemical Society*, **121**(33), 7469–7475 (1999).
201. S. Siebenborn, R. Marquard and B. Honermeier, Investigations on domestication and quality improvement of the dye plant *Rubia tinctorum* L. (madder), *Pflanzenbauwissenschaften*, **5**(2), 49–57 (2001).
202. B. Dittmann, L. Adam and K. H. Karabensch, Anbau und Ernte von Färber-Resede und Krapp, Abstract, in Symposium on *Naturfarben – Chance für Produktinnovationen*, Potsdam, Germany, 18–19 September 2001.
203. W. Maltry and L. Adam, Temperatureinfluß auf die Farbinhaltsstoffe von Färberresede und Krapp beim Trocknen, Abstract, in Symposium on *Naturfarben – Chance für Produktinnovationen*, Potsdam, Germany, 18–19 September 2001.
204. L. Adam and C. Müller, Schaderregerauftreten bei Färberresede und Krapp in Brandenburg, Poster, in Symposium on *Naturfarben – Chance für Produktinnovationen*, Potsdam, Germany, 18–19 September 2001.
205. L. Adam and B. Dittmann, *Färberpflanzen – Krapp, Anbau, Ernte und Nacherntebehandlung, Landesanstalt für Landwirtschaft, Abteilung Acker- und Pflanzenbau*, Güterfelde, Germany, 2001.
206. B. Sterk, *Rubia* pigmenta naturalia, *Textile Forum*, (2), 32–33 (2007).
207. H. Baydar and T. Karadogan, Agronomic potential and industrial value of madder (*Rubia tinctorum* L.) as a dye crop, *Turkish Journal of Agriculture and Forestry*, **30**(4), 287–293 (2006).
208. I. Boldizsar, A. Laszlo-Bencsik, Z. Szucs and B. Danos, Examination of the anthraquinone composition in root-stock and root samples of *Rubia tinctorum* L. plants of different origins, *Acta Pharmaceutica Hungarica*, **74**(3), 142–148 (2004).
209. S. Siebenborn, R. Marquard, I. Turgut and S. Yuce, Evaluation of different madder genotypes (*Rubia tinctorum* L.) for dyestuff production, *Journal of Herbs, Spices and Medicinal Plants*, **9**(4), 281–287 (2002).
210. R. Farhoudi, M. T. Makkizadeh, F. Sharifzadeh, M. Kochak Por and S. Rashidi, Study of dormancy-breaking of madder seed (*Rubia tinctorum*), *Seed Science and Technology*, **35**(3), 739–743 (2007).
211. Z. A. Toth, O. Raatikainen, T. Naaranlahti and S. Auriola, Isolation and determination of alizarin in cell cultures of *Rubia tinctorum* and emodin in *Cortinarius sanguineus* using solid-phase extraction and high-performance liquid chromatography, *Journal of Chromatography*, **630**, 423–428 (1993).
212. V. D. Busto, J. Rodriguez-Talou, A. M. Giulietti and J. C. Merchuk, Effect of shear stress on anthraquinones production by *Rubia tinctorum* suspension cultures, *Biotechnology Progress*, **24**(1), 175–181 (2008).
213. N. Orban, I. Boldizsar, Z. Szucs and B. Danos, Influence of different elicitors on the synthesis of anthraquinone derivatives in *Rubia tinctorum* L. cell suspension cultures, *Dyes and Pigments*, **77**(1), 249–257 (2008).
214. J. Jakab and I. Kiraly, Comparison of the effect of different fungal elicitors on *Rubia tinctorum* L. suspension culture, *Biologia Plantarum*, **45**(2), 281–290 (2002).
215. L. J. P. Van Tegelen, R. J. M. Bongaerts, A. F. Croes, R. Verpoorte and G. J. Wullems, Isochorismate synthase isoforms from elicited cell cultures of *Rubia tinctorum*, *Phytochemistry*, **51**(2), 263–269 (1999).

216. O. V. Mantrova, M. V. Dunaeva, I. N. Kuzovkina, B. Schneider and F. Muller-Uri, Effect of methyl jasmonate on anthraquinone biosynthesis in transformed madder roots, *Russian Journal of Plant Physiology*, **46**(2), 248–251 (1999).
217. I. N. Kuzovkina, O. V. Mantrova, I. E. Al'terman and S. A. Yakimov, Culture of genetically transformed hairy roots derived from anthraquinone-producing European madder plants, *Russian Journal of Plant Physiology*, **43**, 291–298 (1996).
218. M. Perassolo, C. Quevedo, V. Busto, F. Ianone, A. M. Giulietti and J. R. Talou, Enhancement of anthraquinone production by effect of proline and aminoindan-2-phosphonic acid in *Rubia tinctorum* suspension cultures, *Enzyme and Microbial Technology*, **41**(1–2), 181–185 (2007).
219. A. Vasconsuelo, S. Morelli, G. Picotto, A. M. Giulietti and R. Boland, Intracellular calcium mobilization: a key step for chitosan-induced anthraquinone production in *Rubia tinctorum* L., *Plant Science*, **169**(4), 712–720 (2005).
220. A. Vasconsuelo, A. M. Giulietti and R. Boland, Signal transduction events mediating chitosan stimulation of anthraquinone synthesis in *Rubia tinctorum*, *Plant Science*, **166**(2), 405–413 (2004).
221. J. J. Shim, J. H. Shin, T. Pai, I. S. Chung and H. J. Lee, Permeabilization of elicited suspension culture of madder (*Rubia akane* Nakai) cells for release of anthraquinones, *Biotechnology Techniques*, **13**(4), 249–252 (1999).
222. S. S. Madaeni and H. Daneshvar, The concentrating of alizarin using a reverse osmosis process, *Journal of the Serbian Chemical Society*, **70**(1), 107–114 (2005).
223. R. Räisänen, P. Nousiainen and P. H. Hynninen, Emodin and dermocybin natural anthraquinones as mordant dyes for wool and polyamide, *Textile Research Journal*, **71**(10), 922–927 (2001).
224. F. A Nagia and R. S. R. El-Mohamedy, Dyeing of wool with natural anthraquinone dyes from *Fusarium oxysporum*, *Dyes and Pigments*, **75**(3), 550–555 (2007).
225. P. H. Hynninen, R. Räisänen, P. Elovaara and E. Nokelainen, Preparative isolation of anthraquinones from the fungus *Cortinarius sanguineus* using enzymatic hydrolysis by the endogenous β-glucosidase, *Zeitschrift für Naturforschung*, **55c**, 600–610 (2001).
226. M. Stalman, A.-M. Koskamp, R. Luderer, J. H. J. Vernooy, J. C. Wind, G. J. Wullems and A. F. Croes, Regulation of anthraquinone biosynthesis in cell cultures of *Morinda citrifolia*, *Journal of Plant Physiology*, **160**(6), 607–614 (2003).

11

Dyes from Lichens and Mushrooms

Riikka Räisänen

11.1 Use of Lichen and Mushroom Dyes in the Past

Lichens and mushrooms form a group of more peculiar natural colorants. They have been widely used in Europe and also in other parts of the world as sources of colorants. Especially lichens were important in dyeing violet and purple colours. In the times of Mediterranean civilizations, which refer to the period from ancient times to the beginning of the Middle Ages, violet and purple colours were most commonly obtained from molluscs. Mollusc and shellfish dyestuffs were laborious to obtain and thus very expensive. Instead, lichens appeared abundantly on coastal areas and were easier to collect. Also the dyeing process for lichen colorants was simpler. The only disadvantage was the poor light-fastness. Hence, lichen dyes were used for cheaper textiles or as an under dye for the molluscs when producing, for example, the royal purple. In that way the use of the precious mollusc colorants was economized [1–3]. Royal purple or Tyrian purple was the name of the colour obtained originally from molluscs. Tyrian purple refers to the place (Tyre, in Lebanon) that was the familiar dyeing centre whereas royal purple refers to the high cost of the colorant, which allowed only the rich and wealthy to wear it [4]. To obtain the purple colour from lichen, a process with ammonia was required. The purple dyeing tradition seemed to be unique for the Mediterranean and European cultures [1]. The dyeing process will be discussed in more detail later in Section 11.3.1.

In Northern Europe, Scotland and Scandinavia yellowish, brownish and reddish brown colours were obtained using a simpler dyeing method in which lichens were boiled in water together with mordanted or nonmordanted wool [1, 2, 5–9]. Similar simple dyeing methods were applied by the American Indians when they used dyestuffs obtained from lichens in their traditional textiles [1, 2].

Handbook of Natural Colorants Edited by Thomas Bechtold and Rita Mussak

Resent investigations reveal that there is more information than suspected about the use of fungi for dyeing textiles. It seems that fungi have been used as a source of colour on all continents, Europe, North Africa and America as well as Asia [1]. The American Indians obtained red colour from the fungus *Echinodontium tinctorium*. Also, in America furs were sometimes dyed with colorants obtained from mushrooms of the *Boletus* species. Dyeing with alum mordants gave furs a luxurious golden hue [10, 11]. In Italy and France fungi belonging to Polyporales were used to dye wool. In some 15th century dyeing books, there are references to a substance called 'popo', 'opoppo' or 'pococco', which was used to obtain a yellow colour. The substance was also used to obtain an orange when dyed together with the red insect dye kermes. Several French books from the 17th century mention 'agaric'. 'Agaric' is described as fungus, a parasite of larch. The recent investigations suggest that 'popo' and 'agaric' could be *Polyporus mori* or *Fomes fomentarius* and *Laricifomes officinalis*, respectively [1]. The dyeing recipes are very similar for 'popo' and 'agaric' and some recipes mention that grey and black shades can be obtained when mordanting with iron. It seems that 'popo' and 'agaric' have been used on an industrial scale for several centuries, from the 15th to the 17th century, in the major European textile centres and goods have been traded internationally. Fungi were obtained from the Alps or Eastern Mediterranean countries [1].

In Northern Europe, several authors refer to lichen and fungal colorants [10–15]. In the book *Svenska Färge-Konst* from 1720 the Swede Johan Linder mentions old *Lactarius* (milk cap) as a source of a yellow colour [13]. Eighty years later, in 1804, Palmstruch mentions several mushrooms in his book *Svensk Botanik* [15]. It seems that mushrooms were mainly used for food and medicine, but a few times authors mention that they also gave colour to cotton and linen. The rare use of mushrooms for dyeing was obviously a result of the fact that fungi were not known very well. Most of the species were not identified at all and that caused doubt and confusion. During the Second World War, some experiments were carried out in Europe aiming at the use of fungi as a source of textile colorants [12], but it was not until the 1970s when the American M. Rice published a book about dyeing with mushrooms that the colorants became popular among craftsmen [16]. Due to the recent debate on diminishing fossil resources and ecological design natural dyes have gained growing interest. Research has been done to promote the use of natural dyes in several fields in industry, e.g. food, cosmetics and textiles.

11.2 Cultivation of Lichens and Mushrooms

Contrary to general belief, lichens are not always rare. In fact some lichens are weedy and they may grow fast [1]. Most lichens grow relatively slowly; thus their collection for dyeing purposes is questionable. Some researchers suggest that lichens could be harvested wisely [1, 3]. According to the public right of access, however, in most countries, lichens may not be collected from nature without the permission of the land owner [17]. From an ecological point of view, fungi serve as a more interesting source of raw material. In contrast to higher plants whose growth rates are rather slow, single cell algae and fungi can grow in high yields within a short period of time, only several hours, and are thus more suitable and economical also for the new technologies like biotechnological production [18, 19]. When collecting mushrooms from nature, one does not need to think about the ethical questions of harming species growth and population, because only the fruit bodies are collected and mycelium remains in the ground. Furthermore, the exploitation of mushrooms could be beneficial for rural areas, promoting extra income for farmers and families. Edible mushrooms are already collected through networks and the same networks could be utilized for collecting fungal material for dyeing purposes.

Due to the symbiotic relationship with their host trees mycorrhizal fungi seem to be rather difficult to cultivate [20, 21]. However, there are successful attempts to obtain aseptic cultures or even fruiting bodies of mycorrhizal fungi like *Cortinarius semisanguineus*, which contain anthraquinones suitable for dyeing, and *Cantharellus cibarius*, a delicious edible mushroom [22–25]. Also several other ectomycorrhizal fungi have been cultivated successfully. The highest mycelial growth rates of ectomycorrhizal fungi were reported to be about half that of the wood-rotting fungi [26]. The amounts of aseptic cultures in agar-containing flasks varied from 1.1 to 94.9 mg/flask, the incubation time being 21 days [26]. Fungi that do not form mycorrhiza, i.e. saprotrophs, are easy to cultivate and the growing procedures are utilized widely in production of edible mushrooms. *In vitro* cultivations have been explored by utilizing, for example, wood shavings, sawdust, paddy husk, sugarcane bagasse, banana leaves, dried moringa stems and paddy straw as culture media. In recent investigations shredded newspapers were found as a cost-effective medium for culturing, the availability of paper waste being abundant [27, 28]. In the case of basidiomycetous saprotrophs, the growth rates expressed as biological efficiency (calculated as the fresh weight of fruit bodies per dry weight of substrate) may vary from an average of 50 % up to even over 130 % [20, 27].

In commercial production of edible mushrooms, the first fungal bodies may usually be recognized within six weeks after the preparation of the culture medium and adapting spores. In nine weeks, one square metre may produce 15–30 kg of mushrooms [29].

11.3 Dyestuffs in Lichens and Mushrooms

Lichens are cryptograms, symbiotic organisms comprised of alga and fungal partners. Due to this symbiotic origin lichens and fungi contain similar types of colorants (Table 11.1). There are colorants or precursors of colorants that may be found from both lichen and fungi; furthermore, there are colorants that may be typical for only lichens or mushrooms. Nevertheless, colorants belong to various classes of chemical compounds, some of which are not found elsewhere in nature [1, 5]. Many colorants in lichens and fungi are benzoquinone derivatives [5]. Particularly, terphenylquinone compounds are found. Some species, e.g. *Sarcodon*, *Phellodon*, *Hydnellum* and *Telephora*, containing these compounds are known to impart a blue colour, which is rare in nature, to wool fibres [1, 5, 10, 12, 16].

11.3.1 Lichen Dyestuffs: Orchils and Litmus

Both orchil and litmus are colorants that have not been found in plants. They are formed from their colourless pre-compounds through consecutive enzymatic hydrolysation, decarboxylation and oxidation processes, respectively. Pre-compounds of orchil and litmus exist in several species of lichen, e.g. *Roccella* and *Lecanora* spp. [5]. The procedure for the

Table 11.1 *Some examples of the colorants occurring in fungi and lichens; l = lichen, f = fungus* [1, 5, 30, 34]

Dye source	*Colouring component and its chemical structure*
Roccella sp. (l), *Ochrolechia* sp. (l) *Evernia* sp. (l)	lecanoric acid 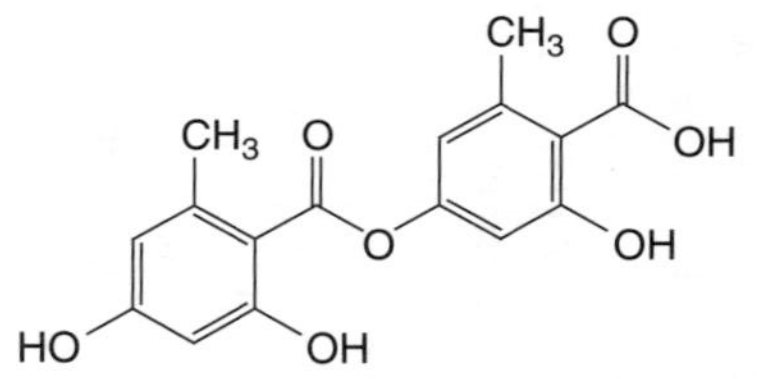
Parmelia sp. (l), *Xanthoria parietina* (l)	atranorin 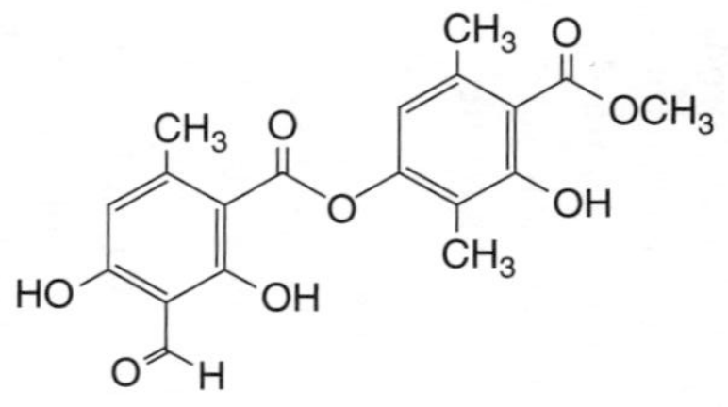
Ochrolechia tartarea (l), *Lasallia pustulata* (l)	gyrophoric acid 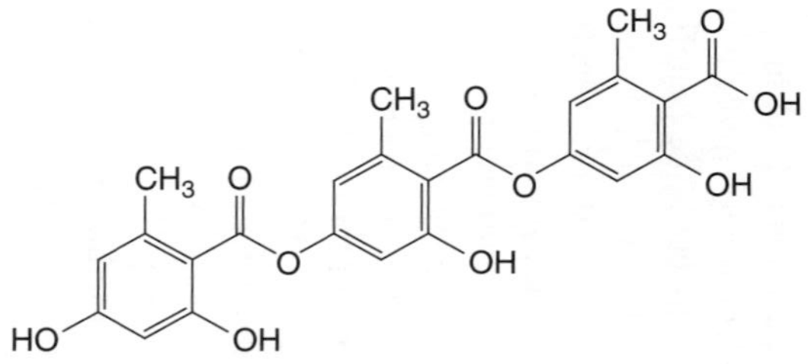

Parmelia sp. (l), *Usnea dasypoga* (l), *Ramalina crassa* (l)	salazinic acid 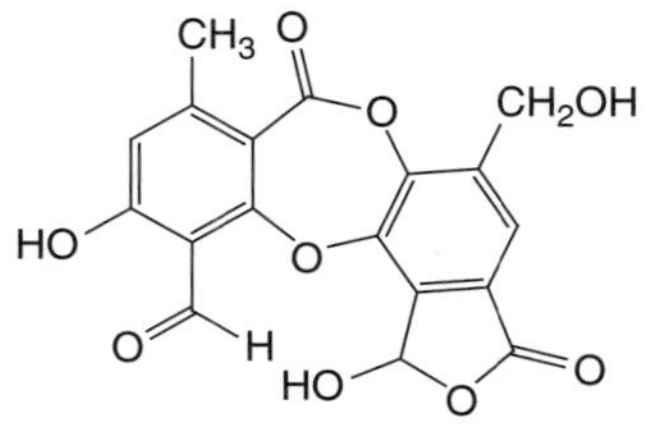
Boletus sp. (f), *Xerocomus* sp. (f), *Paxillus atrotomentosus* (f) *Sarcodon squamosus* (f)	atromentin 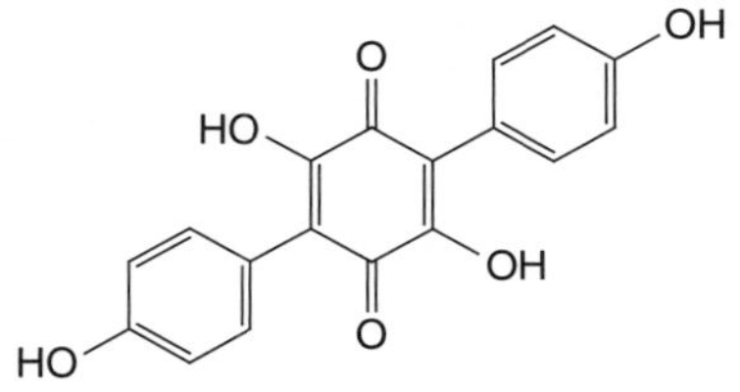
Sticta sp. (l), *Polyporus* sp. (f), *Hapalopilus nidulans* (f)	polyphoric acid 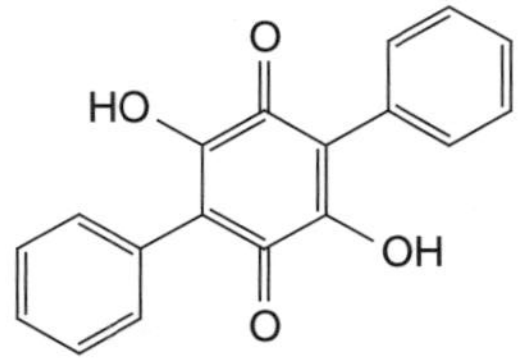

(continued)

Table 11.1 *(Continued)*

Dye source	*Colouring component and its chemical structure*
Lobaria sp. (l), *Trametes* sp. (f), *Polyporus* sp. (f), *Hydnum* sp. (f). *Sarcodon squamosus* (f)	telephoric acid 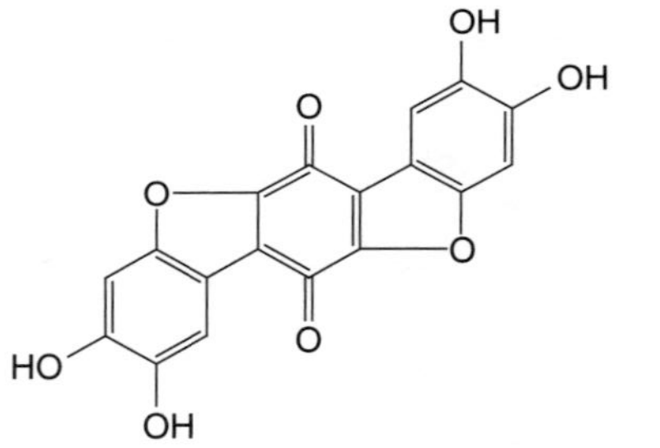
Boletales *sp. (f),* *Suillus* sp. (f)	grevillin A: R1, R2 =H grevillin B: R1=H, R2=OH grevillin C: R1, R2=OH 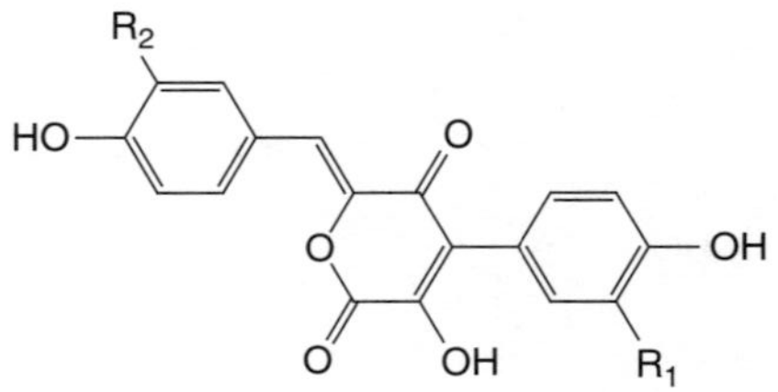
Xanthoria sp. (l) *Dermocybe* sp. (f)	physcion 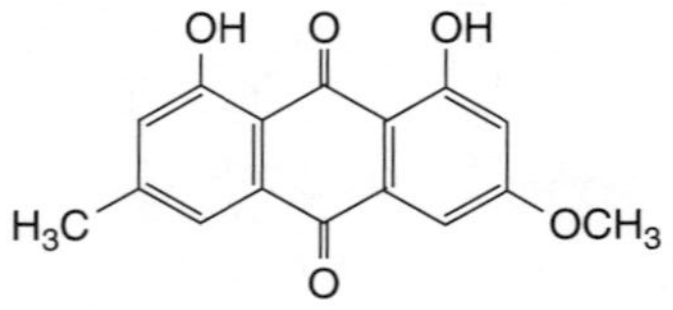

Dermocybe sp. (f), *Cortinarius sanguineus* (f)	emodin	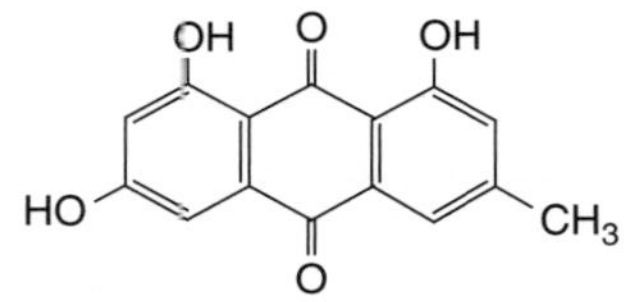
Dermocybe sp. (f), *Cortinarius sanguineus* (f)	dermocybin	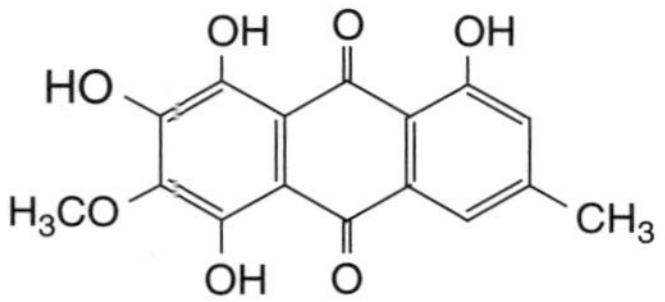
Boletales *sp. (f),* *Sticta aurata* (l)	pulvinic acid	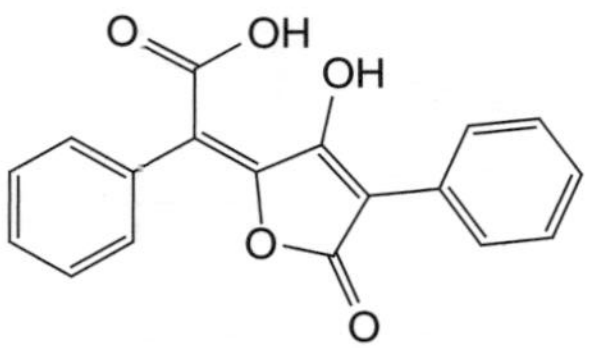

(*continued*)

Table 11.1 *(Continued)*

Dye source	*Colouring component and its chemical structure*
Boletus sp. (f), *Pisolithus arhizus* (f)	norbadione A 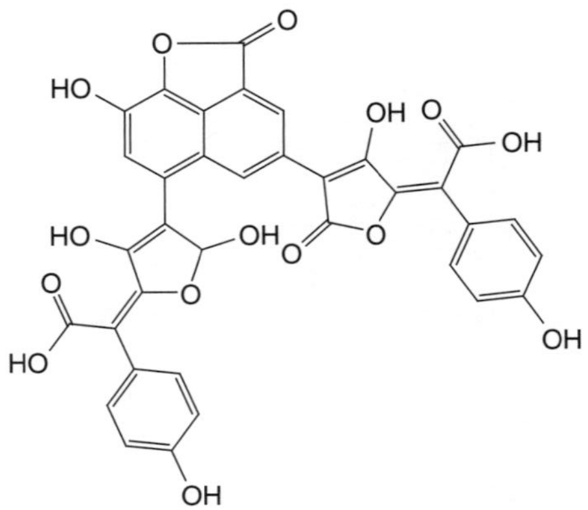

ancient lichen purple was to prepare a water extract with ammonia and to let the liquid stand in a warm place for several days. The reaction occurred in a form of enzymatic hydrolysis, where the noncoloured compounds of lecanoric acid were first hydrolysed to orsellinic acid and decarboxylated further on to orcinol. Orcihol oxidized in the air and formed the purple orceins, e.g. α-aminoorcein and α-hydroxyorcein, or litmus (Figure 11.1) [1, 5]. Orchil is a complex of several other orceins among those mentioned earlier, i.e. β- and γ-aminoorcein, β- and γ-hydroxyorcein and β- and γ-aminoorceimine [1]. The colour of both litmus and orchil are pH-dependent. In an acidic environment the dyestuffs form red cations while in the basic environment they are bluish violet in their anionic forms [1, 5]. Lichens from *Rocella* sp. contain lecanoric acid from 3 to 4 % of the dry weight [5].

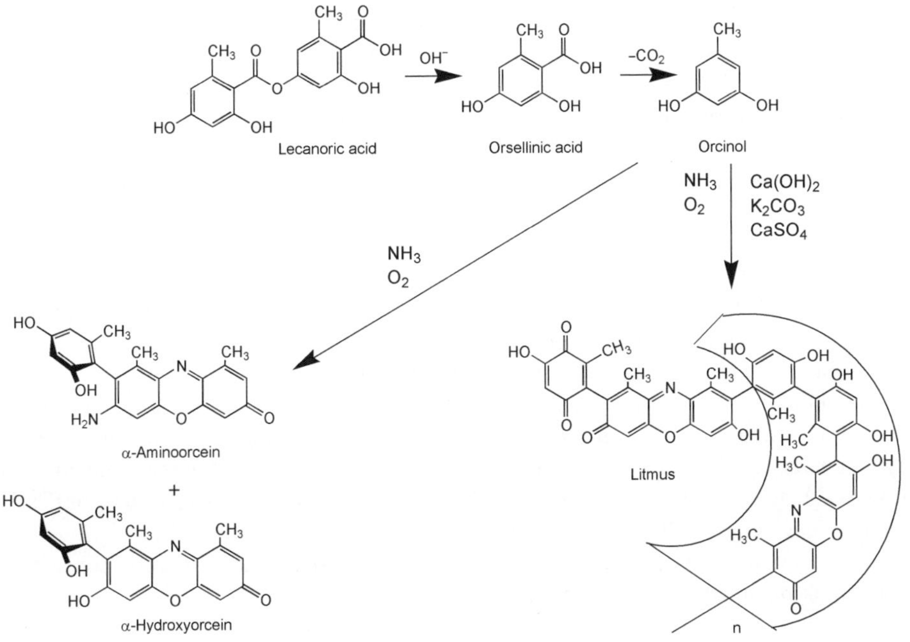

Figure 11.1 Chemical reactions occurring in the purple dyeing procedure with lichens of Roccella and Lecanora species [1, 5]. Reproduced by permission of Archetype Publications

11.3.2 Yellowish, Brownish and Reddish Colorants from Lichen

Atranorin and gyrophoric acid are found only in lichens, e.g. the former in *Parmelia* sp. and *Xanthoria parietina* and the latter in *Ochrolechia tartarea* and *Lasallia pustulata*. Salizinic acid, the derivative of atranorin, is found in several *Parmelia* spp., e.g. in *Parmelia saxatilis*. These lichens have been used to dye yellowish, reddish and brownish hues for the woollen materials used in tweeds and tartans in the Scottish highlands [1, 5, 6]. Dyeing with these lichens is also common in Scandinavia. For the colorants two types of dyeing methods are used: the process with ammonia, which considers the consecutive enzymatic hydrolysation, decarboxylation and oxidation processes, and the water boiling method, where the lichens are treated in water together with the textile material [1, 2, 6]. With the ammonium procedure more violet tones are obtained.

11.3.3 Benzoquinone Derivatives

11.3.3.1 Blue Colours from Terphenylquinones

Terphenylquinone compounds are the main colorants used to produce the blue colour in dyeing processes with mushrooms. Blue hues may be obtained from *Sarcodon*, *Phellodon* and *Hydnellum* as well as from *Telephora* spp. *Hapalopilus nidulans* contains polyporic acid in 20–40 % of its fresh weight. *Paxillus atrotomentosus* and *Telephora palmata* contain pigments in notably less amounts. *P. atrotomentosus* contains atromentin and telephofic acid together from 2 to 4 %, whereas *T. palmata* contains telephoric acid in under 1 % of its dry weight [5, 30].

It may be recognized that there are two species of *Sarcodon*, *S. imbricatus* and *S. squamosus*, which are very often confused with each other [31]. Earlier it was thought that *Sarcodon imbricatus* imparted the blue colour to wool fibres. However, very often the blue colour was not obtained and that confused dyers. The recent DNA investigations reveal that there are two different *Sarcodon* species, one of which grows with pines, *Pinus sylvestris*, and the other with spruces, *Picea abies*. Mushrooms collected from a pine forest seem to give stronger blue colours than those collected from spruce forests [31].

The dyestuff quantities mentioned above and the dyeing experiments show that to obtain strong blue colours great amounts of fresh mushrooms are needed. Greyish and light blue colours may be obtained when the fresh weight of mushrooms is 10 or 20 times greater than the weight of the textile material used. For stronger colours an amount at least 30 times the weight of the fibre is needed [32]. Thus, even though the blue colours of the mushrooms are interesting and in dyeing they produce shades that are rare in natural dyestuffs, it seems that the amounts of colorants are so low that utilization of them in large-scale dyestuff production does not seem very economical. However, natural compounds often have other unique, e.g. medicinal, properties that make them valuable. Terphenylquinones have been found to have antimicrobial properties, and those are the ones worth considering when thinking about future uses of these compounds [33].

11.3.3.2 Yellows from Grevillines

Benzoquinone derivatives, grevillines, together with terphenylquinones are important colorants of Boletales sp. (Table 11.1). Grevillines produce hues from yellow to red in these mushrooms [1, 5, 33].

11.3.4 Anthraquinones

Some of the best mushrooms for dyeing seem to be the ones in *Cortinarius* sp. The richly coloured appearances of the fruit bodies already reveal them as containing plenty of colorants. Mushrooms in *Cortinarius* sp. appear in brown, red, olive green and even violet colours.

The main colorants in *Cortinarius*, subgroup *Dermocybe* sp., are anthraquinones. Anthraquinones are one of the most important classes of dyestuffs among commercial colorants. Anthraquinones are relatively stable and light-fast and they give a bright colour. The best ancient red colours obtained from madder, cochineal and kermes were also anthraquinone-based dyestuffs. In recent years, natural anthraquinones have been used as colouring agents for beverages, sweets and other foods [17, 35, 36]. In addition, the cosmetic industry utilizes natural colorants more and more in all kinds of products, including hair colorants [17, 37, 38].

11.3.4.1 Blood-red webcap, Cortinarius sanguineus

In the fungus *C. sanguineus*, emodin- and dermocybin-1-β-D-glycopyranosides are the most abundant anthraquinones. In the fresh fungi as much as 90 % of the pigments exist as glycosides [39]. The most important biosynthetic route to anthraquinones seem to be the acetate–malonate pathway, which includes suitable folding and condensation of a polyketide chain derived from acetate units (Figure 11.2). According to the acetate–malonate pathway, the biosynthetic relationships show that the yellow compounds emodin, physcion and endocrocin exist in the beginning of the synthesis pathway whereas the red compounds dermocybin, dermorubin and 5-chlorodermorubin are more complicated in structure and occur in the latter part of the synthesis pathway [30]. This may be verified in dyeing experiments, which reveal that young mushrooms give more yellowish and orange colours whereas old mushrooms give dark red and reddish brown colours.

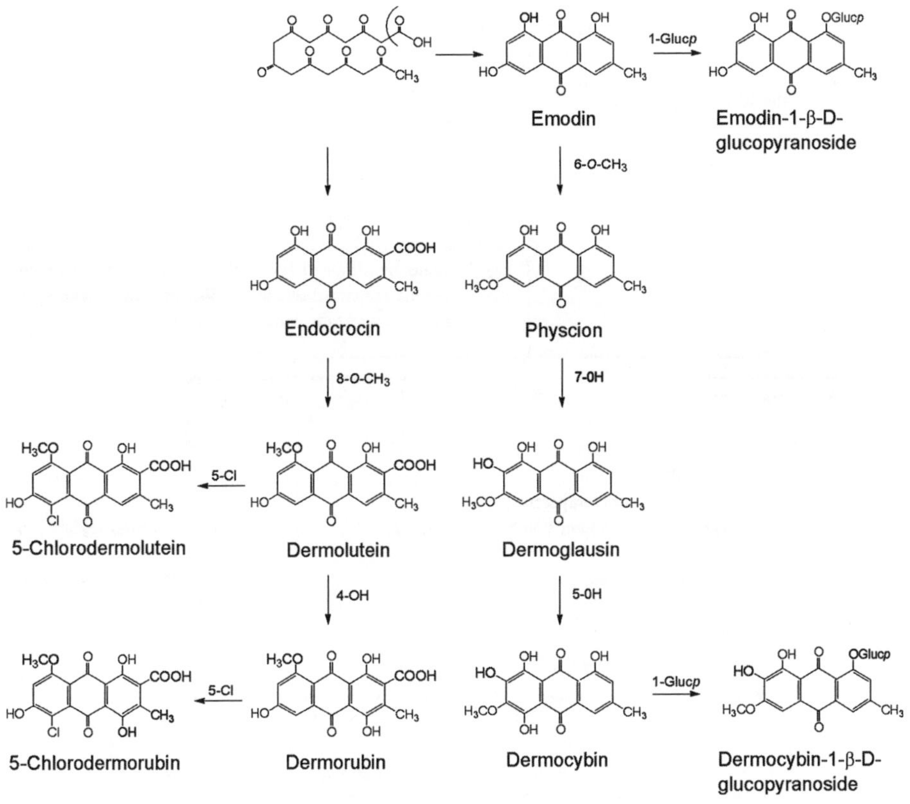

Figure 11.2 *Biosynthetic relationships between pigments of Cortinarius sanguineus [30]. Reproduced by permission of Springer-Verlag*

Since fungi are structurally and functionally, e.g. they do not have photosynthesis, simpler organisms than higher plants, the isolation of secondary metabolites, i.e. pigments, from them seems to be quite easy. Hynninen *et al.* have developed a preparative scale isolation method for the anthraquinone colorants from wild grown mushrooms [40]. In the procedure fresh fungal bodies were used (Figure 11.3). As known, most natural colorants exist as glycosides in living

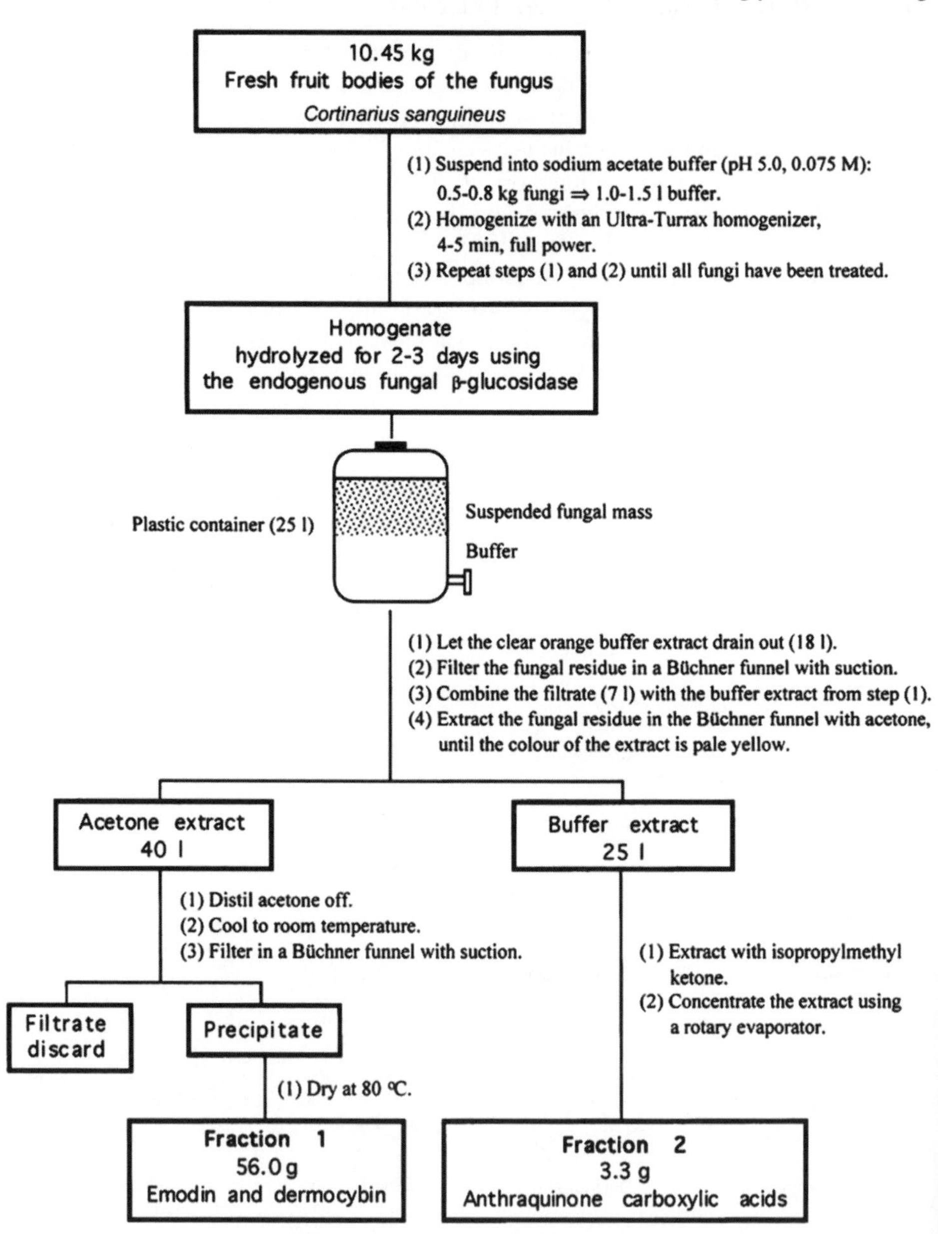

Figure 11.3 *Protocol for the isolation of the anthraquinone aglycones from the fungus Cortinarius sanguineus applying hydrolysis catalysed by the endogenous β-glucosidase. Reproduced by permission of Verlag der Zeitschrift fuer Naturforschung*

organisms. Glycosidic compounds are in most cases so water soluble that they are inconvenient for dyeing – the dyestuffs remain rather in the dye solution than find their way to fibres. In the isolating procedure the endogenous β-glucosidase enzyme of the fungus was employed to break down the glucosidase linkage and thus the colorants were obtained as free anthraquinones, which are in the most appropriate form for dyeing purposes. In addition, it was noticed in other experiments that during the drying process of the fruit bodies the glycoside linkages partly broke down by themselves. It seems that during the evaporation of water the cell membranes were disrupted, enabling the β-glucosidase enzymes and anthraquinone glycosides to meet. Thus, in the preparative scale colorant isolation procedures it seems to be advisable to dry the mushrooms as the primary step. Drying is cost-effective; it reduces the mass and enables easy storage for years. Also the dry fruit bodies can be crushed, which decreases the space required for storage. From 10 kg of fresh mushrooms, which is about 1 kg in dry weight, 60 g of anthraquinone dyestuff powder was obtained [40]. This means 6 % of pigment content, which is fairly good. For plants, the colorant content from the dry mass is often under 5 %; e.g. *Rubia tinctorum* contains anthraquinones of around 2–3 % of the dry weight [41].

Due to the isolation process, pigments were obtained in powder form. This is useful when storing the colorants as well as when actually using them, because the concentration of the dyestuff in the dye solution can be measured. This enhances the repeatability of dyeings.

Räisänen *et al.* applied pure anthraquinone dyestuffs, which were obtained by the previously described isolation procedure, as mordant, acid and disperse dyes both for natural and synthetic fibres using modern dyeing techniques in laboratory scale. The results showed that the dyeing procedures were accomplished successfully. The dye uptakes of natural anthraquinones were in most cases over 70 % and for the disperse dyeing processes even nearly 100 % [42–44]. The colour fastness of the dyed materials varied from low to excellent

Table 11.2 *Color-fastness properties of fungal anthraquinones in different textile fibres. Samples 1–5 are wool and silk dyed with the crude extract of C. sanguinea, 6–15 wool and polyamide dyed with pure emodin and dermocybin using the mordant dyeing procedure; 16 and 17 refer to polyester dyed with emodin and dermocybin using the disperse dyeing method. The washing-fastness results in the colour change are given in two ways: as grey-scale rating converted from ΔE values [45] and as visually assessed grey-scale rating. LF = lightfastness, vis ass = visual assessment, r = redder, bl = bluer*

No.	Fabric/dye mordant	LF	Washingfastness							
			Color change		Staining					
			from ΔE	vis ass	WO	CA	SE	PA	CO	CTA
	WO / *C. sanguinea*									
1	No mordant	5	4–5	2	2				3	
2	$KAl(SO_4)_2$	5	4	3, r	3/4				3/4	
	SE / *C. sanguinea*									
4	No mordant	2	2–3	2, bl	4	4	4	4	3/4	4
5	$KAl(SO_4)_2$	3	3	2–3, r	4	4/5	4/5	4	4	4
	WO / emodin									
6	No mordant	3	1	3–4, r	3				3/4	

(continued)

Table 11.2 *(Continued)*

No.	Fabric/dye mordant	LF	Washingfastness							
			Color change		Staining					
			from ΔE	vis ass	WO	CA	SE	PA	CO	CTA
7	$KAl(SO_4)_2$	3	1	3, r	3				3/4	
8	$FeSO_4$	2	1	3, r	3/4				4	
	PA / emodin									
9	No mordant	2	1–2	1–2	2/3	4/5	4	2/3	4	3/4
10	$KAl(SO_4)_2$	2	1	1	3	3/4	3/4	2/3	4	3/4
11	$FeSO_4$	2	1	1	3	4/5	4/5	3	4	3/4
	WO / dermocybin									
10	No mordant	2	4–5	4	3/4				3/4	
11	$KAl(SO_4)_2$	1	3–4	3	4/5				4/5	
12	$FeSO_4$	3	4	3–4, r	4/5				4/5	
	PA / dermocybin									
13	No mordant	1	4–5	4, bl	4/5	4	4/5	4/5	4	4/5
14	$KAl(SO_4)_2$	2	2–3	2–3, r	4/5	4	4/5	4/5	4	4/5
15	$FeSO_4$	2	3–4	3, r	4/5	4/5	4/5	4/5	4/5	4/5
16	PES / emodin	7, r	4	5	5/PES				5	
17	PES / dermocybin	6	4–5	5	5/PES				5	

among the different fibres and different dyeing techniques (Table 11.2). Excellent results were obtained, especially in the case of high-temperature dyed polyester. As a conclusion, it was stated that natural anthraquinone compounds can produce bright hues with good colour-fastness properties in both natural and synthetic textile materials. Natural resources could be utilized more widely for anthraquinone dyestuff production, because the total synthesis is quite cumbersome, frequently requiring toxic reagents, such as heavy metal salts. Also, utilization of natural processes diminishes the formation of toxic side products [46].

11.3.5 Other Colorants of Fungi

11.3.5.1 Pulvinic Acid Derivatives: the Yellow-Orange Colorants

Pulvinic acid and its derivatives form yellow and orange-red colours in fungi of Boletales species and lichen *Sticta aurata*. Dimers of pulvinic acids have been found in the brown outer layer of the cap of *Boletus erythropus* [34]. Pulvinic acid derivatives also cause the characteristic blueing of the flesh of many fungi of Boletales sp. when the mushroom is cut and the flesh is in touch with air [1, 5, 34].

11.3.5.2 Brown from Hadiones

The famous dyeing mushroom *Pisolithus arhizus*, which produces brown colours, contains dyestuffs of the badione group. Norbadione A occurs in the fungus as potassium salts and exists in amounts of over 25 % of the dry weight of the fungus. Badione A, bisnorbadio-quinone A and pisoquinone may also occur as minor compounds [1].

11.4 Colour-fastness of Lichen and Mushroom Dyes

The properties of the products have to fulfil the requirements of consumers. Even in past centuries, the textile trade required quality control, testing and regulations [47]. Consumers are aware of factors, such as cost, expected wear of life and preferred methods of cleaning, that affect their satisfaction with a product, but these factors are often connected with the level of colour-fastness exhibited by the item. One of the important properties of dyed material is the fastness of the colour. The level of colour-fastness desired in a product is determined primarily by its intended use. Low light-fastness may be acceptable for items that are inexpensive and whose lifetime is short, e.g. paper. Also textiles that are not directly exposed to light, e.g. underwear and tights, may require lower light-fastness. In contrast, excellent light-fastness is demanded especially for furnishing fabrics, since they are expected to last for many years.

Light-fastness is among the most important properties that must be satisfactory for textile dyes. On fading, most samples show loss and dulling of the colour, accompanied by a change of hue [48]. Most natural dyes initially fade rapidly and then continue to fade at a slower rate. Most of the light-fast natural dyes, such as the anthraquinones, fade gradually at a constant rate [49]. A significant number of natural dyes, when treated in weakly alkaline solution, show marked changes in hue. Natural compounds contain OH groups, the deprotonation of which is pH-dependent. In many cases, ionization has an effect on the colour of the compound, as it causes a bathochromic shift in the electron absorption spectrum of the dye molecule [50, 51]. The tendency to change colour makes it necessary to have accurate knowledge of the pH of alkaline solutions for the cleaning of textiles dyed with natural dyestuffs. Apparently, repeated washings under mildly alkaline conditions do not have a cumulative effect and therefore after several washings textiles reach a state of stability [48].

The ISO (International Organization for Standardization) has published several standards for colour-fastness tests. The most commonly used method to test the light-fastness is the ISO 105-B02, in which the dyed samples are exposed to a xenon arc lamp together with the standard blue scale. The rating values vary in the range 1–8, 8 being the best value. The light-fastness properties of the woollen materials dyed with the blue terphenylquinone colorants of *Sarcodon squamosus* varied from 3 to 5, which is rather low for textile dyeing purposes [32]. The low fastness properties of terphenylquinone compounds had been noticed in ancient times among lichen dyes [1].

Natural compounds contain plenty of O or OH groups through which colorants may attach directly to the textile fibres. The other possibility is to form coordination linkages with the metal atoms of the mordants. It seems that the low light- and washing-fastness of many lichen and fungal dyes is not due to lack of appropriate functional groups, but the general three-dimensional structure of the compound. The colorants have one or more single bonds in their chemical structure and these single bonds enable the molecules to twist. Due to twisting, the functional groups end up in a position where they are unable to form linkages between each other or between the molecule and the fibre. Hence, molecules are not flat. It is known that for dyeing purposes flat and firm dyestuffs, like anthraquinones, are better than very three-dimensional ones. Flat molecules may approach the fibre surface more closely and thus attach to the fibres more strongly. It is evident that anthraquinones have outstanding fastness properties compared to other chemical groups in fungal and lichen dyestuffs. Dyeing is a manifold phenomenon in which physical and chemical interactions exist. The close attachment of dye molecules with the fibres enable charge-transfer forces, van der Waals forces, hydrophobic interaction and even π–π interaction to occur. In addition to covalent, coordination, ionic and hydroxyl bonds, these have a remarkable effect for the fixation [52].

11.5 New Approaches to Lichen and Fungal Natural Dyes

Recently, more focus has been laid on studies to exploit other unique properties of natural compounds. Fungi and lichens have been used for centuries as medicines and now these medicinal properties have been recognized as a growing opportunity in textile markets [18].

Emodin is probably the most widely distributed anthraquinone, being present in moulds, higher fungi, lichens, flowering plants and insects [53]. For pharmaceutical studies, emodin has been obtained from fungi that have been cultivated on a culture medium. *Penicilliopsis clavariaeformis* Solms was used in the experiments, but some *Cortinarius* species were mentioned as being worth investigation [54].

Recently, natural compounds have been found to have UV-protective properties and when applied to textiles could protect human skin from hazardous solar ultraviolet radiation [55]. It seems that cleavage of hydrogen bonds in the molecules of the natural dyes contributes to their capacity to absorb UV radiation [55]. Also the antimicrobial activity of natural colorants brings new ideas to be considered when developing protective clothing or home textiles such as bedlinen, towels and carpets. According to Singh *et al.*, the concentration of dye increased the bactericidal properties of dyed textile. The antimicrobial properties seem to be related to the dye structure, especially the presence of functional groups in it [33]. Modern textiles are required to have high degrees of chemical and photolytic stability in order to maintain their structure and colour. Antimicrobial agents are added particularly to natural fibres to make them resistant to biological degradation. In this respect, the antimicrobial activity of natural colorants could be taken into account [33].

It is important to test the dyes also for their cytotoxicity and carcinogenic properties. For example, some anthraquinones, such as emodin and its chloroderivatives, have been found to have slightly toxic or carcinogenic properties [56–58]. Unfortunately, only a few natural anthraquinones have been tested. On the other hand, emodin, for example, has been used as a laxative and thus its toxic properties seem to be mild.

References

1. D. Cardon, *Natural Dyes: Sources, Tradition, Technology and Science*, Archetype Publications, London, 2007.
2. K. D. Casselman, *Lichen Dyes and Dyeing: The New Source Book*, Dover Publications, Mineola, New York, 2001.
3. K. D. Casselman, Magic, mystery and mayhem: lichen dyes old and new, in *Seminar 12.8.2002*, Publication of the Natural Dyes Product Research and Development Project, EVTEK Institute of Art and Design, Vantaa, 2002.
4. C. J. Cooksey, Tyrian purple: 6,6′-dibromoindigo and related compounds, *Molecules*, **6**, 736–769 (2001), available at http://mdpi.org/molecules/papers/60900736.pdf, 29 January 2008.
5. H. Schweppe, *Handbuch der Naturfarbstoffe*, Ecomed, Landsberg/Lech, 1993.
6. S. Grierson, D. G. Duff and R. S. Sinclair, Natural dyes of the Scottish Highlands, *Textile History*, **16**, 23–43 (1985).

7. G. W. Taylor, Reds and purples: from the Classical World to Pre-Conquest Britain, in P. Walton and J.-P. Wild (eds), *Textiles in Northern Archaeology*, NESAT III: Textile Symposium in York, 6 September 1987, Northern European Symposium for Archaeological Textiles, Monograph 3, IAP Archetype Publications, London, 1990.
8. A. Wallert, Fluorescent assay of quinone, lichen and redwood dyestuffs, *Studies in Conservation*, **31**, 145–155 (1986).
9. A. R. Hall and P. R. Tomlinson, Dyeplants from Viking York, *Antiquity*, **58**, 58–60 (1984).
10. K. Høiland, Garnfargning med sopp. Historikk, *Ottar*, **152**, 3–4 (1985).
11. E. Sundström, *Värjäämme Yrteillä, Sienillä ja Jäkälillä*, Kustannus-Mäkelä, Karkkila, 2003.
12. C. Sundström and E. Sundström, *Sienivärjäys*, Otava, Helsinki, 1983.
13. J. Linder, *Svenska Färge-konst. Med Inlåndska Orter, Grås, Blommor, Blad, Låf, Barkar, Rötter, Wårter och Mineralier*, 2nd edition, Lars Salvius, Stockholm, 1749.
14. J. P. Westring, *Svenska Lafvarnas Färghistoria eller Sättet att Använda dem till Färgning och annan Hushållsnytta*, Carl Delén, Stockholm, 1805.
15. J. W. Palmstruch, *Svensk botanik,* C. Delén, Stockholm, 1802–1819.
16. M. Rice, *Mushrooms for Color*, 2nd edition, Mad River Press, Eureka, California, 1980.
17. O. Saastamoinen, Forest policies, access rights and non-wood forest products in northern Europe, *Unasylva* 198. Non-wood Forest Products and Income Generation. FAO – Food and Agriculture Organization of the United Nations, in http://www.fao.org/docrep/x2450e/x2450e00.HTM, 7 August 2007.
18. N. Durán, M. F. S. Teixeira, R. De Conti and E. Esposito, Ecological-friendly pigments from fungi, *Critical Rev. Food Sci. Nutrition*, **42**, 53–66 (2002).
19. K. Perumal, E. Sumathi and S. Chandrasekarenthiran, Microbial pigments for textile dyeing, Presentation at the International Symposium/Workshop on *Natural Dyes*, 5–12 November 2006 Hyderabad, India, 2006.
20. J. W. Deacon, *Modern Mycology*, 3rd edition, Blackwell Science, London, 2000.
21. T. E. Brandrud, H. Lindström, H. Marklund, J. Melot and S. Muskos, *Cortinarius Flora Photographica*, Cortinarius HB, Matfors, 1990.
22. V. Hintikka, Some types of mycorrhizae in the humus layer of conifer forests in Finland, *Karstenia*, **14**, 9–11 (1974).
23. E. Danell and N. Fries, Methods for isolation of *Cantharellus* species and the synthesis of ectomycorrhizae with *Picea abies, Mycotaxon*, **38**, 141–148 (1990).
24. E. Danell, Formation and growth of the ectomycorrhiza of *Cantharellus cibarius, Mycorrhiza*, **5**, 89–97 (1994).
25. E. Danell and F. Camacho, Successful cultivation of the golden chantarelle, *Nature*, **385**, 303 (1997).
26. A. Ohta, Ability of ectomycorrhizal fungi to utilize starch and related substrates, *Mycoscience*, **38**, 403–408 (1997).
27. Q. A. Mandeel, A. A. Al-Laith and S. A. Mohamed, Cultivation of oyster mushrooms (*Pleurotus* spp.) on various lignocellulosic wastes, *World Journal of Microbiology and Biotechnology*, **21**, 601–607 (2005).
28. K. Perumal, Cost effective process of artificial cultivation of *Ganoderma lucidum* basidiomata on sugarcane bagasse using polythene bags, 30 December 2005, submitted to Provisional Specification in the Patent Office Government of India, IPR Buildings, Chennai, Patent Application 1969/CHE/2005.
29. Finfood, Oppimateriaali, Ylli, Viljellyt sienet, available at http://www.finfood.fi/finfood/ffom.nsf, 11 February 2008.
30. M. Gill and W. Steglich, Pigments in fungi (Macromycetes), in W. Herz, H. Grisebach, G. W. Kirby and Ch. Tamm (eds), *Progress in the Chemistry of Organic Natural Products*, **51**, 125–174 (1987).
31. H. Johannesson, S. Ryman, H. Lundmark and E. Danell, *Sarcodon imbricatus* and *S. squamosus* – two confused species, *Mycol. Res.*, **103**, 1447–1452 (1999).
32. S. Hiltunen, *Blue Colours from Mushrooms* (in Finnish), Master's Thesis, Department of Home Economics and Craft Science, University of Helsinki, 2005.
33. R. Singh, A. Jain, S. Panwar, D. Gupta and S. K. Khare, Antimicrobial activity of some natural dyes, *Dyes Pigm.*, **66**, 99–102 (2005).

34. G. Gruber, *Isolierung und Strukturaufklärung von Chemotaxonomisch Relevanten Sekundärmetaboliten aus Höheren Pilzen, Insbesondere aus der Ordnung der Boletales*, PhD Thesis, Faculty of Chemistry and Pharmacy, Ludwig-Maximilians University Munich, 2002.
35. P. R. N. Carvalho and C. H. Collins, HPLC determination of carminic acid in foodstuffs and beverages using diode array and fluorescence detection, *Chromatogr.*, **45**, 63–66 (1997).
36. F. E. Lancaster and J. F. Lawrence, High-performance liquid chromatographic separation of carminic acid, α- and β-bixin, and α- and β-norbixin, and the determination of carminic acid in foods, *J. Chromatogr.*, **A 732**, 394–398 (1996).
37. H. Hoeffkes and B. Bergmann, Natural hair dye powder for mixing with water prior to hair application, German Patent 19600225, 1997.
38. J. F. Grollier, G. Rosenbaum and J. Cotteret, Use of hydroxyanthraquinones for the coloration of human keratin fibres, US Patent 4602913, 1986.
39. W. Steglich and W. Lösel, Pilzpigmente X. Anthrachinon-glucoside aus *Cortinarius sanguineus* (Wulf. ex Fr.) Wünsche, *Chem. Ber.*, **105**, 2928–2932 (1972).
40. P. H. Hynninen, R. Räisänen, P. Elovaara and E. Nokelainen, Preparative isolation of anthraquinones from the fungus *Cortinarius sanguineus* using enzymatic hydrolysis by the endogenous β-glucosidase, *Z. Naturforsch.*, **55c**, 600–610 (2000).
41. *Forum Färberpflanzen*, Gulzower Fachgespräche, Vol. 18. Fachagentur Nachwachsende Rohstoffe e.V. Gulzow, 2001.
42. R. Räisänen, P. Nousiainen and P. H. Hynninen, Emodin and dermocybin natural anthraquinones as high-temperature disperse dyes for polyester and polyamide, *Textile Res. J.*, **71**, 922–927 (2001).
43. R. Räisänen, P. Nousiainen and P. H. Hynninen, Emodin and dermocybin natural anthraquinones as mordant dyes for wool and polyamide, *Textile Res. J.*, **71**, 1016–1022 (2001).
44. R. Räisänen, P. Nousiainen and P. H. Hynninen, Dermorubin and 5-chlorodermorubin natural anthraquinone carboxylic acids as dyes for wool, *Textile Res. J.*, **72**, 973–976 (2002).
45. K. J. Smith, Colour-order systems, colour spaces, colour difference and colour scales, in R. McDonald (ed), *Colour Physics for Industry*, 2nd edition, Society of Dyers and Colourists, Bradford, UK, 1997.
46. R. Räisänen, *Anthraquinones from the Fungus Cortinarius sanguineus as Textile Dyes*, PhD Thesis, Department of Home Economics and Craft Science Research, Report 10, University of Helsinki, Dark, Vantaa, 2002.
47. J. Zahn, 'Bei Leibesstraff sollen die Tücher...' Ein (zweiter) Blick zurück in Sachen Qualität und Qualitätssicherung, *Textilveredlung*, **27**, 358–365 (1992).
48. D. G. Duff, R. S. Sinclair and D. Stirling, Light-induced colour changes of natural dyes, *Studies in Conservation*, **22**, 161–169 (1977).
49. P. C. Crews, The fading rates of some natural dyes, *Studies in Conservation*, **32**, 65–72 (1987).
50. P. F. Gordon and P. Gregory, *Organic Chemistry in Colour*, Springer-Verlag, Heidelberg, 1983.
51. W. Barz and J. Köster, Turnover and degradation of secondary (natural) products, in E. E. Conn (ed.), *The Biochemistry of Plants. A Comprehensive Treatise,* Vol. 7, *Secondary Plant Products*, Academic Press, London, 1981.
52. E. R. Trotman, *Dyeing and Chemical Technology of Textile Fibres*, 6th edition, Edward Arnold, Sevenoaks, 1990.
53. R. H Thomson, *Naturally Occurring Quinones*, 2nd edition, Academic Press, London, 1971.
54. L. Roth, C. Roth, H. K. Frank, H. T. Bruester and P. Wernet, Use of fungi that produce anthraquinone or dianthrone dyestuffs – for the manufacture of antiviral preparations, German Patent 3913040, 1990.
55. X. X. Feng, L. L. Zhang, J. Y. Chen and J. C. Zhang, New insights into solar UV-protective properties of natural dye, *J. Clean. Prod.*, **15**, 366–372 (2007).
56. M. Gross, R. Levy and H. Toepke, Zum vorkommen und zur Analytik des Mycotoxins Emodin, *Nahrung*, **28**, 31–44 (1984).
57. P. A. Cohen, J. B. Hudson and G. H. N. Towers, Antiviral activities of anthraquinones, bianthrones and hypericin derivatives from lichens, *Experientia*, **52**, 180–183 (1996).
58. A. von Wright, O. Raatikainen, H. Taipale, S. Kärenlampi and J. Mäki-Paakkanen, Directly acting geno- and cytotoxic agents from a wild mushroom *Cortinarius sanguineus, Mutation. Res.*, **269**, 27–33 (1992).

12

Tannins and Tannin Agents

Riitta Julkunen-Tiitto and Hely Häggman

12.1 Introduction

Tannins are plant phenolics, which is one of the most widespread and complex group of chemical compounds in the plant kingdom. Tannins are constitutive polymeric polyphenols, and fairly recently were also found to be induced by damage (e.g. Constabel, 1999). They have relatively high molecular mass with a typical aromatic ring stucture with hydroxyl substituents. These chemically reactive components easily form conjugations with other biomolecules via their hydroxyl moieties. They are able to precipitate proteins, enzymes, carbohydrates and alkaloids, and are used industrially for millenia to convert raw animal hides into leather. The most effective molecular weight for a tanning process is 500–2000 (3000) and the binding activity is dependent on the tannin structure (Seigler, 1998).

In plants two different groups of tannins are found: hydrolysable tannins (HT) and proanthocyanidins (syn. condensed tannins, CT). Brown algae produce phlorotannins, which consist of units of phloroglucinol (e.g. Porter, 1989; Waterman and Mole, 1994). Evolutionarily older CTs (about 370 million years) preceded hydrolysable tannins and are found in ferns, gymnosperms and angiosperms, especially in dicotyledonous grasses, shrubs and trees (Harborne, 1984; Seigler, 1998). CTs tend to be correlated with a woody growth form of higher plants (e.g. Waterman and Mole, 1994). HTs are found widely in angiosperms and also in some green algae (Harborne, 1984; Waterman and Mole, 1994). Several species also contain complex mixtures of both tannins, such as *Acasia*, *Acer*, *Quercus* and *Betula* (e.g. Reed and Mueller-Harvey, 1987; Julkunen-Tiitto *et al.*, 1996; Tikkanen and Julkunen-Tiitto, 2003).

Tannins are found most often in plant cell vacuoles, and are concentrated in epidermal tissues (e.g. Siegler, 1998). Depending on the species, a high proportion of tannins are also found in weakly soluble or insoluble forms, obviously mainly bound into cell wall material

Handbook of Natural Colorants Edited by Thomas Bechtold and Rita Mussak

(e.g. Strack *et al.*, 1989; Keski-Saari *et al.*, 2005, 2007). Tannins are found in wood, bark, leaves, buds, floral parts, seeds and roots, and usually at high constitutive levels by dry weight of plants. HTs are especially abundant in roots of several woody species (e.g. Keski-Saari and Julkunen-Tiitto, 2003; Keski-Saari *et al.*, 2007). A high content of tannins is found in galls induced in plants by herbivores (e.g. Nyman and Julkunen-Tiitto, 2001). The content of tannins will vary over plant phenology and ontogeny. Often young juvenile plants/organs will contain higher amounts of soluble HTs while CTs predominate to the

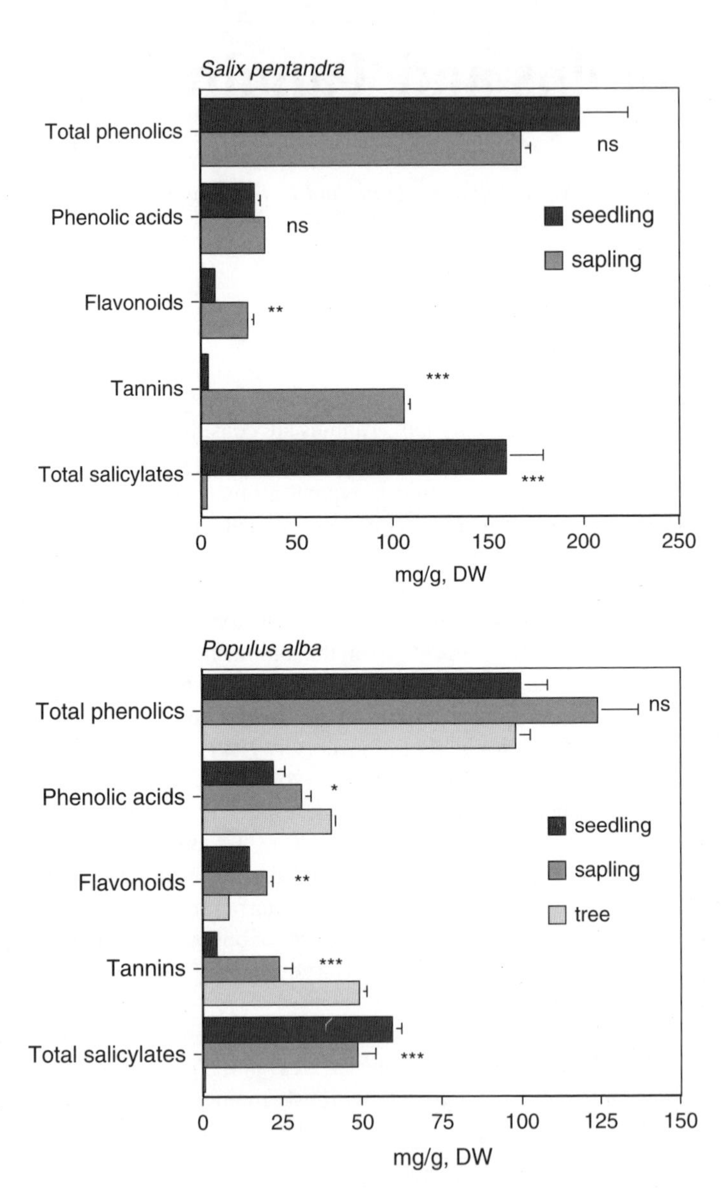

Figure 12.1 *Ontogenetic change of phenolics in the leaves of Salix pentanda and Populus alba*

Plate 1 *Medieval Portuguese illumination, dating from a 12–13th century, Lorvão 15, fl. 50 kept at Torre do Tombo (Lisbon). Dark blues were painted with indigo, whereas for the backgound the inorganic and precious pigment lapis-lazuli was used (See Figure 1.1)*

Plate 2 *Textile object 2045 from the Coptic period in the Ismaillia Museum, Egypt (photo picked up by the author) (See Figure 3.1)*

Plate 3 *Textile Ottoman 12014 in the Islamic Art Museum, Cairo, Egypt (photo picked up by the author)* *(See Figure 3.2)*

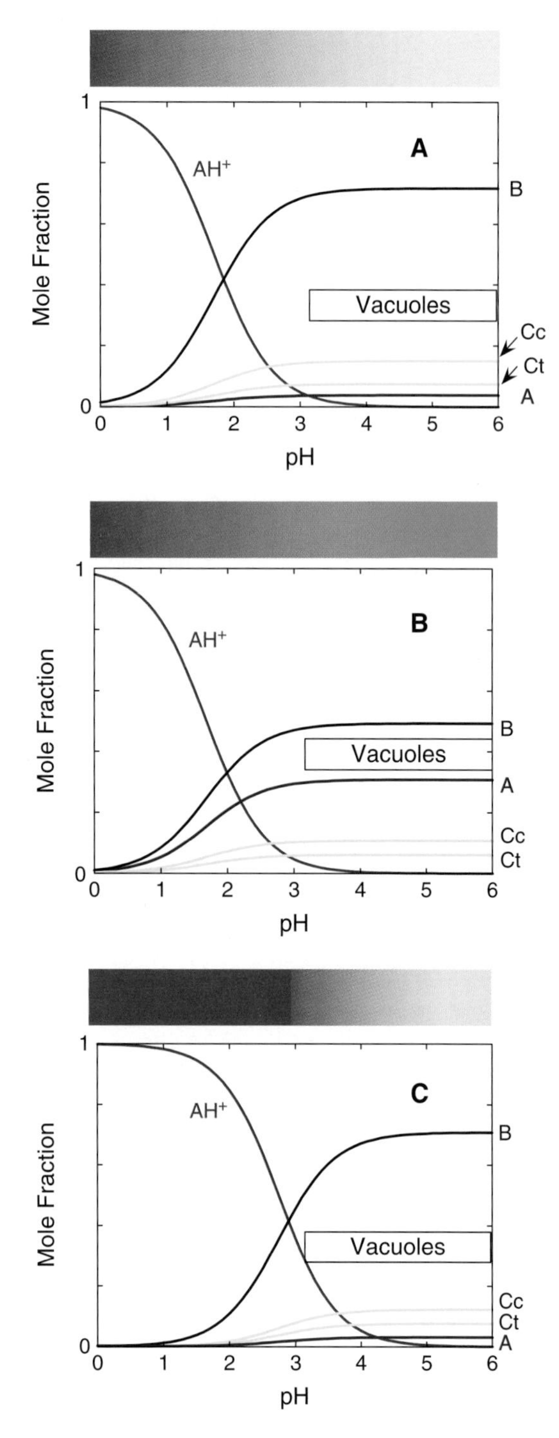

Plate 4 *Mole fraction distribution of the several species of malvidin 3,5-diglycoside:* ***A****-free* ($pK'_a = 1.7$); ***B****-simulation in the case of co-pigmentation with A for a copigment concentration of 0.1 M and an association constant,* $K'_{cp} = 100\ M^{-1}$; ($pK'_a = 1.7 = 1.68$) ***C****-the smae as in B but when exclusive copigmentation with* $\boldsymbol{AH^+}$ *occurs,* $K_{cp} = 100\ M^{-1}$ ($pK'_a = 1.7 = 2.74$). *The band at the top simulates the colour that is obtained at different pH values. Reprinted with permission from Reference [17a], © 2000, Elsevier Science (See Figure 9.2)*

Plate 5 *Wood species that had more or less been successfully dyed with indigotin. Species are arranged according to increasing blue components (b* gets more negative); from left to right in each row: oak, nut, cherry, ash, beech heartwood, beech, Douglas fir, alder, pine, poplar, maple and birch (See Figure 17.5)*

Plate 6 *Wood species that had more or less been successfully dyed with chlorophyll sodium salt. Species are arranged according to increasing green components (a* gets lower); from left to right in each row: beech heartwood, beech, alder, birch, maple and poplar (See Figure 17.6)*

Plate 7 *Variation of black alder (Alnus glutinosa Gaertn.) wood color due to six independent convection drying processes (See Figure 17.7)*

Plate 8 *Color impression of black alder (Alnus glutinosa Gaertn.) steamed for 0, 1, 2, 3, 4, 8 (first row) and 10, 12, 14, 16, 18 and 30 days (second row) (See Figure 17.8)*

Plate 9 *Thermally treated black alder (Alnus glutinosa Gaertn.) after 1, 2 and 3 days duration (top down) (See Figure 17.9)*

Plate 10 *Untreated (left) and ammoniated (right) oak (Quercus sp.) wood from the same individual tree (See Figure 17.10)*

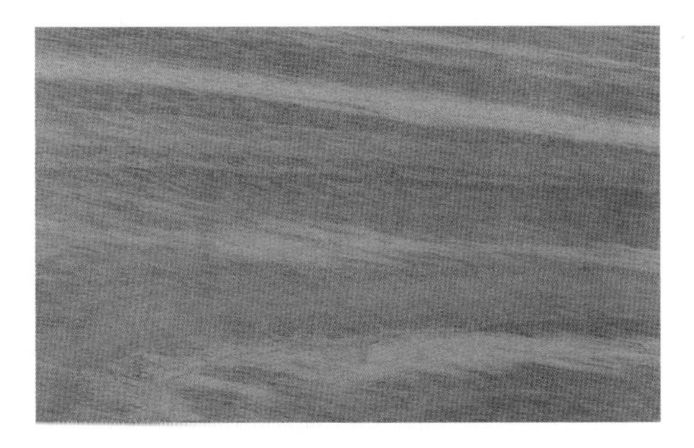

Plate 11 *Ammoniated cherry (Prunus avium L.) wood (See Figure 17.11)*

Plate 12 *Comparison of bleached (first and third rows) and untreated wood (second and fourth rows). Species from left to right: poplar, maple, spruce, beech, beech heartwood, birch, alder, ash, pine, cherry, oak and nut (See Figure 17.13)*

Plate 13 *Color impression of black alder (Alnus glutinosa Gaertn.) steamed for 0, 1, 2, 3, 4, 8 (first row) and 10, 12, 14, 16, 18 and 30 days (second row) after 24 h UV radiation (See Figure 17.15)*

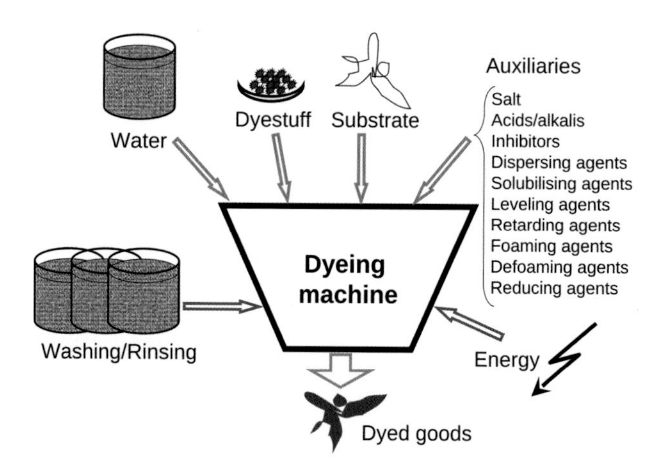

Plate 14 *Dyeing process: requirements for a general dyeing process (See Figure 18.1)*

end of the growing season and in later ontogenetic phases (e.g. Feeny, 1970; Scalbert and Haslam, 1987; Salminen *et al.*, 2001; Tikkanen and Julkunen-Tiitto, 2003). Even a linear increase was found in CTs in willows (*Salix sericea* and *S. eriocephala*) with the age of the seedlings (Fritz *et al*,, 2001). Interestingly, the content of soluble CTs is increased in the leaves of two to three year old deciduous *S. pentandra* and *Populus alba* (Salicaceae) saplings, followed by a strong decrease of smaller molecular weight salicylates and flavonoids (Figure 12.1). Moreover, the genotype-dependent tannin metabolism in plants is modified by different environmental factors, such as soil nutrients (e.g. Mansfield *et al.*, 1999; Keski-Saari *et al.*, 2005). For instance, nitrogen fertilization is found to decrease CTs in several woody species (Mattson *et al.*, 2004), while CO_2 elevation alone enhances the accumulation of CTs in *Betula pendula* and *S. myrsinifolia* (e.g. Julkunen-Tiitto *et al.*, 1993; Mattson *et al.*, 2004). The enhancement of temperature alone reduces CTs in the leaves *of B. pendula* and *B. pubescens* and are unaffected in the leaves of *S. myrsinifolia* (Veteli *et al.*, 2002; Veteli, 2003). The interaction effect between different environmental factors has been reported for CTs. In *B. pendula* and *S. myrsinifolia* leaves, CO_2 elevation enhanced CT levels mostly at the highest level of nitrogen fertilizer (Julkunen-Tiitto *et al.*, 1993; Mattson *et al.*, 2004), while temperature elevation had an opposite effect on CTs at elevated CO_2 and fertilizer level. Though condensed tannins have some absorption in the region of ultraviolet-B (UV-B) radiation they seem not to have a response in the leaves of *Salix* sp. and *Betula* sp. leaves on enhanced UV-B radiation (Tegelberg *et al.*, 2003; Keski-Saari *et al.*, 2005).

Generally, tannins have a great importance at cellular level as well as in plant physiology and ecology. They may be fytotoxic, antimicrobic, antimycotic, antiviral, allelopathic, phytoalexins, signal molecules and also free radical scanvengers.

12.2 Chemical Structure, Biosynthesis and Degradation

Plants produce a high variety of tannins, leading to an unlimited number of different chemical profiles in different plant organs. The variation in structure of tannins within plant genera may be as great as the variation among genera (Ayres *et al.*, 1989). Of three major types of tannins HTs are grouped as gallotannins or ellagitannins, the former are mainly gallic (**1**) and the latter ellagic acids (**2**), yielding on hydrolysis tannins having Mr 500–2000 (Figure 12.2). Both gallic and ellagic acids or other degradation products have also been detected upon hydrolysis of some HTs. Gallic acids in gallotannins are esterified with the hydroxyl groups of central polyol moiety, which is very often glucose but also in some species such as shikimic acid, quinic acid, glucitol, quercitol or cyclitol (Gross, 1992; Salminen, 2002). The simplest and most widespread gallotannin is pentagalloylglucose (ß-1,2,3,4,6-pentagalloyl-*O*-D-glucopyranoside (PGG, **3**), which is proposed to be the precursors of both higher gallo- and ellagitannins. Ellagitannins are esters of hexahydroxydiphenic acids (HHDP, **4**), and are formed by oxidative C–C coupling (most common between C-4/C-6 and C-2/C-3) between the adjacent galloyl residues of related gallotannins (**5**). Hexahydroxydiphenic acid occurs together with gallic acids in certain hydrolysable tannins (in oak and chestnut). In gallotannins additional galloyl residues are attached by depside bonds, usually at the C-3 position of gallic acid (Gross, 1992; Siegler 1998).

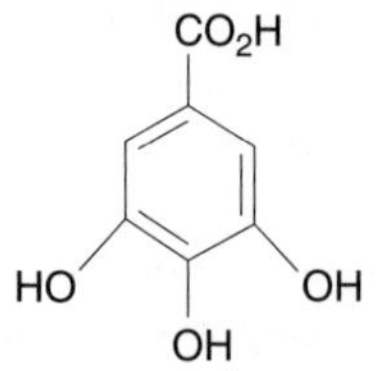

gallic acid (1)

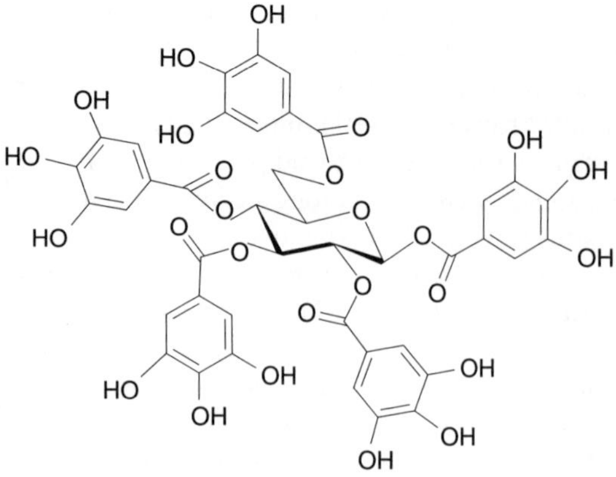

pentagalloylglucose (3)

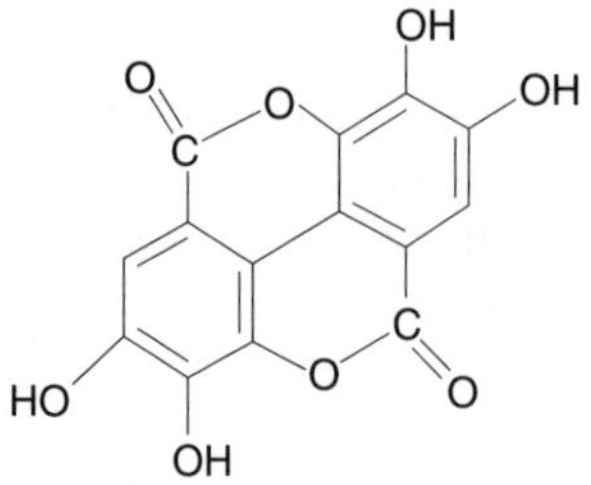

ellagic acid (2)

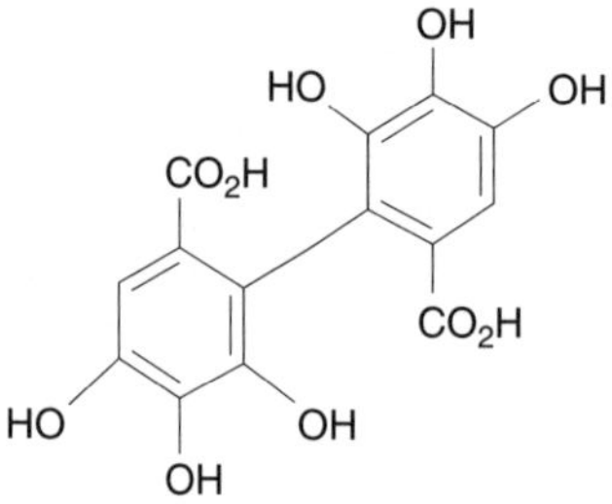

hexahydroxydiphenic acid (4)

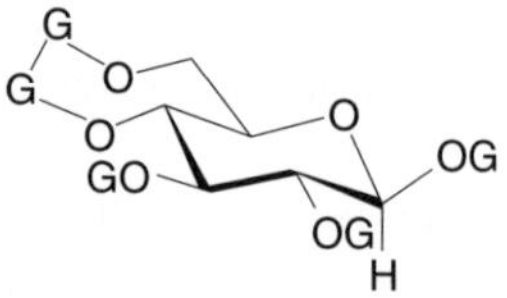

ellagitannin eugeniin (5)
G = gallic acid
G-G = 4

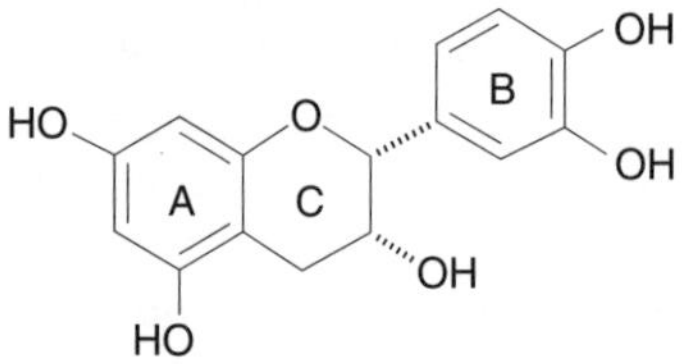

(–)-epicatechin (6)

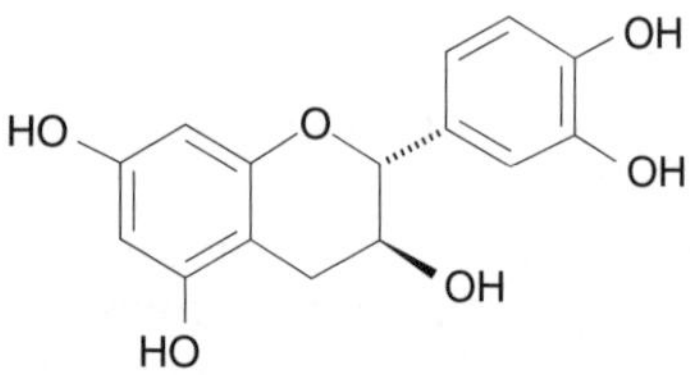

(+)-catechin (7)

Figure 12.2 *Some monomers of hydrolysable and condensed tannins and β-pentagalloylglucose*

CTs are oligomers or polymers of 15-carbon polyhydroxyflavan-3-ol monomer units (mostly (-)-epicatechin or (+)-catechin) (**6, 7**) linked by an acid labile C–C bond between C-4 and C-8, which gives linear structures, or between C-4 and C-6, which gives a branched globular structure (Figure 12.2) (Porter, 1992). CTs with C-4/C-8 linkages are more widespread that those of C-4/C-6 (Hagerman and Butler, 1991; Haslam, 1998).

Recently, Cheynier (2005) has reported ether bonds between O-7 and C-2 and between O-5 and C-2. C-3 hydroxylated monomers are the most common (Porter, 1992). CTs are able to be distinguished based on their hydroxylation pattern in the A- and B-rings. The procyanidin unit (3′,4′-hydroxyl substitution in the flavan B-ring) is the most widespread, while propelargonidin polymers (4′-hydroxyl substitution in the flavan B-ring) are rare (Haslam, 1998). Mixed procyanidin and prodelfinidin have been found to co-occur when procyanidin units predominate. Generally, proanthocyanidin polymers are polydisperse and may contain molecules with degrees of polymerization of 2 to more than 50. They are able to bind with ellagitannins as well as gallotannins (Porter, 1992; Haslam, 1998).

Due to their complex chemical nature (oligomers, polymers) to reveal the biosynthesis and polymerization pathways of tannins has been and still is a challenging task. However, most of the main biosynthetic pathways leading to tannins are now well established and many of the enzymes have been cloned (e.g. Peter and Constabel, 2002). Current knowledge indicates that precursors for the biosynthesis of hydrolysable tannins are supplied most probably via the shikimic acid pathway as 3-dehydroshikimic acids (e.g. Gross, 1992; Ossipov *et al.*, 2002). The direct aromatization of 3-dehydroshikimic acid gives rise to gallic acid (3,4,5-trihydroxyl benzoic acid), which is esterified to a core polyol. Monogalloylglucose is the starting compound, which is activated by UDP glucose. UDP glucose activated gallic acid, ß-*O*-glucogallin, functions as both donor and acceptor in the sequential addition of galloyl units to the hydrolysable tannin (Gross, 1992) (Figure 12.3).

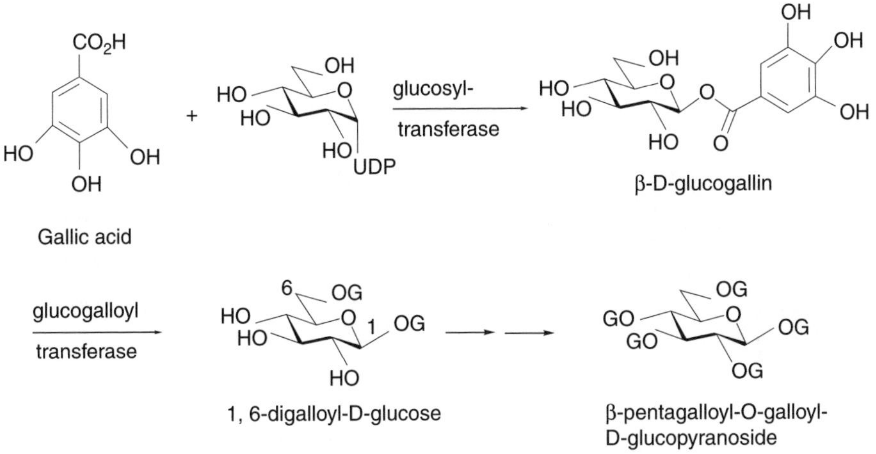

Figure 12.3 *Biosynthesis of hydrolysable tannins*

The biosynthesis of CTs occurs through two different pathways: the phenylpropanoid pathway and the polyketide pathway. The polyketide pathway supplies malonyl moieties for A-ring formation in flavonoid biosynthesis. Via the phenylpropanoid pathway aromatic amino acid, L-phenylalanine is nonoxidatively deaminated to E-cinnamate by phenylalanine ammonia-lyase (PAL). Chalcone synthase catalyses the formation of the chalcone by condensing 4-coumaroyl-CoA with three molecules of malonyl-CoA. This C_{15} skeleton product possesses a monohydroxyl B-ring, typical for nearly all flavonoids, giving rise to

CTs. The chalcone isomerase catalyses the ring closure of chalcone to give rise to flavanone, which is transformed by flavanone 3-hydroxylase to dihydroflavonol and further to leucoanthocyanidin (flavan-3,4-*cis*-diol) catalysed by dihydroflavonol 4-reductase. Flavan-3,4-diol is finally reduced by flavan-3,4-diol 4-reductase to catechin (flavan-3-ol) or via anthocyanidin by anthocyanidin reductase to epicatechin (flavan-3-ol) (e.g. Strack, 1997; Petersen *et al.*, 1999; Xie *et al.*, 2003). Flavan-3-ols serves as a starting unit for still weakly clarified polymerization for CTs (Waterman and Mole, 1994; Haslam, 1998) (Figure 12.4).

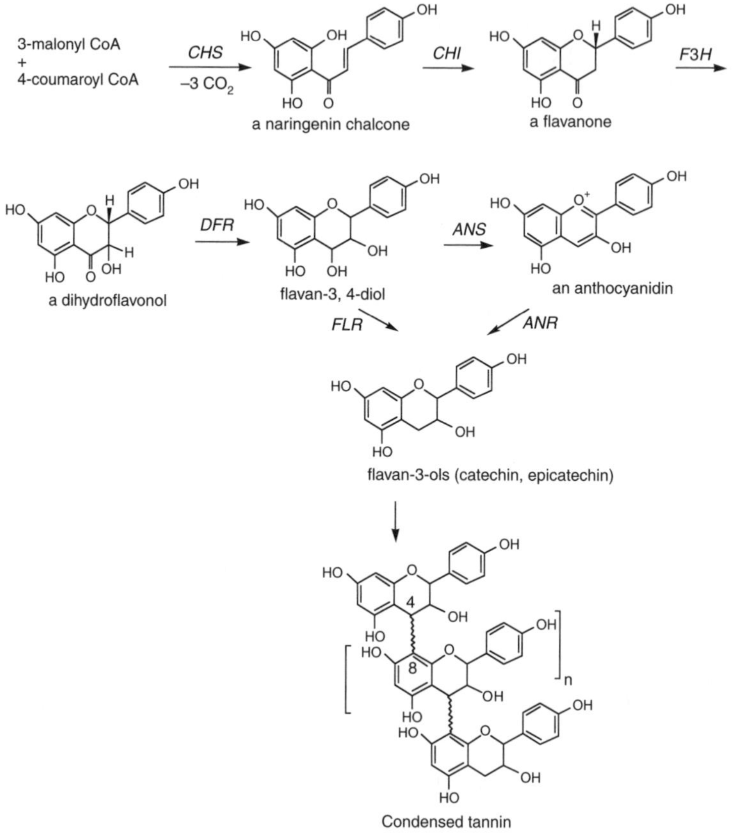

Figure 12.4 *Biosynthesis of condensed tannins*

According to Xie *et al.* (2003) the critical step in the synthesis of flavan-3-ol (epicatechin, leading to CTs production) is that it is catalysed by anthocyanidin reductase (ANR) enzyme encoded by the *Banyuls* gene. This discovery made the epicatechin-based CT production also possible in plant tissues, which do not usually synthesize these compounds, assuming that the tissue contains the necessary precursors, i.e. a functional anthocyanin biosynthetic pathway. Anthocyanin and/or CT pathways have been found to be controlled by over 25 transcription factors belonging to MYB, bHLH, WD40, WKRY, WIP, homeodomain and bMADS protein families (Dixon *et al.*, 2005). Recently, Xie *et al.* (2006) succeeded in metabolic engineering of CT in tobacco through co-expression of anthocyanidin reductase gene and the PAP1 MYB transcription factor.

CT and its precursor, flavan-3-ols, readily ionize in aqueous solution, and solubility of CTs is dependent on phenolate ion formation (Porter, 1992). In mild acid solutions, CTs decompose easily to give flavan-3-ols and quinone methide, while in stronger acid solutions quinone methide is in equilibrium with a carbocation. Carbocation will change to anthocyanidins in hot alcohol solutions (Porter, 1992). CTs are relatively unstable polymers, which may even decompose at natural pH of their aqueous solutions (pH 3–4) (Porter, 1992). Similarly, HTs are susceptible to oxidation at high pH. The ester bonds in HTs are also base and enzyme labile (Halsam, 1989). HTs are readily degraded to gallic acid or hexahydroxydiphenic acids (HHDP) and the core polyol. HHDP will spontaneously form the internal ester ellagic acid (Hagerman and Butler, 1989).

12.3 Properties of Tannins

The presence and intensity of activities underlined for tannins are determined by the structure of each tannin and structural type of tannin (Okuda *et al.*, 1989). The use of the chemical properties of tannins obviously dates back to the unwritten history of mankind. Tanning processes to make leather from animal skin was well developed in ancient civilizations from Babylon to China, and leather making belongs to the oldest human activities. Historically tannin use to cure leather dates back to 5000–6000 BC. One example is a leather funeral tent of an Egyptian queen dated at 1100 BC in a museum in Cairo (Hegert, 1989). Tannins (such as hemlock, chestnut, oak, quebracho, mangrove, wattle, etc.) are still important today in the leather industry.

Both CTs and HTs are able to bind and precipitate proteins. The astringent sensation (puckeriness and dryness, loss of lubrication in the mouth and throat) produced in the mouth is due to binding of tannins with mucoproteins (Bate-Smith, 1973; Haslam, 1989, 1998). The formation of soluble and insoluble tannin–protein complexes is caused by crosslinking the separate protein molecules by the phenol (association with hydrogen bonding between hydroxyls of tannins and the polar group of proteins).

Tannins seem to be most astringent to mammals, leading to antinutritional or even toxic effects; tannin–protein complexes are less digestible than free proteins. Tannins may also act by interacting with the gut wall and so affect the passage of nutrients (Hagerman *et al.*, 1992). Herbivores show adaptation to tannins; they are able to modify the gut environment to inhibit tannin reactions (high pH and surfactants, or even degrade the compounds). If gut reactions are not enough or have failed, tannins are able to be degraded after being absorbed in the liver (McArthur *et al.*, 1991; Buttler, 1993). One of the best-characterized defence mechanisms in animals is tannin-binding protein (TBP) of herbivore saliva (Mehansho *et al.*, 1987). Binding of tannins by TBP in the digestive tract could protect the inhibition of digestive enzymes and other essential physiological activities (Mole *et al.*, 1990). TBP is shown to contain an extraordinarily high content of proline, as well as glutamine and glycine, and is found in root vole (*Microtus oeconomus*), European moose and North American moose (*Alces alces*) (Juntheikki *et al.*, 1996).

The protein precipitating ability of different tannins varies widely and determines most of the physiological activities of herbal medicines, the taste of foodstuffs and beverages and the nutritional value of feeds and herbivore food selection and physiology (e.g. Harborne, 1988). Condensed tannins apparently function against the action of microbes and hydrolysable tannins act as a deterrent against chewing phytophagous insects or animals (e.g. Seigler, 1998). Tannins may also have positive effects as phagostimulants and feeding cues (Schults, 1989). Moreover, tannins are able to inhibit the activity of pathogens by complexing extracellular enzymes produced by pathogens or may interfer with the metabolism of the pathogen itself (Laks and McKaig, 1988; Seigler, 1998). In soil, tannins are reported to have a central role in humus formation by producing complexes that are resistant to microbial decay (Ya *et al.*, 1989).

CTs also have an important agronomic trait which protects ruminant animals from potential pasture bloat caused by high-leaf-protein-containing forages. By binding proteins in the rumen condensed tannins reduce the rate of fermentation and subsequent reduction in methane production and better nitrogen nutrition. The recent genome and transcriptome level information of CT biosynthesis and regulation may potentially produce biotechnological approaches to breeding CT-containing forages (such as alfalfa) (Xie *et al.*, 2006). However, the recent study failed to reduce enteric methane emissions from growing cattle when the feed contained up to 2 % of the dietary DM as quebarcho tannin extract, although the protein-binding effect of quebracho tannin extract was evident (Beauchemin *et al.*, 2007).

In addition to protein, CTs and HTs are able to bind alkaloids, carbohydrates and metals, such as iron, aluminium, copper, vanadium and calsium (Gaffney *et al.*, 1986; Porter, 1992; Haslam, 1998). Metal complexes will form readily in an aqueous solution of tannins. Metal complexation of tannins is a distinctive feature and in the past it has been a very important, but poorly investigated, feature. Over many centuries, the blue-black iron tannate complex was the main source of writing inks and also ancient Egyptians used the complex as a hair dye (Slabbert, 1992). Ferric ion complexes with *o*-diphenols are stable, highly coloured and may have negative features in leather manufacture where tannins are used. Contact with iron surfaces, iron-contaminated water or other tannin chemicals results in unsightly blue/grey stains (Slabbert, 1992). Aluminium (III) and boron (III) complexes with tannins (wattle) are colourless. Boric acid complex is reported to be water soluble while iron and aluminium complexes are insoluble. The complex of titanium (IV) with *o*-diphenols is highly yellow-coloured and is used to caste a golden colour on leather.

12.4 Chemical Activities of Tannins

Chemical activities of CTs are determined by the properties of A- and B-rings of the molecules. The A-ring is primarily responsible of the lability of the interflavonoid bond and the use of adhesives and the B-ring determines the properties of formation of metal complexes and antioxidant characteristics of tannins (Laks, 1989). Especially, a 5,7-dihydroxy A-ring is susceptible to interflavonoid bond cleavage under acidic or basic conditions (Hemingway, 1989). Also the interflavonoid bond between C-4 and C-8 is more easily cleaved (**5**) than that between C-4 and C-6, and CTs with the upper units having the

2,3-*cis* configuration (**6**) are more readily decomposed compared with their 2,3-*trans* analogues (Scalbert, 1992). Hydroxylation in the B-ring reflects functionally phenol, catechol and pyrogallol having one, two or three hydroxyls, respectively. Often the reactivity and complexity of reactions increase with addition of hydroxyls to B-rings. Acidity and reactivity to electrophiles and reducing power will increase if comparing phenol to catechol or pyrogallol (e.g. Laks, 1989).

12.5 Analysis of Tannins

12.5.1 Sample Preservation

The research of chemically and biochemically unstable tannins has suffered from methodological difficulties. Care should be taken in sample collection, prehandling, preservation and also quantification of tannin content found in plants. There are several studies available where preservation methods of plant material for tannin analyses have been screened. Generally, whenever possible, intact fresh plant material should be preferred for tannin extraction and analyses (e.g. Harborne, 1984; Julkunen-Tiitto, 1985; Julkunen-Tiitto and Sorsa, 2001; Mueller-Harvey, 2001). However, if analytical procedures necessitate sample preservation, Hagerman (1988) has recommended freeze-dryed or frozen material for CTs, while Orians (1995) preferred vacuum drying. Both freeze drying and vacuum drying were found more appropriate methods than air or oven drying for HTs of *Betula* leaves (Salminen, 2003). The degree of chemical changes during different drying methods is also strongly dependent on plant organs. Oven drying (less than 50 °C) did not change CT content of *Salix* bark but was deleterious for whole twig drying (Julkunen-Tiitto and Tahvanainen, 1989). Instead, heat drying (at higher than 60 °C) of *Salix* leaves induced a strong decrease in soluble CTs and (+) catechins (Julkunen-Tiitto, 1985; Julkunen-Tiitto and Sorsa, 2001), as did oven drying (at 40 °C) for CTs of *Crassula argentea* (succulent species) (Bate-Smith, 1977). If air drying is the only possibility direct sunlight should be avoided (Okuda *et al.*, 1989). Moreover, the storage time of dried material will have a strong effect on the extractable yield of tannins. For instance, Scalbert (1992) has reported that the storage of dried sorghum grain for 18 days decreased 15 % of the CT content and 4 years of storage of oak wood decreased 40 % of the ellagitannins yield. Opposite results were obtained for the outbuilding storage of *S. myrsinifolia* dried shoots: there were no change in CT content after one year of storage (Julkunen-Tiitto *et al.*, 2005).

12.5.2 Extraction and purification

Generally, tannins are regarded as water soluble. However, the solubility of the higher molecular weight tannins are difficult for any solvent (Hagerman and Butler, 1991). The complex formation is suggested to be one reason for the difficulties of isolation or insolubility of tannins (Porter, 1992). The soluble and insoluble fractions vary among species, sometimes the insoluble fraction may be even higher than the soluble one and is seen as giving a high anthocyanidin (butano/HCl method) or ellagic acid (acid degradation) yield on hydrolysis of the extracted residue (Porter, 1992; Scalbert, 1992; Keski-Saari *et al.*, 2007). However, the extraction conditions, especially temperature, will have a marked effect on tannin extractability. Though most tannins are water soluble, water is not the best extraction solvent for tannins (Scalbert, 1992). Aqueous organic solvent mixtures are recommended for extraction of tannins (e.g. Hagerman and Butler, 1994). The most often used are acetone (e.g. 50 or 70 %) and methanol (e.g. 50 %) as extractants. Aqueous acetone is often shown to be a better extractant than aqueous methanol (Julkunen-Tiitto, 1985; Scalbert, 1992; Hagerman and Butler, 1994). Especially, methanol is not recommended for gallotannins because it may cleave the pepside bonds, even at room temperature, but not in acidified methanol (Haslam *et al.*, 1961). Instead, methanol is reported to be a better solvent for tannins of low molecular mass and if plant tissues contain a high amount of enzymes (Mueller-Harvey, 2001). Generally, tannin yields obtained by the solvent mixtures are variable and are dependent on plant species, prehandling of plant material and also on individual tannin structures (Salminen, 2003).

Tannin extracts often contain complex mixtures of other phenolics and nonphenolics (polysaccharides and proteins) (Scalbert, 1992). Solvent fractionation with diethyl ether has been used to separate small molecular mass phenolics (such as gallic acid or catechin) from HTs or CTs (Singleton, 1974). The more feasible method is to use sorption of tannins by Sephadex LH20, Toyopearl HW-40 or Diaion HP-20 (Okuda *et al.*, 1989; Hagerman and Butler, 1994). However, it has been reported that on Sephadex, labile oligomeric HTs may be degraded and large molecular mass CTs and HTs may be bound so strongly that they cannot be eluted with acetone (Okuda *et al.*, 1989; Mueller-Harvey, 2001). After isolation HTs are fairly stable at room temperature (in the air), while CTs (as well as catechin and epicatechin units) will slowly oxidize to give reddish colours (Okuda *et al.*, 1989).

12.5.3 Quantification of Tannins

There are general and specific functional group assays and binding assays to measure the content of tannins in plant material (Hagerman and Butler, 1989, 1991; Porter, 1989). When using colorimetric tests their nonspecificity for tannins should be considered, as recommended by Hagerman and Butler (1989). When the mixture of CTs is measured by colorimetric tests this will at best give results only for semi-quantitative comparisons (Mueller-Harvey, 2001).

General assays for CTs and HTs include the Folin–Dennis (1912), Folin–Ciocalteau (1927) and Prussian blue (Price and Butler, 1977) assays, which are nonspecific and detect all other phenolics and other oxidizable compounds in samples. The Folin–Ciocalteau method gives about 30 % more colour than the Folin–Dennis assay (e.g. Julkunen-Tiitto, 1985), while proteins interfere only weakly in the Prussian blue method (Hagerman and Butler, 1991). Both of the latter methods are recommended for total phenolic determination (Waterman and Mole, 1994).

The most often used and the more specific functional group assay for CTs is anthocyanidins assay (Porter *et al.*, 1986) and vanillin addition assay (Broadhurst and Jones, 1978). In anthocyanidin assay (or acid butanol assay) proantocyanidins are depolymerized oxidatively via the interflavan bond by hydrochloric acid in butanol, and subsequent autooxidation of formed carbocations yields red antocyanidins (Porter *et al.*, 1986). Colour formation is sensitive to water and acetone in the reaction mixture (Waterman and Mole, 1994). Transition metals (such as Fe(III)) as autooxidation elicitors enhance colour intensity (Scalbert, 1992). The colour yield is roughly proportional to the nonterminal flavonoid units in CTs, and it also is dependent on the stability of the interflavan bond (Hagerman and Butler, 1991, 1994).

In the vanillin (aromatic aldehyde) test, vanillin forms a red-coloured complex with the meta-substituted ring of CTs and flavanols (such as catechin and epicatechin). Vanillin forms a complex via the flavonol A-ring at C-6 and essential for the reaction is the double bond between C-2 and C-3 and the absence of carbonyl at C-4 (Sarkar and Howarth, 1976). Colour formation is very sensitive to temperature, light and the presence of water. A small amount of water and less than 1.5 °C difference in temperature may drastically alter the colour yield of the reaction (Porter, 1989; Hagerman and Butler, 1994; Waterman and Mole, 1994). Waterman and Mole (1994) recommended the use of some other methods in addition to the vanillin test for CTs.

Rhodanine and potassium iodate tests have been developed for quantification of gallotannins. The rhodanine test is specific for gallic acid and gives a red colour (Inoue and Hagerman, 1988). The amount of gallic acid is determined before and after hydrolysis of the gallotannins. The weakness of the method is that different gallotannins in plants contain different numbers of gallic acid units, and ellagitannins may also contain gallic acids (Mueller-Harvey, 2001). The potassiun iodate test yielding a pink colour (Bate-Smith, 1977) is shown to be unreliable for complex mixtures of tannins or other polyphenols (Hagerman *et al.*, 1997). Hartzfeld *et al.* (2002) have improved the method and introduced a methanolysis step followed by oxidation of formed methyl gallate with KIO_3. The colour test for ellagitannin is based on the hydrolysis and reaction of ellagic acid with sodium nitrite to give a red nitrosylated chromophore (Wilson and Hagerman, 1999). The method is oxygen, glassware purity and time sensitive.

The ability of tannins to precipitate proteins has also been exploited in quantitative determinations and in biological activity assays of the tannin (containing samples) (Hagerman and Buttler, 1989). The quantification is based on the amount of tannin precipitated by standard protein protocols and in biological activity assays; either the amount of precipitated protein is determined or the quality of the complexes formed is analysed (e.g. Makkar *et al.*, 1987). Protein-binding responses are dependent on tannin and protein size and structure and reaction conditions (pH, temperature, reaction time, among others) (Hagerman and Butler, 1978; Martin and Martin, 1982; Scalbert, 1992). Proline-rich proteins have a very high affinity for tannins (e.g. Seigler, 1998). The precipitation recovery is also dependent on protein/tannin ratios in the reaction mixture (Hagerman and Robbins, 1987). According to Hagerman and Butler (1989), precipitation assays must be interpreted cautiously. Instead, the simple radial diffusion method developed by Hagerman (1987) is recommended for determining insoluble tannin–protein (bovine serum albumin) complexes where the tannin sample diffuses through protein-containing agar gel. The area of the ring is proportional to the amount of tannin used.

Modern high-performance liquid chromatography (HPLC) accompanied with diode array (DAD) and electrospray ionization (ESI) mass spectrometry (MS) are the powerful methods used for quantification of hydrolysed units of tannins (such as gallic acid, ellagic acid and procyanidins) and also HTs (e.g. Bianco *et al.*, 1988; Salminen *et al.*, 1999).

Normal (NP) and reversed phase (RP) columns can be used for the separation of mixtures of tannins. RP-HPLC is an appropriate method only for shorter-chain CTs, up to trimers. RP-HPLC used for CTs of *Acacia* sp., *Euclea schimperi* and *Pterolobium stellatum* showed poorly resolved hills eluting between 10 and 20 min (Mueller-Harvey, 2001). Recently, Karonen *et al.* (2006) have developed a quantitative NP-HPLC method of CTs for *Betula* leaf extracts as a late eluting peak in the chromatogram. Salminen (2003) successfully used RP HPLC-ESI-MS (positive and negative modes) for the quanification of HTs in *B. pubescens* leaves. He also compared the rhodanine colour test with the HPLC-MS method and found that the colour test will give an underestimated total concentration of galloylglucoses in an extract. The high selectivity of HPLC-ESI-MS also effectively excludes possible false detection in a complex mixture where HTs may form as a minor fraction together with other more abundant phenolics (Salminen, 2002).

12.6 Use, Toxicology and Safety Aspects of Tannins

The role of tannins in plants is under active research, which has widened their role from the originally presented metabolic by-products to more specific roles. For instance, different tannins may result in different responses within the same herbivore species and the same tannin structure may cause different responses in different herbivores (Rautio *et al.*, 2007). These results underline the importance of considering the potential toxicological effects of tannins or tannin residues on a case by case basis.

Tannins are also a part of our everyday life. They are present in many common plant foods such as sorghum, barley, cranberry, wine and tea (Chung *et al.*, 1998; Foo *et al.*, 2000; Dixon *et al.*, 2005). They are important quality factors with a gastronomic value, e.g. in wines, but they are also considered because of their therapeutic or functional food properties. Tannins are strong antioxidants protecting against oxidative tissue damage (Shi *et al.*, 2003, and references therein), they protect animal cells against UV radiation induced damage (Carini *et al.*, 2000) and they have also been found effective in urinary tract infections (Foo *et al.*, 2000). Chung *et al.* (1998) have pointed out that the overall effects of tannins on human health is vast, but sometimes also conflicting. In the review of Chung *et al.* (1998), the authors indicate that the incidences of certain cancers have been related to consumption of tannin-rich foods, while other reports indicate that carcinogenic activity might rather be related to components associated with tannins. Vice versa, CTs have also been shown to prevent the growth of human cancer cells *in vitro* (Tamagawa and Fukushima, 1998; Ye and Krohn, 1999). In addition, the health-promoting role of tannins has been increasingly studied and it has been shown that tannins are able to reduce the levels of serum lipids and blood pressure as well as to improve brain health (Jerez *et al.*, 2007). These findings indicate a wide commercial potential of plant-derived tannins.

Tannins are used for very many applications in the pharmaceutical (as gallic acid, pyrogallic acid, sulfanilamide drugs), food (as clarifying agent and flavour stability for beer, wine), feed, dyestuff (as fixing agents), cosmetic and chemical (slurry-treating agents in the petrochemical industry, in treating agents for various industrial manufactures) industries. There are a number of commercial suppliers for different tannin preparations (e.g. http/www.biomatnet.org/secure/Fair/S340.htm, http://www.ferco-dev.com/home.htm, http//www.indofinechemical.com/). Furthermore mimosa bark (*Acacia mearnsii*), quebracho wood (*Schinopsis balansae*), pine bark (*Pinus radiata*), oak bark (*Quercus* sp.), grape (*Vitis vinifera*) and wattle (*Acacia mearnsii*), are used for industrially produced tannin extracts and preparations (Cadahía *et al.*, 1998, Garnier *et al.*, 2000; Vivas *et al.*, 2003). At present the most intensive use of tannins is in the manufacture of wood adhesives and leather (Taiwo and Ogunbodede, 1995; Garnier *et al.*, 2000; López-Suevos and Riedel, 2003; Li *et al.*, 2004).

Natural tannin-based adhesives for wood composites have been commercially available for several decades and, especially, the antimicrobial role of tannins has been exploited in the wood composite industry and adhesives (Chow, 1977; Pizzi, 1982; Li and Maplesden, 1998). Building materials such as plywood, other laminated veneer products, fibreboard and fibreglass insulation are major uses of adhesives (Lu and Shi, 1995). Tannin binders are hardened by formaldehyde or other chemicals to form resins (e.g Chow, 1977; Taiwo and Ogunbodede, 1995; Pizzi, 2006). Petrochemical-derived formaldehyde-based wood adhesives are still today the dominating ones in European markets. However, formaldehyde emissions, their suspected carcinogeny (Sowumni *et al.*, 2000; Li *et al.*, 2004) and increasing costs due to high petroleum prices and natural gas are directing a search to find alternative adhesive materials. A formaldehyde-free adhesive has been constructed by mixing condensed tannin and polyethyleneimine together, providing a suitable adhesive for the ply-wood composite having properties of high water resistance (Li *et al.*, 2004). Moreover, new technologies such as nanotechnology are developing in order to solve critical problems related to adhesion of tannins in wood material.

The use of natural tannins in leather production has an ancient history and the old methods are still valid in the present-day industry. In addition to natural leather tanning agents, toxic chrome salts, aldehydes, sulfonated polymers and other resins are also in use (http//www.bionatnet.org/secure/FP5/S1180.htm; Haslam, 1998). Based on the conclusions of an EU project, chrome salts and natural tannins cover more than 90 % of the total leather manufactured, of which the former covers about 70 %. Chrome salts are generally used to manufacture soft leather. Natural tannins have a high protein precipitation ability and strong astringent effect, which give possibilities to manufacture the most fine leather for different kinds of accessories, among other things (Haslam, 1998).

Employment in the leather and leather manufacturing products industry has been associated with various diseases caused by biological, toxicological and carcinogenic agents. During the tanning process, infections may occur as the hide serves as a medium for micro-organisms and yeasts (http://www.ilo.org/encyclopedia/). Skin disorders and allergic dermatitis have also been diagnosed. Tannery workers have the potential for exposure to numerous known or suspected occupational carcinogens (reviewed in http://www.ilo.org/encyclopedia/), and, for instance, excess risks for pancreatic, testicular, sinonasal and bladder cancers have been reported. In the leather industry especially, tannins have been regarded as possible etiological agents for nasal cancer (Battista *et al.*, 1995).

Considering the safety aspects, chemical usage as well as waste disposal practices have been improved since the early history of the leather industry (Durant *et al.*, 1990), the wastewater treatment plants have improved practices and chemical regulations are more stringent than they used to be. Also technological improvements are in progress. One of the attempts has been to prepare tanning agents by copolymerization of vegetable tannins and/or waste lignocellulosic materials with acrylics and to apply these pre-polymerized tanning agents for leather fabrication. During the research a novel, high-performance melamine–urea–formaldehyde (MUF) resin was successfully used in copolymerization, producing leather products of a quality corresponding to that of the chrome-treated leather (http://www.biomatnet.org/secure/FPS/F1180.htm). Also the biological wastewater treatment has found to minimize environmental damage and to cause a decrease in overall effluent toxicity (Vijayaraghavan and Murthy, 1997).

In the textile industry heavy metal ions or natural or synthetic tannins have been used as assisting chemicals or auxiliary substances to be attached to the fibres permanently (so-called mordant dyes). The mordants allow many natural dyes, which would otherwise wash out, to be utilized by the industry. In recent work Koyunluoglu *et al.* (2006) studied the effect of pre-ozonation of commercial textile tannins on their biodegradability and toxicity. The results indicated no significant changes in acute toxicity for natural tannin whereas the inhibitory effect of synthetic tannin could be completely eliminated after 40 min ozonation at a rate of 1000 mg/h and a fair biodegradability improvement was found.

Similar to the leather industry, occupational exposure to tannins and tannin agents potentially leading to health problems also may occur in the wood industry from wood dust, in such work as mill work and furniture, cabinet and pattern making. The health problems may include inhalation and oral intake of wood dust or dermal exposure (e.g. Mark and Vincent, 1986). Hardwood dust exposure is associated with the risk of sinonasal cancer (Nylander and Dement, 1993) and in the European Union all hardwood dust is considered carcinogenic (Council of European Union, Directive 1999/38/EC, 1999).

In conclusion, plant tannins, utilized for a variety of applications, are derived from diverse plant sources characterized by different chemical structures. In many applications synthetic tannins are also utilized. In the leather tannin industry most of the toxic effects of the tanning procedures are caused especially by chrome and other hazardous ions needed by the industry. Therefore, when considering the toxicity and safety aspects of the applications where either tannins or tannin agents are used, they should be evaluated case by case.

References

Ayres, M. P., Clause Jr, T. P., MacClean, S. F., Redman, A. M. and Reichardt, P. B. (1989) Diversity of structure and antiherbivore activity in condensed tannins, *Ecology*, **78**, 1696–1712.

Bate-Smith, E. C. (1973) Haemanalysis of tannins – the concept of relative astringency, *Phytochemistry*, **12**, 907–912.

Bate-Smith, E. C. (1977) Astringent tannins from *Acer* species, *Phytochemstry*, **16**, 1421–1426.

Battista, G., Compa, P., Orsi, D., Norpoth, K. and Maier, A. (1995) Nasal cancer in leather workers: an occupational disease, *J. Cancer Res. Clin. Ongol.*, **121**, 1–6.

Beauchemin, K. A., McGinn, S. M., Martinez, T. F. and McAllister, T. A. (2007) Use of condensed tannin extract from quebarcho trees to reduce methane emissions from cattle, *Journal of Animal Science*, **85**(8), 1990–1996.

Bianco, M. A., Handaji, A. and Savolainen, H. (1988) Quantitative analysis of ellagic acid in hardwood samples, *Sci. Total Environ.*, **222**, 123–126.

Broadhurst, R. B. and Jones, W. T. (1978) Analysis of condensed tannins using acidified vanillin, *J. Sci. Food Agric.*, **29**, 788–794.

Buttler, L. G. (1993) Polyphenol and herbivore diet selection and nutrition, in A. Scalbert (ed.), *Polyphenolic Phenomena*, INRA Editions, Paris, pp. 149–154.

Cadahía, E., Conde, E., Fernández de Simón, B. and Concepción Garcá-Vallejo, M. (1998) Changes in tannic composition of reproduction cork *Quercus suber* throughout industrial processing, *J. Agric. Food Chem.*, **46**, 2332–2336.
Carini, M., Aldini, G., Bombardelli, E., Morazzoni, P. and Maffei Facino, R. (2000) UVB-induced hemolysis of rat erythrocytes: protective effect of procyanidins from grape seeds, *Life Sci.*, **67**, 1799–1814.
Cheynier, V. (2005) Polyphenols in foods are more complex than often thought, *Am. J. Clin. Nutr.*, **81**, 223S–229S.
Chow, S. (1977) Adhesives development in forest products, *Wood Sci. Technol.*, **17**, 1–11.
Chung, K.-T., Wong, T. Y., Wei, C.-I., Huang, Y.-W. and Lin, Y. (1998) Tannins and human health: a review, *Critical Reviews in Food Science and Nutrition*, **38**, 421–464.
Constabel, C. P. (1999) A survay of herbivore-inducible defensive proteins and phytochemicals. Pathogens, in A. A. Agrawaal, S. Tuzun and E. Bent (eds), *Induced Plant Defenses against Herbivores*, APS Press, St Paul, pp. 137–166.
Council of European Union (1999) Council Directive 1999/38/EC of 29 April 1999, *Official Journal of the European Communities*, **L138**, 66–69.
Dixon, R. A., Sharma, S. B. and Xie, D. (2005) Proanthocyanidins – a final frontier in flavonoid research? *New Phytol.*, **165**, 9–28.
Durant, J. L., Zemach, J. J. and Hemond, H. F. (1990) The history of leather industry waste contamination in the Aberjona watershed: a mass balance approach, *Civil Engineering Practice*, **5**, 41–66.
Feeny, P. (1970) Seasonal changes in oak leaf tannins and nutrients as a cause of spring feeding by winter month caterpillars, *Ecology*, **51**, 565–581.
Folin, O. and Ciocalteu, V. (1927) On tyrosine and tryptophane determination in proteins, *J. Biol. Chem.*, **27**, 627–650.
Folin, O. and Dennis, W. (1912) On phosphotungstic–phosphomolybdic compounds as colour reagents, *J. Biol. Chem.*, **12**, 239–343.
Foo, L. Y., Lu, Y., Howell, A. B. and Vorsa, N. (2000) The structure of cranberry proanthocyanidins which inhibit adherence of uropathogenic P-fimbriated *Escherichia coli in vitro, Phytochemistry*, **54**, 173–181.
Fritz, R. S., Hochwender, C. G., Lewkiewicz, D. A., Bothwell, S. and Orians, C. M. (2001) Seedling herbivory in slugs in a willow hybrid system: developmental changes in damage, chemical defense, and plant performance, *Oecologia*, **129**, 87–97.
Gaffney, S. H., Martin, R., Lilley, T. H., Haslam, E. and Magnolato, D. (1986) The association of polyphenols with caffeine and alpha- and beta-cyclodextrin in aqueous media, *J. Chem. Soc. Comm.*, 107–109.
Garnier, S., Pizzi, A., Vorster, O. C. and Halasz, L. (2000) Comparative rheological characteristics of industrial polyflavonoid tannin extracts, *J. Applied Polymer Sci.*, **81**, 1634–1642.
Gross, G. G. (1992) Enzymatic synthesis of gallotannins and related compounds, in H. A. Stafford and R. K. Ibrahim (eds), *Phenolic Metabolism in Plants*, Plenum Press, New York, pp. 297–324.
Hagerman, A. E. (1987) Radial diffusion method for determining tannin in plant extracts, *J. Chem. Ecol.*, **14**, 437–449.
Hagerman, A. E. (1988) Extraction of tannins from fresh and preserved leaves, *J. Chem. Ecol.*, **14**, 453–462.
Hagerman, A. E. and Butler, L. G. (1978) Protein precipitaion method for quantitative determination of tannins, *J. Agr. Food Chem.*, **26**, 809–812.
Hagerman, A. E. and Butler, L. G. (1989) Choosing appropriate methods and standards for assaying tannins, *J. Chem. Ecol.*, **15**, 1795–1810.
Hagerman, A. E. and Butler, L. G. (1991).Tannins and lignins, in G. A. Rosenthal and M. R. Berenbaum (eds), *Herbivores: Their Interactions with Secondary Plant Metabolites*, Vol. 1, Academic Press, New York, pp. 355–388.
Hagerman, A. E. and Butler, L. G. (1994) Assay of condensed tannins or flavonoid oligomers and related flavonoids in plants, *Meth. Enzymol.*, **234**, 429–437.
Hagerman, A. E. and Robbins, C. T. (1987) Implications of soluble tannin–protein complexes for tannin analysis and plant defense mechanisms, *J. Chem. Ecol.*, **13**, 1243–1259.
Hagerman, A. E., Robbins, C. T., Weerasuriya, Y., Wilson, T. C. and McArthur, C. (1992) Tannin chemistry in relation to digestion, *J. Range Management*, **45**, 57–62.

Hagerman, A. E., Shao, Y. and Johnson, S. (1997) Methods for determination of condensed and hydrolyzable tannins, in ACS Symposium Series 662, pp. 209–222.
Harborne, J. B. (1984) *Phytochemical Methods: A Guide to Modern Techniques of Plant Analysis*, Chapman & Hall, New York.
Harborne, J. B. (1988) *Introduction to Ecological Biochemistry*, 3rd edition, Academic Press, New York.
Hartzfeld, P. W., Forkner, R., Hunter, M. D. and Hagerman, A. E. (2002) Determination of hydrolyzable tannins (gallotannins and ellagitannins) after reaction with potassium iodate, *J. Agric. Food Chem.*, **50**, 1785–1790.
Haslam, E. (1989) *Plant Polyphenols, Vegetable Tannins Revisited*, Cambridge University Press, Cambridge.
Haslam, E. (1998) *Practical Polyphenols, From Structure to Molecular Recognition and Physiological Action*, Cambridge University Press, Cambridge.
Haslam, E., Howarth, R. D., Mills, S. D., Rogers, H. J., Armitage, R. and Searle, T. (1961) Gallotannins, Part II. Some esters and depsides of gallic acid, *J. Chem. Soc.*, 1836–1842.
Hegert, H. L. (1989) Hemlock and spruce tannins: an Odyssey, in R. W. Hemingway and J. J. Karchesy (eds), *Chemistry and Significance of Tannins*, Plenum Press, New York, pp. 3–20.
Hemingway, R. W. (1989) Reactions at the interflavavonoid bond of proanthocyanidins, in R. W. Hemingway and J. J. Karchesy (eds), *Chemistry and Significance of Tannins*, Plenum Press, New York, pp. 265–283.
Inoue, K. H. and Hagerman, A. E. (1988) Determination of gallotannins with rhodanine, *Analytical Chemistry*, **169**, 363–369.
Jerez, M., Selga, A., Sineiro, J., Torees, J. and Numez, M. J. (2007) A comparison between bark extracts from *Pinus pinaster* and Pinus radiata: antioxidant activity and procyanidin composition, *Food Chemistry*, **100**, 439–444.
Julkunen-Tiitto, R. (1985) Phenolic constituents in the leaves of Northern willows: methods for the analysis of certain phenolics, *J. Agric. and Food Chem.*, **33**, 213–217.
Julkunen-Tiitto, R. and Sorsa, S. (2001) Testing the drying methods for willow flavonoids, tannins and salicylates, *J. Chem Ecol.*, **27**(4), 779–789.
Julkunen-Tiitto, R. and Tahvanainen, J. (1989) The effects of sample preparation methods of extractable phenolics of Salicaceae species, *Planta Medica*, **55**, 55–58.
Julkunen-Tiitto, R., Tahvanainen, J. and Silvola, J. (1993) Increased CO_2 and nutrient status changes affect phytomass and the production of plant defensive secondary chemicals in *Salix myrsinifolia (Salisb.), Oecologia*, **95**, 495–498.
Julkunen-Tiitto, R., Rousi, M., Bryant, J., Sorsa, S., Keinänen, M. and Sikanen, H. (1996) Chemical diversity of several Betulaceae species: comparison of phenolics and terpenoids in northern birch stems, *Trees*, **11**, 16–22.
Julkunen-Tiitto, R., Rousi, M., Meier, B., Tirkkonen, V., Tegelberg, R., Heiska, S., Turtola, S. and Paunonen, R. (2005) Herbal medicine production: breeding and cultivation of salicylates producing plants in A. Jalkanen and P. Nygren (eds), *Sustainable Use of Renewable Natural Resources – From Principles to Practices*, University of Helsinki, Department of Forest Ecology Publications 34, e-publication.
Juntheikki, M.-R., Julkunen-Tiitto, R. and Hagerman, A. E. (1996) Salivary tannin-binding proteins in root vole (*Microtus oeconomus Pallas), Biochemical Systematics and Ecology*, **24**, 25–35.
Karonen, M., Ossipov, A., Sinkkonen, J., Loponen, J., Haukioja, E.and Pihlaja, K. (2006) Quantitative analysis of polymeric proanthocyanidins in birch leaves with normal-phase HPLC, *Phytochemical Analysis*, **17**, 149–156.
Keski-Saari, S. and Julkunen-Tiitto, R. (2003) Resource allocation in different parts of juvenile mountain birch plants: effect of nitrogen supply on seedling phenolics and growth, *Physiologia Plantarum*, **118**, 114–126.
Keski-Saari, S., Pusenius, J. and Julkunen-Tiitto, R. (2005) Phenolic compounds in seedlings of *Betula pubescens* and *B. pendula* are affected by enhanced UV-B radiation and different nitrogen regimes during early ontogeny, *Global Change Biol.*, **11**, 1180–1194.
Keski-Saari, S., Flack, M., Heinonen, J., Zon, J. and Julkunen-Tiitto, R. (2007) Phenolics during early development of *Betula pubescens* seedlings: inhibition of phenylalanine ammonia lyase, *Trees – Structure and Function*, **21**, 263–272.

Koyunluoglu, S., Aíslan-Alaton, I., Eremektar, G. and Gemirli-Babuna, F. (2006) Pre-ozonation of commercial textile tannins: effects on biodegradability and toxicity, *J. Environ. Science and Health, Part A*, **41**, 1873–1886.

Laks, P. E. (1989) Chemistry of the condensed tannin B-ring, in R. W. Hemingway and J. J. Karchesy (eds), *Chemistry and Significance of Tannins*, Plenum Press, New York, pp. 249–263.

Laks, P. E. and McKaig, P. A. (1988) Flavonoid biocides: wood preservatives based on condensed tannins, *Holzforschung*, **42**, 299–306.

Li, J. and Maplesden, F. (1998) Commercial production of tannins from radiata pine bark for wood adhesives, *IPENZ Transactions*, **25**(1), 46–52.

Li, J., Simonsen, G. J. and Karchesy, J. (2004) Novel wood adhesives from condensed tannins and polyethylenimine, *Int. J. Adhesion and Adhesives*, **24**, 327–333.

López-Suevos, F. and Riedel, B. (2003) Effect of *Pinus pinaster* bark extracts content on the cure properties of tannin-modified adhesives and on bonding of exterior grade MDF, *J. Adhesion Science and Technol.*, **17**(11), 1507–1522.

Lu, Y. and Shi, Q. (1995) Larch tannin adhesive for particleboard, *Holz als Roh- und Werkstoff*, **53**(1), 17–19.

McArthur, C., Hagerman, A. E. and Robbins, C. T. (1991) Physiological strategies of mammalian herbivores against plant defences, in R. T. Palo and C. T. Robbins (eds), *Plant Defences Against Mammalian Herbivory*, CRC Press, Boca Raton, Florida, pp. 103–114.

Makkar, H. P.S., Dawra, R. K. and Singh, B. (1987) Protein-precipitation assay for quantitation of tannins: determination of protein in tannin–protein complex, *Anal. Biochem.*, **166**, 435–439.

Mansfield, J. L. Curtis, P. S., Zak, D. R. and Pregitser, K. S. (1999) Genotypic variation for condensed tannin production in trembling aspen (*Populus tremuloides*, Salicaceae) under elevated CO_2 and high- and low-fertility soil, *American Journal of Botany*, **86**(8), 1154–1159.

Mark, D. and Vicent, J. (1986) A new personal sampler for airborne total dust in workplaces, *Annals of occupational Hygiene*, **30**, 89–102.

Martin, J. S. and Martin, M. M. (1982) Tannin assays in ecological studies: lack of correlation between phenolics, proanthocyanidins and protein-precipitating constituents in mature foliage of six oak species, *Oecologia*, **54**, 205–211.

Mattson, W. J., Kuokkanen, K., Niemelä, P., Julkunen-Tiitto, R., Kellomäki, S. and Tahvanainen, J. (2004) Elevated CO_2 alters birch resistance to *Lagomorpha* herbivores, *Global Change Biology*, **10**, 1402–1413.

Mehansho, H., Buttler, L. G. and Carlson, D. M. (1987) Dietary tannins and salivary proline rich proteins: interactions, inductions and defence mechanisms, *A. Rev. Nutr.*, **7**, 432–440.

Mole, S., Buttler, L. G. and Iason, G. (1990) Defence against dietary tannin in herbivores: a survey for proline rich proteins in mammals, *Biochem. Syst. Ecol.*, **18**, 287–293.

Mueller-Harvey, I. (2001) Analysis of hydrolysable tannins, *Animal Feed Science and Technology*, **91**, 3–20.

Nylander, L. A. and Dement, J. M. (1993) Carcinogenic effects of wood dust: review and discussion, *American Journal of Industrial Medicine*, **24**, 619–647.

Nyman, T. and Julkunen-Tiitto, R. (2001) Manipulation of phenolic chemistry of host plants by gall-inducing sawflies, *Proc. Natl Acad. Sci. USA*, **97**, 13184–13187.

Okuda, T., Yoshida, T. and Hatano, T. (1989) New methods in tannin analysis, *J. Nat. Prod.*, **52**, 1–31.

Orians, C. M. (1995) Preseerving leaves for tannin and phenolic glycoide analyses: a comparison of methods using three willow taxa, *J. Chem. Ecol.*, **21**, 1235–1243.

Ossipov, V., Salminen, J.-P., Ossipova, S., Haukioja, E. and Pihlaja, K. (2002) Gallic acid and hydrolyzable tannins are formed in birch leaves from an intermediate compound of the shikimate pathway, *Biochemical Systematics and Ecology*, **31**, 3–16.

Peter, D. J. and Constabel, C. P. (2002) Molecular analysis of herbivore-induced condensed tannin synthesis: cloning and expression of dihydroflavonol reductase from trembling aspen (*Populus tremuloides*), *The Plant Journal*, **32**, 701–712.

Petersen, M., Strack, D. and Matern, U. (1999) Biosynthesis of phenylpropanoids and related compounds, in M. Wink (ed.), *Biochemistry of Plant Secondary Metabolism*, Vol. 2, Sheffield Academic Press, pp. 150–221.

Pizzi, A. (1982) Pine tannin adhesives for particle boards. *Holz als Roh- und Werkstoff*, **40**, 293–301.

Pizzi, A. (2006) Recent developments in eco-efficient bio-based adhesives for wood bonding: opportunities and issues. *J. Adhesion Science and Tehcnology*, **20**(8), 829–846.

Porter, L. J. Hrstich, L. N. and Chan, B. C. (1986) The conversion of procyanidins and prodelphinidins to cyanidin and delphinidin, *Phytochemistry*, **25**, 223–230.

Porter, L. J. (1989) Tannins, in J. B. Harborne (ed.), *Methods in Plant Biochemistry*, Vol. 1, Academic Press, London, pp. 389–419.

Porter, L. J. (1992) Structure and chemical properties of the condensed tannins, in R. W. Hemingway and P. E. Laks (eds.), *Plant Polyphenols*, Plenum Press, New York.

Price, M. L. and Butler, L. G. (1977) Rapid visual estimation and specrtophotometric determination of tannin content of sorghum grain, *J. Agr. Food Chem.*, **25**, 1268–1273.

Rautio, P., Bergvall, U.A., Karonen, M. and Salminen, J.-P. (2007) Bitter problems in ecological feeding experiments: commercial tannin preparation and common methods for tannin quantifications. *Biochem. Syst. Ecol.*, **35**, 257–262.

Reed, J. D. and Mueller-Harvey, I. (1987) Charecterization of phenolic compounds, including flavonoid and tannins, of 10 Ethiopian browse species by high-performance liquid chromatography, *J. Sci. Food Agric.*, **39**, 1–14.

Salminen, J.-P. (2002) Birch leaf hydrolyzable tannins: chemical, biochemical and ecological aspects, *Annales Universitatis Turkuensis*, Ser AI, 290.

Salminen, J.-P. (2003) Effect of sample drying and storage, and choice of extraction solvent and analysis method on the yield of birch leaf hydrolysable tannins, *J. Chem. Ecol.*, **29**, 1289–1305.

Salminen, J.-P., Ossipov, V., Loponen, J., Haukioja, E. and Pihlaja, K. (1999) Characterization of hydrolyzable tannins from leaves of *Betula pubescens* by high-performance liquid chromatography–mass spectromentry, *J. Chromatogr.*, **A864**, 283–291.

Salminen, J.-P., Ossipov, V., Haukioja, E. and Pihlaja, K. (2001) Seasonal variation in the content of hydrolyzable tannins in leaves of *Betula pubescens, Phytochemistry*, **57**, 15–22.

Sarkar, S. K. and Howarth, R. E. (1976) Specificity of vanillin test for flavanols, *J. Agr. Food Chem.*, **24**, 317–320.

Scalbert, A. (1992) Quantitative methods for the estimation of tannins in plant tissues, in R. W. Hemingway and P. E. Reed (eds), *Plant Polyphenols*, Plenum Press, New York, pp. 259–279.

Scalbert, A. and Haslam, E. (1987) Polyphenols and chemical defence of the leaves of *Quercus rubur, Phytochemistry*, **26**, 3191–3195.

Schults, J. C. (1989) Tannin–insect interaction, in R. W. Hemingway and J. J. Karchesy (eds), *Chemistry and Significance of Tannins*, Plenum Press, New York, pp. 417–433.

Seigler, D. S. (1998) *Plant Secondary Metabolism*, Kluwer Academic Publishers, Boston.

Shi, J., Yu, J., Pohorly, J. E. and Kakuda, Y. (2003) Polyphenolics in grape seed –biochemistry and functionality, *J. Med. Food*, **6**, 291–299.

Singleton, V. L. (1974) Analytical fractionation of the phenolic substances of grapes and wine and some practical uses of such analyses, *Am. Chem. Soc., Adv. Chem. Ser.*, Washington, DC, **137**, 184–196.

Slabbert, N. (1992) Complexation of condensed tannins with metal ions, in R. W. Hemingway and P. E. Laks (eds), *Plant Polyphenols*, Plenum Press, New York, pp. 421–436.

Sowumni, S., Ebewele, R. O., Peters, O. and Conner., A.,H. (2000) Differential scanning calorimetry of hydrolyzed mangrove tannin, *Polymer International*, **49**, 574–578.

Strack, D. (1997) Phenolic metabolism, in P. M. Dey and J. B. Harborne (eds), *Plant Biochemistry*, Academic Press, London, pp. 387–416.

Strack, D., Heilman, J., Wray, V. and Dirks, H. (1989) Structures and accumulation patterns of soluble and insoluble phenolics from Norway spruce needles, *Phytochem.*, **28**, 2071–2978.

Taiwo, E. A. and Ogunbodede, R. A. (1995) Production of tannin adhesives from bark of Nigerian trees, *Wood Science and Technol.*, **29**,103–108.

Tamagawa, K. and Fukushima, S. (1998) Proanthocyanidins from barley bran potentiate retinoic acid-induced granulolytic and sodium butyrate-induced monocytic differentiation of HL-60 cells, *Biosci. Biotechnol. Biochem.*, **62**, 1483–1487.

Tegelberg, R., Veteli, T., Aphalo, P. J. and Julkunen-Tiitto, R. (2003) Clonal differences in growth and phenolics in willows exposed to elevated ultraviolet-B radiation, *Basic and Applied Ecology*, **4** (3), 219–228.

Tikkanen, O.-P. and Julkunen-Tiitto, R. (2003) Phenological variation as protection against defoliating insects: the case of *Quercus robur* and *Operophtera brumata*, *Oecologia*, **136**, 244–251.
Veteli, T. O. (2003) Global atmospheric change and herbivory. Effects of elevated levels of UV-B radiation, atmospheric CO_2 and temperature on boreal woody plants and their herbivores, *University of Joensuu, PhD Dissertation of Biology*, **19**, 107.
Veteli, T. O., Kuokkanen, K., Julkunen-Tiitto, R., Roininen, H. and Tahvanainen, J. (2002) Effects of elevated CO_2 and temperature on plant growth and herbivore defensive chemistry, *Global Change Biology*, **8**, 1240–1252.
Vijayaraghavan, K. and Murthy, D. V. S. (1997) Effect of toxic substances in anaerobic treatment of tannery wastewater, *Bioprocess Engng*, **16**, 151–155.
Vivas, N., Nonier, M. F., Vivas de Gaulejac, N., Absalon, C., Bertrand, A. and Mirabel, M. (2003) Differentiation of procyanidin tannins from seed, skin and stems of grapes (*Vitis vinifera*) and heartwood of quebracho (*Schinopsis balansae*) by matrix assisted laser desorption/ionization time-of-flight mass spectrometry and thioacidolysis/liquid chromatography/electrospray ionization mass spectrometry, *Analytica Chimica Acta*, **523**, 247–256.
Waterman, P. G. and Mole, S. (1994) *Analysis of Phenolic Plant Metabolites*, Blackwell Scientific Publications, Oxford.
Wilson, T. C. and Hagerman, A. E. (1999) Quantitative determination of ellagic acid, *J. Agric. Food Chem.*, **38**, 1678–1683.
Xie, D.-Y., Sharma, S. B., Paiva, N. L., Ferreira, D. and Dixon, R. A. (2003) Role of anthocyanidin reductase, encoded by *Banyuls*, in plant flavonoid biosynthesis, *Science*, **299**, 396–399.
Xie, D.-Y., Sharma, S. B., Wright, E., Wang, Z.-Y. and Dixon, R. A. (2006) Metabolic engineering of proanthocyanidins through co-expression of anthocyanidin reductase and the PAP1 MYB transcription factor, *The Plant Journal*, **45**, 895–907.
Ya, C., Caffney, S. H., Lilley, T. H. and Haslam, E. (1989) Carbohydrate-phenol complexation, in R. W. Hemingway and J .J. Karchesy (eds), *Chemistry and Significance of Tannins*, Plenum Press, New York, pp. 307–322.
Ye, X.and Krohn, R. L. (1999) The cytotoxic effects of a novel IH636 grape seed proanthocyanidin extract on cultured human cancer cells, *Molec. Cell. Biochem.*, **196**, 99–108.

13

Carotenoid Dyes – Properties

U. Gamage Chandrika

13.1 Introduction

13.1.1 Occurrence of Carotenoids

Carotenoids are red, orange or yellow pigments and are found in many plants and animals. They are the most widespread of all groups of naturally occurring pigments. Approximately 750 carotenoids (not including *cis–trans* isomers) have been isolated and characterized from natural sources [1]. They share common structural features, such as the polyisoprenoid structure and a series of centrally located bonds [2]. Most of the carotenoids are xanthophylls and contribute to the autumnal tint in leaves of deciduous trees and the impressive feathers of birds such as the wild cock bird, the flamingo and many others. The attractive colours of many red and yellow fruits and vegetables are also attributable to their carotenoid content. Carotenoids occur in algae, fungi, yeasts, moulds, mushrooms and bacteria and in all classes of plants and animals, including mammals. No animal is able to synthesize carotenoids. They obtain it from food they consume. Some carotenoids may be slightly altered by oxidative metabolism during digestion and absorption [3].

13.1.2 Chemistry of Carotenoids

All carotenoids are polyisoprenoids and possess an extensive system of conjugated double bonds. This serves as the light absorbing chromophore responsible for the yellow, orange and red colour of the plant sources and provides the visible absorption spectrum that is the basis for their identification and quantification. The basic carotenoid structure is symmetrical, linear and usually contains 40 carbon atoms and often have one or two cyclic structures at the end of their conjugated chains (see Figures 13.1, 13.2 and 13.3). The most important feature

Handbook of Natural Colorants Edited by Thomas Bechtold and Rita Mussak

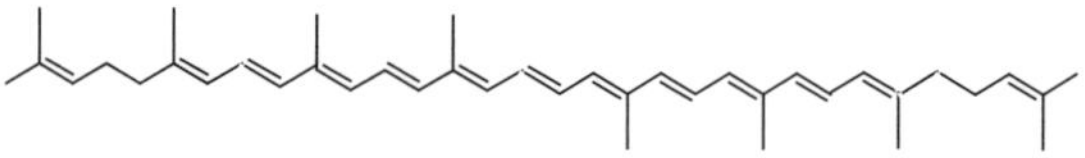

Lycopene

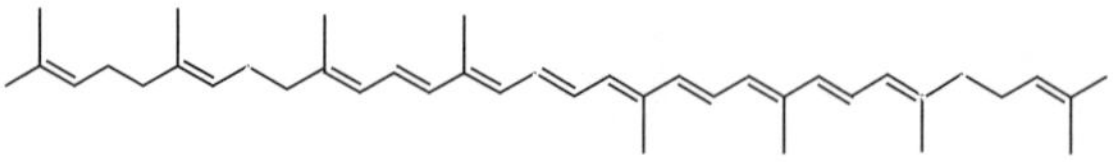

Neurosporene

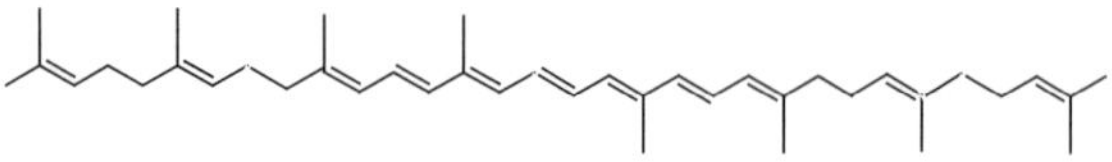

ζ-Carotene

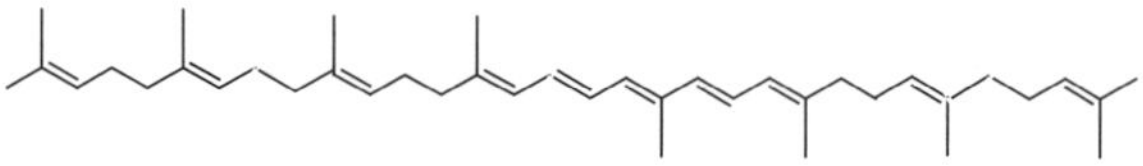

Phytofluene

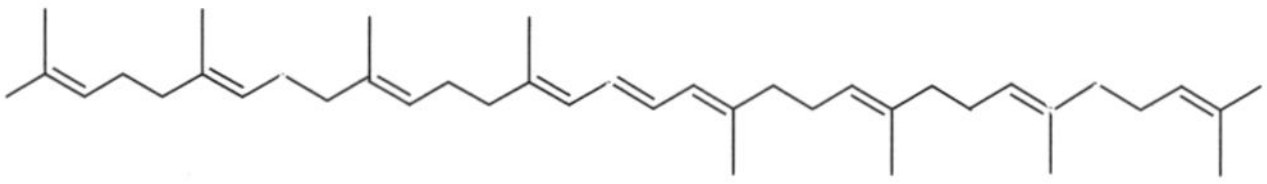

Phytoene

Figure 13.1 Acyclic carotenes

of carotenoids is the long system of alternating double and single bonds. It gives them their distinctive molecular shape, chemical reactivity and light absorbing properties.

Carotenoids are generally classified into two groups: (1) hydrocarbon carotenoids are collectively termed carotenes (see Figures 13.1 and 13.2); (2) those containing oxygen are called xanthophylls (see Figure 13.3). Modifications, such as rearrangements, hydrogenation, dehydrogenation, double bond migration, isomerization, chain shortening or extension, introduction of oxygen functions or a combination of these processes, result in many carotenoid structures. Carotenoids, whether carotenes or xanthophylls, may be acyclic, monocyclic or bicyclic. Cyclization occurs at one or both ends of the molecule, forming one or two six-membered β-rings (termed β-ionone rings) or ε-rings (referred to as α-ionone rings). Thus, the monocyclic γ-carotene has one β-ring while the bicyclic β-carotene, β-cryptoxanthin, zeaxanthin and astaxanthin have two of these rings. The bicyclic α-carotene and lutein each have one β-ring and one ε-ring.

13.1.3 Chemical Characteristics of Natural Carotenoids

Of the acyclic carotenes (Figure 13.1), lycopene and ζ-carotene are the most common. Phytoene and phytofluene are colourless carotenoids and have been reported in many carotenogenic fruits. Neurosporene has limited occurrence and is normally found in small amounts.

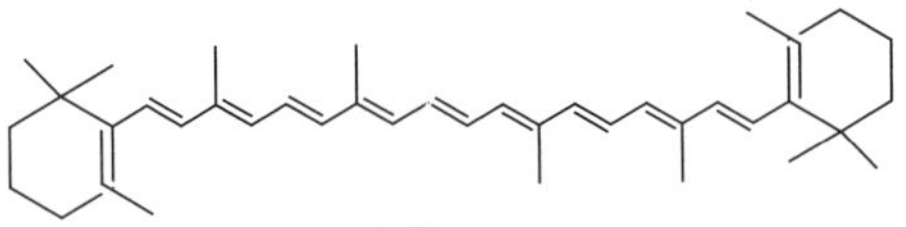

β-Carotene

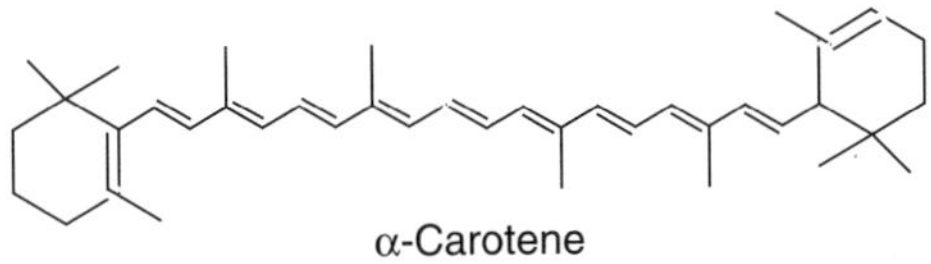

α-Carotene

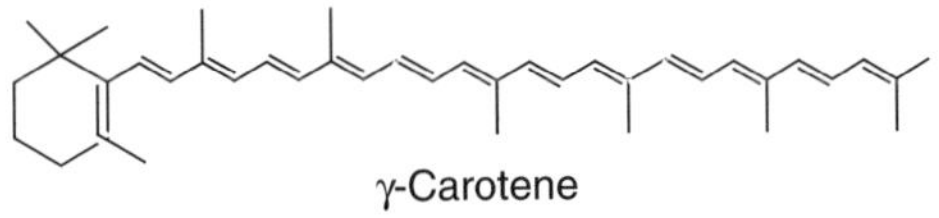

γ-Carotene

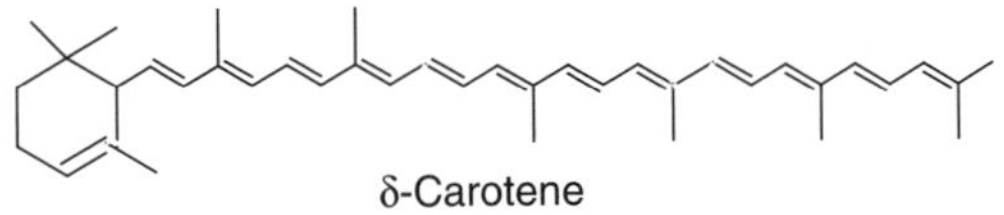

δ-Carotene

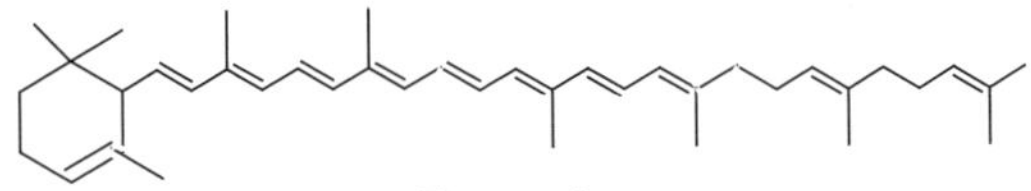

α-Zeacarotene

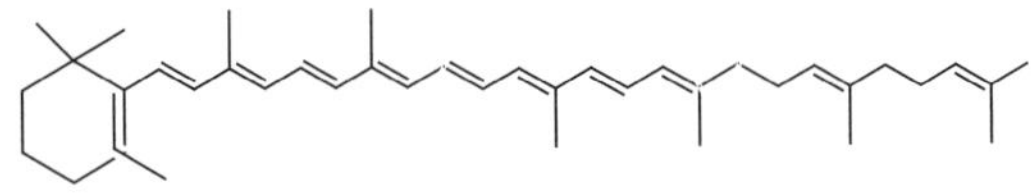

β-Zeacarotene

Figure 13.2 *Cyclic carotenes*

The bicyclic β-carotene (Figure 13.2) is the most widespread of all carotenoids in foods. The bicyclic α-carotene and monocyclic γ-carotene sometimes accompany β-carotene, generally at much lower concentrations. A great variety of xanthophylls, which are oxygenated carotenoids, are found in plant foods (Figure 13.3). Epoxy carotenoids (Figure 13.4) comprise a major group of xanthophylls, but the physiological significance of its presence is unclear. The epoxide derivatives of zeazanthin, antheraxanthin, mutatoxanthin, vioalxanthin, luteoxanthin, auroxanthin and neoxanthin are widely distributed. Xanthophylls, which are epoxides of β-carotene, β-carotene-5,6-epoxide, β-carotene-5,6,5′,6′-diepoxide, mutatochrome, luteochrome and β-cryptoxanthin, especially β-cryptoxanthin-5,6-epoxide and cryptoflavin, are frequently encountered. Less frequently encountered is δ-carotene. Although green leaves contain unesterified hydroxy carotenoids, most carotenols in ripe fruit are esterifies with fatty acids. Some examples of plant sources of common carotenoids are shown in Table 13.1.

The existence of uncommon or species-specific carotenoids has also been reported, capsanthin and capsorubin, the main carotenoids of red pepper, being an example (Figure 13.5). There are two unique carotenoids that have been commonly used as textile dyes: crocetin, the yellow pigment found in the stigmata of saffron (*Crocus stativus*) flowers,

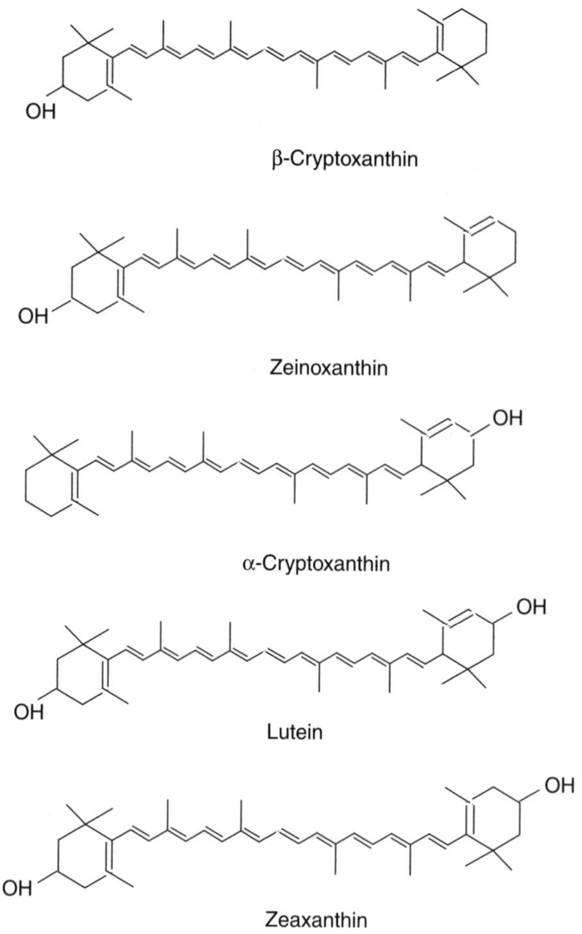

β-Cryptoxanthin

Zeinoxanthin

α-Cryptoxanthin

Lutein

Zeaxanthin

Figure 13.3 Xanthophylls

Indian mahogany (*Cedrela toona*) and harshingar (*Nyctanthes arbor-tistis*) flowers, and the apocarotenoid bixin, found in high concentrations in the seed coat of *Bixa orellana*. Bixin and its hydrolysis product, the bicarboxylic acid norbixin, are extensively employed as food colorants (annatto) (Figure 13.5).

Astaxanthin and tunaxanthin are principal carotenoids of some fish, such as salmon and trout, and most crustaceans (e.g. shrimp, lobster and crab). Echinenone and canthaxanthin are the intermediates in the transformation of dietary carotenoids and can be detected as accompanying minor carotenoids.

Although most carotenoids exist in nature in the more stable *trans* form, the occurrence of natural *cis* isomers has also been reported. Prolycopene and bixin are classical examples of natural *cis* isomers. The *cis* isomer may sometimes be an artefact because *trans–cis* isomerization can easily happen during carotenoid analysis.

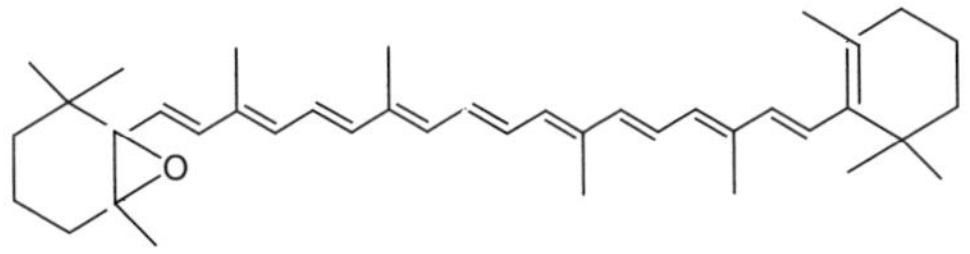

β-Carotene-5, 6-epoxide

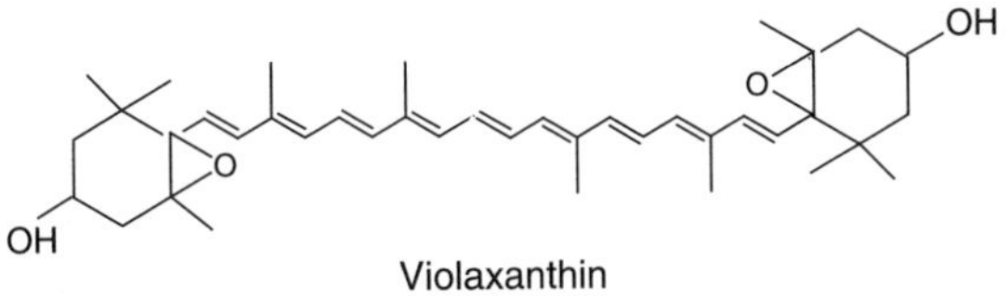

Violaxanthin

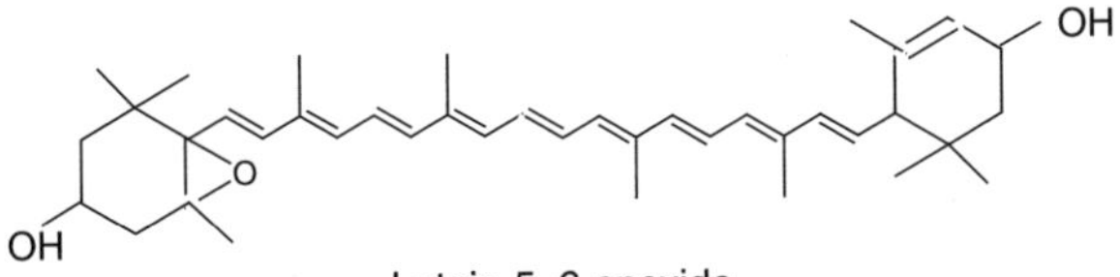

Lutein-5, 6-epoxide

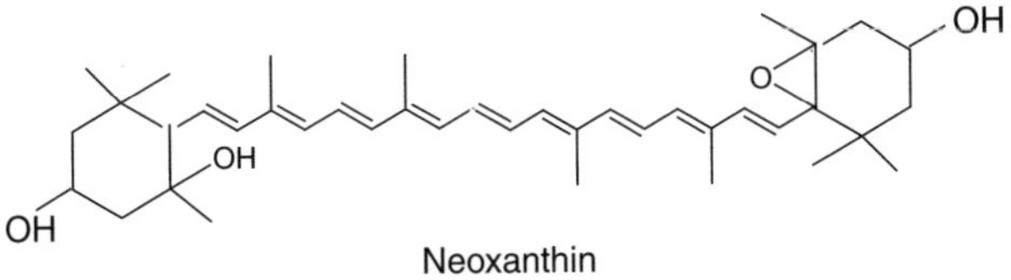

Neoxanthin

Figure 13.4 *Epoxy carotenoids*

Table 13.1 *Chemical characteristic of plant sources of common carotenoids*

Carotenoids	*Plant sources*
α-Carotene	*Mauritia vinifera*, *Cucurbita moschata*, carrots, red plam oil, *Cucurbita maxima*
β-Carotene	Green leafy vegetables, cantaloup melon, jack fruits (*Artocarpus heterophyllus*) *Cucurbita maxima*, yellow and orange fleshed sweet potatoes, mango, carrots, broccoli, apricot
β-Cryptoxanthin	Caja (*Spondias lutea*), pitanga (*Eugenia uniflora*), yellow and orange fleshed papaya
χ-Carotene	Passion fruits
Lycopene	Red-fleshed guava, pink grapefruit, red fleshed papaya, tomato, red-fleshed watermelon
Lutein	Green leafy vegetables, broccoli, Brussel sprout, corn, *Cucurbita maxima*
Zeaxanthin	Buriti, corn

References: Rodriguez-Amaya [4], Markovic *et al.* [5], Niizu and Rodriguez–Amaya [6], Holden *et al.* [7], Chandrika *et al.* [8–10], Priyadharshani and Chandrika [11]. The scientific names are given for foods that are not so common.

13.2 Properties and Functions of Carotenoids

13.2.1 Carotenoids Role as Pro-vitamin A

Certain carotenoids are capable of being converted to vitamin A. This is known to occur in most insects, fish, reptiles, birds and mammals. This ability is not shared by members of the

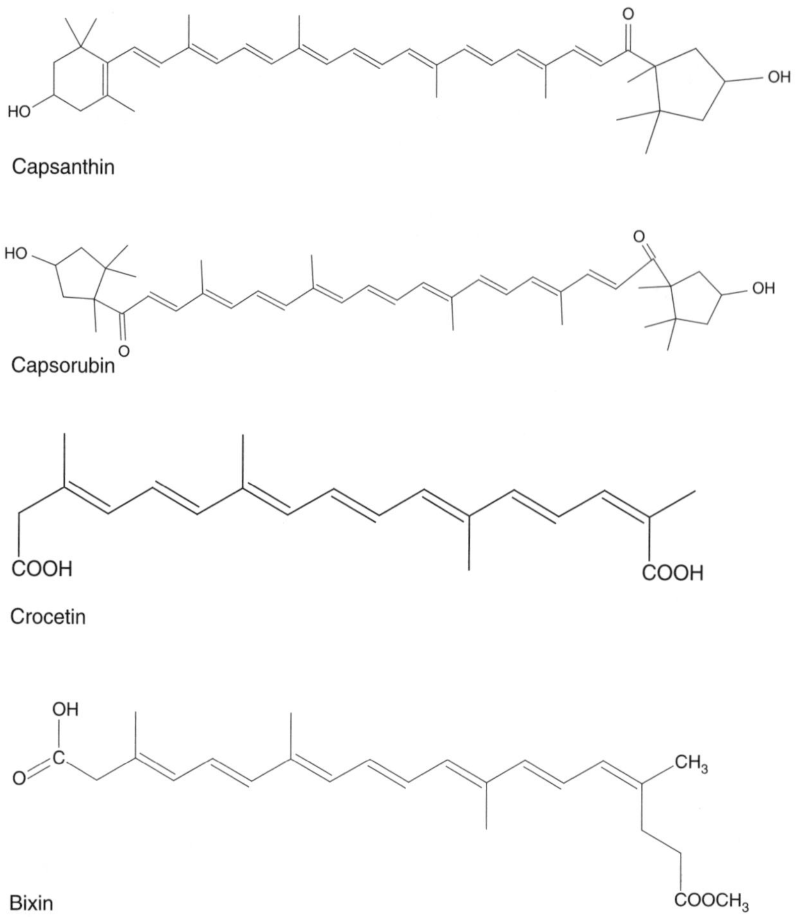

HO Capsanthin O OH

HO Capsorubin O O OH

COOH COOH Crocetin

OH C O CH3 COOCH3 Bixin

Figure 13.5 *Principal carotenoids of natural dye (colorants)*

carnivorous cat family [12]. The enzyme responsible for the conversion is 15,15′-dioxygenase. It is known that the concentration of this enzyme in the intestine of humans is very low due to a low human requirement for vitamin A. Further, excess vitamin A is toxic to humans. The cleavage enzyme is present at such low levels in humans that even oral ingestion of 200 mg/day of β-carotene does not lead to vitamin A toxicity but does lead to awkward orange coloration [13, 14]. The vitamin A (retinol) molecule is essentially one half of the molecule of all-*trans*-β-carotene with an added molecule of water at the end of the lateral polyene chain. For vitamin A activity to occur, carotenoids must include at least one unsubstituted β-ionone ring and a polyene side chain. The other end of molecule may have a cyclic or acyclic structure. Thus, all-*trans*-β-carotene makes by far the largest contribution to vitamin A activity in foodstuffs, while α-carotene, γ-carotene and β-cryptoxanthin contribute to a lesser extent. Lutein, lycopene and zeaxanthin cannot be converted to vitamin A.

13.2.2 Use of Carotenoids as Markers of Dietary Practices

β-Carotene, α-carotene, β-cryptoxanthin, lycopene, lutein and zeaxanthin are the principle carotenoids found in human blood [15] and except for zeaxanthin, the carotenoids are most frequently found in foods. In the case of pro-vitamin A carotenoids usual levels in plasma are significantly higher in women than in men [16].

13.2.3 Carotenoids as Antioxidants

There are several distinct mechanisms by which carotenoids function as antioxidants. One mechanism is singlet oxygen quenching, which is the ability of carotenoids to quench the highly reactive form of oxygen known as singlet oxygen ($^1O_2$) [17]. Singlet oxygen is a highly reactive species, capable of oxidizing nucleic acids, various amino acids in proteins and unsaturated fatty acids. Fortunately for plants as well as humans, carotenoids (CAR) are the most effective quencher of singlet oxygen found in nature, via the following reaction:

$$^1O + CAR \longrightarrow {}^3O_2 + {}^3CAR$$

$$^3CAR \longrightarrow CAR + heat$$

Further, due to the long conjugated polyene nature of these molecules, they lose the excess energy in the excited state (^{3}CAR) via vibrational and rotational interactions with the solvent system, ultimately reforming the ground state carotenoid, ready to begin another cycle of $^1O_2$ quenching. The acyclic lycopene was found to be a more efficient antioxidant than the cyclic β-carotene [18], although both have eleven conjugated double bonds. There have been many reports of the ability of various carotenoids to interfere with radical-initiated reactions, particularly with those that result in lipid peroxidation [17, 19]. Studies of *in vitro* antioxidant activities of β-carotene and other carotenoids demonstrated excellent activity, in some cases showing a more powerful antioxidant action than α-tocopherol [20].

13.2.4 Carotenoids in the Macular Region of the Retina

The macula is a structure at the centre of the human retina that is composed primarily of densely innervated cone photoreceptors. The macula often deteriorates beyond the age 70, leading to serious visual impairment and handicaps [21]. It has long been noted that this region is yellow (thus its name, macula lutea or yellow spot). Bone *et al.* [22] and Handelman *et al.* [23] found that the pigment is a mixture of two very similar carotenoids, lutein and zeaxanthin. In the very centre, at the point of highest concentration, the pigment consists mostly of zeaxanthin, whereas the peripheral retina contains mostly lutein. The physiological significance of this selective accumulation is primarily based on filtration of potentially damaging blue light and quenching of photochemically induced reactive oxygen species. It is believed that via these mechanisms lutein and zeaxanthin can contribute to reducing the risk of age-related macular degeneration, which is the leading cause of irreversible loss of vision in the elderly population in the West [24]. Furthermore, intake of these carotenoids, either with food or supplements, can increase their levels in macula [25]. Lutein is more widely distributed in foods than zeaxanthin. The main food sources of lutein are green leafy vegetables and pumpkins (see Table 13.1).

13.2.5 Carotenoids as Anticancer Agent

Lycopene has aroused considerable interest as a health-promoting phytochemical in recent years [26, 27]. There is scientific evidence that it may protect humans against prostate, lung and oesophagus cancers, and coronary heart disease [28]. It is found that lycopene can synergize with other phytonutrients in the inhibition of cancer cell growth. The mechanism of the inhibitory effects of lycopene and other carotenoids involves interference in several pathways related to cancer cell proliferation and includes changes in the expression of many proteins participating in these processes, such as connexins, cyclins, cyclin-dependent kinases and their inhibitors. These changes in protein expression suggest that the initial effect involves modulation of transcription by ligand-activated nuclear receptors or by other transcription factors [29].

13.2.6 Carotenoids as a Natural Colorant

The most commonly used natural carotenoid extracts in food are annatto, paprika, saffron, tagetes, carrot extracts, lycopene and β-carotene. More important chemical physical properties of natural carotenoid colorants are summarized in Table 13.2.

13.3 General Procedure for Carotenoid Analysis

This section deals with the analysis of known carotenoids in plant sources. There are two main objectives in carotenoid analysis. They are: (1) to establish the identity of the compound that has been isolated to characterize its structure and (2) to determine the content of the isolated carotenoid in the biological sample. Carotenoid analysis usually consists of (a) sampling and sample preparation, (b) extraction, (c) partition to a solvent compatible with the subsequent chromatographic steps, (d) saponification and washing, (e) concentration or evaporation of the solvent, (f) chromatographic separation, (g) identification and (h) quantification [37].

13.3.1 Sampling

Special attention should be given to the sampling procedure. The content of catrotenoids is controlled by many factors, as described in Section 14.1 in Chapter 14. To obtain reliable analytical data, the sample must be representative of the entire materials under investigation. The factors that should be considered in the sample plan consist of

Table 13.2 *Chemical and physical properties of some natural carotenoid colorants*

Name English	*Botanical*	*Chemical basis*	*Relevant representative*	*Content % of weight*	*Analytics/ references*	*Solubility*	*Complex formation*	*Chemical properties/ sensitivity*	*Toxicology/Ref.*
Annatto (seeds)	*Bixa orellana*	Carotenoids	Bixin, norbixin	3.5–5.2 %	Solvent extraction [30]	Oil, water		Light sensitive, stable in alkaline mediums, degrades in acids, degrades by oxidation, resistant to microbial attack	Not showing any carcinogenic effect [31]
Paprika seed	*Capsicum annuum*	Carotenoids	Capsanthin, capsorubin, β-carotene			Oil		Antioxidative activity	
Saffron (flower)	*Crocus sativus*	Carotenoids	Crocetin, zeaxanthin, β-carotene, crocin	70 % of total volatiles	HPLC [32]	Oil, water		Fairly stable to light, oxidation and microbial attack, changes in pH	Saffron is nontoxic in animal studies (LD_{50} – 20.7 g/kg), noncytotoxic *in vitro* studies (LD_{50} – 200 m kg/ml) [33]
Tagetes	*Tagetes erecta* L.	Carotenoids	Lutein, zeazanthin	8.7 % of total carotenoids	HPLC [34]	Hexane		Lutein is a free-flowing orange-red powder	Generally recognized as safe (GRAS) [35]
Carrot extract	*Daucus carota*	Carotenoids	α-Carotene, β-carotene	29–64 % of total carotenoids	HPLC [11]	Oil		Antioxidative activity	Not showing any adverse effect [31]
Tomato	*Lycopersicon esculentum*	Carotenoids	Lycopene	85–90 % of total carotenoids	HPLC	Water		Antioxidant properties, thermal degradation of naturally occurring antioxidants takes place	No observed effect level at 1 % [36]

(a) the purpose of analysis, (b) the nature of the population to be studied, (c) the nature of the substance to be measured, (d) the distribution of the analyte within the population and (e) statistical considerations.

13.3.2 Extraction

Samples of tissues are homogenized prior to extraction at low temperature and in the presence of antioxidant. A good extraction procedure should release all the carotenoids from the tissue and bring them into the solution without altering their structures. There is no standard procedure for carotenoid extraction because of the wide spectrum of analysed materials (foodstuffs, plants, animal and human samples) and the wide range of carotenoids present. Extraction of carotenoids from plant sources in our laboratory is carried out according to Rodriguez-Amaya [37].

13.3.3 Saponification of Carotenoids

Saponification not only hydrolyses the carotenol esters in sample like fruits, but also removes chlorophylls and unwanted lipids, which may interfere with the chromatographic separation and shorten the column's life. Saponification and the subsequent washing can result in losses of carotenoids, especially xanthophylls. Therefore it should be omitted from the analytical procedure whenever possible. For example, the saponification step can be avoided in carotenoid analysis of leafy vegetables as these carotenoids have not been esterified. When it is necessary, saponification should be carried out after transferring carotenoids to petroleum ether and saponified for 16 h (in the dark, at room temperature) with an equal volume of 10% potassium hydroxide in methanol containing the antioxidant butylated hydroxytoluene (1% in petroleum ether). The alkali is washed with water into a separatory funnel until neutral. For samples containing high concentrations of lipids, such as vegetable oil, avocado and palm fruit, a better procedure for eliminating lipids should be pursued before saponification. In our laboratory such samples are dissolved in acetone and left at −20 °C for 4–5 hours to solidify the lipids. The solidified lipids are then filtered through a sintered glass funnel in the freezer compartment. The carotenoid solution is then dried with anhydrous sodium sulfate and concentrated in a rotary evaporator (35 °C) for chromatographic separation.

13.3.4 Chromatographic Separation

Reverse phase HPLC is the current method of choice for carotenoid separation. In these systems, a nonpolar mobile phase, such as various mixtures of methanol, acetonitrile, tetrahydrofuran and ethyl acetate, are recommended [38]. Antioxidants such as BHT are added to the mobile phase and the temperature of the mobile phase should be maintained low and constant (around 20 °C) to prevent decomposition of carotenoids during the HPLC analysis and improve the reproducibility of quantitative analysis [39]. Isocratic separation is sufficient for the determination of food samples containing few principle carotenoids. An example would be the analysis of carotenoids in cooked nonleafy vegetables [40] (lutein, α-carotene and β-carotene) on a reversed phase C_{18} column using a mobile phase consisting of acetonirile:methanol:triflouroacetic acid (58:35:7). Gradient elution has the advantage of greater resolving power and elution of strongly retained compounds. Thus this should be employed when analysing foods containing several principle carotenoids. An example is the analysis of carotenoids in green leafy vegetables on a reversed phase C_{18} column and gradient elution using the ternary mobile phase consisting of acetonirile:methanol: 0.05% triethylamine in ethyl acetate. An example is shown in Figure 13.6. Triethyl amine minimizes the effects of acidity generated by the free silanol group present on the silica support. Applications of the C_{30} stationary phase to food analyses have been developed especially for carotenoid analyses. The resolution of geometrical isomers in this type of column is excellent. This column with an isocratic solvent system consisting of methyl-*tert*-butyl ether:methanol:-water (56:40:4) was used for the separation of *cis–trans* isomers of β-carotene in cooked green leafy vegetables [41].

Open column chromatography is the classical method for separating carotenoids for quantitative analysis. Usually carotenoids are separated on an MgO:celite (1:1) column. The absorbent is activated for 2 h at 100 °C, cooled in a desiccator prior to use and the column is developed with increasing concentration of acetone in petroleum ether to give the characteristic separation pattern. Advantages of this technique are that the colour enabled visual monitoring of the separation and to isolate standards that are not commercially available.

13.3.5 Chemical Tests

Chemical tests are conducted to verify the type and position of substituents in xanthophylls and an iodine-catalysed isomerization reaction is carried out to verify the geometric configuration as summarized in Table 13.3 [37].

13.3.6 Detection and Identification of Carotenoids

Carotenoids exhibit very high extinction coefficients between 400 and 500 nm. Only a few hydrophobic natural compounds absorb at these wavelengths. Therefore carotenoids with known structure can be conclusively identified by the combined use of chromatographic behaviour, UV–visible absorption spectra (wavelengths of maximum absorptions and the shape of the spectral fine structure) and specific group chemical reactions to confirm the type, location and number of functional groups in xanthophylls, HPLC retention time, co-chromatography with authentic carotenoid standards, the UV–visible spectra obtained with a photodiode array detector and mass spectra obtained with a mass detector.

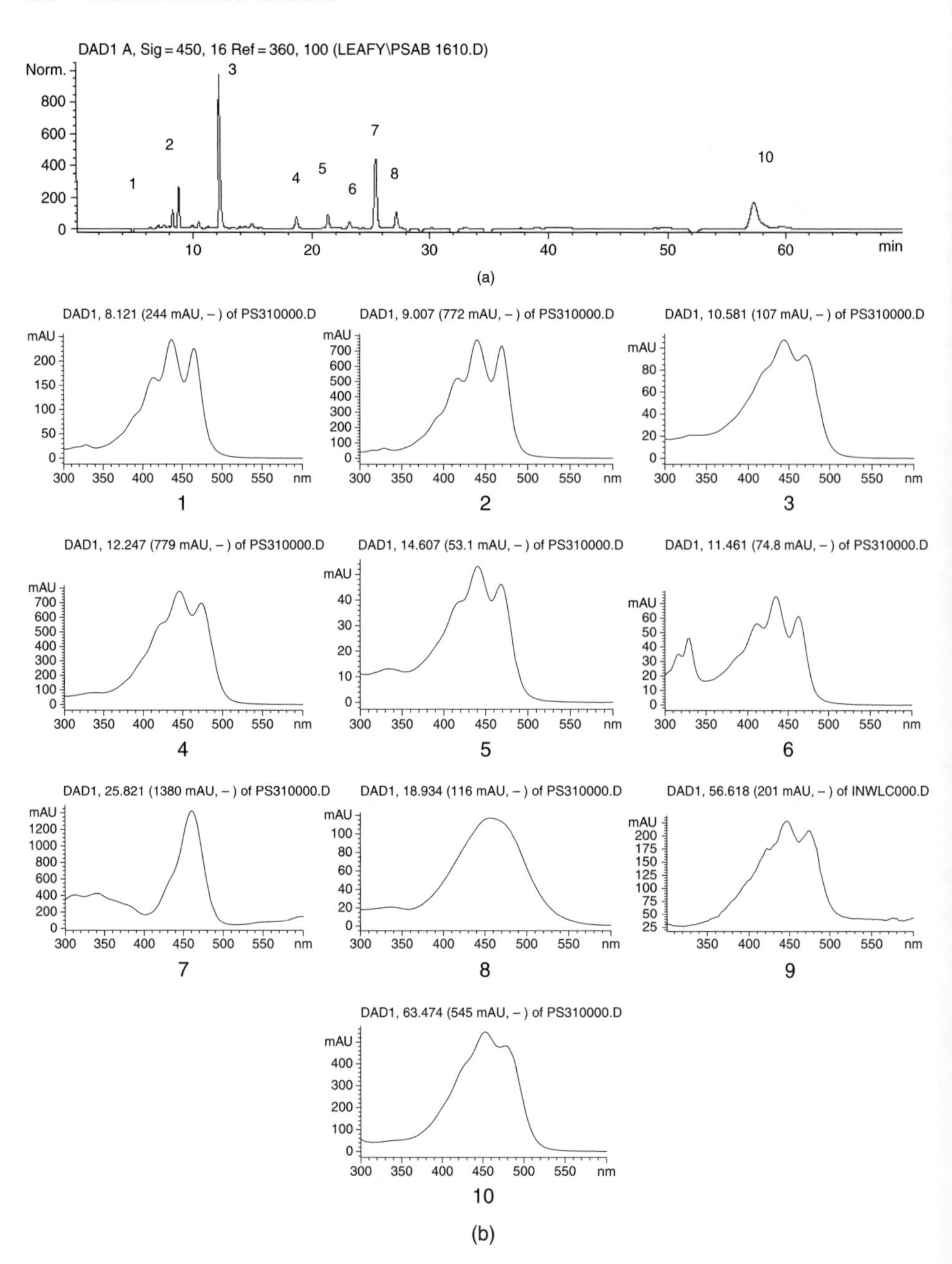

Figure 13.6 *(a) RP-HPLC chromatogram and (b) photodiode array spectra of the carotenoids of Boerhavia diffusa. Column – Agilent 5 μm, RP – C_{18} (250 mm × 4.6 mm); mobile phase: acetonitrile:methanol:ethyl acetate containing 0.05 % of triethylamine at a flowrate of 0.5 ml/min, A gradient of 95:05:00 to 60:20:20 in 70 minutes. Peak identification: (1) neoxanthin, (2) violaxanthin, (3) lutein, (4) β-cryptoxanthin, (5) α-cryptoxanthin, (6) β-carotene-5,6,5′,6′-diepoxide, (7) internal standard, (8) chlorophyll peak, (10) β-carotene. Reproduced from U. G. Chandrika and N. Colombagama (unpublished data)*

Table 13.3 Summary of chemical tests use in the identification of carotenoids

Chemical tests	If positive, possible carotenoids
Acetylating of primary and secondary hydroxyl group	α-Cryptoxanthin, β-cryptoxanthin, β-cryptoxanthin monepoxides, neoxanthin (3 OH groups), vioaxanthin (2 OH groups) zeaxanthin (2 OH groups), lutein, zeaxanthin (3 OH group), rubixanthin (1 OH group)
Methylation of allylic hydroxyl groups	α-Cryptoxanthin, lutein
Epoxide-furanoid rearrangement and HCl fumes	β-Catorotene-5,6-epoxide, β-cryptoxanthin monepoxides, neoxanthin, lutein
Reduction with sodium borohydride	Canthaxanthin, echinenone

A specific problem is the identification of the geometrical isomers. Some of the *cis* isomers have identical spectra as compared to the *trans* isomer, for example all-*trans* and 5-*cis*-lycopene. Others exhibit characteristic UV–visible absorption with an additional '*cis* band' in the spectrum, as shown in Figure 13.7.

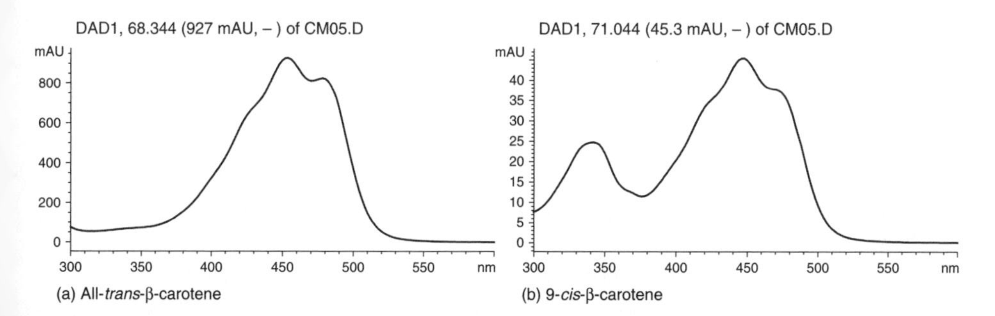

Figure 13.7 *UV–visible spectrums of (a) all-trans β-carotene and (b) 9-cis- β-carotene*

13.3.7 Quantification of Carotenoids

For the quantification of carotenoids by HPLC chromatography the external or internal standard method may be applied. Both require the availability of pure standards. The concentrations of the various carotenoids can also be determined spectrophotometrically using molar extinction coefficients as described by Rodriguez-Amaya [37].

13.4 Problems in Carotenoid Analysis

In carotenoid analysis various factors have to be considered:

(a) There are a large number of naturally occurring carotenoids; thus correct identification of carotenoids is essential for accurate quantification. The limited commercial availability of carotenoid standards is also a serious problem. Carotenoid standards can be isolated by open column chromatography, which is time consuming and requires experience and patience.

(b) The carotenoid composition of foods varies qualitatively and quantitatively. Thus, the analytical procedures have to be adapted to the carotenoids and other nutrient compositions of each of the food samples. For example, some fruits (e.g. avocado, palm fruits) have high fat contents and hence need to use specialized extraction methods to extract carotenoids. On the other hand, some fruit carotenoids are esterified; thus a sponification step should be included in the analysis.
(c) There are wide variations in carotenoid concentrations in any given carotenoid source. Normally, one to four main carotenoids are present, with a series of minor carotenoids at low or trace levels. The analysis of these minor carotenoids is a very difficult task.
(d) The highly unsaturated carotenoid molecule is susceptible to isomerization and oxidation during analysis and during storage of samples prior to analysis.
(e) There are variations in the nature of the matrix [42, 43]. Thus sample preparation, extraction and storage conditions should be established for each carotenoid source type.

A serious problem in this type of analysis arises from the carotenoid instability. Whatever the analytical method chosen, precautionary measures to avoid artefact formation and quantitative losses must be part of the analytical process, as described by Rodriguez-Amaya [37]. They are: (a) completion of the analysis within the shortest possible time, (b) exclusion of oxygen, e.g. by using a vaccum, nitrogen or argon atmosphere, (c) protection from light, (d) avoiding high temperatures, (e) avoiding contact with acids and (f) use of reagent-grade or distilled solvents free from harmful impurities, such as peroxide-free diethyl ether and tetrahydrofuran or acid-free chloroform.

The general procedure for carotenoid analysis consists of eight steps as described above in Section 13.3.4. Errors can be introduced in any of these steps. Common sources of errors are: (a) nonidentification of the variety/cultivar properly, (b) samples not representing the food collections under investigation, (c) incomplete extraction, (d) physical losses during different steps, (e) incomplete chromatographic separation, (f) misidentification of carotenoids, (g) faulty quantification and calculation and (h) isomerization and oxidation of carotenoids during analysis and/or during storage of samples before analysis.

Specifically for HPLC, currently the method of choice for carotenoid analysis, the following sources of errors have been observed: (a) incompatibility of the injection solvent and mobile phase, resulting in peak splitting, and distortion [44], (b) quantification of overlapping peaks, (c) varying purity [45, 46] and instability of carotenoid standards, (d) inconclusive or erroneous identification and (e) inaccuracies in the preparation of standard solutions and in the calibration and erroneous calculation.

References

1. G. Britton, S. Liaaen-Jensen and H. Pfrander, *Carotenoids Handbook*, Birkhäuser Verlag, Basel, Switzerland, 2004.
2. J. A. Olson, in *Modern Nutrition in Health and Disease–Vitamin A, Retinoids, and Carotenoids*, 8th edition, Lea and Febiger, Philadelphia, Pennsylvania, pp. 287–307, 1994.
3. T. W. Goodwin, Metabolism, nutrition, and function of carotenoids, *Annu. Rev. Nutr.*, **6**, 273–297 (1986).

4. D. B. Rodriguez-Amaya, Latin American food sources of carotenoids, *Arch Latinoamer. Nutr.*, **49**, 745–845 (1999).
5. M. Markovic, U. Mülleder and H. Neunteufl, Carotenoids content of different verities of pumpkin, *J. Food Comp. Anal.*, **15**, 633–638 (2002).
6. P. Y. E. Niizu and D. B. Rodriguez-Amaya, A melancia como fonte de licopeno, *Rev. Inst. Adolfo Lutz.*, **62**(3), 195–199 (2003).
7. J. M. Holden, A. L. Eldridge and C. R. Beecher, I. M. Buzzard, S. Bhagwat, C. S. Davis, L.W. Douglass, S. Gebhardt, D. Haytowitz and S. Schakel, Carotenoids content of U.S. foods: an update of the database, *J. Food Comp. Anal.*, **12**, 169–196 (1999).
8. U. G. Chandrika, E. R. Janz, S. M. D. N. Wickramasinghe and N. D. Warnasuriya, Carotenoids in yellow and red-fleshed papaya, *J. Sci. Food Agric.*, **83**(12), 1279–1282 (2003).
9. U. G. Chandrika, E. R. Jansz and N. D. Warnasuriya, Carotenoid composition of ripe jakfruits (*Artocarpus heterophyllus*) kernel and its bioconversion, *J. Sci. Food Agric.*, **85**, 186–190 (2005).
10. U. G. Chandrika, K. S. S. P Fernando and K. K. D. S. Ranaweera, Carotenoids content and *in vitro* bioaccessibility of lycopene from guava (*Psidium guajava*) and watermelon (*Citrullus lanatus*) by HPLC-DAD, *Int. J. Food. Sci. Nutr.* (2008, in press).
11. A. M. B. Priyadarshani and U. G. Chandrika, Content and composition of selected non-leafy vegetables from Sri Lanka, *J. Nat. Sci. Found.*, **35**(4), 251–253 (2007).
12. F. J. Schweigert, J. Raila, B. Wichert and E. Kienzle, Cats absorb β-carotenes, but it is not converted to vitamin A, J. Nutr., **132**, 1610S–1612S (2002).
13. A. Bendich and J. A. Olson, Biological action of carotenoids, *FASEB J.*, **3**, 1927–1932 (1989).
14. R. McGowan, J. Beattie and P. Galloway, Carotenaemia in children is common and benign: most can stay at home, *Scott. Med. J.*, **49**(3), 82–84 (2004).
15. K. S. Epler, R. G. Zeigler and N. E. Craft, Liquid chromatographic method for the determination of carotenoids, retinoids and tocopherols in human serum and in foods, *J. Chromatogr.*, **619**, 37–48 (1993).
16. B. Olmedilla, F. Granado and I. Blanco, Seasonal and sex-related variations in six serum carotenoids, retinol and alpha-tocopherol, *Am. J. Clin. Nutr.*, **60**, 106–110 (1994).
17. P. Palozza and N. I. Kinsky, Antioxidant effects of carotenoids *in vivo* and *in vitro*: an overview, *Meth. Enzymol.*, **213**, 403–420 (1992).
18. P. Di Mascio, S. Kaiser and H. Sies, Lycopene as the most efficient biological carotenoid singlet oxygen quencher, *Arch. Biochem. Biophys.*, **274**(2), 532–538 (1989).
19. N. I. Krinsky, Antioxidant functions of carotenoids, *Free. Radic. Biol. Med.*, **7**, 617–635 (1989).
20. N. J. Miller, J. Simpson, L. P. Candeias, P. M. Bramley and C. A. Rice-Evans, Antioxidant activities of carotenes and xanthophylls, *FEBS J.*, **384**, 240–246 (1996).
21. R. Klein, B. E. Klein, S. C. Jensen and S. M. Meuer, The five-year incidence and progression of age-related maculopathy: the Beaver Dam Eye Study, *Ophthalmology*, **104**, 7–12 (1997).
22. R. A. Bone, J. T. Landrum, L. Femandez and S. L. Tarsis, Analysis of the macular pigment by HPLC: retinol distribution and age study, *Invest. Ophthalmol. Vis. Sci.*, **29**, 843–849 (1988).
23. G. J. Handelman, E. A. Dratz, C. C. Reay and F. J. G. M. Van Kuijk, Carotenoids in the human macula and whole retinas, *Invest. Ophthalmol. Vis. Sci.*, **29**(6), 850–855 (1988).
24. N. I. Krinsky, J. N. Landrum and R. A. Bone, Biologic mechanisms of the protective role of lutein and zeaxanthin in the eye, *Annu. Rev. Nutr.*, **23**, 171–201 (2003).
25. J. T. Landrum, R. A. Bone, H. Joa, M. D. Kilburn, L. L. Moore and K. Sprague, A one-year study of the macular pigment: the effect of 140 days of a lutien supplement, *Expl Eye Res.*, **65**, 57–62 (1997).
26. W. Stahl and H. Sies, Lycopene: a biologically important carotenoid for humans? *Arch. Biochem. Biophys.*, **336**, 1–9 (1996).
27. P. M. Bramley, Is lycopene benefical to human health? *Phytochem.*, **54**, 233–236 (2000).
28. L. Arab and S. Steck, Lycopene and cardiovascular disease, *Am. J. Clin. Nutr.*, **71**(Suppl.), 1691–1695 (2000).
29. Y. Sharoni, M. Danilenko, S. Walfisch, H. Amir, A. Nahum, A. Ben-Dor, K. Hirsch, M. Khanin, M. Steiner, L. Agemy, G. Zango and J. Levy, Role of gene regulation in the anticancer activity of carotenoids, *Pure. Appl. Chem.*, **74**, 1469–1477 (2002).

30. Dominic W. S. Wong, *Mechanism and Theory in Food Chemistry*, J. J. Offset Press, Wazirpur, Delhi, 1996.
31. S. S. Deshpande, *Handbook of Food Toxicology*, Marcel Dekker, New York, (2002).
32. V. Sujata, G. A. Ravishankar and L.V. Venkataraman, Methods for the analysis of the saffron metabolites crocin, crocetins, picrocrocin and safranal for determination of the quality of the spice using thin-layer chromatography, high-performance liquid chromatography and gas chromatography, *J. Chromatogr.*, **624**, 497–502 (1992).
33. F. Abdullaev, Biological properties and medicinal use of saffron (*Crocus sativus* L.), *Acta Hort. (ISHS)*, **739**, 339–345 (2007); http://www.actahort.org/books/739/739_44.htm.
34. F. Khachik, A. Steck and H. Pfander, Isolation and structural elucidation of (13Z,13′Z,3R,3′R,6′R)-lutein from marigold flowers, kale, and human plasma, *J. Agric. Food Chem.*, **47**(2), 455–461 (1999).
35. A. Alves-Rodrigues and A. Shao, The science behind lutein, *Toxicol. Lett.*, **150**(1), 57–83 (2004).
36. D. Jonker *et al.*, Ninety day oral toxicity study of lycopene from *Blakeslea trispora* in rats, *Regul. Toxicol. Pharmacol.*, **37**, 396–406 (2003).
37. D. B. Rodriguez-Amaya, *A Guide to Carotenoid Analysis in Foods*, ILSI Press, Washington, 1999.
38. G. Su, K. G. Rowley, N. D. H. Balazs, Carotenoids: separation methods applicable to biological samples, *J. Chromatogr. B.*, **781**, 393 (2002).
39. K. J. Scott, D. J. Hart, Further observations on problems associated with the analysis of carotenoids by HPLC-2: column temperature, *Food Chem.*, **47**, 403 (1993).
40. A. M. B. Priyadarshani and U. G. Chandrika, Content and *in-vitro* accessibility of pro-vitamin A carotenoids from Sri Lankan cooked non-leafy vegetables and their estimated contribution to vitamin A requirement, *Int. J. Food. Sci. Nutr.*, **58**(8), 659–667 (2007).
41. U. G. Chandrika, U. Svenberg and E. R. Jansz, Content and *in vitro* accessibility of β-carotene from cooked Sri Lankan green leafy vegetables and their estimated contribution to vitamin A requirement, *J. Sci. Food Agric.*, **86**, 54–61 (2006).
42. D. B. Rodriguez-Amaya, Critical review of pro-vitamin A determination in plant foods, *J. Micronutr. Anal.*, **5**, 191–225 (1989).
43. D. B. Rodriguez-Amaya, Carotenoids and Food Preparation. The Retention of Pro-vitamin A Carotenoids in Prepared, Processed, and Stored Foods, OMNI Press, Washington, 1997.
44. F. Khachik, G. R. Beecher, J. T. Vanderslice and G. Furrow, Liquid chromatographic artifacts and peak distortion. Sample solvent interaction in the separation of carotenoids, *Anal. Chem.*, **60**(8), 807–811 (1988).
45. F. W. Quackenbusch and R. L. Smallidge, Non-aqueous reverse phase liquid chromatographic system for separation and quantification of pro-vitamins A, *J. Assoc. Anal. Chem.*, **69**, 767–772 (1986).
46. N. E. Craft, L. C. Sander and H. F. Pierson, Separation and relative distribution of all *trans*-β-carotene and its *cis*-isomers in β-carotene preparations, *J. Micronutr. Anal.*, **8**, 209–221 (1990).

14

Carotenoid Dyes – Production

U. Gamage Chandrika

14.1 Factors Influencing Carotenoid Composition in Plant Sources

The composition of carotenoids differs qualitatively as well as quantitatively among plant sources. The composition of carotenoids varies in a given plant source due to factors such as stage of maturity, cultivar or varietal differences, climatic or geographic effects, part of the plant used, conditions during agricultural production, post-harvest handling, processing and storage.

14.1.1 Stage of Maturity

Carotenoid levels in plant sources vary depending on the maturity at harvest. Leafy vegetables do not follow a defined carotenoid pattern. For example, mature leaves of kale [1], cabbage [2], lettuce and endive [3] typically have a higher β-carotene content than young leaves, but in New Zealand spinach, young leaves have higher carotenoid levels than the mature leaves [4]. In carrots the carotenoid content increases up to three weeks before harvest and then decreases as the vegetable matures [5].

In fruits where the colour at the ripe stage is due to pigments other than carotenoids, e.g. anthocyanins such as yellow cherry [6], red currant strawberry [7] and olive fruits [8], the carotenoid content decreases during ripening. In contrast, in most carotenoid-containing fruits and fruit vegetables, ripening is accompanied by enhanced carotenoid biosynthesis, as seen in Algerian date varieties [9]. Furthermore, Mercandanate and Rodriguez-Amaya [10] found a three- to fourfold increase in two varieties of mango respectively from unripe to ripe fruit. Ripeness was judged by means of visual colour and texture.

Handbook of Natural Colorants Edited by Thomas Bechtold and Rita Mussak

14.1.2 Cultivar or Varietal Differences

Variability in carotenoid data in plant sources reported by different authors may be due to the use of different items/varieties, since the same common name is often applied to items with different scientific names and vice versa [11]. Food should be unambiguously named and described and all local and scientific names up to the subspecies or variety level should be included. A detailed description of the plant source (e.g. flesh colour, peel type/colour, stage of maturity, pH) and any other informative data, if available, such as origin, soil and climate, size, cultivar, harvest date, preservation and packing medium, will improve the applicability and comparability of data. Cultivar/varietal difference could be quantitative, as in the case of Brazilian mangoes [10], red-fleshed Brazilian papayas [12], citrus cultivars from Florida [13], red tomatoes [14], pandanus fruit [15] and yellow-fleshed sweet potatoes and pumpkin [16]. An example is shown in Table 14.1 for varieties of sweet potatoes and pumpkins found in Sri Lanka. However, pronounced qualitative and quantitative variations can also be observed as in squashes and pumpkins [16, 18, 19] (Table 14.1), orange and red-fleshed papaya [12, 20] and emberella [21].

Table 14.1 *Concentrations of the major carotenoids of some selected Sri Lankan nonleafy vegetablesa*[a]

Vegetable	*β-Carotene*	*α-Carotene*	*Lutein*
Carrots	43.8 ± 5.6	20.5 ± 1.7	3.8 ± 0.4
Pumpkin varieties			
Arjuna	50.9 ± 5.7	27.3 ± 3.1	39.1 ± 4.7
Ruhunu	8.7 ± 1.2	6.2 ± 1.0	8.2 ± 1.2
Meemini	6.2 ± 2.1	11.8 ± 3.1	30.8 ± 4.7
Janani	3.0 ± 0.9	1.2 ± 0.3	31.2 ± 2.8
Samsun	5.1 ± 0.8	2.0 ± 0.4	45.8 ± 4.8
Squashes	6.0 ± 0.8	5.1 ± 1.1	
Sweet potato varieties			
CARI-426	42.8 ± 4.3	–	–
P_2-20	35.4 ± 5.2	–	–
CIP-440060	37.5 ± 3.8	–	–
420027	59.0 ± 6.2	–	–
187617-1	14.7 ± 2.3	–	–

[a] Data presented as μg/g fresh weight ± SD (n = 6).
Reference: Priyadharshani and Chandrika [17].

14.1.3 Climatic or Geographic Effects

Geographic effects are generally hard to assess since some cultivar difference is also usually involved. Significant differences in carotenoid composition (geographic effect) can be seen in 'Formosa' papayas (same cultivar) produced in two Brazilian states with different climates. As compared with those from Sao Paulo, papaya from the hot state of Bahia presented higher β-carotene, β-cryptoxanthin and lycopene content [12]. The β-carotene varied in 10 leafy vegetables collected at different times during the year [3]. In two 'kale' cultivars produced in the same farm and analysed at the mature stage, the β-carotene, lutein, violaxanthin and total carotenoid levels were significantly higher in the winter than in the summer [22]. Considerable variation of carotenoid content was observed in fruits and vegetables obtained from five cities in the United States [23].

14.1.4 Post-harvest Storage and Packing

Post-harvest changes in carotenoid content depend on the plant source, the length of time of storage and the conditions, such as light, temperature, relative humidity and irradiation. Plant sources are prone to deleterious changes induced by respiratory, metabolic and enzymatic activities. In addition to pest and microbial spoilage, temperature-induced injury can also be observed. To extend carotenoid shelf life in carrots, several approaches such as reducing water activity and removing oxygen have been used [24]. Blanching may not reduce carotenoid content in plant sources significantly and is recommended as a primary step in the processing and storing as it gives the benefit of inactivating enzymes involved in carotenoid destruction [25]. It was shown that freezing after blanching results in negligible changes in carotenoid content [26]. An example is shown in Table 14.2.

Table 14.2 *Effect of freezing after blanching on β-carotene content of green leafy vegetables*

Green leaves	*Total amount (μg/g DW) Mean ± SD*[a]	
	Raw	*Blanched*
Manioc	519 ± 176.8	517.3 ± 7.1
(*Manihot esculenta Crantz*)	(88%)[b]	(71%)[c]
Katurumurunga	486.7 ± 39.7	437.6 ± 13.6
(*Sesbania grandiflora* (L.))	(72.5%)[b]	(78.2%)[c]
Mukunuwanna	252.7 ± 48.2	149.3 ± 6.8
(*Alternanthera sessilis* (L.))	(75%)[b]	(74%)[c]
Gotukola	297.4 ± 48.2	275.5 ± 31.9
(*Centella asiatica* (L) Urban)	(84%)[b]	(77.4%)[c]

[a] Mean ± SD of all samples ($n = 3$ for total content).
[b] Moisture content of raw green leaves.
[c] Moisture content of blanched green leaves.
Reference: Chandrika *et al.* [27, 28].

14.1.5 Changes in Processing/Cooking

Most of the research done on the influence of processing had been based on the total carotene or carotenoid content. The effect on processing on pro-vitamin A carotenoids were reviewed by Rodriguez-Amaya [29]. Plant sources are frequently cooked to some extent before consumption. In addition to enhancing the preservation of food, processing/cooking allows compaction of foods, thereby increasing the value of the product [24]. During processing/cooking, isomerization of carotenoids from the *trans* form to the *cis* form occurs, with consequent alteration of the carotenoid bioavailability and bioconversion. Despite a reduction in carotene content due to cooking, heat processing has the potential of increasing the bioavailability/bioaccessibility of carotenoids during cooking of vegetables [17, 30]. The low *in vitro* accessibility obtained for raw carrots compared to cooked carrots could be attributed to the fibrous matrix of these vegetables that might have hindered the release of pro-vitamin A during *in vitro* digestion (Table 14.3). Mild cooking could increase the bioavailability of carotenoids by disrupting or softening plant cell walls and carotenoid–protein complexes [28, 31]. The differences in all-*trans*-β-carotene retained among vegetables cooked by the same method (see Table 14.4) is possibly due to differences in the degree of susceptibility of leaves to heat treatment or as a result of differences in varieties [32].

Table 14.3 *In vitro accessible pro-vitamin A in the nonleafy vegetable preparations (per 1 g dry weight)*

Vegetable preparations	β-carotene (μg/g DW) Mean ± SD[a]	In vitro accessibility (%)[b]	α-carotene (μg/g DW) Mean ± SD[a]	In vitro accessibility (%)[b]
Carrots				
Carrot salad	244.3 ± 26.0	60	88.4 ± 11.6	48.1
Carrot curry	309.0 ± 42.7	74.7	159.7 ± 25.4	78.1
Boiled carrot	301.0 ± 65.6	73.8	119.3 ± 10.7	60.6
Raw carrot	55.2 ± 8.5	12.1	20.0 ± 3.2	9.4
Pumpkins				
Pumpkin curry	91.0 ± 10.3	32.3	46.2 ± 6.4	29.3
Boiled pumpkin	55.1 ± 9.6	18.7	27.5 ± 10.1	17.7
Boiled pumpkin with scraped coconut	74.2 ± 7.9	25.2	37.6 ± 6.2	24.2
Squashes				
Squash curry	6.3 ± 1.4	14.1	4.5 ± 0.92	11.3
Sweet potatoes				
Sweet potato curry	47.6 ± 7.3	23.3	–	–
Boiled sweet potato	23.1 ± 3.8	11.0	–	–
Boiled sweet potato with scraped coconut	30.8 ± 5.4	14.6	–	–

[a] Mean ± SD in six samples in duplicate.
[b] As a percentage of the amount present in the vegetable preparations.
Reference: Priyadharshani and Chandrika [17].

Table 14.4 *Total content, retention and in vitro accessibility of all-trans-β-carotene in traditionally cooked vegetables (stir-fried[b])*

Green leaves	Total content (μg/g DW) Mean ± SD[a]		Retention on cooking[c] (%)	In vitro accessibility	
	Blanched	Stir-fried[b]		(μg/g DW) Mean ± SD[a]	(%)[d]
Manioc (*Manihot esculenta Crantz*)	517.3 ± 7.1 (71%)[e]	254.0 ± 37.0 (15%)[f]	50.7 ± 2.6	83.2 ± 11.7	32
Katurumurunga (*Sesbania grandiflora* (L.)	437.6 ± 13.6 (78.2%)[e]	323.2 ± 5.4 (14%)[f]	76.0 ± 3.2	155.4 ± 6.8	36
Mukunuwanna (*Alternanthera sessilis* (L.))	149.3 ± 6.8 (74%)[e]	91.8 ± 7.9 (21%)[f]	60.3 ± 2.6	17.2 ± 2.6	19
Gotukola *Centella asiatica* (L) Urban	275.5 ± 31.9 (77.4%)[e]	76.3 ± 8.6 (14%)[f]	29.6 ± 6.2	9.1 ± 1.0	12

[a] Mean ± SD of all samples ($n = 3$ for total content, $n = 4$ for *in vitro* content), equivalent to 1 g dry weight of the green leaves.
[b] Stir-fried green leafy vegetables (finely cut green leafy vegetables stir-fried in a coconut oil for 5–10 min).
[c] Compare with amounts in blanched leaves.
[d] Compare with amounts in stir-fried preparation.
[e] Moisture content of blanched green leaves.
[f] Moisture content of stir-fried green leaves.
Reference: Chandrika *et al.* [28.]

The variation in *in vitro* availablility of all-*trans*-β-carotene between vegetables cooked by similar methods (Table 14.4) was likely to be due to differences in leaf matrices and the rate at which the carotenoid–protein complexes are disrupted during cooking [30]. The low *in vitro* availability obtained for cooked *A. sessilis* and *C. asiatica* could be attributed to the fibrous matrix of these edible portions, which might have hindered the release of all-*trans*-β-carotene during *in vitro* digestion (Tables 14.3 and 14.4).

14.1.6 Effect of Agrochemicals

Only a few studies have been carried out on the effect of agrochemicals on carotenoid content. There are reports of its decrease [33], its increase [34] and its remaining the same [33]. Mercandante and Rodriguez Amaya reported in 1991 [22] that the carotenoid content in 'kale' samples from a natural farm is greater than that from a neighbouring farm using agrochemicals.

References

1. C. H. Azevedo-Meleiro and D. B. Rodriguez-Amaya, Carotenoid composition of kale as influenced by maturity, season and minimal processing, *Journal of Science of Food and Agriculture*, **85**, 591–597 (2005).
2. R. M. Welch, Agronomic problems related to provitamin A carotenoid-rich plants. *European Journal of Clinical Nutrition*, **51**, S34–S38 (1997).
3. D. M. R. Ramos and D.B. Rodriguez-Amaya, Determination of the vitamin A value of common Brazilian leafy vegetables, *Journal of Micronutritional Analysis*, **3**, 147–155 (1987).
4. C. H. Azevedo-Meleiro and D. B. Rodriguez-Amaya, Carotenoids of endive and New Zealand spinach as affected by maturity, season and minimal processing, *Journal of Food Composition and Analysis*, **18**, 845–855 (2005).
5. C. Y. Lee, Changes in carotenoid content of carrots during growth and post-harvest storage, *Food Chemistry*, **20**, 285–293 (1986).
6. J. Gross, Carotenoid pigments in the developing cherry (*Prunus avium*), *Gartenbauwiss.*, 50 (1985).
7. J. Gross, Chlorohyll and carotenoid pigments in *Ribes* fruits, *Scientific Horticulture*, **18**, 131–136 (1982/83).
8. M. I. Minguez-Mosquera and J. Garrido-Fernandez, Chlorophyll and carotenoid presence in olive fruit, *Journal of Agricultural Food Chemistry*, **37**, 1–7 (1989).
9. H. Boudries, P. Kefalas and D. Hornero-Mendez, Carotenoid composition of Algerian date varieties (*Phoenix dactylifera*) at different edible maturation stages, *Food Chemistry*, **101**(4), 1372–1377 (2007).
10. A. Z. Mercandanate and D. B. Rodriguez-Amaya, Effects of ripening, cultivar differences, and processing on the carotenoid composition of mango, *Journal of Agricultural Food Chemistry*, **46**, 128–130 (1998).
11. F. Granado, B. Olimedilla, Y. Blanco, E. Gil-Martinez and E. Rojas-Hidalgo, Variability in the intercomparison of food carotenoid content data: a users point of view, *Critical Review of Food Sciences and Nutrition*, **37**, 621–633 (1997).
12. M. Kimura, D. B. Rodiguez-Amaya and S. M. Yokoyama, Cultivar differences and geographic effects on the carotenoid composition and vitamin A values of papaya, *Lebensmittel-Wissenchaft and Technologies*, **24**, 415–418 (1991).
13. R. L. Rouseff, G. D. Sadler, T. J. Putnam and J. E. Davis, Determination of β-carotene and other hydrocarbon carotenoids in red grapefruit cultivars, *Journal of Agricultural Food Chemistry*, **40**, 47–51 (1992).

14. D. J. Hart and K. J. Scott, Development and evaluation of an HPLC method for the analysis of carotenoids in foods, and the measurement of the carotenoid content of vegetables and fruits commonly consumed in the UK, *Food Chemistry*, **54**, 101–111 (1995).
15. L. Englberger, W. Aalbersberg, J. Schierle, G. C. Marks, M. H. Fitzgerald, F. Muller, A. Jekkein, J. Alfred and N. Velde, Carotenoid content of different edible pandanus fruit cultivars of the republic of the Marshall Islands, *Journal of Food Composition and Analysis*, **19**, 484–494 (2006).
16. A. M. B. Priyadarshani and U. G. Chandrika, Content and *in-vitro* accessibility of pro-vitamin A carotenoids from Sri Lankan cooked non-leafy vegetables and their estimated contribution to vitamin A requirement, *International Journal of Food Sciences and Nutrition*, **58**(8), 659–667 (2007).
17. A. M. B. Priyadarshani and U. G. Chandrika, Content and composition of selected non-leafy vegetables from Sri Lanka, *Journal of National Science Foundation*, **35**(4), 251–253 (2007).
18. H. K. Arima and D. B. Rodriguez-Amaya, Carotenoid composition and vitamin A value of squash and pumpkin from North-eastern Brazil, *Archives Latinoamricano Nutrition*, **40**, 284–292 (1990).
19. M. Murkovic, U. Mulleder and H. Neunteufl, Carotenoid content in different varieties of pumpkins, *Journal of Food Composition and Analysis*, **15**, 633–638 (2002).
20. U. G. Chandrika, E. R. Jansz, S. M. D. N. Wickramasinghe and N. D. Warnasuriya, Carotenoids in yellow and red-fleshed papaya, *Journal of the Science of Food and Agriculture*, **83**(12), 1279–1282 (2003).
21. K. S. S. P Fernando, U. G. Chandrika and K. K. D. S. Ranaweera, Carotenoids from two different varieties of emberella (*Spondias ceytherea* and *Sphondias dulcis*), *Proceedings of the Sri Lanka Association for the Advancement of Science*, **62**(1), 200 (abstract) (2006).
22. A. Z. Mercandanate and D. B. Rodriguez-Amaya, Carotenoid composition of a leafy vegetable in relation to some agricultural variables, *Journal of Agricultural Food Chemistry*, **39**, 1094–1097 (1991).
23. J. L. Bureau and R. J. Bushway, HPLC determination of carotenoids in fruits and vegetables in the United States, *Journal of Food Science*, **51**, 128–130 (1986).
24. S. A. Desobry, F. M. Netto and T. P. Labuza, Preservation of β-carotene from carotene from carrots, *Critical Review of Food Sciences and Nutrition*, **38**, 381–397 (1998).
25. C. G. Edwards and C. Y. Lee, Measurement of pro-vitamin A carotenoids in fresh and canned carrots and green peas, *Journal of Food Science*, **51**, 534–535 (1986).
26. M. J. Oruna-Concha, M. J. Gonzalez-Castro, J. Lopez-Hernandez and J. Simal-Lazano, Effects of freezing on the pigment content in green beans and pardon peppers, *Z. Lebensm. Unters Forsch. A*, **205**, 148–152 (1997).
27. U. G. Chandrika, E. R. Jansz and N. D. Warnasuriya, Identification and HPLC quantification of green leafy vegetables commonly used in Sri Lanka, *Journal of National Science Foundation Sri Lanka*, **33**(2), 141–145 (2005).
28. U. G. Chandrika, U. Svenberg and E. R. Jansz, Content and *in vitro* accessibility of β-carotene from cooked Sri Lankan green leafy vegetables and their estimated contribution to vitamin A requirement, *Journal of Science of Food and Agriculture*, **86**, 54–61 (2006).
29. D. B. Rodriguez-Amaya, Effect of processing and storage on food carotenoids, *Sight and Light News Letter*, **3/2002**, 25–35 (2002).
30. K. H. van het Hof, C. Gartner, C. E. West and L. B. M. Tijburg, Potential of vegetable processing to increase the delivery of carotenoids to man. *International Journal of Vitamin and Nutritional Research*, **68**, 366–370 (1998).
31. L. Hussein and M. El-Tohamy, Vitamin A potency of carrot and spinach carotenes in human metabolic studies, *International Journal of Vitamin Nutrition Research*, **60**, 229–235 (1989).
32. M. I. Gomes, Carotene content of some green leafy vegetables of Kenya and effects of dehydration and storage on carotene retention, *Journal of Plant Foods*, **3**, 231–244 (1981).
33. J. P. Sweeney and A. C. J. Marsh, Effect of selected herbicides provitamin A content of vegetables, *Journal of Agricultural Food Chemistry*, **19**, 854–856 (1971).
34. J. Rouchand, C. Moons and J. A. Meyer. Effects of pesticide treatments on the carotenoid pigments of lettuce, *Journal of Agricultural Food Chemistry*, **32**, 1241–1245 (1984).

15

Chlorophylls

Ursula Maria Lanfer Marquez and Daniela Borrmann

15.1 Introduction

Natural chlorophylls are the most abundant pigments in nature, whose principal function is photosynthesis, the fundamental process for life on Earth. Nevertheless, although being so obviously present in nature, their use as colorants is highly restricted due to their chemical instability and high production costs. Promising alternatives with a real potential of usage are water-soluble, metal-chelated chlorophyll derivatives. These pigments have been proposed to be used as food colorants and for a wide range of other industrial applications based on unique physicochemical and photochemical characteristics. Chlorophylls have also been investigated for their outstanding biological activities, which created a shift of emphasis to exert potential beneficial effects on human health. Whole green plant extracts and natural chlorophyll-derived pigments are marketed in different pharmaceutical forms for a variety of dietary and medicinal uses. Popular therapeutic purposes include anti-inflammatory activity, accelerant of wound healing, immune modulator properties and body deodorization in geriatric and ileostomy patients, all of which have been documented in the literature for more than 50 years [1].

Most of the information relates to chlorophyll as part of a vegetable extract, but not as an isolated compound. Sodium copper chlorophyllin, on the other hand, has been studied separately as an individual synthetic compound, with respect to its antimutagenic and cancer-preventative properties, as already reported in original and review articles [2–7]. Both chlorophylls and chlorophyllins are able to reduce the mutagenic activity of a number of dietary and environmental carcinogens, but, in general, copper-complexed chlorophyllins seem to be more effective. A comparative study showed chlorophyllin to be a more effective antimutagen than retinol, β-carotene, vitamins C and E against certain dietary and environmental carcinogens [8]. Chlorophyllin can also neutralize several physically

Handbook of Natural Colorants Edited by Thomas Bechtold and Rita Mussak

relevant oxidants *in vitro*, and limited data from animal studies suggest that chlorophyllin supplementation may decrease oxidative damage induced by chemical carcinogens and radiation [9–11]. However, still controversial results reported in the literature may be related to the high variability in composition and purity of commercial preparations. It is reasonable to expect further investment in research for future development of stable chlorophylls as either complexes with metals or natural molecules embedded in a proper matrix and the establishment of adequate methods for their quality control. In addition, despite natural chlorophylls being considered safe for intake due to their long known usage in foods, there are still some uncertainties regarding toxicological aspects of synthetic chlorophyll derivatives that should be addressed in the future.

15.2 Chlorophylls as Colorants

Chlorophylls are essential pigments for photosynthesis and can be found on land in all plants, ferns, mosses and also in algae and some photosynthetic bacteria from aquatic and marine environments. Life on Earth relies on photosynthesis and besides chlorophylls some related tetrapyrroles and carotenoids are responsible for the complex regulatory function of converting solar energy into chemical energy and also to build up material. Photosynthesis occurs in two stages, a light-dependent and a light-independent stage. In the light-dependent reaction, a chlorophyll molecule absorbs one photon and loses one electron. This electron is passed to pheophytin, which passes the electron to a quinone molecule, allowing the start of a flow of electrons down an electron transport chain that leads to the reduction of NADP into NADPH. In addition, it serves to create a proton gradient across the chloroplast membrane. Its dissipation is used by the enzyme ATP-synthase for the concomitant synthesis of ATP. The chlorophyll molecule regains the lost electron by taking one from a water molecule through a process called photolysis, which releases oxygen. In the light-independent reaction the enzyme ribulose-1,5-bisphosphate carboxylase captures CO_2 from the atmosphere and, with the help of the newly formed NADPH, releases triose phosphates. Those are generally considered the prime end-products of photosynthesis. They can be used as immediate food nutrients or rearranged to energy-rich carbohydrates, like disaccharide sugars such as sucrose and fructose, which can be transported to other cells or packaged for storage as insoluble polysaccharides such as starch [12].

Chlorophylls are noncovalently connected to proteins and are always accompanied by carotenoids that protect the instable molecules from degradation. They present a closed circuit of ten conjugated double bonds, allowing them to absorb light in the blue and red regions of the spectrum and to reflect light, which is perceived as green.

Several natural chlorophylls in plants and photosynthetic organisms have already been described and are known under their trivial names as chlorophyll *a*, *b*, *c*, *d* and *e* (Figure 15.1), but it is likely that more structures will be discovered as trials of less known photosynthetic organisms progress. Higher plants contain only chlorophylls *a* and *b* and their respective breakdown metabolites like pheophytins, chlorophyllides and pheophorbides. In addition to chlorophylls *a* and *b*, chlorophylls *c*, *d* and *e* are found in algae and there are also bacteriochlorophylls present in photosynthetic bacteria [13].

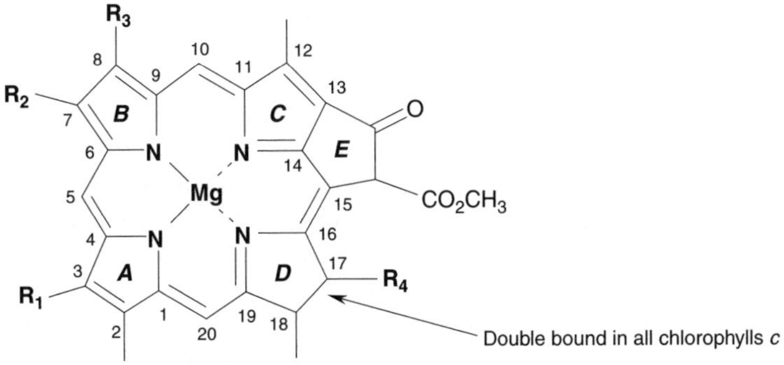

	R_1	R_2	R_3	R_4
Chlorophyll *a*	$CH{=}CH_2$	CH_3	CH_2CH_3	$(CH_2)_2{-}COO{-}C_{20}H_{39}$
Chlorophyll *b*	$CH{=}CH_2$	CHO	CH_2CH_3	$(CH_2)_2{-}COO{-}C_{20}H_{39}$
Chlorophyll c_1	$CH{=}CH_2$	CH_3	CH_2CH_3	$CH_2{=}CH{-}COOH$
Chlorophyll c_2	$CH{=}CH_2$	CH_3	$CH{=}CH_2$	$CH_2{=}CH{-}COOH$
Chlorophyll c_3	$CH{=}CH_2$	$COOCH_3$	$CH{=}CH_2$	$CH_2{=}CH{-}COOH$
Chlorophyll *d*	CHO	CH_3	CH_2CH_3	$(CH_2)_2{-}COO{-}C_{20}H_{39}$

Figure 15.1 *Molecular structure of natural chlorophylls*

Chlorophylls have received attention for a long time, not only because of their significance in living systems but also because of their potential relevance as natural pigments in a limited range of applications as colorants. The intense green colour of natural chlorophylls suggests that they may be useful as commercially available lipid-soluble additives in foods, pharmaceutical and cosmetic products and toiletries. Natural chlorophylls are permitted as food colorants throughout Western Europe and can be extracted from an assortment of green leaves and algae. However, in practice natural chlorophylls are rarely used as colorants for several reasons. Firstly, carotenoids, phospholipids and other oil-soluble substances are co-extracted, resulting in products with diversified composition and variable levels of pigments, which makes subsequent purification steps indispensable. Secondly, endogenous plant enzymes and extraction conditions employed can easily promote chemical modification on the sensitive chlorophylls, yielding unattractive brownish-green degradation products like pheophytins and pheophorbides. Consequently, the production costs considering the mentioned difficulties are very high and therefore a more widespread application of natural chlorophylls as colorants is limited. To overcome some of these limitations, semi-synthetic, metal-chelated and water-soluble chlorophyll derivatives, called chlorophyllins, have been produced as promising alternatives to generate colorants with a higher stability and tinctorial strength.

The most successful commercial product of metal-chelated chlorophyllins corresponds to sodium copper chlorophyllin and its commercial production has been described in details by Hendry [14]. It is prepared from a crude chlorophyll extract, usually derived from alfalfa (*Medicago sativa*), or a few genera of pasture grasses. The peripheral phytyl group (a 20C-diterpene) attached by an ester linkage to the macro cycle of chlorophyll *a* and *b* and some other derivatives is hydrolysed by weak alkali or weak acid or even by the action of chlorophyllase, yielding complex mixtures of water-soluble chlorophyllins. Then, in an acidic environment the central magnesium is replaced by hydrogen and the molecule reacts with copper to form a stable blue-green complex. The two main components found in these preparations are trisodium copper chlorin e6 and disodium copper chlorin e4, depicted in Figure 15.2. There are several commercial chlorophyllin products available that have been developed in a broad range of colours, and each batch has been shown to present a different profile of chemical constituents. Some of these have different metal ions and contents, while others have slightly modified chromophores. Sodium copper chlorophyllin is available as a powder or in liquid form and has been used in Europe as a colorant in sweets, ice cream, desserts, cheese and cucumber relishes, mainly in combination with carotenoids or curcumin. In Western Europe, copper chlorophyllins are permitted as food colorants, while their use in the United States of America is limited to dentifrice and as a colour additive in citrus-based dry beverage mixes [15].

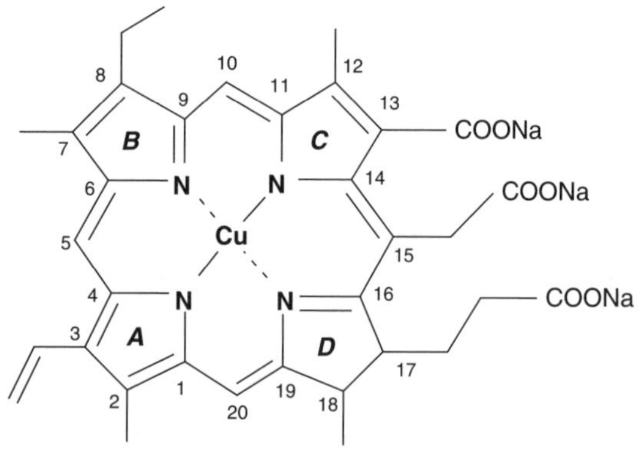

Figure 15.2 *Structure of trisodium copper chlorin e6. The substitution of Na by H at C-13 produces disodium copper chlorin e4. Both substances belong to the main constituents of commercial sodium copper chlorophyllin*

Recently, copper chlorophyllin has been proposed to be applied as a colorant in textile and leather industries as well as for paper production. As synthetic colorants have been questioned because of their toxicity and some of them have already been forbidden in many European countries, the use of natural or semi-synthetic colorants has become attractive. In 2004, the first paper with colours extracted from natural and water-based colorants was introduced into the market. Chlorophyllins have shown good interaction with other additives of papermaking and therefore may be considered an ecologically appealing alternative for dyeing paper products [16].

15.3 Other Applications of Chlorophylls and their Derivatives

Some applications of chlorophylls and chlorophyllins, which are not related to their green colour but to their photochemical characteristics, have been investigated. For example, an interesting application of a chlorophyll *a* derivative (methyl 3-carboxy-3-devinyl-pyro-pheophorbide *a*), in which the carboxyl group is directly attached to the conjugated macrocycle to facilitate efficient electron transfer, would be its integration in titania-based Grätzel-type dye-sensitized solar cells. Experiments performed in the laboratory showed satisfying results with this derivative, whereas natural chlorophylls and pheophorbides tested did not show sufficient photo-conversion efficiency [17]. This potential energy source deserves to be exploited more precisely in the future.

Copper chlorophyllin has also been reported to have an influence on health. It has been traded as an over-the-counter drug for controlling body odours in geriatric and ileostomy patients, as an anti-inflammatory agent and an accelerant in wound healing [18]. However, the most encouraging investigations in *in vitro* and *in vivo* studies reported in the medical area are related to potential health benefits of chlorophylls and chlorophyllins, due to antioxidant and antimutagenic activities, modulation of metabolizing enzymes, reduction of the bioavailability of mutagens and carcinogens and radioprotective properties [7, 9, 10, 19]. These biological activities are consistent with the prevention of human cancers, but mechanisms that may be involved are not yet completely elucidated. Nevertheless, few other studies report adverse effects, such as genotoxicity, clastogenicity and tumor-promotion [3, 20]. One of the mechanisms of cancer preventative action has been attributed to the formation of strong noncovalent complexes with planar aromatic carcinogens from dietary and environmental sources, like aflatoxin, nitrosamines and benzo[*a*]pyrene. When co-administered with mutagens or carcinogens, chlorophyllin and chlorophyll reduce the toxicant-DNA binding by acting as an inhibitor or interceptor molecule, itself forming reversible complexes in the gut and tissues and thereby interfering with the absorption from the intestinal tract [21].

More recently, there has been disputed evidence that chlorophylls have also been used for phototherapy, taking advantage of their unique physicochemical characteristics. In the screening for substances that might be useful for photodynamic therapy of cancer, chlorophylls and related tetrapyrroles have been found to possess the ability to cause the death of tumour cells. However, these substances after absorbing light quanta are poorly selective for tumour cells and photodynamic damage can be propagated in healthy cells by direct reaction with substrates like lipid fatty acids. Therefore there has been a certain concern regarding phototoxicity against normal cells. Until now, there is a limited picture regarding side effects and additional studies regarding the biological and toxicological effects of applying chlorophyllin topically are warranted. The relatively low extinction coefficient of chlorophyllins demands a high pigment concentration to promote photosensitization and applied light presents poor skin penetration, which does not reach more than five millimetres [22].

15.4 Chemical Structures and Physicochemical Properties

The knowledge of the array of the different, but closely related, chemical structures of natural porphyrins and the several steps in their biosynthesis and interrelationships is useful for a better understanding of their physicochemical properties.

Briefly, the biosynthesis of the tetrapyrrole macrocycle starts from 5-aminolevulinic acid (ALA), which itself is formed by the condensation of succinyl-CoA and glycine. The ALA is subsequently transformed into uroporphyrinogen III, which is the first tetrapyrrole macrocycle and precursor of all natural tetrapyrroles (chlorophylls, haem and bilins). Enzymatic modification yields protoporphyrin IX, and from this step on, biosynthesis differs between the tetrapyrrole groups. Depending on which metal is inserted in the centre of the porphyrin, either haem is formed by inclusion of iron or chlorophyll by insertion of magnesium (Figure 15.3).

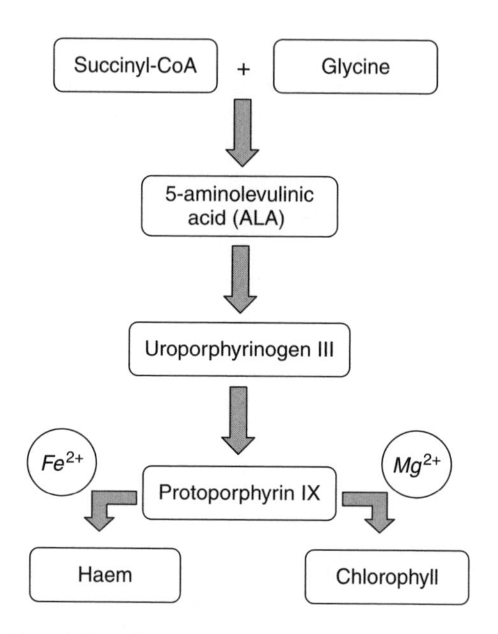

Figure 15.3 *Scheme of chlorophyll synthesis*

Natural chlorophyll molecules are basically square planar arrangements of four pyrrole rings, conjugated by methine bridges, to which a fifth cyclopentanone ring has been added. The four nitrogen atoms of the macrocyle are coordinated with a centrally located magnesium ion and at C-17 a propionic acid residue is added, usually esterified with phytol, a long-chain alcohol. The whole chlorophyll molecule is hydrophobic, but acquires a hydrophilic character after hydrolysing the ester bond and releasing the phytol.

Porphyrins can be found in three states of oxidation: true porphyrins that have the highest number of double bonds; dihydroporphyrins, in which the carbon between C-17 and C-18 (ring D) is saturated, and tetrahydroporphyrins, which have an additional reduction between C-7 and C-8 at ring B. Therefore, chlorophylls *a* and *b* are not true porphyrins, but dihydroporphyrins, to which a fifth ring, a cyclopentanone, has been added. Chlorophyll *b* differs from chlorophyll *a* only by having an aldehyde group at C-7, instead of a methyl group. Chlorophyll *c*, despite being named chlorophyll, is a true porphyrin due to a double bond between C-17 and C-18 of ring D. Chlorophyll *d* seems to be originated from chlorophyll *a*, being structurally similar except for the vinyl group at C-3, which is oxidized to a formyl group [23]. Figure 15.1, previously shown, summarizes the chemical constitution of the chlorophylls *a*, *b*, *c* and *d*. Tetrahydroporphyrins, the most reduced forms, also called bacteriochlorophylls, are only found in some photosynthetic bacteria.

Natural chlorophylls and their derivatives absorb strongly in the red and blue regions of the visible spectrum and therefore reflect and transmit green light due to the presence of several conjugated carbon–carbon double bonds arranged as resonance hybrids. Therefore, absorbance spectroscopy is the simplest way roughly to identify and to quantify chlorophylls in an extract. The whole chromophore is raised to an excited state that can be detected by typical absorbance bands. Reflection of light is highest at the absorption minimum around 550 nm and lowest at the absorption maximum around 440 nm, which also includes carotenoids. Small structural differences allow each type of natural chlorophyll to absorb light at different wavelengths and their absorbance spectra are vaguely different, though overlapping. Figure 15.4 shows the typical absorption spectra of chlorophylls *a* and *b*. The nature of the central atom, peripheral groups and the environment of the pigment, like the solvent and temperature, may strongly influence the position of maximum absorbance and the shape of the spectrum.

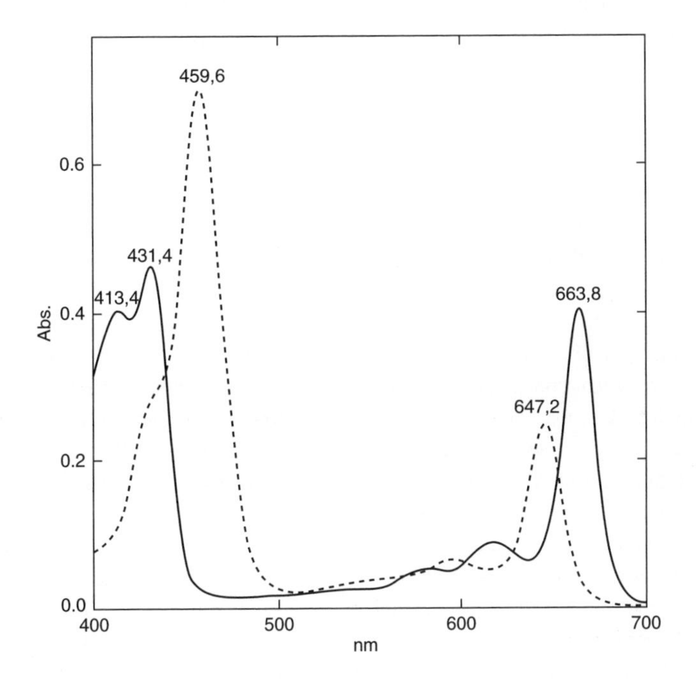

Figure 15.4 *Absorbance spectra of isolated chlorophyll a (continuous line) and b (dashed line)*

Commercially available sodium copper chlorophyllin, the most investigated metal-chelated derivative, is not a single compound but a mixture of several water-soluble chlorin compounds. Depending on the raw material and the manufacturing process employed, saponification can result in hydrolysis of the phytyl and methyl ester groups, cleavage of the cyclopentanone ring and several other secondary reactions. Although the major components are known to be copper(II)pheophorbide *a*, copper chlorin e6 and e4 and copper(II)rhodin g7 (an aldehyde analogue of chlorin e6), other derivatives and degradation products can be found, some of them lacking metal. The corollary is that commercial products may have a batch-to-batch variability in composition that makes any quality control extremely difficult [24].

15.5 Stability and Analysis

Natural chlorophylls are highly instable and subject to colour changes, being sensitive against low pH, heat, light, oxygen and the simultaneous action of enzymes. Numerous studies have been conducted to investigate the colour changes from bright blue-green to olive-brown of many vegetables, and a first-order kinetic has already been established for formation of pheophytins *a* and pheophytins *b* from their respective chlorophylls [25]. Pheophytins and pheophorbides are the most stable pigments resulting from chlorophyll degradation among those naturally found in vegetables. Chlorophylls are easily degraded, unless they are located inside the chloroplasts attached to carotenes, xantophylls and a highly structured internal system of membranes, called thylakoids. These envelopes serve to isolate and to protect the chlorophylls. However, cutting vegetables, subjecting vegetables to heating treatments and isolating the pigments causes disruption of this protective environment and chlorophylls are quickly degraded.

Conversely, metallochlorophylls, in which the magnesium at the centre of the porphyrin ring has been replaced with a transition element like copper, are stable to moderate light and heat and are more resistant to low pH and oxidative reactions in general. For instance, the fat-soluble copper chlorophyll was lost only to 42 % after 500 h of exposure to light, while the magnesium chlorophyll lost 95 % during only 45 h of light exposure [26]. The presence of 3d orbitals in the copper molecule seems to increase the covalent character of the linkage between the ligand and the copper acceptor, stabilizing the copper chlorophyll structure.

Analysis of natural chlorophylls, semi-synthetic metallochlorophyllins and their derivatives constitutes an analytical challenge, due to the chemical heterogeneity and different solubilities of the compounds. Chlorophylls and metallochlorophylls are rather hydrophobic and can be extracted with a single solvent or a mixture of organic solvents. The main strategies for determination of chlorophylls are based on spectrophotometric and chromatographic methods and there are many literature references available [27]. Absorbance spectroscopy appears to be the simplest way to identify major pigments, and comparisons of their spectra with those of pure pigments allow their identification. However, single absorbance measurements are ineffective for measurements of individual pigment concentration. Individual components, their eventual degradation products and metallochlorophylls have been successfully separated by RP-HPLC and can be detected either with DAD, spectrophotometric or fluorimetric detectors coupled or not to MS, due to the lack of proper standards [28].

15.6 Sources, Storage and Handling

Chlorophylls are the most abundant pigments in plants. Dark green leafy vegetables like spinach, parsley and garden cress are rich sources of natural chlorophylls but are not explored for commercial production of isolated pigments. Algae, silkworm excreta, alfalfa, pine needles, other pasture grasses and plant harvest by-products have been used to extract chlorophylls. However, chlorophyll concentrations in plant sources are highly influenced by climate conditions, like temperature, humidity, oxygen and nitrogen concentrations in water and air, sun exposure and geographical location, and therefore harvested amounts differ in a wide range.

Mass cultivation of microalgae has received a lot of attention during the last decades for diverse applications. However, the use of algae for commercial production of pigments is restricted to a few plants, mainly *Dunaliella* in Australia and Israel and *Spirulina* and *Haematococcus* in the USA [29].

Especially the photoautotrophic and unicellular microalga, *Spirulina platensis*, that is present in aqueous and saline habitats is rich in polyunsaturated fatty acid and has high contents of vitamins, especially B_{12} and several pigments, like carotenoids, xantophylls, phycobilins and chlorophylls. Carotenoids from *Spirulina* have already found commercial application in the food, cosmetic and pharmaceutical industries, but extraction of chlorophylls from algae is still hampered due to the high costs of extraction and drying and the inherent difficulty to disrupt the cellulose membrane. *Spirulina* fresh biomass is one of the largest sources of chlorophyll in nature, containing a tenfold higher amount than fresh spinach [30, 31]. The incorporation of the whole biomass instead of isolation of individual pigments may be of particular interest for food application, because it will provide green colour and increase its nutritional value [32].

The green colour is mainly due to the pigments chlorophyll *a* (blue-green) and chlorophyll *b* (yellow-green), but degradation products can be formed depending on the stage of maturity or senescence of the raw material, so precautions must be taken during storage and handling. In the early stages of catabolism, the replacement of magnesium by two hydrogen atoms takes place in an acidic environment and enzymes are responsible for the cleavage of the phytol chain. Nevertheless, the degradation products, pheophytins, chlorophyllides and pheophorbides, are still greenish since the tetrapyrrole ring has not been affected. Only late oxidative degradation steps characterized by an opening of the tetrapyrrole are responsible for the complete loss of coloration [33]. The disappearance of chlorophyll is a natural process related to a planned slowing down of photosynthesis in metabolically active tissues.

Almost any type of handling, processing and storage of raw material leads to a fading of the original colour and pheophytins are the predominant degradation products found. For that reason, harvest followed by processing should be carried out in the shortest possible time since, once plants are harvested, chlorophylls are consistently chemically and enzymatically degraded. Exposure to acidic media should be avoided and enzymes should be inactivated by blanching.

15.7 Purity, Standardization and Quality Control

Spectrophotometric assays, widely used for determination of natural chlorophylls, are inadequate for correct determination of each component as all components have similar absorption spectra. Spectrometric analytical results disagree with data obtained by elemental analysis. There are no chlorophyllin standards for quantification and there are no officially approved analytical methods to monitor the quality control. In studies where spectrophotometric analysis was accompanied by elemental analysis of copper and nitrogen, a significant overestimation by the spectrophotometric analysis was found. This can be explained by the optical properties of the chlorophyllin in the blue region of the spectrum where measurements take place. The absorption increases significantly if other colouring compounds like carotenoids and metal-free chlorophyll analogues are present in the sample. The ratio between copper and nitrogen contents has been used frequently as an indicator of the purity of chlorophyllin; however, studies have also shown inconsistent results. For an optimized quality control, the copper/nitrogen index should be accompanied by measurement of the percentage of absolute copper and nitrogen. The interference of carotenoids can be avoided by obtaining the total absorption in the red region of the spectrum [34, 35]. This heterogeneity of commercial samples of chlorophyll–copper complexes may explain the different biological responses observed when administered to laboratory animals in *in vitro* and *in vivo* tests. In addition, proper methods of analysis are needed for quality control [36, 37].

15.8 Toxicological and Safety Aspects

Natural chlorophylls are not known to be toxic, and no acute toxic effects have been reported for sodium copper chlorophyllin despite more than 50 years of use as an over-the-counter drug. Nevertheless, reported anticarcinogenic, antimutagenic, antioxidant and radioprotective effects have created an expectancy regarding potential health benefits, and as a consequence, safety and toxicological aspects have arisen. This seems especially true for the copper-complexed chlorophyllins derived from natural chlorophylls.

Information on the absorption of these compounds into the bloodstream that could justify post-absorptive biological effects is still limited, partially because of the general supposition that those chlorophylls are not absorbable. However, cumulative scientific evidence is strengthening the assumption that after intake by mammals, at least a minimal absorption of both natural chlorophylls and semi-synthetic water-soluble derivatives occurs [38–40]. Recently, Egner *et al.* [41] provided concrete evidence that commercial copper chlorophyllin composed mainly of copper-chlorin e4, copper-chlorin e6 and copper-chlorin e4 ethyl ester can be absorbed by humans.

Concerning toxicological aspects, the Joint FAO/WHO (Food and Agriculture Organization/World Health Organization) Expert Committee on Food Additives (JECFA) [42] evaluated in 1975 the original study performed by Harrison *et al.* in 1954 [38], and due to the lack of reported adverse effects in rats, recognized 1500 mg of copper-chlorophyllin per kg body weight/day as the no observed effect level (NOEL). By applying a 200-fold safety factor to this NOEL, the acceptable daily intake was calculated as 7.5 mg/kg body weight/day.

The American Food and Drug Administration also reviewed the study published by Harrison *et al.* [38] and permits the over-the-counter use of chlorophyllin copper complex as an internal deodorant for incontinent individuals in doses up to 300 mg/day [43].

Since the safety of chlorophyll or chlorophyllin supplements has not been tested in pregnant or lactating women, they should be avoided during these periods. In view of the toxicological data, considerable research should still be invested for risk assessment in order to identify adverse health effects related to semi-synthetic chlorophyll derivatives.

References

1. J. C. Kephart, Chlorophyll derivatives – their chemistry, commercial preparations and uses, *Economic Botany*, **9**, 3–38 (1955).
2. R. H. Dashwood, V. Breinholt and V. Bailey, Chemopreventive properties of chlorophyllin: inhibition of aflatoxin B1 (AFB1)-DNA binding *in vivo* and anti-mutagenic activity against AFB1 and two heterocyclic amines in the *Salmonella* mutagenicity assay, *Carcinogenesis*, **12**(5), 939–942 (1991).
3. D. Sarkar, A. Sharma and G. Talukder, Chlorophyll and chlorophyllin as modifiers of genotoxic effects, *Mutation Research*, **318**, 239–247 (1994).
4. V. Breinholt, J. Hendricks, C. Pereira, D. Arbogast and G. Bailey, Dietary chlorophyllin is a potent inhibitor of aflatoxin B_1 hepatocarcinogenesis in rainbow trout, *Cancer Research*, **55**, 57–62 (1995).
5. M. G. Ferruzzi, V. Böhm, P. D. Courtney and S. J. Schwartz, Antioxidant and antimutagenic activity of dietary chlorophyll derivatives determined by radical scavenging and bacterial reverse mutagenesis assays, *Journal of Food Science*, **67**, 2589–2595, 2002.
6. P.A. Egner, A. Muñoz, and T. W. Kensler, Chemoprevention with chlorophyllin in individuals exposed to dietary aflatoxin, *Mutation Research*, **523–524**, 209–216 (2003).
7. M. G. Ferruzzi and J. Blakeslee, Digestion, absorption and cancer preventative activity of dietary chlorophyll derivatives, *Nutrition Research*, **27**, 1–12 (2007).
8. T. M. Ong, W. Z. Whong, J. Stewart and H. E. Brockman, Comparative antimutagenicity of five compounds against five mutagenic complex mixtures in *Salmonella typhimurium* strain TA98, *Mutation Research*, **222**, 19–25 (1989).
9. J. P. Kamat, K.K. Boloor and T. P. A. Devasagayam, Chlorophyllin as an effective antioxidant against membrane damage *in vitro and ex vivo, Biochimica e Biophysica Acta*, **1487**, 113–127 (2000).
10. K. K Boloor, J. P. Kamat and T. P. A. Devasagayam, Chlorophyllin as a protector of mitochondrial membranes against gamma-radiation and photosensitization, *Toxicology*, **155**(1–3), 36–71 (2000).
11. S. S. Kumar, B. Shankar and K. B. Sainis, Effect of chlorophyllin against oxidative stress in splenic lymphocytes *in vitro* and *in vivo, Biochimica et Biophysica Acta*, **1672**(2), 100–111 (2004).
12. Govindjee, J. T. Beatty, H.Gest and J. F. Allen (eds), *Advances in Photosynthesis and Respiration – Discoveries in Photosynthesis*, Vol. **20**, Springer, Berlin, 2005.
13. G. A. F. Hendry, Chlorophylls, in G. J. Lauro and F. J. Francis (eds), *Natural Food Colorants – Science and Technology*, IFT Basic Symposium Series, Marcel Dekker, New York, 2000.
14. G. A. F Hendry, Chlorophylls and chlorophyll derivatives, in G. A. F Hendry and J. D. Houghton (eds), *Natural Food Colorants*, 2nd edition, Chapman & Hall, London, 1996.
15. US Food and Drug Administration, Listing of color additives exempt from certification: sodium copper chlorophyllin, 21 CFR Part 73, 67 Federal Register, 35429, May 20, 2002.
16. E. M. D. Frinhani and R. C. Oliveira, The applicability of natural colorants in papermaking, *Tappi Journal*, **5**(7), 3–7(2006).
17. X. F. Wang, J. Xiang, P. Wang, Y. Koyama, S. Yanagida, Y. Wada, K. Hamada, S. Sasaki and H. Tamiaki, Dye-sensitized solar cells using a chlorophyll a derivative as the sensitizer and carotenoids having different conjugation lengths as redox spacers, *Chemical Physics Letters*, **408**, 409–414 (2005).
18. S. A. Chernomorsky and A. B. Segelman, Biological activities of chlorophyll derivatives, *New Jersey Medicine*, **85**(8), 669–673 (1988).
19. S. S. Kumar, R. C. Chaubey, T. P. A. Devasagayam, K. I. Priyadarsini and P. S. Chauhana, Inhibition of radiation-induced DNA damage in plasmid pBR322 by chlorophyllin and possible mechanism(s) of action, *Mutation Research*, **425**(1), 71–79 (1999).
20. L. Romert, M. Curvall and D. Jenssen, Chlorophyllin is both a positive and negative modifier of mutagenicity, *Mutagenesis*, **7**, 349–355 (1992).
21. V. Breinholt, M. Schimerlik, R. Dashwood and G. Bailey, Mechanisms of chlorophyllin anti-carcinognenesis against aflatoxin B1. Complex formation with the carcinogen, *Chemical Research in Toxicology*, **8**, 506–514 (1995).
22. D. K. Kelleher, O. Thews, A. Scherz, I. Salomon and P. Vaupel, Combined hypothermia and chlorophyll-based photodynamic therapy: tumour growth and metabolic microenvironment, *British Journal of Cancer*, **89**, 2333–2339 (2003).

23. U. M. L. Marquez and P. Sinnecker, Chlorophylls: properties, biosynthesis, degradation and functions, in C. Socaciu (ed.), *Food Colorants – Chemical and Functional Properties*, Chemical and Functional Properties of Food Components Series, CRC Press, Boca Raton, Florida, 2008.
24. H. Inoue, H. Yamashita, K. Furuya, Y. Nonomura, N. Yoshioka and S. Li, Determination of copper(II) chlorophyllin by reversed-phase high-performance liquid chromatography, *Journal of Chromatography A*, **679**, 99–104 (1994).
25. C. A. Weemaes, V. Ooms, A. M. Van Loey and M. E. Hendrickx, Kinetics of chlorophyll degradation and color loss in heated broccoli juice, *Journal of Agricultural and Food Chemistry*, **47**, 2404–2409 (1999).
26. P. A. Bobbio and M. C. Guedes, Stability of copper and magnesium chlorophylls, *Food Chemistry*, **36**, 165–168 (1990).
27. B. Schoefs, Chlorophyll and carotenoid analysis in food products. Properties of the pigments and methods of analysis, *Trends in Food Science and Technology*, **13**, 361–371 (2002).
28. M. J. Scotter, L. Castle and D. Roberts, Method development and HPLC analysis of retail foods and beverages for copper chlorophyll (E141[i]) and chlorophyllin (E141[ii]) food colouring materials, *Food Additives and Contaminants*, **22**(12), 1163–1175 (2005).
29. L. Gouveia, A. Reis, V. Veloso and J. A. Empis, Microalgal biomass as a sustainable alternative raw material, *International Journal for Green Chemistry*, **May/June**, 28–34 (1996).
30. F. Khachik, G. R. Beecher and N. F. Whittaker, Separation, identification and quantification of the major carotenoid and chlorophyll constituents in extracts of several green vegetables by liquid chromatography, *Journal of Agricultural and Food Chemistry*, **34**, 603–616 (1986).
31. R. Henrikson, *Earth Food Spirulina*, Ronore Enterprises Inc, California, 1989.
32. E. D. G. Danesi, C. O. Rangel-Yangui, J. C. M. Carvalho and S. Sato, An investigation of effect of replacing nitrate by urea in the growth and production of chlorophyll by *Spirulina platensis, Biomass and Bioenergy*, **23**, 261–269 (2002).
33. S. Hörtensteiner, K. L. Wüthrich, P. Matile, K. H. Ongania and B. Kräutler, The key step in chlorophyll breakdown in higher plants: cleavage of pheophorbide *a* macrocycle by a mono-oxygenase, *Journal of Biological Chemistry*, **273**(25), 15335–15339 (1998).
34. S. Chernomorsky, Chlorophyllin copper complex: quality control, *Journal of the Society of Cosmetic Chemists*, **44**, 235–238 (1993).
35. S. Chernomorsky, Variability of the composition of chlorophyllin, *Mutation Research*, **324**(4), 177–178 (1994).
36. R. H. Dashwood, The importance of using pure chemicals in (anti) mutagenicity studies: chlorophyllin as a case in point, *Mutation Research*, **381**(2), 283–286 (1997).
37. S. Chernomorsky, R. Rancourt, D. Sahai and R. Poretz, Evaluation of commercial chlorophyllin copper complex preparations by liquid chromatography with photodiode array detection, *Journal of AOAC International*, **80**(2), 433–435 (1997).
38. J. W. E. Harrison, S. E. Levi and B. Trabin, The safety and fate of potassium sodium copper chlorophyllin and other copper compounds, *Journal of the American Pharmaceutical Association*, **43**, 722–737 (1954).
39. K. K Park and Y. J. Surh, Chemopreventive activity of chlorophyllin against mouse skin carcinogenesis by benzo[a]pyrene and benzo[a]pyrene-7,8-dihydrodiol-9,10-epoxide, *Cancer Letters*, **102**, 143–149 (1996).
40. T. M. Fernandes, B. B. Gomes and U. M. L. Marquez, Apparent absorption of chlorophyll from spinach in an assay with dogs, *Innovative Food Science and Emerging Technologies*, **8**, 426–432 (2007).
41. P. A. Egner, K. H. Stansbury, E. P. Snyder, M. E. Rogers, P. A. Hintz and T. W. Kensler, Identification and characterization of chlorine-4 ethyl ester in sera of individuals participating in the chlorophyllin chemoprevention trial, *Chemical Research in Toxicology*, **13**, 900–906 (2000).
42. FAO/WHO, Toxicological evaluation of some food colours, enzymes, flavour enhancers, thickening agents and certain other food additives, Joint FAO/WHO Expert Committee on Food Additives, *WHO Food Additives Series 6*, pp. 74–77, Geneva, 1975.
43. Federal Register, R., Deodorant drug products for over-the-counter use, Final Monograph, Vol. **55**(92), pp.19862–19865.

Part IV

Application in Technical Use and Consumer Products

16

Flavonoids as Natural Pigments

M. Monica Giusti and Taylor C. Wallace

16.1 Introduction

Consumers first assess the freshness and quality of food by its appearance, which makes color a vital characteristic for the ultimate initial acceptance of a product. The flavonoid family comprises the most widely distributed group of secondary plant products consumed by humans, and are responsible or contribute to the color of many foods. Many fruits and vegetables owe their attractive colors to the presence of flavonoid pigments that range from white/cream to red to purple to blue. Frequently, yellow producing flavonoids (chalcones, aurones and some flavonols) co-occur with carotenoids mainly in flower species [1]. However, anthocyanins represent the most common group of natural flavonoid colorants on the market, being responsible for most of the orange-red to purple colors in nature. In processed foods, natural flavonoid pigmentation can be lost easily during manufacturing and enhancement of the product is obtained by the addition of other colorants. Recently, food manufacturers have begun to re-introduce many natural flavonoid colors to foods as opposed to synthetic dyes because of their functional health properties and findings of new colorant extract sources that exhibit an improved performance.

As a component of the human diet, flavonoids are known to have health promoting activities because of their antioxidant properties *in vitro* and *in vivo* [2]. Flavonoids have the capability to induce protective enzyme systems, guarding the body from many chronic age related diseases including cardiovascular disease, macular degeneration and cancer [2], as well as some pathological disorders such as gastric and duodenal ulcers, vascular fragility and viral infections [3]. Biological effects of flavonoids on mammalian metabolism and antiallergenic, anti-inflammatory, and anti-proliferative activities have recently been reported [4].

There are over 6000 naturally occurring flavonoids that have been isolated from plants [5], most of which are pale, yellow or colorless. Conversely, a group of around 600

Handbook of Natural Colorants Edited by Thomas Bechtold and Rita Mussak

flavonoids known as the anthocyanins are responsible for producing intense red, blue and purple colors in many plants. Anthocyanins extracts/colorants in food may impart color, in addition to their potential influence of taste, antioxidant power and protection of vitamins and enzymes. The bioavailability of anthocyanins and other flavonoids to humans seems to be limited and dependent upon the plant source and food matrix. Because this book deals directly with natural colorants, the focus of this chapter will pertain to anthocyanins and other color producing flavonoids.

16.2 Role of Localized Flavonoids in the Plant

The flavonoid family comprises the most widely distributed group of secondary plant products consumed by humans. In the average human diet, intake of flavonoids has been estimated to be between 100 and 1000 mg/day [6]. The highest levels of flavonoids in the human diet include soy isoflavones, flavonols, flavones and anthocyanins. Total flavonoid concentration in foods is dependent upon genetic factors of the plant species and environmental conditions such as light, ripeness and post-harvest processing preparation [7].

Plants are localized and thus are increasingly vulnerable to a variety of harmful environmental stress factors including frost hardiness, drought, fungal/microbial symbiosis, metal composition of soil and most notably exposure to UV radiation. Production of flavonoids is a mechanism that aids in the plant's protection by providing photo-protective [8], antioxidant [9], antifungal [10] and antimicrobial [11] functions. Several species of herbaceous insects are deterred or are sensitive to flavonoids [12]. However, the most obvious role of many color producing flavonoids is providing beautiful visual appeal to flowers, fruits and vegetables, while serving as pollination attractants and encouraging seed dispersal in most angiosperms.

Flavonoids accumulate in the epithelial cells of a mature plant's leaves, pollen, stigmata, and/or floral primordia [13]. Build-up of these compounds in a young plant's stem is also common. Although more commonly present in the plant cell vacuole, anthocyanins can accumulate in vesticles, eventually coalescing into a structure known as an anthocyanoplast in a number of species [14]. Similar vesticle formation can be found with proanthocyanidin phytoalexins in the cytoplasm of sorghum cells near a fungal infection site [15]. Once attached to a fungal infection site, the phytoalexins are released near the pathogen to aide in the protection and maintenance of the cell [15]. These antimicrobial effects of flavonoids have been described, but the roles and mode of action is not yet fully understood [16].

16.3 General Flavonoid Chemical Structure

The chemical nature of a flavonoid depends on its structural class and the degree of hydroxylation/methoxylation and conjugation. Flavonoids contain a C_6—C_3—C_6 carbon structure, which varies around the characteristic heterocyclic oxygen ring (Figure 16.1). All flavonoid compounds are derivatives of a 2-phenylchromone structure composed of three phenolic rings known as the A-, B- and C-rings [17]. These rings can exhibit different patterns of sugar, acid and R-group conjugations, which may play a significant role in the bioactivity of the compound.

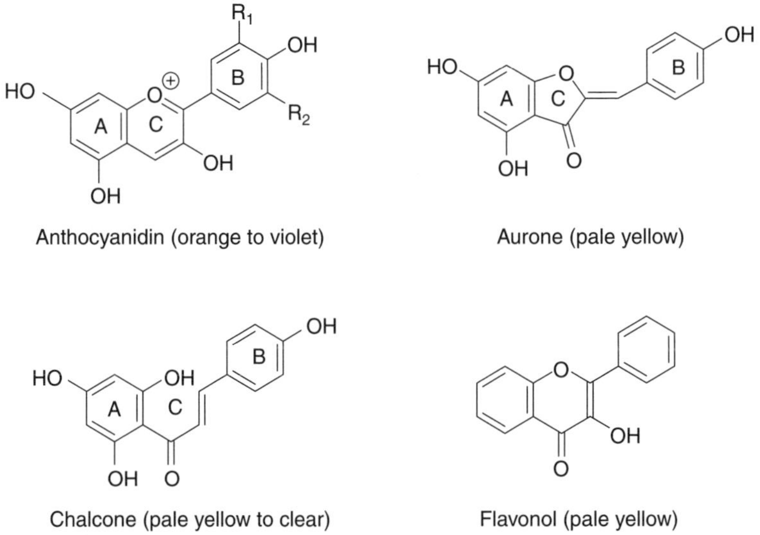

Anthocyanidin (orange to violet)

Aurone (pale yellow)

Chalcone (pale yellow to clear)

Flavonol (pale yellow)

Figure 16.1 *General base structures of color producing flavonoids.* R_1/R_2 *= H, OH and/or* OCH_3

Most flavonoids exhibit two major absorption bands, those being from 250 to 285 nm (A-ring) and 320 to 385 nm (B-ring) [18]. Flavonoids that absorb light in the visible spectrum between 400 and 700 nm are known as pigments. Anthocyanins produce the most intense flavonoid pigment color, absorbing light between 490 and 550 nm [19]. Pigments contain unbound or 'loosely bound' electrons, which require less energy to become excited and thus allow the compounds to absorb light in the visible region [18]. The color produced is not only characteristic of the absorbed light but also that which is reflected or scattered as well. An increase in double bond conjugation and/or hydroxylation of the pigment allows the compound to absorb at a higher wavelength and produce color.

16.4 Biosynthesis of Flavonoids

Flavonoids are produced via the phenylpropanoid pathway, which is also responsible for producing many plant secondary metabolites including lignans, lignins, stilbenes and hydroxycinnamic acids [1]. These compounds are first produced in the cytosol of the plant cell and later transferred to the vacuole [20]. Final conjugation of the flavonoid aglycon with a sugar glucoside is thought to be involved in the transfer of these compounds into the vacuole [21]. All flavonoids are derived from the amino acid phenylalanine and the citric acid cycle derivative acetyl-CoA, but there are many different paths in the production of various groups (Figure 16.2). The flavonoid structure consists of two phenyl rings that are connected by a three-carbon bridge, usually forming the heterocyclic oxygen or C-ring. The biosynthesis of color producing flavonoids, those being anthocyanins, chalcones, aurones and flavonols, which have the ability to absorb light in the visible spectrum of 400 to 700 nm, are shown in Figure 16.2. We recommend a more extensive review of general flavonoid biosynthesis in the book of Andersen and Markham [22].

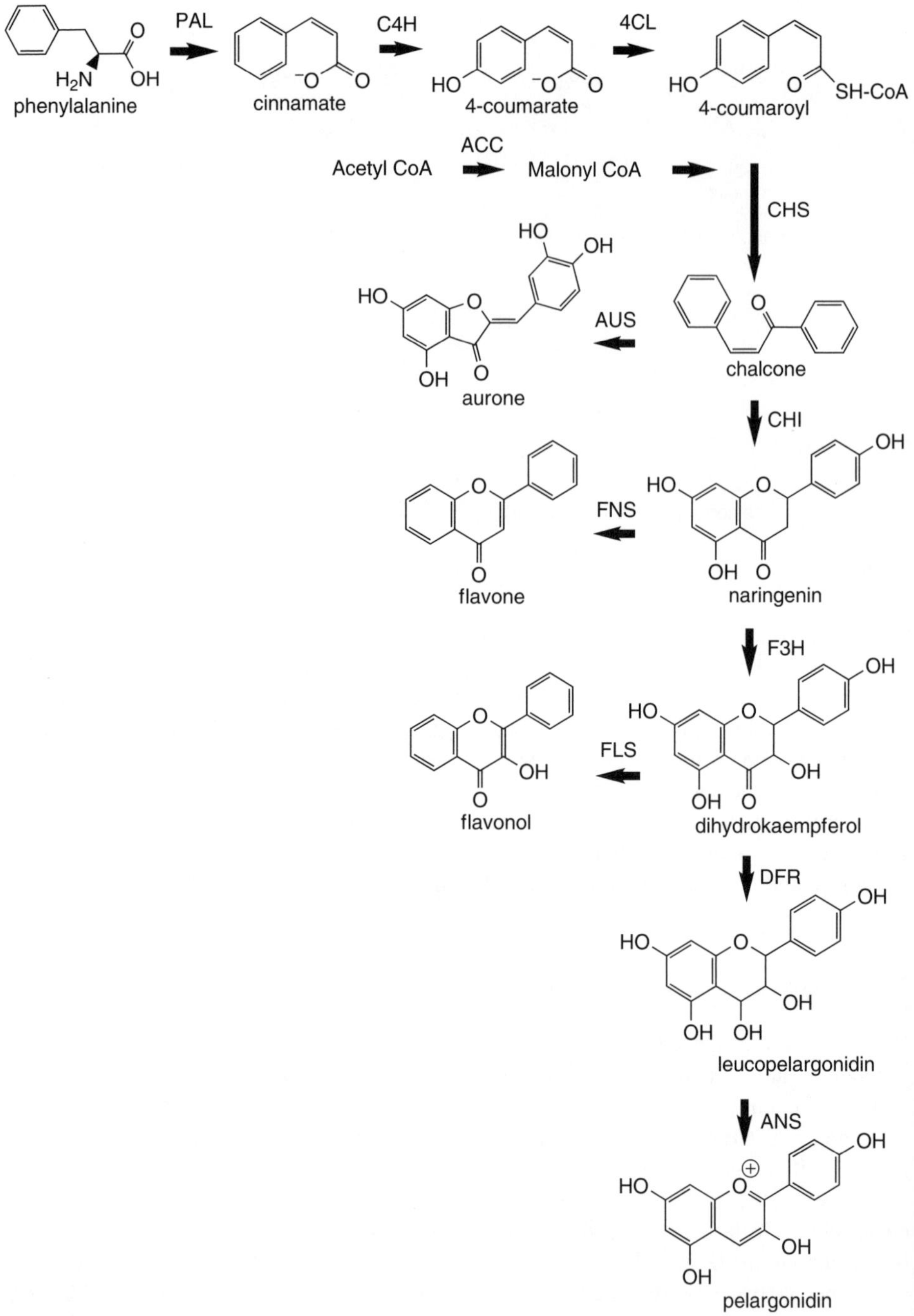

Figure 16.2 *Schematic of the biosynthetic pathway for color producing pigments. PAL: phenylalanine ammonia-lyase; C4H: cinnamate 4-hydroxylase; 4-coumarate CoA ligase; ACC: acetyl-CoA carboxylase; CHS: chalcone synthase; CHI: chalcone isomerase; F3H: flavanone 3β-hydroxylase; DFR: dihydroflavonol 4-reductase; ANS: anthocyanidin synthase; AUS: aureusidin synthase; FNS: flavone synthase; FLS: flavonols synthase. From Schwinn and Davies [1], in Plant Pigments and Their Manipulation, Blackwell Publishing, Oxford, 2004, p. 92*

16.5 Anthocyanins as Natural Colorants

16.5.1 Color Stability

There is a large consumer demand for natural red colors which may offer an alternative to the use of synthetic dyes. This interest has drastically increased with findings and increasing consumer perception and awareness of the potential functional benefit from natural plant ingredients. Anthocyanins are responsible for the attractive orange to red to violet colors exhibited by many flowers, berries and vegetables. Anthocyanins absorb light in the visible region because they are protonated at the heterocyclic oxygen ring under acidic conditions. Other factors such as R-group conjugations (Figure 16.1 and Table 16.1) also exert a great impact on the absorbance and visible appearance of these compounds. Anthocyanin pigments have generally been thought to be relatively unstable except in low pH environments. Recently, researchers have shown the capability of anthocyanins to become increasingly stable in food, drug and cosmetic matrices by co-pigmentation and complexation reactions. Degradation of an anthocyanin's color in foods occurs mainly during processing. Increased anthocyanin concentration has been shown to support higher color stability. Knowledge of an anthocyanin's chemistry can lead to the increased stabilization of the pigment. With this in mind, the major factors affecting the color produced and pigment stability are the individual anthocyanin structure as well as the environment that surrounds them: pH, temperature, oxygen, ascorbic acid, light, enzymes, sugars, sulfur dioxide, co-pigmentation and metal complexation. Color and pigment degradation rates vary greatly among the anthocyanins because of their diverse structural characteristics.

Table 16.1 *Anthocyanidins more commonly found in nature, their B-ring conjugations and maximum absorbance [23]*

Anthocyanidins	R_1	R_2	λ_{max} *(nm) visible color*
Pelargonidin	H	H	494 nm/orange
Cyanidin	OH	H	506 nm/orange-red
Peonidin	OMe	H	506 nm/orange-red
Delphinidin	OH	OH	508 nm/red
Petunidin	OMe	OH	508 nm/red
Malvidin	OMe	OMe	510 nm/bluish-red

16.5.2 Structure

At the primary level the degree of hydroxylation/methoxylation of the anthocyanidin B-ring (Table 16.1) and the nature of sugar and/or acid conjugations have the greatest effect on the color produced by these pigments. An increased number of hydroxyl and/or methoxyl groups on the B-ring of an anthocyanidin results in a bathochromic shift of the visible absorption maximum, which has a bluing effect on the color produced. Substitutions on the R-groups of the B-ring may also affect the stability of the pigments: hydroxylation of the B-ring has been reported to decrease the stability of the anthocyanin, while methoxylation increases stability [24]. Foods containing the aglycon malvidin and petunidin seem to show increased color and pigment stability as compared to other hydroxylated anthocyanins. Sugar substitution of the anthocyanidin may increase the visible absorption maximum of the pigment, producing a more red-orange color [23]. Acylation of the sugar substitutions and/or individual anthocyanidins may also produce bathochromic (increased wavelength) and/or hyperchromic (increased absorption) shifts, altering the spectra of a compound (Figure 16.3).

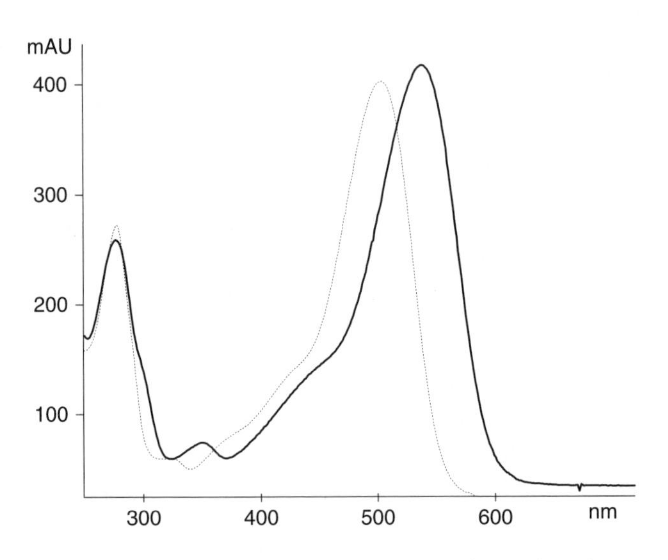

Figure 16.3 *Bathochromic and hyperchromic shifts produced by acylation of purple carrot anthocyanins at 520 nm: Cy-3-gal-xyl-glu (---) and Cy-3-gal-xyl-glu + p-coumaric acid (). Bathochromic shift indicates an increase in the visible absorption maximum (nm). Hyperchromic shift indicates an increase in the peak intensity (mAU)*

Acylation of the anthocyanin, mainly with aromatic acids, can drastically improve the pigment stability possibly through intermolecular/intramolecular co-pigmentation, and self-association reactions making acylated anthocyanins more desirable as colors from natural sources [23]. This effect is more noticeable when there are two or more acylating cinnamic acids attached to the anthocyanin molecule. Because of such interactions, acylated anthocyanins and other flavonoids may exhibit increased color and pigment stability to pH changes, heat treatment, and light exposure [23]. The concentration and varying structures of the sugars/acids conjugating an anthocyanin have a significant effect on the color and pigment stability of the molecule.

Small differences in chemical structure have a clear impact on color and tinctorial strength of anthocyanins and other flavonoids [25]. The colors of non- and monoacylated anthocyanins have been largely determined by the substitution on the B-ring of the aglycon, while acylations may play a major role on color and tictorial strength, particularly on diacylated anthocyanins and anthocyanins with multiple acylations. Anthocyanins and other flavonoids have also been known to interact with components of the food matrix, including proteins, fat, inorganic salts, metals and other phenolic compounds.

16.5.3 Structural Transformation and pH

At any given pH level, anthocyanins exist in an equilibrium of different chemical forms. Anthocyanins typically exhibit an absorption maximum at a pH of 1.0 when the anthocyanidin is in its most stable form, known as oxonium or the flavylium cation (Figure 16.4). In this form, the pigment produces a bright orange-red to violet color, attractive for many applications. However, at a pH of 4.5, anthocyanins become colorless in its chalcone and/or hemiketal forms. A cyanidin-3-glucoside molecule will only retain 50 % of the orange-red color, producing an oxonium ion at a pH of 3.0 [26]. When the anthocyanin is in a pH environment of 7.0 the pigment produces a dull blue to green color when the predominant chemical structure is in the form of a quinonoidal base (Figure 16.4).

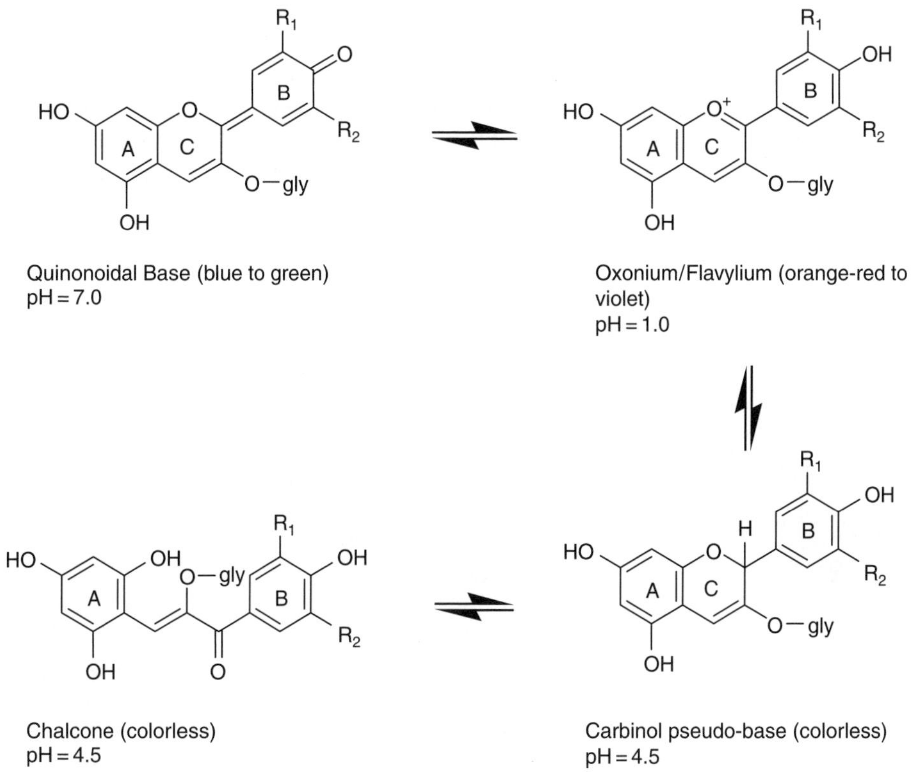

Quinonoidal Base (blue to green)
pH = 7.0

Oxonium/Flavylium (orange-red to violet)
pH = 1.0

Chalcone (colorless)
pH = 4.5

Carbinol pseudo-base (colorless)
pH = 4.5

Figure 16.4 *Anthocyanin structural transformations with a change in pH*

The food matrix may enhance or decrease the stability of anthocyanins to pH changes. Wallace and Giusti [27] showed increased anthocyanin color and pigment stability in fat-containing yogurt matrices (pH ~ 4.3) because of anthocyanin polymerization and an interaction between the fat matrix of the yogurt and the acylating components of anthocyanins. Similar findings were also reported in milk matrices containing high fat content [28]. Such interactions may also be present between acylated anthocyanins and other food, drug and cosmetic matrices whose environmental conditions (pH, temperature, etc.) are not suitable to anthocyanin color and pigment stability. Kinetic degradation of anthocyanins due to pH changes have been reported to follow largely a first-order reaction. Zero-order degradation of anthocyanins has been reported; however, statistical differences between zero- and first-order kinetic reactions in quality related systems have been shown to be potentially insignificant [29].

16.5.4 Temperature

Heat treatment is one of the most important processes in food manufacturing and has a detrimental influence on the stability of anthocyanin compounds. A logarithmic rise in anthocyanin degradation has been reported by an arithmetic increase in temperature [30–32] (Figure 16.5). Anthocyanins first undergo hydrolysis of the glycosidic bond when exposed to heat treatment, which leads to loss of color since anthocyanidins are much less stable than their glycosylated forms. It is postulated that the first step in anthocyanidin degradation is conversion of the molecule to chalcone, which eventually produces an ά-diketone [26]. The result is the production of a dull dark brown to yellow pigment that is unsuitable as a natural colorant. In general, structural characteristics that lead to increased stability to pH changes also lead to increased thermostability. Thermodegradation of anthocyanins has been largely reported to follow first-order kinetics [33]. Highly acylated anthocyanins seem to be more stable in the food matrix [27]. For example, diacylated anthocyanins are typically more thermostable as compared to mono- and nonacylated anthocyanins. Complex sugar residues of red cabbage anthocyanins were also proposed to be protective against thermal degradation [34].

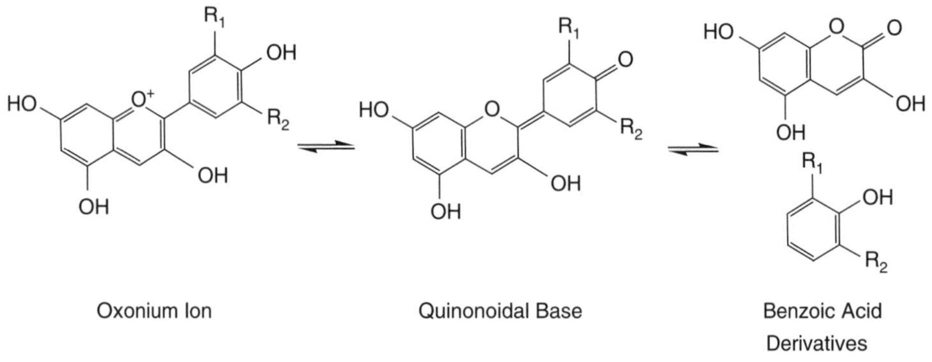

Oxonium Ion ⇌ Quinonoidal Base ⇌ Benzoic Acid Derivatives

Figure 16.5 Degradation of the oxonium ion at pH 3.7 accelerated by heat

16.5.5 Oxygen and Ascorbic Acid

Oxygen amplifies the impact of other anthocyanin degradation processes. The unsaturation of the anthocyanin structure makes it increasingly susceptible to degradation by oxygen and ascorbic acid. The presence of oxygen accompanied with elevated temperature was the most detrimental combination of many factors tested against color deterioration of different berry juices and isolated anthocyanins in a study conducted by Nebesky *et al.* [35]. It has long been known in the food industry that packaging anthocyanin-containing products such as juices under a vacuum or in a nitrogen atmosphere can increase the shelf life of the pigments significantly. Reaction of anthocyanins with oxygen compounds typically yields a dull yellow-brown oxidized color. Damaging effects of oxygen on the anthocyanin pigments can take place through a direct oxidative mechanism and/or through indirect oxidation, where the oxidized components of the media further react with anthocyanins, giving rise to colorless or yellow-brown products [36].

Fortification of food with ascorbic acid is a common practice used to inhibit oxidation while nutritionally enhancing the product. When anthocyanins are in the presence of ascorbic acid both compounds have been known to simultaneously disappear. This is most likely due to the formation of hydrogen peroxide and complete degradation of the anthocyanin during ascorbic acid oxidation [30]. This reaction can be catalyzed by the presence of copper [30]. Other theories suggest that direct condensation of ascorbic acid with anthocyanins destroys both molecules [37]. Contrary to thermal degradation, galactosides were more stable as compared to arabinosides of cranberry anthocyanins against degradation induced by both oxygen and ascorbic acid [38]. In some cases, anthocyanins have been shown to be protected by ascorbic acid against enzymatic degradation [39].

16.5.6 Light

Light exerts two contrasting effects on anthocyanins by favoring their biosynthesis in the plant, but accelerating their degradation particularly after extraction and incorporation into a food matrix as a colorant. Light-induced degradation is dependent on the concentration of molecular oxygen present [40]. The most vigorous anthocyanin loss can be experienced when the pigments are exposed to florescent light [41]. As with most natural pigments that lack the structural stability and tinctorial strength, anthocyanins are predisposed to photo-oxidation by light. Furtado *et al.* [42] found the end-products of light-induced degradation of anthocyanins to be the same as the ones produced by thermal degradation. However, the kinetic pathways of the two reactions are different [42].

Acylated anthocyanins, once again, seem to exhibit increased stability to light exposure conditions. The stability of acylated pelargonidin derivatives from radish in a maraschino cherry application was compared when the product was stored in the dark versus the same product exposed to light. This comparison showed only a slight increase in the rate of pigment degradation when they were exposed to light [43]. This was attributed to the presence of cinnamic acid acylations attached to the anthocyanin molecules.

16.5.7 Enzymes and Sugars

Sugars and enzymes are naturally present in fruits, and during processing are often added to enhance a food product. Pectolytic enzymes are often used during juice processing to increase the juice yield and color extraction. Under certain conditions enzyme preparations may degrade anthocyanins and other pigments present in fruit by hydrolyzing glycoside substituents [44]. Hydrolyzed anthocyanins in their pure aglycon form are extremely unstable and degrade quickly, losing their coloring properties. Inactivation of enzymes generally improves an anthocyanin's stability in the food matrix [45]. The most common anthocyanin degrading enzymes are glycosidases, which break the glycosidic bond between an anthocyanin and its residual sugar, resulting in the more unstable anthocyanidin [46]. This is of great importance because β-glucosides are the most prevalent pigments in anthocyanin-colored fruits [47]. Similar findings have also been found in anthocyanin-galactoside rich products containing β-galactosidase enzymes [44]. Mold is a common contributor of these enzymes to fruit products [44].

Other enzymes present in fruits such as peroxidases and phenolases also commonly degrade anthocyanins [48]. Interestingly, cyanidin was observed to react directly with polyphenol oxidase, but pelargonidin did not react at all [49]. Generally, enzymes degrade other phenolic compounds present in the media, after which their corresponding quinones react with anthocyanins, leading to brown condensation products. This has been noted in many studies with pure anthocyanins including cyanidin-3-glucoside [50], pelargonidin-3-glucoside [51] and in wine model solutions [52].

Sugars as well as their degradation products at lower concentrations are known to decrease the stability of anthocyanins. In a study by Daravingas and Cain [53], all of the tested sugars (fructose, sucrose, glucose and xylose) increased anthocyanin degradation in the same way. The reaction of anthocyanins with the degradation products of sugar generally yields a brown polymerized complex. Sugars at high concentrations have also been known to protect anthocyanins from degradation because of their ability to lower a product's water activity (a_w). Sugar solutions also help to stabilize anthocyanins during frozen storage by the inhibition of enzymatic reactions [54].

16.5.8 Sulfur Dioxide

Sulfur dioxide has been used extensively in the fruit and vegetable industry, chiefly as an inhibitor of microbial growth and of enzymatic and nonenzymatic browning. The presence of sulfur dioxide can also lead to the color loss of anthocyanins by a reversible bleaching mechanism, which generally occurs when fruits are treated with 500 to 3000 ppm of SO_2. Sulfite bleaching of anthocyanins has been attributed to a nucleophilic attack of the oxonium ion's (flavylium cation) C-4 position by the negatively charged bisulfate ion [36]. This reaction is thought to disrupt the conjugated double bond system, resulting in loss of color (Figure 16.6).

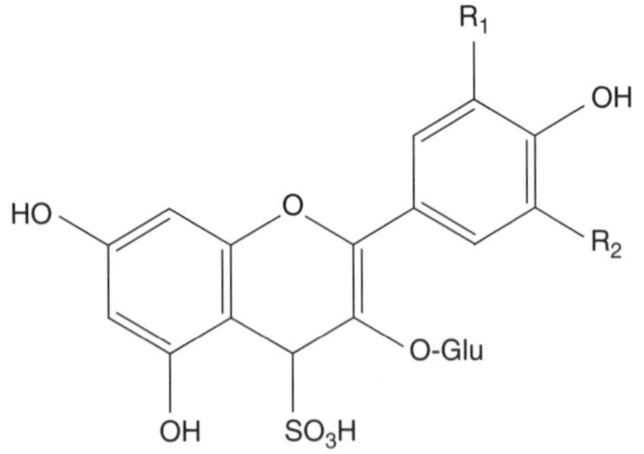

Figure 16.6 Colorless anthocyanin–sulfate complex

The extraction efficiency of anthocyanins from sources such as grape skins can be greatly increased by the addition of sulfur dioxide or its equivalent in bisulfite or *meta*-bisulfite [55]. Anthocyanin color loss from sulfite bleaching can be restored by washing before further processing. Color loss in anthocyanin-containing fruits and vegetables is a result of a structural complex formation between SO_2 and the C-4 position (Figure 16.6) of the anthocyanidin [56]. This, in turn, shifts the maximum absorbance of the complex outside of the visible spectra. Anthocyanins that are resistant to the sulfur bleaching effect generally have a conjugated C_4 position [57].

16.5.9 Co-pigmentation and Metal Complexation

Co-pigmentation is a valuable and natural tool for enhancing and stabilizing the color of anthocyanin-rich products. Co-pigmentation can take place through several interactions: intermolecular complex formations, intramolecular complex formations, self-association mechanisms and metal complexation (Figure 16.7). Co-pigmentation is observed as a bathochromic shift in the visible range towards higher wavelength, which is also called the bluing effect, since the color of an anthocyanin changes from a red to a more blue hue [58] or as a hyperchromic shift in which the intensity of the anthocyanin color increases [23]. Anthocyanin co-pigmentation reactions are strongly affected by pH, temperature, concentration and molecular structure. Co-pigmentation reactions are much weaker at very low pH ranges as compared to pH values between 2 and 5 [59].

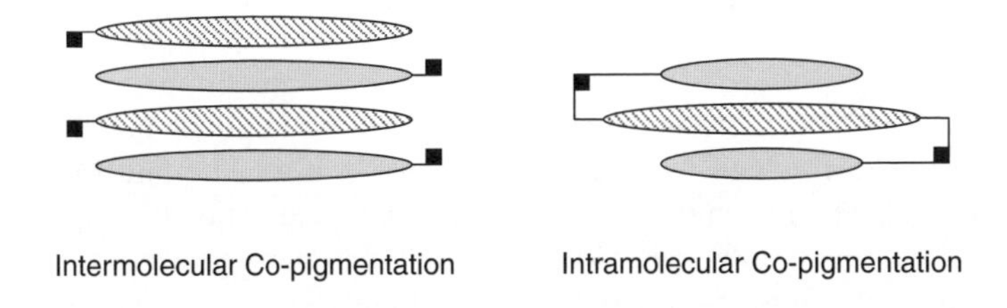

Figure 16.7 *Anthocyanin interactions: anthocyanidin (≡), co-pigment (#), sugar conjugate (■)*

Intermolecular interaction is a more prominent means of co-pigmentation in fruits, which contain nonacylated anthocyanins. Intermolecular co-pigmentation is defined as the interaction between a colored anthocyanin to a colorless co-pigment that is not bound covalently to the anthocyanin molecule [60]. Hydrogen bonding, hydrophobic interactions and electrostatic interactions have been speculated as the main driving force for intermolecular co-pigmentation, resulting in a 1:1 complex formation [61] (Figure 16.7).

Intermolecular co-pigmentation is loosely defined such that the co-pigment is part of the anthocyanin molecule. The covalent bond between the acylation of the anthocyanin molecule and co-pigment stabilizes the complex [30]. Intramolecular co-pigmentation is thought to be stronger and more effective in stabilizing anthocyanin color, probably due to the strength of the covalent bonds present [62]. This type of co-pigmentation is mostly associated with anthocyanins derived from flowers and vegetables, which generally contain acylation (Figure 16.7).

The mechanism of self-association has been described as stacking like interactions [63] (Figure 16.8). Self-associations of anthocyanins have been observed to take place during wine aging and it is assumed that they may partially contribute to the color of aged wines.

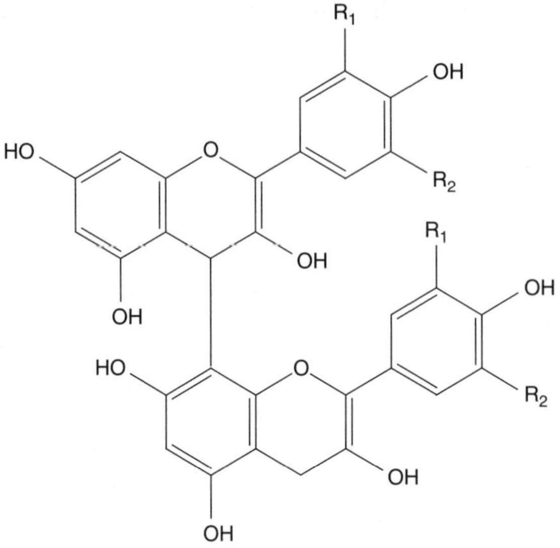

Figure 16.8 Self-association of anthocyanins at the C-4 position

Unlike polyphenolic copigments, metal ions seem to be rarely involved in color stabilization. However, some highly charged metal ions such as Al^{3+} and Mg^{2+} have been reported to possibly strengthen the pigment–co-pigment interaction leading to hyperchromic and/or bathochromic shifts [64]. Goto and Kondo [65] have utilized X-ray technology to show a natural six pigment–six co-pigment complex around two magnesium ions in a crystalline state, producing a hyperchromic and bathochromic shift in flower petals. The most common metals and anthocyanin complexes are with tin (Sn), copper (Cu), iron (Fe), aluminum (Al), magnesium (Mg) and potassium (K) [66]. Only cyanidin, delphinidin and petunidin based anthocyanins, which have more than one free hydroxyl group in the B-ring are capable of metal chelation on the aglycon [67].

16.6 Other Flavonoids as Natural Colorants

16.6.1 Yellow Flavonoid Pigments

The yellow flavonoid pigments found in nature include chalcones, aurones and some flavonols. Most yellow flavonoid pigments in nature co-occur with carotenoids, serving as pollination attractants or UV-protecting pigments. Aurones produce the most intense yellow color of these pigments, but represent one of the smallest classes of flavonoid compounds identified [1]. Aurones can be formed by oxidation of chalcones during extraction and purification [1]. Chalcones act as an intermediate for many color producing pigments in the biosynthetic process. Chalcones with a free or unsubstituted hydroxyl group at the C-6 position spontaneously cyclize to colorless flavanones [68]. Flavonols produce the least intense color of the yellow producing flavonoid pigments because of their absorption at a lower wavelength. Most flavonols have an additional hydroxyl group attached to the A-ring, commonly on the C-8 position. All yellow producing flavonoids degrade under similar conditions (pH, temperature, oxygen, etc.) as anthocyanins.

An abundant yellow flavonoid colorant is that of the safflower *Carthamus* sp., which is utilized primarily in beverages [69]. The use of these extracts to color foods is permitted in the United States. However, a number of odor and flavor compounds exist that are not easily separated, which makes utilizing the colorant in high quantities difficult.

16.6.2 Tannins

Tannins commonly found in grains, legumes, fruits and herbs are naturally occurring phenolic compounds that have the ability to combine with proteins and other polymers such as polysaccharides. These compounds are water soluble and have a molecular weight ranging between 500 and 3000. Tannins range from light yellow to light brown in color and contribute to the color of a number of foods and beverages, including fruit juices, whiskey/bourbon spirits and whole grain bread products. They are noted for their ability to precipitate proteins such as alkaloids and gelatin [70]. It is estimated that the average consumption of tannins among humans is about 1 g/day [71]. Three types of tannins exist: condensed tannins (proanthocyanidins), hydrolyzable tannins and the phlorotannins, which are generally found in marine algae and are not utilized by humans. Proanthocyanidins reduce voluntary intake and absorption of protein and other minerals in monogastric animals. Commonly, proanthocyanidins are extracted from barley because of their role in precipitating protein during beer production, known as 'chilling haze'. Typically proanthocyanidins are colorless, but can be converted to colored products during food processing. The presence of proanthocyanidins was first noticed in cocoa beans, when upon heat treatment under acidic conditions the compounds hydrolyze into cyanidin and epicatechin (Figure 16.9). Other proanthocyanidins have shown the ability to hydrolyze into common anthocyanidins (pelargonadin, cyanidin, delphinidin and petunidin) under similar conditions.

Proanthocyaninidin (Colorless) $[H^+]$ cyanidin (orange-red) + epicatechin (brown)

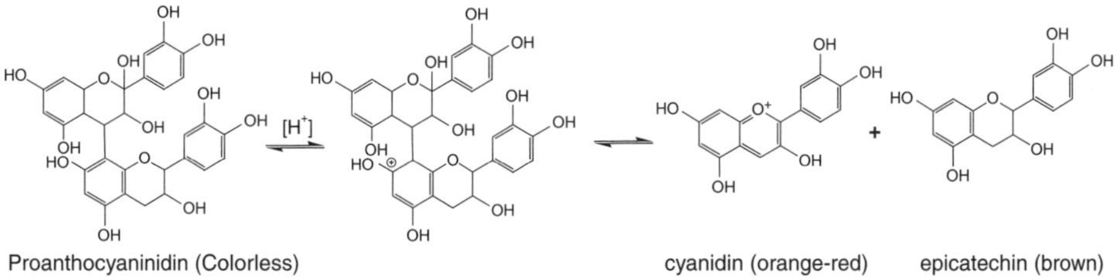

Figure 16.9 *Hydrolysis of proanthocyanidin. From Forsyth and Roberts [72],* Chem. Ind. (Lond.), *755 (1958)*

16.7 Therapeutic Effects of Flavonoids in the Diet

Flavonoids are thought to play an essential role in human health by demonstrating strong antioxidant activities [2]. It is thought that flavonoids are absorbed by passive diffusion after the cleaving of glycoside groups by gut microflora [73]. The compounds in general have been noted to help aide in the prevention of almost every chronic disease known because of these functional properties. Flavonoids that are most notable to human health include: flavanones, flavones, flavans (flavanols) and flavonols. Flavonols are the most abundant flavonoids found in foods, with quercetin, kaempferol and myricetin being the most prominent [18].

Recent findings of the many versatile health benefits of flavonoid pigments has led to a rapid increase in industry and consumer demand for these value-added colors from natural sources. Many flavonoids have been shown to contain free radical scavenging and antioxidant activity both *in vitro* and *in vivo* [2]. Flavonoids in general exhibit potential in the suppression and prevention of many chronic age-related diseases such as coronary heart disease [74], cancer [75], capillary fragility [76], atherosclerosis [77] and stroke [3]. Current research suggests that flavonoids show biological properties that may account for cancer chemoprotection [4]. Recent attention has been given to the ability of these compounds to inhibit the cell cycle, cell proliferation and oxidative stress and to induce detoxification enzymes, apoptosis and thoroughly activate the human immune system [18]. This is in addition to their positive influence on visual acuity, anti-inflammation and cognitive functions [78].

Flavonoids are generally absorbed in the gastrointestinal tract of humans or animals and excreted intact or as metabolites in the urine and feces [2]. Bioavailability of the flavonoid aglycon absorbed can range between 0.1 % for anthocyanins and tea catechins to 20 % for quercetin and isoflavones. Bioavailability may also depend on the food source. For example, absorption of quercetin has been known to be about four times greater when consumed from onions as opposed to apples [73]. Colonic bacteria are largely responsible for the breakdown of flavonoids into phenyl acids before absorption [73]. Measurement of flavonoids in the bloodstream and plasma after ingestion shows rapid absorption of the antioxidants [79]. This thus might reduce the risk of oxidative damage of DNA and thus potentially prevent the formation of many cancers [80].

Interest in anthocyanin pigments has increased because of their beautiful color and functionality in food products. Possible improvement of anti-inflammatory effects, visual acuity and cognitive function make anthocyanins an attractive alternative to typical synthetic red dyes in food matrices. There are numerous anecdotal accounts for the bioactive properties of anthocyanins. British RAF pilots were fed bilberries to enhance their night vision during the Second World War, while Chernobyl victims were directed to consume chokeberry products to alleviate the effects of high-dosage radiation [26]. Anthocyanin pigments have been known to contain a high oxygen radical absorbing capacity (ORAC) value in many *in vitro* analyses. Numerous studies have found that anthocyanin absorption into the blood and serum is minimal. Aside from their intact form, additional methylated, glucoronidated and sulfated anthocyanin metabolites have been identified in the blood serum of rats, suggesting a potentially active form of anthocyanins *in vivo* [81]. Typically, nonacylated anthocyanins seem to be more bio-available to humans and animals. Of the most common occurring anthocyanidins, cyanidin derivatives seem to be more easily absorbed *in vivo* and contain higher ORAC values [82]. For a more extensive review of the various health benefits of flavonoids we recommend the book of Andersen and Markham [22].

16.8 Regulations on the Use of Flavonoid Colorants

Anthocyanins are the most popular flavonoid colorants on the market. Many anthocyanin-rich extracts (e.g. grape skin, red radish and red cabbage) have had commercial success in the market. In the United States, The Color Additives Amendment of the Food, Drug and Cosmetic Act establishes general regulations for the use of food colorants in the United States. US regulations classify food colorants into Certified Colorants, requiring approval and certification by the Food and Drug Administration (FDA) for every batch produced, and Colorants Exempt from Certification. Colorants obtained from natural sources as well as nature-identical compounds may follow under the 'exempt' category. The flavonoid colorants approved for use in food in the US fall therefore into the 'colors exempt from certification' category. US approved flavonoid colorants include fruit and vegetable juice, grape skin extract and grape color extract. According to the Unites States Nutrition Labeling and Education Act of 1991, all food companies are required to list color additives by their common or usual name on the product's ingredient statement. Colors derived from natural sources or those that are considered nature identical can be listed by their common/usual name or generally as 'natural colors added' (US Code of Federal Regulations, Title 21, 2007).

Consumption of colors from natural sources has significantly increased in the last decade throughout the European Union because of their role as a healthy alternative to synthetic dyes. The European Union has tried to achieve a uniform additive legislation for Common Market Countries. Each permitted color has been assigned an E (E = Europe) number. Anthocyanins have been classified as a general group under the E163 category. Because countries in the European Union (EU) can interpret and adopt different regulations, the regulations on food colorants can be variable from country to country. Italy is the most restrictive of the EU countries on colors derived from natural sources. However, Norway does not allow synthetic colorants in food and greatly promotes the use of natural alternatives [55]. Anthocyanins are generally accepted as safe food colorants in the European Union as well as in most other countries around the world.

The hundredfold safety factor known commonly as NOEL (no observed effect level) is widely accepted worldwide to determine the maximum concentration for food/beverage colorants. This safety factor is derived from a factor of 10 to change animal data to human conditions and another factor of 10 to account for the variability among humans [55].

The findings of the powerful antioxidant power and the potential health benefits of anthocyanins and other flavonoids as well as increasing consumer demand for natural products and ingredients makes it likely that anthocyanin and other flavonoid based colorants will increase in use by the food industry as natural and safe food colorants. Findings of anthocyanin sources with increased stability as well as an increased understanding of the co-pigmentation reactions and other factors that may enhance their natural stability are critical and will allow wider application of these colorants in our food supply.

References

1. K. E. Schwinn and K. M. Davies, Flavonoids, in K. M. Davies (ed.), *Plant Pigments and Their Manipulation*, Blackwell Publishing, Oxford, 2004.
2. N. C. Cook and S. Samman, Review: flavonoids – chemistry, metabolism, cardio protective effects, and dietary sources, *J. Nutr. Biochem.*, **7**, 66–76 (1996).
3. R. S. R. Zand, D. J. A. Jenkins and E. P. Diamandis, Flavonoids and steroid hormone-dependent cancers, *J. Chromatogr. B*, **777**, 219–232 (2002).
4. W. Ren, Z. Qian, H. Wang, L. Zhu and L. Zhang, Flavonoids: promising anticancer agents, *Medicinal Res. Rev.*, **23**(4), 519–534 (2003).
5. W. Watjen, G. Michels, B. Steffan, P. Niering, Y. Chovolou, A. Kampkotter, Q. Tran-Thi, P. Proksch and R. Kahl, Low concentrations of flavonoids are protective in rat H411E cells whereas high concentrations cause DNA damage and apoptosis, *J. Nutr.*, **135**, 525–531 (2005).
6. A. S. Aherne and N. M. Obrien, Dietary flavonols: chemistry, food content, and metabolism, *Nutr.*, **18**, 75–81 (2002).
7. Y. H. Chu, C. L. Chang and H. F. Hsu, Flavonoid content of several vegetables and their antioxidant acitivity, *J. Sci. Food Agric.*, **80**, 561–566 (2000).
8. K. G. Ryan, E. E. Swinny, K. R. Markham and C. Winefield, Flavonoid gene expression and UV photoprotection in transgenic and mutant *Petunia leaves, Phytochem.*, **59**, 23–32 (2002).
9. M. Tattini, C. Galardi, P. Pinelli, R. Massai, D. Remorini and G. Agati, Deferential accumulation of flavonoids and hydroxycinnamates in leaves of *Ligustrum vulgare* under excess light and drought stress, *New Phytologist.*, **163**, 547–561 (2004).
10. R. J. Grayer and J. B. Harborne, A survey of antifungal compounds from higher plants, 1982–1993, *Phytochem.*, **37**, 19–42 (1994).
11. C. H. Chou, Roles of allelopathy in plant biodiversity and sustainable agriculture, *Crit. Rev. Plant Sci.*, **18**, 609–636 (1999).
12. O. Thoison, T. Sevenet, H. M. Niemeyer and G. B. Russell, Insect antifeedant compounds from *Nothofagus dombeyi* and *N. pumilio, Phytochem.*, **65**, 2173–2176 (2004).
13. W. A. Peer, D. E. Brown, B. W. Tague, G. K. Muday, L. Taiz and A. S. Murphy, Flavonoid accumulation patterns of transparent testa mutants of Arabidopsis, *Plant Physiol.*, **126**, 536–548 (2001).
14. R. A. Dixon and C. L. Steel, Flavonoids and isoflavonoids – a gold mine for metabolic engineering, *Trends in Plant Sci.*, **4**, 394–400 (1999).
15. B. A. Snyder and R. L. Nicholson, Synthesis of phytoalexins in sorghum as a site specific response to fungal ingress, *Science*, **248**, 1637–1639 (1990).
16. D. Treutter, Significance of flavonoids in plant resistance and enhancement of their biosynthesis, *Plant Biology*, **7**, 581–591 (2005).
17. A. H. Clifford and S. L. Cuppett, Review: anthocyanins-nature, occurrence, and dietary burden, *J. Sci. Food Agric.*, **80**, 1063–1072 (2000).
18. L. H. Yao, Y. M. Jiang, J. Shi, F. A. Tomas-Barberan, N. Datta, R. Singanusong and S. S. Chen, Flavonoids in food and their health benefits, *Plant Foods for Human Nutr.*, **59**, 113–122 (2004).
19. M. M. Giusti and R. E. Wrolstad, Characterization and measurement of anthocyanins by UV–visible spectroscopy, in R. E. Wrolstad and S. J. Schwartz (eds), *Handbook of Food Analytical Chemistry*, John Wiley & Sons Inc., New York, 2005.
20. B. Winkle-Shirley, Biosynthesis of flavonoids and effects of stress, *Current Opinion in Plant Biology*, **5**, 218–223 (2002).
21. W. Barz and U. Mackenbrock, Constitutive and elicitation induced metabolism of isoflavones and pterocarpans in chickpea (*Cicer arietinum*) cell suspension cultures, *Plant Cell Tissue Organ. Cult.*, **38**, 199–211 (1994).
22. O. M. Andersen and K. R. Markham, *Flavonoids Chemistry, Biochemistry, and Applications*, Taylor & Francis Group, New York, 2006.
23. M. M. Giusti and R. E. Wrolstad, Acylated anthocyanins from edible sources and their applications in food systems, *Biochem. Engng J.*, **14**, 217–225 (2003).

24. G. Mazza and E. Miniati, *Anthocyanins in Fruits, Vegetables, and Grains*, CRC Press, Boca Raton, Florida, 1993.
25. O. Dangles, N. Saito and R. Brouillard, Anthocyanin intramolecular copigment effect, *Phytochem.*, **34**, 119–124 (1993).
26. R. E. Wrolstad, Anthocyanin pigments – bioactivity and coloring properties, *J. Food Sci.*, **69**(5), 419–421 (2004).
27. T. C. Wallace and M. M. Giusti, Determination of color, pigment, and phenolic stability in yogurt systems colored with non-acylated anthocyanins from *Berberis boliviana* L. as compared to other natural/synthetic colorants, *J. Food Sci.*, **73**, c241–c248 (2008).
28. P. Jing and M. M. Giusti, Characterization of anthocyanin-rich waste from purple corncobs (*Zea mays* L.) and its application to color milk, *J. Agric. Food Chem.*, **53**, 8775–8781 (2005).
29. T. P. Labuza and D. Riboh, Theory and application of arrhenius kinetics to the prediction of nutrient losses in foods. *Food Technol.*, **36**, 66–74 (1982).
30. F. J. Francis, Food colorants: anthocyanins, *Crit. Rev. Food Sci. Nutr.*, **28**, 273–314 (1989).
31. L. Havlikova and K. Mikova, Heat stability of anthocyanins, *Z. Lebensm. Unters Frosch.*, **181**, 427–432 (1985).
32. J. Rhim, Kinetics of thermal degradation of anthocynain pigment solutions driven from red flower cabbage, *Food Sci. Biotechnol.*, **11**, 361–364 (2002).
33. J. Ahmed, U. S. Shivhare and G. S. V. Raghavan, Thermal degradation kinetics of anthocyanin and visual colour of plum puree, *Eur. Food Res. Technol.*, **218**, 525–528 (2004).
34. M. Dyrby, N. Westergaard and H. Stapelfeldt, Light and heat sensitivity of red cabbage extract in soft drink model systems, *Food Chem.*, **72**, 431–437 (2001).
35. E. A. Nebesky, W. B. Esselen, J. E. W. McConnell and C. R. Fellers, Stability of color in fruit juices, *Food Res.*, **14**, 261–274 (1949).
36. R. L. Jackman, R. Y. Yada, M. A. Tung and R. A. Speers, Anthocyanins as food colorants – a review, *J. Food Biochem.*, **11**, 201–247 (1987).
37. M. S. Poei-Langston, Color degradation in an ascorbic acid–anthocyanin–flavanol model system, *J. Food Sci.*, **46**(4), 1218 (1981).
38. M. S. Starr and F. J. Francis, Oxygen and ascorbic acid effect on the relative stability of four anthocyanin pigments in cranberry juice, *Food Technol.*, **22**, 1293–1295 (1968).
39. S. T. Talcott, C. H. Brenes, D. M. Pires and D. del Pozo-Insfran, Phytochemical stability and color retention of copigmented and processed muscadine grape juice, *J. Agric. Food Chem.*, **51**, 957–963 (2003).
40. E. L. Attoe and J. H.Von Elbe, Photochemical degradation of betanine and selected anthocyanins, *J. Food Sci.*, **46**, 1934–1937 (1981).
41. N. Palamidis and P. Markakis, Stability of grape anthocyanin in carbonated beverages, *Semana Vitivinicola*, **33**, 2633–2639 (1978).
42. P. Furtado, P. Figueiredo, H. Chaves das Neves and F. Pina, Photochemical and thermal degradation of anthocyanidins, *J. Photochem. Photobiol. A*, **75**, 113–118 (1993).
43. M. M. Giusti and R. E. Wrolstad, Radish anthocyanin extract as a natural red colorant for maraschino cherries, *J. Food Sci.*, **61**(4), 688–694 (1996).
44. J. D. Wightman and R. E. Wrolstad, β-glucosidase activity in juice-processing enzymes based on anthocyanin analysis, *J. Food Sci.*, **61**(3), 544–548 (1996).
45. A. Garcia-Palazon, W. Suthanthangjai, P. Kajda and I. Zabetakis, The effects of high hydrostatic pressure on b-glucosidase, peroxidase and polyphenoloxidase in red raspberry (*Rubus idaeus*) and strawberry (*Fragaria * ananassa*), *Food Chem.*, **88**, 7–10 (2004).
46. H. T. Huang, The kinetics of the decolorization of anthocyanins by fungal anthocyanase, *J. Am. Chem. Soc.*, **78**, 2390–2393 (1956).
47. J. J. Macheix, A. Fleuriet and J. Billot, *Fruit Phenolics*, CRC Press, Boca Raton, Florida, 1990.
48. F. Kader, B. Rovel, M. Girardin and M. Metche, Mechanism of browning in fresh highbush blueberry fruit (*Vaccinium corymbosum* L.). Partial purification and characterization of blueberry polyphenol oxidase, *J. Sci. Food Agric.*, **73**, 513–516 (1997).
49. P. Wesche-Ebeling and M. W. Montgomery, Strawberry polyphenol oxidase: its role in anthocyanin degradation, *J. Food Sci.*, **55**, 731–734 (1990).

50. F. Kader, M. Irmouli, N. Zitouni, J. Nicolas and M. Metche, Degradation of cyanidin 3-glucoside by caffeic acid *o*-quinone. Determination of the stoichiometry and characterization of the degradation products, *J. Agric. Food Chem.*, **47**, 4625–4630 (1999).
51. F. Kader, M. Irmouli, J. P. Nicolas and M. Metche, Proposed mechanism for the degradation of pelargonidin 3-glucoside by caffeic acid *o*-quinone, *Food Chem.*, **75**, 139–144 (2001).
52. P. Sarni, H. Fulcrand, V. Souillol, J. Souquet and V. Cheynier, Mechanisms of anthocyanin degradation in grape must-like model solutions, *J. Sci. Food Agric.*, **69**, 385–391 (1995).
53. G. Daravingas and R. F. Cain, Thermal degradation of black raspberry anthocyanin pigments in model systems, *J. Food Sci.*, **33**, 138–142 (1968).
54. R. E. Wrolstad, G. Skrede, P. Lea and G. Enersen, Influence of sugar on anthocyanin pigment stability in frozen strawberries, *J. Food Sci.*, **55**, 1064–1065 (1990).
55. F. J. Francis, *Colorants*, Eagan Press Handbook, St Paul, Minneapolis,1999.
56. C. F. Timberlake and P. Bridal, Effect of substituents on the ionization of flavylium salts and anthocyanins and their reactions with sulfur dioxide, *Chem. Ind.*, **2**, 1965–1966 (1966).
57. C. F. Timberlake and P. Bridal, Flavylium salts resistant to sulfur dioxide, *Chem. Ind.*, **1**, 1489 (1968).
58. S. Asen, R. N. Stewart and K. H. Norris, Copigmentation of anthocyanins in plant tissues and its effect on color, *Phytochem.*, **11**, 1139–1144 (1972).
59. M. Williams and G. Hrazdina, Anthocyanins as food colorants: effect of pH on the formation of anthocyanin–rutin complexes, *J. Food Sci.*, **44**, 66–68 (1979).
60. R. Brouillard, The *in vivo* expression of anthocyanin color in plants, *Phytochem.*, **22**, 1311–1323 (1983).
61. Y. Cai, T. H. Lilley and E. Haslam, Polyphenol–anthocyanin copigmentation, *J. Chem. Soc. Chem. Commun.*, 380–383 (1990).
62. R. Brouillard, Chemical structure of anthocyanins, in Pericles Markakis (ed.), *Anthocyanins as Food Colors*, Academic Press Inc., New York, 1982.
63. T. Hoshino, U. Matsumoto, N. Harada and T. Goto, Chiral excitation coupled stacking of anthocyanins: interpretation of the origin of anomalous CD induced by anthocyanin association, *Tetrahedron Lett.*, **22**, 3621 (1981).
64. M. Elhabiri, P. Figueiredo, K. Toki, N. Saito and R. Brouillard, Anthocyanin–aluminum and – gallium complexes in aqueous solution, *J. Chem Soc. Perkin Trans.*, **2**, 355–362 (1997).
65. T. Goto and T. Kondo, Angew, *Chem. Int. Ed. Engl.*, **30**, 17 (1991).
66. P. Markakis, Stability of anthocyanins in foods, in *Anthocyanins as Food Colors*, CRC Press, Boca Raton, Florida, 1982.
67. Y. Osawa, Copigmentation of anthocyanins, in P. Markakis (ed.), *Anthocyanins as Food Colors*, Academic Press Inc., New York, 1982.
68. C. O. Miles and L. Main, Kinetics and mechanism of the cyclisation of 2,6-dihydroxy-4,4-dimethoxychalcone; influence of the 6-hydroxyl group on the rate of cyclisation under neutral conditions, *J. Chem. Soc. Perkin Trans.*, **2**, 1639–1642 (1985).
69. C. Yukawa, T. Ichi, K. Onishi and H. Sato, US Patent 6936292 (2005).
70. A. E. Hagerman, Y. Zhao and S. Johnson, Methods for determination of condensed and hydrolysable tannins, in F. Shahadi (ed.), *Antinutrients and Phytochemicals in Foods*, American Chemical Society, Washington, DC, 1997.
71. W. S. Pierpoint, Flavonoids in human food and animal feed – stuffs: amounts and consequences, in N. P. Das (ed.), *Flavonoids in Biology and Medicine III. Current Issues in Flavonoids Research*, National University of Singapore, Singapore, 1990.
72. W. G. C. Forsyth and J. B. Roberts, *Chem. Ind. (Lond.)*, 755 (1958).
73. P. C. H. Hollman, J. H. M. De Vries, S. D. D. Van Leeuwen, M. J. B. Mengelers and M. B. Datan, Absorption of dietary quercetin in healthy ileostomy volunteers, *Am. J. Clin. Nutr.*, **62**, 1276–1282 (1995).
74. D. Bagchi, M. Bagchi, S. J. Stohs and D. K. Das, Free radicals and grape seed proanthocyanidin extract: importance in human health and disease prevention, *Toxicology*, **148**, 187–197 (2000).
75. Z. Juranic and Z. Zizak, Biological activities of berries: from antioxidant capacity to anti-cancer effects. *Biofactors*, **23**, 207–211 (2005).

76. E. Mian, S. B. Curri, A. Lietti and E. Bombardelli, Anthocyanosides and the walls of the microvessels: future aspects of the mechanism of action of their protective effect in syndromes due to abnormal capillary fragility, *Minerva Med.*, **68**(52), 3565–3581 (1977).
77. A. Kadar, L. Robert, M. Miskulin, J. M. Tixier, D. Brechemier and A.M. Robert, Influence of anthocyanoside treatment on the cholesterol-induced atherosclerosis in the rabbit, *Paroi Arterielle*, **5**(4), 187–205 (1979).
78. S. Zafra-Stone, T. Yasmin, M. Bagchi, A. Chatterjee, J. A. Vinson and D. Bagchi, Berry anthocyanins as novel antioxidants in human health and disease prevention, *Molec. Nutr. Food Res.*, **51**, 675–683 (2007).
79. I. F. F. Benzie, Y. T. Szeto, J. J. Strain and B. Tomlinson, Consumption of green tea causes rapid increase in plasma antioxidant power in humans, *Nutr. Cancer*, **34**, 83–87 (1999).
80. Q. Feng, Y. Torii, K. Uchida, Y. Nakamura, Y. Hara and T. Osawa, Black tea polyphenols, theaflavins, prevent cellular DNA damage by inhibiting oxidative stress and suppressing cytochrome P450 1A1 in cell cultures, *J. Agric. Food Chem.*, **50**, 213–220 (2002).
81. M. Dell Agli, A. Busciala and E. Bosisio, Vascular effects of wine polyphenols, *Cardiovascular Res.*, **63**(4), 593–602 (2004).
82. H. Wang, G. Cao and RL. Prior, Oxygen radical absorbing capacity of anthocyanins, *J Agric. Food Chem.*, **45**, 304–309 (1997).

17

Application of Natural Dyes in the Coloration of Wood

Martin Weigl, Andreas Kandelbauer, Christian Hansmann, Johannes Pöckl, Ulrich Müller and Michael Grabner

17.1 Introduction

Within this chapter several methods are presented related to the utilization of natural plant dyes for the coloration of wood and wood-based products. Two completely different concepts are shown:

- attachment or inclusion of additional colorants;
- modification of wood components and their natural color.

While some of these natural staining methods are well known and have been commonly used for centuries others have appeared in the last decades. Additionally, basic principles of new methods are described, which might be used in the future as new green staining technologies.

The motivation for staining of wood depends on socioeconomic, geographical, ecological, artistically, scientific or even other considerations. Often consumers are interested to use wood because of its appealing material properties but they may not be satisfied with the range of the native color spectrum or the inhomogeneity of the product. On the other hand, the furniture and flooring industry has established very high standards with respect of the color regularity of wood surfaces. Beside advanced grading and sorting methods new staining methods and modification of the native color are also demanded to guarantee a certain wood color. Completely changing the natural color of wood while retaining the natural structural and haptic properties of wood is another motivation. Depending on the applied modification or staining method, product usability might be limited to interior use

Handbook of Natural Colorants Edited by Thomas Bechtold and Rita Mussak

such as for floorings or furniture or especially enabled for exterior use such as for facades, terrace floorings or fences.

After a brief introductory section especially related to optical wood properties, different methods for the alteration of color of wood and wood-based products in order to meet the customer requirements will be discussed. Thereby, the major focus will be on the modification of wood components, i.e. cell wall components and extractives. Although classical treatments such as coating or impregnation with dye solutions are covered as well, these processes are usually based on synthetic dye applications and thus differ not fundamentally from coating or impregnation procedures for other types of substrates. Printing techniques, in principle similar to those used in the textile industry may be applied on wood substrates as well, certainly with technological modification of the involved machining (Voigt and Lentner, 2007).

However, utilization of untreated wood and modification of wood components and their natural color seems to reflect much more the unique property profile of wood-based materials. Colored phenolic compounds of ligno-cellulosic matter may be considered as herbal dyes, which are already immobilized on the substrate and thus represent an interesting example for natural dyeing procedures. A wide range of sorting techniques and modification methods are industrially established and worldwide in use. In comparison, attachment or inclusion of additional natural plant dyes are more theoretical applications as they are industrially not widespread.

17.1.1 General Basics

Wood is commonly used for different purposes ranging from high-performance constructions to filigree art work. In most applications, the appearance of the surface is crucial for costumer decisions. The different wood species cover a wide range of natural colors. However, these colors might change due to modification, treatment or even daily use. Basic knowledge about color and an overview about presented methods will be given in the following section.

17.1.1.1 Surface Color

Visible light (VIS) is a small range within the electromagnetic spectrum, which ranges from 380 to 780 nm. Electromagnetic radiation below 380 nm is called ultraviolet (UV) and radiation above 780 nm infrared (IR). Radiation covering the complete VIS spectrum is white. However, if the VIS spectrum is incomplete the healthy human eye recognizes a certain color, no matter whether emitted by a light source or a body surface.

The surface color depends strongly on the chemical composition and the surface structure as well as on the irradiating spectrum. Surfaces emitting the complete VIS spectrum are perceived as white. In contrast, surfaces absorbing the whole VIS spectrum appear black. Colored objects do absorb certain wavelengths and reflect the rest. The spectral composition of the reflected light is noticed as the surface color.

Absorption of specific wavelengths is caused by the absorption of light by different types of functional groups, such as unsaturated bonds (e.g. C=C, C=O, etc.), as their electrons are easily excited to higher energetic levels or so-called excited states. Atomic groups containing π-electronic systems of sufficient degree of conjugation resulting in a coloration of the object are called chromophores (Hon and Minemura, 2001). Unsaturated bonds can be found in different wood components, which results in the natural color of wood.

17.1.1.2 The Color of Wood

Wood innate color is basically defined by its chemical composition. However, other factors such as specific anatomically formations or physical properties further affect the optical impression (Hon and Minemura, 2001). Due to the wide range of potential variations of wood color between and within the tree species, a huge quantity of natural wood colors exists. However, the naturally and commercially available color assortment does not always meet customer requirements.

Wood Chemical Composition Wooden cell walls are basically formed by the three different structural macromolecular components cellulose, hemicelluloses and lignin. A schematic model for the arrangement of these substances is given in Figure 17.1. From a functional point of view, cellulose acts, for example, as columns, lignin prevents them from deformation and hemicelluloses are the matrix. Due to the lack of chromophore structures within cellulose, hemicelluloses and lignin, no or just small parts of VIS are absorbed and they appear to be slightly yellowish or even white. Therefore, in general, the basic color of the wooden structure is very bright (Hon and Minemura, 2001). Deviations from this basic color are due to the existence of extractable substances such as phenolic compounds (e.g. flavonoids, stilbenes, tannins). Common colors such as brown, red or even black, as well as rarer ones such as green or violet, are due to the existence of such extractives. Concentration of extractives depends strongly on the position within the tree and also varies strongly between trees (Willför *et al.*, 2003; Ters *et al.*, 2006). For example, concentration of phenolic extractives is highest within knots (i.e. the branch base inside the tree stem; Willför *et al.*, 2003). The composition of extractable substances is species specific and most of them do not contribute to the wooden color. Depending on the solvents, different quantities of hydrophilic and hydrophobic organic extractives as well as inorganic compounds can be found in wood. The chemical composition of bark extractives is in general similar to that of wood but is higher in concentration (Fengel and Wegener, 1984). A general classification of organic extractives could be: (a) phenolics, (b) resin acids, (c) sterols in free form and esterified, (d) aliphatic alcohols and waxes and (e) free fatty acids and triglycerides. Phenolics, the main part of chromophore extractives, could be further subsummarized into (a) stilbens, (b) lignans, (c) flavonoids and (d) other monomeric phenolic compounds. In the case of coniferous species, flavanoids are rare in wood but can be found more often within bark, being usually highly condensed. More or less condensed flavonoids as well as hydrolysable tannins (derivates of gallic acid) are often summarized as tannins. Tannins are probably one of the most widely used group of dyeing agents based on wood or bark extractives. However, the color of wood depends strongly on its overall chemical composition.

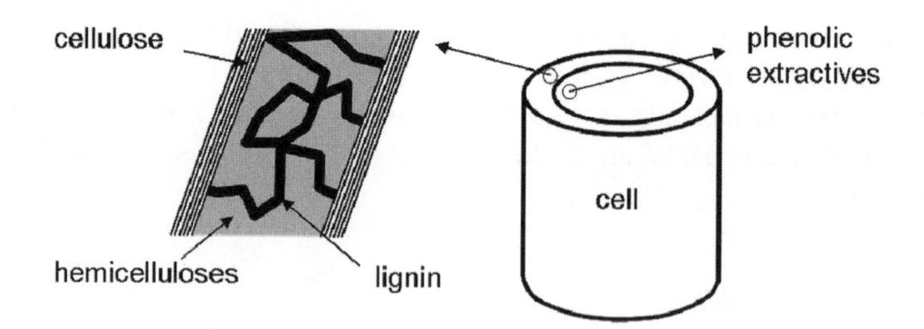

Figure 17.1 Schematic model for the arrangement of cellulose, hemicelluloses and lignin within the cell wall. Extractable substances (e.g. phenolic ones) are mostly associated with the lumen

Wood Anatomical Appearance Figure 17.2 gives an impression of the most important wood anatomical basics that will be referred to within this and the following subsections. The proportions within the figure can deviate strongly from natural conditions. However, parameters on a macro-scale (tree, wood) as well as on a micro-scale (wood, cells) are illustrated. The general appearance of a wooden surface as well as the final optical impression after a color modification process depends – besides other factors – on the wooden anatomy. Coniferous and broad-leaved trees show substantial differences in their wood anatomy and tree-ring structures. Due to differences in the tree physiology, some wood species are characterized by the formation of obligatory or facultative colored heartwood.

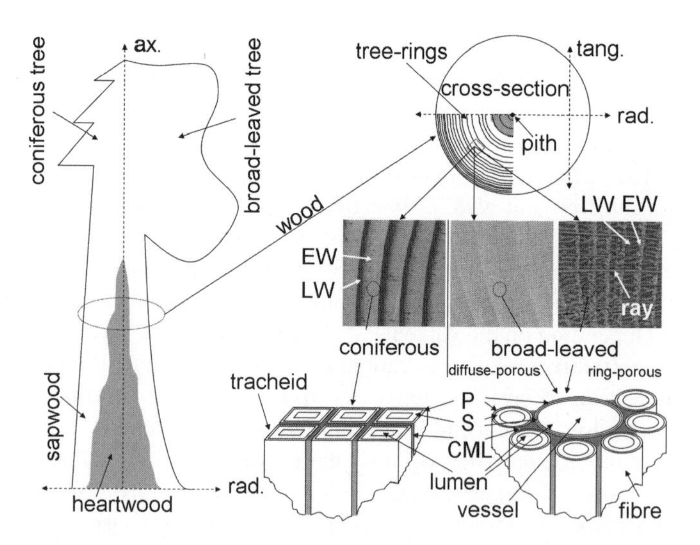

Figure 17.2 Principle wood anatomical structures on the macro (tree and wood) and micro (wood structure and cells) scale (ax. = axial growth direction, rad. = radial growth direction, tang. = tangential growth direction, EW = earlywood, LW = latewood, CML = compound middle lamella, P = primary cell wall, S = secondary cell wall)

Tree rings are the most important anatomical effect influencing the texture of wood surfaces. A separation into the four different wood anatomical basic types (i.e. coniferous, ring porous, diffuse porous and semi-diffuse porous) is based upon intra-annual tree-ring structures (Hoadley, 1990).

Conifers such as spruce, pine or fir have the simplest anatomy. Tree rings are usually very obvious due to changing colors within the tree ring. Earlywood is formed from the beginning of the vegetation period towards midsummer, showing rapid growth in the radial direction, low density, low extractive concentrations and a bright appearance. The subsequently formed latewood shows much lower radial extension, higher density, higher extractive concentrations and a darker appearance.

The arrangement of different cell types within the tree ring of broad-leaved trees can be very variable, but is usually species specific. Ring-porous trees (e.g. mostly oak, ash or elm) do form at least one ring of earlywood vessels. This earlywood ring shows low density due to thin cell walls and wide lumen. For example, earlywood vessel diameters for oak wood can reach values of up to 360 μm (Wagenführ and Scheiber, 1985), which is actually visible with the naked eye. A low degree of lignification, low density and the wide lumen usually give ring-porous earlywood a bright color. In contrast, latewood shows an extremely dense structure and often high contents of extractives, leading to dark colors. Diffuse-porous tree species (e.g. maple, birch or beech) do not show a characteristic density or other kind of pattern within the tree ring. Furthermore, tree rings do not dominate their appearance as they are often hardly visible. The distribution of vessels is almost homogeneous. Therefore optical properties of diffuse-porous tree species are mainly dominated by the content of extractives. Finally, also so-called semi-diffuse-porous tree species exist (e.g. walnut or cherry), being somewhat in between ring- and diffuse-porous species.

In general, variations of wood density will always reflect any kind of coloration process in different ways. While homogeneous wood material without visually evident tree rings will always give more homogeneous coloration, clearly structured tree species will display, for example, dye gradients strongly related to their density profiles and morphological features like cell type or lumen.

Colored heartwood is another anatomical characteristic of wood with a strong influence on the color. Sapwood is the primary appearance of wood being produced by cambial differentiation. Depending on the species, sapwood rings may become modified to heartwood rings. Heartwood shows a lower water content and often increased extractive contents (e.g. oxidized phenolic substances such as flavonoids, stilbens or lignans) coupled with a significant darkening and a shift to a brown or reddish color. The formation of heartwood is often coupled with the formation of tylosis. These are formally living parenchyma cells that grew into the inactive vessels in order to seal them and disable fungal growth. However, in the case of impregnation of wood with dyes such tylosis also strongly influences coloration processes, as the uptake of dyes is reduced.

The ratio between sapwood and heartwood is more or less species characteristic, but it also depends on environmental conditions and tree age. For example, black locust (*Robinia pseudacacia* L.) usually shows extremely narrow sapwood structures consisting usually of about three tree rings. On the other hand, black pine (*Pinus nigra* Arnold) produces the colored heartwood very late so that about two-thirds of the cross-section belongs to the sapwood. However, even within one species a strong variability of color within sapwood and heartwood can occur due to geographical origin or tree height (Grekin, 2007).

Physical Properties of the Wooden Surface Corresponding to Hon and Minemura (2001), the irradiation direction, the moisture content and the surface roughness have a further impact on the wooden color. These parameters significantly influence the accuracy of color measurement. Consequently, they must be equal when comparing wood surface colors.

Irradiation direction influences the amount of reflected, scattered or transmitted light due to the cellular structure of wood. Lightness increases with increasing deviation between the irradiation direction and the axial direction. Color saturation shows exactly the inverted trend with lower amplitude compared to lightness, without any change of the hue.

Moisture content has a certain impact on wooden color as wet wood appears darker than dry wood (Hon and Minemura, 2001). Above fibre saturation (lumen are partly or completely filled with water) light penetrates deeper into the cellular structure and is scattered. Small differences in moisture content below fibre saturation in the range of infinite percent points, such as due to seasonal fluctuations of relative humidity, do not visually influence the wooden color (Pöckl, 2007).

Surface roughness related effects are similar to those described for the irradiation direction. In the later case the effect is due to the anatomical orientation of truncated cells. Rough surfaces additionally show truncated structures independent from the anatomical direction. Especially reflection and scattering are increased on such surfaces. Treated wooden surfaces (e.g. varnished) often appear besides others to be darker due to the reduction of such effects.

17.1.2 Color Measurement

A quantitative description of color is often performed by using the $L^*a^*b^*$ color space values (CIE 1976 color system). It is nowadays the most widely used color system within the wood industry and the majority of authors cited in the following sections refer to it.

The $L^*a^*b^*$ system describes color as a rotational body-shaped space where each spatial location can be described by three coordinates and thus quantitatively represents a certain color. The three coordinates are L^*, the luminance, a^*, the green-red axes and b^*, the blue-yellow axes. Values for L^* reach from 0 to 100 while higher values represent brighter and lower ones darker colors. The extreme values 100 and 0 represent absolute white and black, respectively. Following the L^* axes between the extreme values all grey nuances are located.

Leaving the L^* axes towards the outer sections of the color space and following the two other axes, color saturation increases. Following the perimeter of the color space in the a^*–b^* layer from $+a^*$ (red) to $+b^*$ (yellow) to $-a^*$ (green) to $-b^*$ (blue) and back to $+a^*$, all the colors can be found arranged as in the spectra of visible light.

In order to compare different colors, the color distance (ΔE^*_{ab}) is commonly used. The distance between two coordinates within the color space can be described by (DIN 6174, 1979)

$$\Delta E^*_{ab} = \sqrt{(\Delta L^*)^2 + (\Delta a^*)^2 + (\Delta b^*)^2} \tag{17.1}$$

where the terms for ΔL^*, Δa^* and Δb^* are defined as

$$\Delta L^* = L^*_{\text{sample}} - L^*_{\text{reference}} \tag{17.2}$$

$$\Delta a^* = a^*_{\text{sample}} - a^*_{\text{reference}} \tag{17.3}$$

$$\Delta b^* = b^*_{\text{sample}} - b^*_{\text{reference}} \tag{17.4}$$

17.1.3 Color Stability

Color stability is of interest during the production process as well as during the whole lifetime of the final product. While in the first case mostly reactions due to changed moisture content, oxidation processes or increased temperature are of interest, radiation is one of the most important reasons for surface discoloration during the product lifetime. Electromagnetic radiation provides an energy input in the wooden surface that facilitates radical formation and subsequent chemical reactions that may lead both to degradation (bleaching) as well as to polymerization (darkening) of wood inherent colored compounds. The different chemical reactions depend on the wood species, exposure time and light frequency. Specific details about processes due to radiation will be discussed later. Sundqvist and Morén (2002) argued that the human eye is sensitive to color differences of $\Delta E^*_{ab} = 2 - 3$. In direct comparison of two differently altered wooden surfaces, e.g. by removing a carpet from wooden parquet, even smaller color differences might be obvious.

17.2 Coatings

The most classical way to colorize surfaces of wood-based products is to apply a coating containing dyes or pigments. However, the major function of coating systems is the protection of surfaces against any kind of injury or degradation. For example, due to the application of an acrylic polymer coating on a wooden surface, radiation-induced lignin degradation next to the wood surface is significantly reduced (MacLeod *et al.*,1995). Due to the low surface hardness of wood compared to other materials such as metals or several artificial polymers, accomplishing protection is particularly difficult. While with interior surfaces the major focus of a colored coating lies in an appealing visual appearance, there are much higher demands for products intended for outdoor use. In exterior applications, besides mechanical stress and moisture, the major environmental factor with respect to color changes is light. Typically, coatings of chemical formulations contain light-stabilizing compounds that may act according to different chemical mechanisms, described in Bieleman (1998). Light-induced color changes will be discussed below in more detail.

In general, two types of surface coatings for wood are distinguished, open porous coatings and closed porous coatings (Wicks *et al.*, 1999). Open porous coatings allow permanent moisture equilibrium of the material with the environment. The wood pores are not completely filled with the coating and remain in their native condition. Thus, they contribute to the haptic behaviour of the material by their three-dimensional structure. With closed porous coatings, the porous space is completely filled with material and a smooth surface is formed where the original texture is not felt any more. They do not support material exchange with the environment. This can lead to severe problems as soon as moisture penetrates into the wood (e.g. from a crack) and becomes locked below the surface. The high moisture content increases the probability of secondary effects such as fungal decay or peeling of the coating. Both effects potentially change dramatically the optical appearance. With light colors such surfaces are normally coated with a transparent topcoat that contains light stabilizers in order to prevent color changes due to irradiation (Rothkamm *et al.*, 2003).

Typically, the coating of wood consists of several subsequent steps. In the first place, the sanded surface is stained with a reactive dye that may be of natural origin. After drying and possibly another sanding step, the surface is coated with a base lacquer, dried and sanded again. Finally, a topcoat is applied. Most systems are based on synthetic dyes and pigments. The actual step of coloration is in fact very similar to the process described below in the section about impregnation, and may also be based on natural pigments. Especially, staining solutions based on natural mordant dyes such as alizarin or anthrachinone derivatives are still important today, although they are no longer isolated from natural resources but are chemically synthesized. Kollmann (1955) actually reported that most of the mordant applications based on plant colorants had been substituted by other methods. However, it is mentioned that hematoxylin and aniline are still used. Especially these two substances will be discussed later on in the section about dyeing for analytical purposes as modern staining agents. Any kind of chemical that gives a color change of wooden surfaces can be seen as a mordant. However, mordants based on metal ions had been of special industrial interest for decades. Kollmann (1955) lists, for example, cobalt chloride, copper chloride, copper sulfate, manganese sulfate, nickel sulfate, potassium dichromate and several iron salts. These salts basically react with tannins that are naturally available in wood or are artificially added.

According to DIN 55945 (2007), natural lacquer systems are coatings consisting of naturally occurring compounds that are neither chemically modified nor changed in their natural structure. They do not contain artificial components such as binder resins or solvents or chemical additives. Typically, such materials are oils, waxes and resins of natural provenance. Often these substances are also used as raw material for coatings. Commercial coating systems consist of a number of components of different functions: (a) binder materials, (b) solvents, (c) additives, (d) dyes or pigments and (e) filler. Natural components may occur as film formation aids (cellulose nitrate), as softeners (stearates, ricinoleic acid) or as a basis for synthetic resins (alkyd resins). In this section, some natural materials used as the basis for coating systems that have been used to protect colored wood surfaces since ancient times and are still in use today are discussed (Schleusener, 2003).

Drying oils and waxes are some of the most important components of natural material based coating solutions. Drying oils have been used as binders for coatings formulations for centuries. They may also influence the appearance of the surface by native color or degradation-induced color changes. Although not often used anymore as native components, they are still important raw materials for the manufacture of alkyd resins, epoxy esters and uralkyds (Wicks *et al.*, 1999). Important sources for drying oils are linseeds, safflower, soybean, sunflower and other herbaceous plants.

Chemically, they are triglycerides consisting of glycerol that is esterified with different types of fatty acids. The fatty acids contain varying numbers of allylic double bonds, which act as the active species in the oxidative crosslinking step during drying. Depending on the source, there is a great variability in the composition of the material. A summary of the most abundant components of natural oils is given in Figure 17.3. They differ in number, position and geometric orientation of the double bonds. The mechanism of drying includes (a) an induction step, (b) a period of rapid oxygen consumption and (c) a complex sequence of autocatalytic radical coupling reactions.

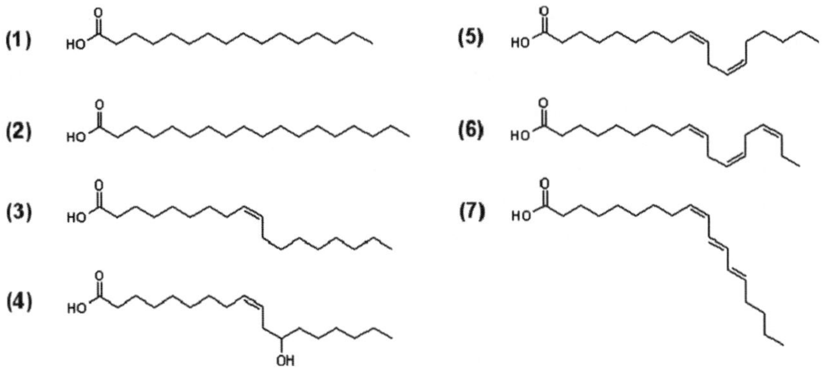

Figure 17.3 *Important naturally occurring fatty acids: (1) palmitic acid, (2) stearic acid, (3) oleic acid, (4) ricinoleic acid, (5) linoleic acid, (6) linolenic acid, (7) α-eleostearic acid*

Waxes are chemically very similar to oils in being esters of long-chained fatty acids with long-chained alcohols. They are solid at room temperature but have a very low melting point. Some other important raw materials are Carnauba wax and bees wax.

Natural resins of relevance for coatings are shellac, colophonium and copals. One major advantage is their low content of organic volatiles. The composition varies strongly depending on the origin of the material. While colophonium and copals are resins from plants, shellac is the only natural resin of commercial importance with an animal origin. Obtained from the excreta of the lac insect (*Kerria lacca* Kerr.), various extraction steps result in a differently colored product depending on the origin and degree of raffination that is easily crosslinked due to a number of reactive functional groups. The term ‘lacquer’ is derived from the insect species that produces shellac. Shellac is easily modified chemically and provides readily formed films of good mechanical stability and good adhesion.

17.3 Dyes

17.3.1 Dyeing for Analytical Purposes

Countless numbers of dyes and mixtures have been established over centuries, aiming for the coloration of wood and other botanical cellular structures for microscopic investigations. Structures within micro-sections showing sample thicknesses in the range of some μm as well as for solid wood investigated under a transmitted respectively reflected light microscope are clearer when samples are dyed. Mixtures or special procedures are often used in order to dye selective anatomical or chemical structures.

While most of the used dyes are synthetic, some do exist with a natural herbal origin. For example, hematoxylin, aniline derivates and reactions including tannins are in use. The following descriptions of dyes refer to Anonymous (1924), Gerlach (1984) and Ruzin (1999).

Hematoxylin is extracted from the heartwood of the tropical logwood (*Haematoxylum campechianum* L.). A similar dyeing agent but less efficient is brazilin, originating from the same botanical family. Pure hematoxylin is colorless and needs to be oxidized either by atmospheric oxygen or by further agents such as $NaIO_3$, KIO_3, $K_2Cr_2O_7$, $KMnO_4$ in order to produce hematein. Several dyeing procedures exist using hematoxylin in an attempt to produce coloration (e.g. dark or brighter variations of black, grey and blue, and partly even yellowish) of, for example, nucleons, nonlignified cell walls, protoplasm, chromatin, chromosomes, mitochondria or fungal mycelia.

Aniline can be extracted from indigotin or from the roots of alkanet (*Alkanna tinctoria* (L.) Tausch). It can be used in a wide range of several mixtures in order to detect lignin, cellulose, callose or mitochondria by development of black, blue, red or yellow color. Even the fluorescent nature of some aniline derivates is partly used for such analyses.

Tannins can be found in several botanical sources such as the wood and bark of oak (*Quercus* sp.) or in grapes and tea. Tannins are well known in wood technology to give black discoloration, for example, of fresh or wet oak wood in contact with iron. The same reaction principle is being used for microtechniques in order to dye meristematic cell walls.

Besides microtechniques, dyes in general, and consequently also herbal colorants, can be used to investigate water and other solvent transport mechanisms within wood and wood-based products. Especially with respect to the impregnation process (see the next section) such investigations are of interest. For example, anatomical barriers like tylosis, pith flecks, rays and others can reduce transport or even stop it. Besides capillary transport pathways within the solid cell wall structure can also be studied using such dyes.

17.3.2 Impregnation

Besides surface coloration, the color of the whole volume of wood pieces or wood-based products can be changed by impregnation with dyes. In comparison to shallow staining methods, the coloration by impregnation allows machine finishing without changing the color. Furthermore, scratches and other injuries do not appear as obvious when deeper layers show the same color as the surface.

17.3.2.1 Technology

Impregnation techniques to be used for natural plant dyes are partly the same as for wood preservatives. Although the aim of wood preservation is often just to impregnate wood with agents against insects or micro-organisms up to a certain depth, these methods can also be used for coloring wood with natural plant dyes. The most important methods developed for wood preservation have been well established for some decades and will be explained here with reference to van Groenou *et al.* (1952) and Langendorf and Eichler (1982). In principle the various impregnation treatments can be divided into two groups aimed at different impregnation depths: (a) pressureless treatments such as submersing, soaking and the osmosis treatment and (b) pressure treatments such as the vacuum/pressure impregnation or the Boucherie treatment.

Submerging and soaking are the simplest treatments of all. Dry wood is immersed and dyeing depth depends on process time, temperature, dye, species and wood dimension. In the case of very short treatments results will be similar to a simple surface treatment.

Osmosis treatment can only be performed on green (not dried) wood that has been carefully debarked or sawn. Cross-sections are completely sealed while all other surfaces are daubed with a dye-containing paste. A package of wood is wrapped in foil and dyes can diffuse into the wood within a duration time of several weeks up to three months. Dyeing depth depends mainly on process time, overall dye concentration, temperature, wood dimension, and wood species. This treatment was developed especially for tree species where impregnation is very difficult due to anatomical barriers, such as closed bordered pits of Norway spruce (*Picea abies* (L.) Karst.).

Vacuum/pressure impregnation is a term for several methods that deviate in processing conditions. Wood subjected to a different number of prior processing steps, such as debarking, splitting, drying etc., is impregnated in a pressurized tank. Often alternating pressures are used. The process is limited by the boiling point of the solvent used during the vacuum phase and the plant design in general. Often wood is evacuated first in order to remove all the entrapped air and the impregnation solution is added during the subsequent pressurized phase. Typical process conditions are in the range of 5 to 20 kPa for the vacuum phase and 800 to 1400 kPa for the pressure phase. By the end of the treatment often a short vacuum phase (approximately 45 kPa) is used in order to remove excessive solution volumes.

The transition between the vacuum and pressure phases can be gradual, stepwise or abrupt. Repeating the vacuum/pressure cycles leads to higher immersion depths. The optimum amount of cycles as well as length of each phase depends on the wood dimensions, wood species, viscosity of the solution, temperature and pressure difference. Modern treatments often contain more than one hundred applications of the vacuum/pressure cycles within a minimum total duration time of ten hours.

The *Boucherie treatment* was developed in order to impregnate wood directly after felling the tree in the forest. Later on the treatment was upgraded to an industrial process. Nondebarked trunks are positioned in such a way that the smaller cross-section is on a lower level than the wider cross-section. The elevated cross-section is connected to a tank containing the impregnation solution. The impregnation process can rely on the hydrostatic pressure difference or on a pumping process. At the other end of the trunk, sap and later on also dyeing solution seep out of the trunk. Depending on the wood anatomical barriers (such as heartwood) eventually a complete coloration is possible.

Independent of the technological process used, basically a piece of wood comes in contact with a solution of plant dyes. Preferably, water will be used as the solvent for environmental reasons. Other solvents would need deviating process parameters. Organic solvents are less desirable due to their toxicological potential but may nevertheless be necessary for dyestuff that is otherwise insoluble. Depending on their polarity, solvents may behave completely differently in the wood matrix. Figure 17.4 shows an example of the influence of the solvent on the transportation mechanism. Comparable samples of black alder (*Alnus glutinosa* Gaertn.) are immersed with their cross-sections in six different solvents (mixtures of water and ethanol) for the same lenth of time. Obviously transportation velocity was the highest for pure ethanol and lowest for pure water. The figure shows the enormous influence of the solvent on the transportation mechanism, which is the limiting factor concerning the use of dyes on solid wood.

Figure 17.4 *Influence of the solvent on axial transport in wood demonstrated for black alder (Alnus glutinosa Gaertn.). Solvents were different mixtures of ethanol and water (100 %, 80 %, 60 %, 40 %, 20 % and 0 % from left to right)*

In general dyeing green wood is more efficient than dry wood as the solvents need to be removed afterwards. Selection of wood dimensions close to the final formats reduces the amount of dyed wood that has to be removed during the following process steps. Furthermore, the impregnation time can be dramatically reduced by choosing smaller wood dimensions. Another general question is whether a complete impregnation is desired. Incomplete colorations to give optical effects or the impregnation up to a certain depth are possible.

17.3.2.2 Color

Laboratory results for small wood samples of different species produced in a vacuum/pressure treatment (20 kPa and 800 kPa) using aqueous solutions (4 g/l) of indigotin and chlorophyll sodium salt respectively are shown in Figures 17.5 and 17.6. The samples show that impregnation is partly limited by anatomical barriers such as the heartwood of beech (*Fagus sylvatica* L.), wood rays and the latewood of oak (*Quercus* sp.) or pith flecks of black alder (*Alnus glutinosa* Gaertn.). However, impregnation was in general more complete for indigotin than for chlorophyll. Table 17.1 shows results of color measurements performed on these samples. Species in the table are arranged due to decreasing b^* values (blue saturation increases downwards) in the case of indigotin. In the case of chlorophyll they are arranged according to decreasing a^* values (green saturation increases downwards).

Figure 17.5 *Wood species that had more or less been successfully dyed with indigotin. Species are arranged according to increasing blue components (b* gets more negative); from left to right in each row: oak, nut, cherry, ash, beech heartwood, beech, Douglas fir, alder, pine, poplar, maple and birch (See Colour Plate 5)*

Figure 17.6 *Wood species that had more or less been successfully dyed with chlorophyll sodium salt. Species are arranged according to increasing green components (a* gets lower); from left to right in each row: beech heartwood, beech, alder, birch, maple and poplar (See Colour Plate 6)*

Table 17.1 Color values of several wood species dyed with indigotin and chlorophyll-sodium salt, respectively

	Indigotin				Chlorophyll-sodium salt		
	L*(D65)	a*(D65)	b*(D65)		L*(D65)	a*(D65)	b*(D65)
Oak	61.71	1	12.56	Beech heartwood	57.1	7.15	15.62
Nut	45.39	2.95	6.07	Ash	70.67	4.37	16.41
Cherry	53.65	−0.73	4.89	Beech	57.05	3.97	14.8
Ash	59.9	−4.56	−0.13	Alder	61.82	2.45	14.01
Beech heartwood	49.1	−3.5	−1.78	Birch	67.71	1.2	12.59
Beech	50.11	−6.66	−3.07	Maple	72.52	1.14	12.51
Douglas fir	51.23	−6.32	−3.97	Poplar	71.74	−0.22	13.23
Alder	54.73	−7.82	−5.04				
Pine	57.36	−8.11	−5.16				
Poplar	58.91	−8.75	−8.79				
Spruce	60.56	−7.7	−9.34				
Maple	60.01	−8.33	−10.2				
Birch	54.18	−7.84	−10.74				

The general appearance of the indigotin dyed samples is blue although *b** values vary strongly. For example, the dominant wood rays and the latewood of oak (*Quercus* sp.) shift color values towards yellow. In the case of chlorophyll, values do not even stand for an expected green coloration. This is due to the fact that dyes applied in this way just give wood an additional color while original wood characters remain almost stable.

17.3.2.3 Products

Such coloring methods using natural plant dyes could be used for solid wood as well as for any kind of wood-based panels. Colorful results can be produced by using veneers (slices of wood, between less than 1 and up to 4 mm in thickness).These veneers can be dyed and glued together into a block, which can be used again to produce veneers. To obtain a striped image the block has to be cut perpendicular to the former direction of the veneers.

Slight coloration of engineered wood products is partly performed in order to classify products with respect to special properties. Completely colored medium density fibre-boards (MDFs) can also be produced showing saturated colors. Such coloration is not due to an impregnation process, as pigments are simply added to the fibres or the glueing system. However, completely colored MDF is nowadays commonly available and accepted by the market (Buchholzer, 2005). It gives the opportunity to millcut structures into boards such as fronts of cupboards without any additional coloration or the need for colorful foils. However, substitution of the existing dyes for such products by natural plant dyes would be technically possible.

17.4 Color Modification

Wood is maybe the only valuable resource used for decorative purposes that originates from plant production. It has a more or less given natural color, which might be changed using methods similar to those described above.

Another method for the use of herbal colorants is to change the color of the wood itself. Wood contains several functional groups within its basic chemical composition that might obtain chromophores due to oxidation or other alterations. Furthermore, the natural chromophore character can also be intensified or even changed due to several modification procedures. Most commonly, color modification methods are based on reactions at increased temperature and relative humidity, such as in the case of technical drying, steaming or thermal modification.

Potential reasons for alteration of the original wooden color are (a) exquisite decorations lead to higher product profitability, (b) exquisite species are rare and often not available, (c) regionally available species in their natural appearance are also used by competing companies and can therefore not be used as a trademark, (d) new decors offer new market opportunities, (e) stringency of resources even appears for the most frequently cultivated species due to forestry management (turnover time, protection of forests, small-sized forests in private holdings, focus on by-products of the forest such as welfare, wild animals or even fish, etc.) and (f) recycling of such products might be much easier as no additional inseparable substances (like coatings) are contained. Further advantages of color-modified wood (such as for dyed wood) are less obvious damage (e.g. scratches) and easier restoration thereafter.

Color modification is always coupled with changes in the chemical structure of wood. These changes might also have a strong impact on other physical or biological properties, such as its mechanics, behaviour against water or the resistance against micro-organisms. Improving one of these properties often leads to decreasing properties in another respect. For example, the good resistance of thermally modified wood against micro-organisms is mainly due to the destruction of hemicelluloses. On the other hand, this leads to significantly reduced mechanical properties (Hill, 2006).

17.4.1 Drying

17.4.1.1 Basics

Due to its chemical structure and the primary ecological function within the tree, wood shows a distinct affinity to water. Thus a major character of wood is its hygroscopicity. Depending on environmental conditions like humidity, temperature and air pressure, the moisture content of wood changes following different physical laws. Under constant environmental conditions the so-called equilibrium moisture content is achieved over time. Changes of moisture content below fibre saturation (approximately 30 % or even less water mass compared to absolute dry wood mass) lead to dimensional changes. These dimensional changes are the major reason for the need of drying. The moisture content needs to be fixed within a small range depending on the conditions during end use (Langendorf and Eichler, 1982). If this is not done accurately warping, bad fit of connections, grooves or even discoloration might appear. Furthermore, material properties depend strongly on the moisture content. For example, strength properties decrease with increasing moisture content (Kollmann and Côté, 1968). Increased moisture content also leads to processing problems, especially concerning glueing, planeing, sanding or other surface treatments. There is also a higher risk of microbiological attack due to better conditions for fungi or bacteria at a higher moisture content. Such attack can also be coupled with a significant color change.

Especially in the case of high-quality wood species such as oak (*Quercus* sp.), higher demands exist on drying quality and product properties than for ordinary species. For example, deformation or cracks are unacceptable and are real problems (Vanek, 1986; Luostarinen and Luostarinen, 2001). As high-quality products are often designed for interior use, decorative aspects and particularly the color are major criteria in quality control. Discoloration might actually appear during the lifetime of the tree but can also occur within several steps of production. In general, discoloration lowers the value of the wood and can lead to complaints. An accurate technical drying process can prevent most storage and process related discolorations.

On the other hand, discoloration might also occur due to drying. For example, Dawson-Andoh *et al.* (2004) showed for hard maple (*Acer saccharum* Marsh.) that lightness decreases while redness and yellowness increase during kiln drying. Drying flaws were defined by Ward and Simpson (1991) as any kind of drying effect on the wood leading to diminished product value. This includes mechanical failures such as cracks as well as discoloration. Inhomogeneous coloration might especially be caused by diffusion processes of resins (Wiberg, 1996), hydrolysis or oxidation of several wood compounds (Schmidt, 1986; Charrier *et al.*, 1995). Schmidt (1986) suggested a pre-drying step for oak wood in the open air in order to achieve just a slight but homogeneous discoloration. Depending on customer requirements, one and the same color after the drying process might be defined as a drying flaw or as increased value.

17.4.1.2 Technology

According to Fengel and Wegener (1984), physical and chemical properties of wood change significantly with increasing temperature and time of impact. In the first place, high temperatures lead to an enormous loss of mass due to evaporation of free and physically bound water as well as highly volatile extractives. Furthermore, chemical changes appear due to elevated temperatures that have a strong impact on wood color and further properties (Hill, 2006). According to Sandermann and Augustin (1963a, 1963b), the rate of wood degradation at temperatures below 100 °C is so slow that it can be neglected. The authors argue that polysaccharides with a more pronounced crystalline degree, such as cellulose, compared to hemicelluloses and lignin are more stable under elevated temperatures.

Besides temperature and processing time, the surrounding treating medium also has an important impact on wood color. According to Sandermann and Augustin (1963a), four methods of heat transfer during drying (and related processes) are usually employed. These are drying in the presence of (a) water or steam, (b) oxygen, (c) inert media such as nitrogen or oils or (d) under evacuated conditions. The use of water leads to hydrolytic reactions and the presence of oxygen to oxidations. Such processes are probably unwanted and can be avoided by employing inert media or vacuum.

Convection drying is industrially the most widespread method. Due to temperature and humidity profiles along several stages of the process, the equilibrium moisture content changes and water is removed by ventilation. Heat transfer is accomplished by convection. Ventilation of the whole system is partially performed. Process temperatures are kept below 100 °C. A characteristic drying pattern consists of a heating, a drying, a conditioning and a cooling phase (Trübswetter, 2006). High temperatures are only applied by the end of the process as soon as wood is below fibre saturation.

An example of some possible color variations due to convection drying of black alder (*Alnus glutinosa* Gaertn.) is given in Figure 17.7. Samples are taken from independent industrial processes. Variability might be due to pre-air-drying, individual wood properties or process parameters such as the drying rate or the temperature.

Figure 17.7 *Variation of black alder (Alnus glutinosa Gaertn.) wood color due to six independent convection drying processes (See Colour Plate 7)*

High-temperature drying is a process operating at temperatures above 100 °C. Heat transfer can, for example, be performed by steam. Vanek (2001) suggested this method for drying wood of large timber dimensions and for problematic species that are known to have difficulties during drying, such as poplar (*Populus* sp.), birch (*Betula* sp.), linden (*Tilia* sp.) or comparable tropical species (Hildebrand, 1979).

Vacuum drying uses the benefit of the reduced boiling point of water due to reduced atmospheric pressure. At pressures of about 10 to 15 kPa the boiling point of water is approximately 50 °C. Consequently temperatures between 50 and 70 °C are used. Major advantages of this process are a reduced drying time and reduction of drying flaws (Trübswetter, 2006) such as discoloration. Prevention of discoloration is due both to the reduced amount of oxygen, prohibiting oxidation reactions, and to the low temperatures, preventing temperature-induced color changes.

High-frequency vacuum drying is a combination of two straightforward techniques. In principle, high-frequency drying could also work without any evacuation. In comparison to the three other methods, heating of wood is not caused by convection or any surrounding media but by dielectric heating of water molecules within the wood in a high-frequency alternating field (Torgovnikov, 1993). This drying procedure is also known for small discolorations (Hansmann, 2002) that result for the same reasons as in vacuum drying. Processing takes place with frequencies in the MHz to GHz range and field intensities of several kV/m.

17.4.1.3 Color

Treating conditions during the drying process have a strong impact on wood color as the following example illustrates (Hansmann, 2007; Seeling *et al.*, 2007). European beech (*Fagus sylvatica* L.) boards of 32 mm thickness were dried by the four above-mentioned methods. A comparison of these four techniques shows that there are significant differences between the drying methods with respect to color change and processing time (Table 17.2). Vacuum drying leads to the lightest and high-temperature drying to the darkest coloration (L*). High-temperature drying shows the highest value in red saturation (a*). Comparatively minor differences were found regarding the yellow saturation (b*). A major difference was found for the drying time. While drying towards the same drying target, moisture content took about 15 days using the convection drying process, vacuum drying took about half that time, high-temperature drying about one-fourth and high-frequency vacuum drying less than 1.5 days. Of course, these results are still species specific and the drying characteristic has to be studied for each different type of wood in order to find the optimum operating conditions for a specific process. For example, Charrier *et al.* (1992) found for two oak species brown discoloration on convection-dried wood and no discoloration using vacuum drying due to prevented oxidation.

Table 17.2 *Comparison of four different drying methods concerning achieved wood color and drying time, based upon 32 mm thick beech (Fagus sylvatica L.) boards (according to Hansmann, 2007, and Seeling et al., 2007)*

	Convection drying	Vacuum drying	High-temperature drying	High-frequency vacuum drying
L*	77.4	79.8	65.5	78.2
a*	7.3	6.1	10.8	6.7
b*	19.2	19.2	19.2	18.8
Time (d)	15.5	7.0	3.5	1.0

Selection of the optimal method depends primarily on the availability of different drying systems and secondarily on the requirements and finally also on production capacities. For example, Luostarinen *et al.* (2002) studied discoloration of birch (*Betula pendula* Roth) wood due to several possible influences. They found that the brightest wood was achieved by winter felling, storage for several weeks and use of convection drying. In comparison with vacuum drying the brightest wood shades were obtained after autumn felling and five-week storage. Differences in discoloration in these two particular cases, however, might be attributed more to very specific drying process properties (e.g. temperatures up to 82 °C for the vacuum process) than to influences of growing site or felling season. Furthermore, there is also a strong variability in color shades within one drying method when different drying schedules are employed (Luostarinen and Luostarinen, 2001).

17.4.2 Steaming

17.4.2.1 Basics

Steaming is one of the most commonly used color modification methods in the wood industry. Further reasons for steaming are the reduction of tension stresses and standardization of wood moisture content. Such benefits also have a positive impact on the subsequent drying process, especially on the drying time (e.g. Simpson, 1975; Harris *et al.*, 1989; Alexiou *et al.*, 1990).

17.4.2.2 Technology

Steaming chambers can operate in a direct or an indirect mode. In the first case, steam is injected directly to the chamber atmosphere while in the second case, injection to a water bath or indirect heating of the water bath is performed. Among others, the direct mode is characterized by a shorter processing time and lower steam pressures (110 kPa instead of 130 to 150 kPa) compared to the indirect mode (Brunner, 1987). However, for conventional processes, pressure within the chamber never exceeds much more than atmospheric pressure. Increased processing time due to slow heating up within the indirect mode causes a gentle wood modification and a reduction of flaws, such as those expressed by discoloration or cracks. For the most widely used steaming processes, air is always almost saturated with steam and temperatures are definitely below the boiling point, commonly in the range of 70 to 90 °C (Brunner, 1987). However, depending on the product requirements (e.g. color, moisture content) and wood dimensions, usual steaming processing times range from about two hours to four days (Brunner, 1987; Trübswetter, 2006).

17.4.2.3 Color

Concerning color modification due to steaming, European beech (*Fagus sylvatica* L.) and black locust (*Robinia pseudacacia* L.) are two of the most prominent middle European species. While European beech achieves a slight pink or reddish color, black locust becomes dark brown. Other species such as alder (*Alnus* sp.), cherry (*Prunus* sp.), oak (*Quercus* sp.) and sometimes conifers are also steamed.

Figure 17.8 gives examples of steamed black alder (*Alnus glutinosa* Gaertn.) wood treated at 95 °C in the presence of air for several time steps. The figure shows the potential of steaming as a color modification method. Compared to the samples shown in Figure 17.7 (the variety of color due to different drying processes), the spectrum of accessible colors is much broader. In comparison to high-temperature treatments (Figure 17.9), similar colors are also developed by steaming. Keeping in mind that the results in Figure 17.8 have just been investigated for one temperature level and at atmospheric pressure and for just one species, it is evident that steaming has a huge potential for giving wood a desired color without adding artificial dyes.

Figure 17.8 *Color impression of black alder (Alnus glutinosa Gaertn.) steamed for 0, 1, 2, 3, 4, 8 (first row) and 10, 12, 14, 16, 18 and 30 days (second row) (See Colour Plate 8)*

For example, Tolvaj *et al.* (2000) concluded that wood of black locust (*Robinia pseudacacia* L.) steamed at 90 to 98 °C for up to 22 days has the potential to compete (according to the color) with exotic species such as African wengé (*Millettia laurentii* De Wild.) or South American sucupira (*Bowdichia* sp.). Furthermore, they showed that steaming leads to a more uniform wood color and that the strongest color reactions take place at the beginning of the process, which can also be seen in Figure 17.8. Strong color changes are described for black walnut (*Juglans nigra* L.) (Brauner and Loos, 1968) or hybrid walnut (*Juglans nigra* L. 23 × *Juglans regia* L.) due to steaming, with the opportunity to almost homogenize the color of sap and heartwood (Burtin *et al.*, 2000). Brauner and Loos (1968) showed for black walnut (*Juglans nigra* L.) that color change can be intensified by increasing temperature, humidity and duration time, whereas the most significant darkening appears within the first hours. Seeling *et al.* (2007) used steaming in order to align color of the discolored red heartwood and non-discolored regions within European beech wood (*Fagus sylvatica* L.). This aim could get achieved, for example, by steaming for 24 h at a temperature of 120 °C.

However, these color changes have their origin in a changed chemical composition of the wood concerning extractable and structural substances. For example, Choong *et al.* (1999) argued that extractives of Southern pine (*Pinus* sp.) move with water due to moisture gradients that automatically appear within the wood during the drying process. Burtin *et al.* (2000) found for steamed walnut strong correlations between the color change and a change in phenolics concentration (decrease of hydrojuglone glucoside, ellagic and gallic acid derivates and accumulation of flavonol and several oxidation products). Degradation of lignin in steamed European beech (*Fagus sylvatica* L.) wood was found by Kürschner and Melcerová (1965a, 1965b).

Besides color change, steaming also leads to a stabilization of wooden color under radiation. An example of this improvement will be discussed later on in Section 17.4.7 on radiation.

17.4.3 Thermal Treatment

17.4.3.1 Technology

Militz (2002) postulated that all the industrially available thermal treatment processes used in Europe have in common a temperature range between 160 and 260 °C. However, treatments using lower temperatures should be dedicated to drying or steaming.

Like the two hydrothermal modifications (drying and steaming) discussed above, product properties depend strongly on similar process parameters. Hill (2006) listed the process time, the temperature, the process atmosphere, the connection of the system to the surroundings, the wood species, the humidity including moisture content of wood, the dimensions of the commodity and the presence of catalysts as crucial for product properties. As described earlier for the drying process (Sandermann and Augustin, 1963a), thermal modification can be achieved in the presence of water or steam, oxygen or inert media (e.g. nitrogen, oils) or under evacuated conditions. Due to the elevated temperatures, such conditions have an even stronger impact on wood color and other properties than actually reported for the drying process.

17.4.3.2 Color

Figure 17.9 gives an impression of the darkening due to thermal modification. Three pieces of the same individual fresh black alder (*Alnus glutinosa* Gaertn.) board had been modified at 160 °C in an open system without any additional water for one, two and three days (top down). Compared to the samples of black alder used for drying and steaming (Figures 17.7 and 17.8), darker samples were produced within a much shorter time. Darkening of wood due to thermal modification depends mostly on temperature, process time and atmospheric composition (Hill, 2006).

Such color changes are due to enormous changes in the chemical composition of wood. According to Hill (2006), changes related to color hue concern extractives (e.g. degradation of tannins) and lignin (e.g. formation of phenolic breakdown products). Disintegration of holocellulose (e.g. formation of furan, furfural and other breakdown products; Fengel and Wegener, 1984) has an impact on lightness. Hill (2006) gave an overview of diffusion processes of lower molecular substances (e.g. fats, waxes, resin acids) towards the surface.

Figure 17.9 *Thermally treated black alder (Alnus glutinosa Gaertn.) after 1, 2 and 3 days duration (top down) (See Colour Plate 9)*

These might lead to changed optical properties, but their existence on the surface depends on the temperature and process time. Further treatment or even more increased temperatures can lead to a complete disintegration and evaporation of such substances. Lignin is thermally much more stable than most of the extractive substances or holocellulose. However, some chemical changes still occur depending on modification conditions (Hill, 2006). For example, the ratio of syringyl and guaiacyl might be changed.

Roffael and Schaller (1971) showed that for spruce sulfite pulp significant darkening appears due to increased temperatures, being also slightly dependent on humidity. Furthermore, a reorganization of α-cellulose, especially concerning the degree of polymerization (dp value), was found to depend on temperature and humidity.

Investigations on sawdust of Scots pine (*Pinus sylvestris* L.) sapwood and European beech (*Fagus sylvatica* L.) heartwood treated at 240 °C under nitrogen atmosphere were performed by Nguila Inari *et al.* (2007). Mass loss in the range between 16 and 18.5 % coupled with a reduction of holocellulose from 77 % to the range of 50 to 60 % were reported. Especially hemicelluloses were considered to be involved in these disintegration processes. Reduction of hemicelluloses leads to a reduced number of free reactive hydroxyl groups, explaining the reduced chemical reactivity of thermally treated wood. Especially for acetic anhydride, maleic anhydride, succinic anhydride and phenyl isocyanate, reduced reactivity was found due to thermal treatment.

The disintegration of hemicelluloses was described by Hill (2006) as a reaction towards volatile heterocyclic substances, methanol and acetic acid. Such decomposition products are the reason for the unsavoury odour of thermally treated wood. The authors' own investigations showed, for example, that independent of the wood species, more than 90 % of test persons classified the odour of thermally treated wood as unsavoury while for wood steamed for up to 30 days less than 50 % gave this classification.

Compared to untreated wood, thermally treated wood showed more color stability (Ayadi *et al.*, 2003), but if exposed to exterior conditions, it still tends to discoloration and greying.

17.4.4 Ammoniation

17.4.4.1 Basics

Ammonia treatment as a method to change the color of wood is known by carpenters as a surface modification (Tinkler, 1921; Flocken *et al.*, 1975) as well as by industry as a process of modifying whole wood pieces. A lot of chemical reactions between the cell wall substances as well as extractives and ammonia are possible and described in the literature, especially for soaking wood with pure ammonia or aqueous solutions and high-pressure treatments.

Bariska (1969) reported that hemicelluloses and low molecular lignin were completely dissolved due to ammonia soaking. Oniško and Matejak (1971) concluded that due to the temporary plasticization of cell walls while soaking, nonbound components such as low molecular sugars, pectin, acids, alcohols, dyes, tannic acids and even inorganic salts can be dissolved. Kalnin'š *et al.* (1967) found about seven times more extractable substances in an aqueous ammonia solution than in water after soaking wood. Parham *et al.* (1971) found a change from cellulose I to cellulose III due to high-pressure treatment of wood with ammonia. Amburgey and Johnson (1979) found indications for increased but still acid pH values and also increased nitrogen concentrations in high-pressure ammonia gas treated wood.

A lot of effort was made between the 1960s and the 1970s to use ammonia for plasticization of solid wood and wood particles with the aim to produce wood and wood products with changed properties. For example, bending of solid wood could be performed after an ammonia treatment and was more accurate than after water vapour treatment (Schuerch, 1964; Bariska, 1969, 1974; Kalnin'š *et al.*, 1969; Davidson and Baumgardt, 1970). Also due to the temporary plasticization of wood, boards with increased density and mechanical properties could be produced (Kalnin'š *et al.*, 1967; Berzin'š *et al.*, 1970; Graf *et al.*, 1971, 1972; Oniško and Matejak, 1971). However, color change of wood had been of lower interest during that time.

17.4.4.2 Color

The described reactions of cell wall substances and extractives with ammonia cause a wide range of modified wood colors. Partial inclusion of nitrogen, oxidation reactions and the presence of tanning agents such as tannin seem to be mainly responsible for color changes due to ammoniation. However, the question of the chemical reasons for these color changes has not yet been answered.

Oak (*Quercus* sp.) is the main wood used for this treatment due to strong and homogeneous darkening and good availability of the species (Figure 17.10). The color reaction of ammonia fumigated oak (*Quercus* sp.) wood was found to be possible only in combination with the availability of atmospheric oxygen (Tinkler, 1921). Furthermore, it is reported there that tannin has a strong affinity to silk and that color changes of silk due to impregnation with tannin as well as a gas phase ammoniation of such silk are similar to the color reactions of oak (*Quercus* sp.) wood. Kalnin'š *et al.* (1969) reported that an ammonia treatment can highlight the texture of inconspicuous hardwoods and has the potential to compete with other species like oak (*Quercus* sp.), walnut (*Juglans* sp.) or mahogany (*Swietenia* sp.). This might be due to the general color change as well as to special color effects, as shown in Figure 17.11 for cherry (*Prunus avium* L.) wood.

Figure 17.10 *Untreated (left) and ammoniated (right) oak (Quercus sp.) wood from the same individual tree (See Colour Plate 10)*

Figure 17.11 Ammoniated cherry (Prunus avium L.) wood (See Colour Plate 11)

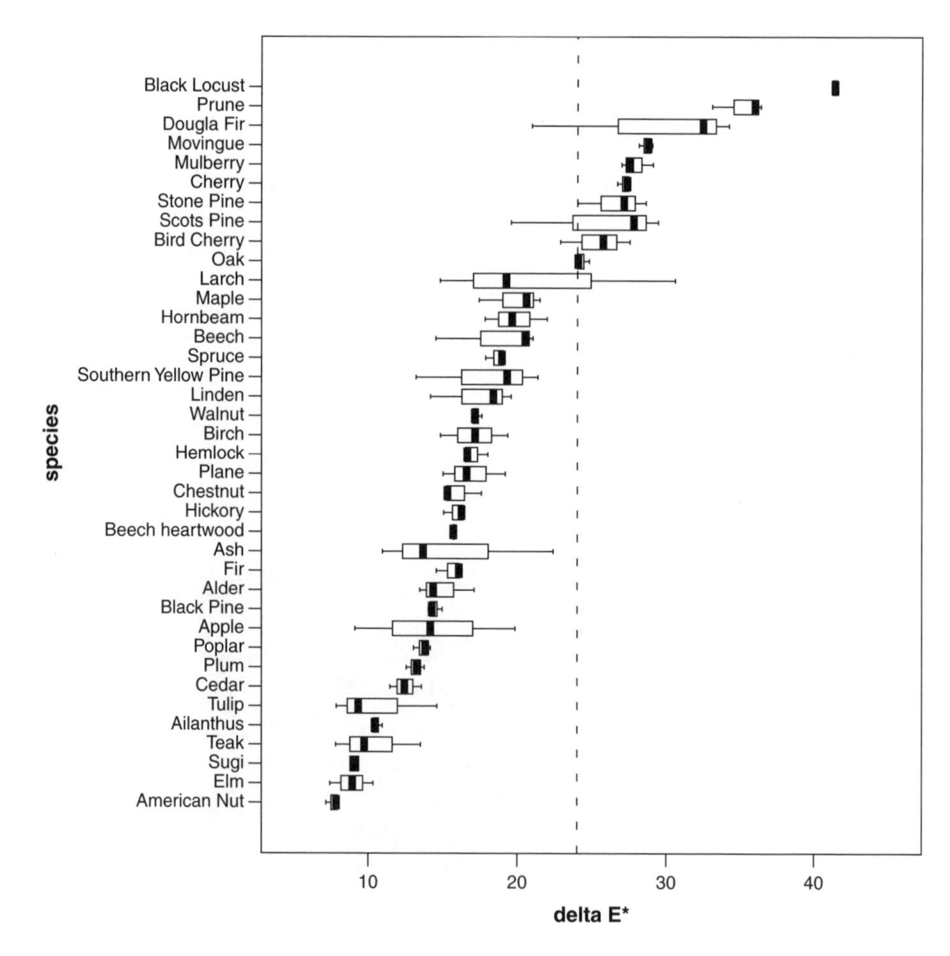

Figure 17.12 Color change (ΔE^*_{ab}) of 38 different wood species due to ammoniation. Broken line indicates the level of oak (Quercus sp.) wood as a reference

Figure 17.12 shows the effect of ammoniation on the color (ΔE^*_{ab}) for 38 wood species. Color change due to ammoniation is the highest for black locust (*Robinia pseudacacia* L.) and the lowest for black walnut (*Juglans nigra* L.) of all the investigated species. As oak (*Quercus* sp.) is the most commonly ammoniated wood species it is taken as a reference. Even nine species (i.e. black locust, prune, Douglas fir, movingue, mulberry, cherry, stone pine, Scots pine and bird cherry) show stronger color changes due to ammoniation than the reference. These color changes (ΔE^*_{ab}) are mainly due to changes in lightness (ΔL^*) as ammoniation leads to significant darkening. However, also changes in the yellow-blue saturation (Δb^*) are evident while red-green saturation (Δa^*) changes just slightly.

In general these wide ranges of color have the potential to substitute tropical wood species as well as thermally modified wood as they are odourless, more cost efficient and physical properties are less negatively influenced.

17.4.5 Bleaching

17.4.5.1 Basics

Bleaching is a major step in pulp and paper production. Oliveira *et al.* (2006) reported the need for high lightness pulp and the problem of lightness reversion due to wood quality, pulping method and bleaching method. For example, lightness reversion of bleached kraft pulp between different clones of *Eucalyptus grandis* × *Eucalyptus urophylla* varied within a range of 100 % according to the lowest value. This variability was also consistent between different bleaching methods and must therefore be attributed to varying wood quality. For example, Wimmer *et al.* (2002) described the relation between the anatomical wood parameter of blue gum (*Eucalyptus globulus* Labill) and the structural and optical paper properties. For example, the micro-fibril angle (angle between the fibre axes and the direction of cellulose micro-fibrils within the secondary cell wall) was positively correlated with light scattering and opacity. Also positive correlations were reported for wood density with light scattering, opacity and roughness.

Agents used within the different bleaching stages can be, for example, sodium hydroxide, calcium hydroxide, sulfuric acid, oxygen, ozone, chlorine dioxide, hydrogen peroxide, sodium chlorite, sodium hypochlorite, hypochlorite acid, silicates of soda, oxalic acid, sodium hypophosphite, potassium permanganate, hydrochloric acid, combinations of these or even others (Douek and Goring, 1976; Loader *et al.*, 1997; Uysal *et al.*, 1999a, 1999b; Oliveira *et al.*, 2006).

17.4.5.2 Color

Bleaching can also be of interest for solid wood. Cherry (*Prunus avium* L.) sometimes shows some greenish tree rings peaking out of the reddish wood. Dark discoloration around iron nails in oak, glue stains or blue stain (due to fungal attack) are further examples of unwished local color variations. Sometimes such kinds of discoloration can be removed by bleaching (Uysal *et al.*, 1999a, 1999b). Furthermore, it can also be used to create a new color assortment, especially when bright wood species are requested by customers.

Figure 17.13 shows a comparison of original wood color and color after hydrogen peroxide bleaching for 12 different wood species. Besides pine (*Pinus sylvestris* L.), all species show a significant brightening due to the process. Even sapwood becomes lighter, which also fits well with the results of Douek and Goring (1976). They found that due to hydrogen peroxide bleaching of Douglas fir (*Pseudotsuga menziesii* E. Murray) parts of lignin (conifer aldehyde groups) are oxidized at their α,β-unsaturated position. While this reaction could cause color changes over the whole wood, the second described reaction (destruction of the flavanoid dihydroquercetin) took place only in heartwood. As this substance is colorless, further reactions must also be responsible for the lightening of heartwood. Even if some heartwood characteristic absorptions within the UV spectrum disappeared due to bleaching, there would still be a distinct color difference between the sapwood and heartwood.

Figure 17.13 *Comparison of bleached (first and third rows) and untreated wood (second and fourth rows). Species from left to right: poplar, maple, spruce, beech, beech heartwood, birch, alder, ash, pine, cherry, oak and nut (See Colour Plate 12)*

The authors' own measurements showed that no significant difference between the color values of bleached black alder (*Alnus glutinosa* Gaertn.), European beech (*Fagus sylvatica* L.), birch (*Betula pendula* Roth), maple (*Acer* sp.), poplar (*Populus* sp.) and Norway spruce (*Picea abies* (L.) Karst.) exist. This means that bleaching gives a certain flexibility to production in times of scarcity of the resource. Also Uysal *et al.* (1999a, 1999b) mentioned that this is one of the major advantages of bleached wood.

Improved lightness and color stability are given in the literature as positive effects of bleaching wood (Douek and Goring, 1976; Uysal *et al.*, 1999a, 1999b). A possible explanation is that due to the bleaching process, possible photo-induced hydroperoxide reactions could become minimized (Hon *et al.*, 1982; Hon and Feist, 1992). However, some disadvantages of bleached wood do exist. Douek and Goring (1976) described hydrogen peroxide in alkaline medium as a good compromise in order to achieve a disintegration of the chromophoric system without significant destruction of the lignin aromatic moiety.

17.4.6 Enzymatic Treatment

17.4.6.1 Basics

In this section, a potential 'green' dyeing procedure is outlined that may be of interest for future applications: the dyeing of wood using oxidative enzymes. Currently, there is no scientific literature available on the targeted coloration of wood and wood-based products by the use of such enzymes. However, enzymes are known to modify the visual appearance of lignocellulosic surfaces dramatically (Ejechi *et al*., 1996). Until now, such enzymatic reactions causing color changes have only been described as unwanted side effects, such as greying, darkening or blueing.

In the natural course of wood decay, fungi degrade selected portions of woody material and thereby cause a massive change in optical appearance (Fackler *et al*., 2006). The excreta of some fungi such as *Trametes* species (*T. versicolor*, *T. hirsuta*, *T. villosa*, etc.), *Pleurotus ostreatus* or *Phanerochaete chrysosporium* contain oxidative enzymes like peroxidases and laccases that readily cause degradation of the lignin polymers in the wood matrix (Ejechi *et al*., 1996). The breakdown of lignin to lower molecular weight compounds facilitates the digestion by the fungi. Removal of the brown binder matrix leaves the cellulosic fibres intact, resulting in a white appearance of the remaining wood residue. Due to this process, the corresponding fungi are labelled 'white-rot' fungi. Alternatively, when hydrolytic enzymes are involved, the cellulosic building material in wood is destroyed, resulting in a brown appearance caused by the remaining lignin polymer. The corresponding fungi are thus so-called brown-rot fungi, such as, for example, *Gloeophyllum* species (Ejechi *et al*., 1996). Of course, the overall process in both cases leads virtually to ultimate degradation of the wood specimen and all mechanical properties are lost, which is highly undesirable.

Although hydrolytic enzymes may have some potential as well in coloration processes, the following subsection will focus on the possible applications of laccases since they are very versatile enzymes.

17.4.6.2 Laccases

Laccases can be isolated from the fermentation broth of a wide variety of white-rot fungi. They are very promising enzymes for the targeted recoloration of wood-based material since in principle they only depend on molecular oxygen as a co-substrate in order to catalyse the desired transformation (Thurston, 1994). In contrast, peroxidases need hydrogen peroxide as an additional substrate (Faber, 1992), which would ultimately result in additional load with chemicals in the dye bath. Laccases catalyse the reduction of molecular oxygen to water by a four-electron transfer reaction. As the electron source, a wide range of aromatic amines and phenols are accepted and oxidized by laccase (Figure 17.14). Typical laccase substrates are monomeric lignin precursors, such as guajacol, vanillic acid, syringic aldehyde and many others. Since laccase catalysed reactions are involved in the synthesis of lignins by trees (Thurston, 1994), it is evident that laccases not only degrade but also polymerize phenolic compounds. This reaction is mainly used when targeted coloration is the objective. By catalysing the oxidative grafting of various kinds of aromatic compounds on to the basic lignin structure, laccases may be very versatile biocatalysts in order to change the visual appearance of wood-based materials. The use of laccase polymerization reactions has been described earlier for other applications, mainly for the textile industry (Kandelbauer *et al.*, 2007a). In the patent literature various hair-dyeing compositions are described that depend on the laccase catalysed coupling of phenols and their subsequent precipitation on to keratinous fibres such as hair (Aaslyng *et al.*, 1999; Onuki *et al.*, 1999; Lang and Cotteret, 2002). A number of chemicals may be grafted by laccase (Schroeder *et al.*, 2007) and by appropriate choice of the monomers different color shades may be generated. By enlarging the π-electron system of aromatic compounds darker color shades are obtained. Different substitution patterns at the aromatic ring lead to the desired shifts in color (Hesse *et al.*, 2005). Since not all aromatic dye precursors are equally reactive with laccase, additional low molecular components that mediate the electron transfer between a grafted molecule and enzyme may be added to the formulation in catalytic amounts. Such mediators are well described in the literature and examples are 2,2′-azino-bis-(3-ethylthiazoline-6-sulfonic acid) (ABTS), hydroxyl benzotriazole (HOBT), violuric acid (VIO) or the 2,2,6,6-tetramethyl-1-piperidinyloxyl radical (TEMPO) (Fabbrini *et al.*, 2001, 2002a, 2002b).

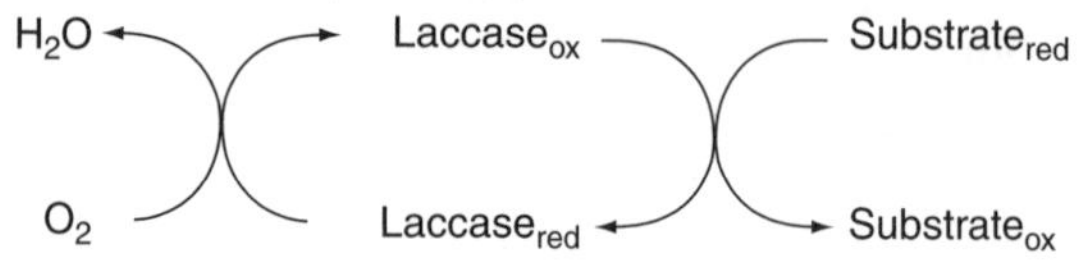

H_2O ← → $Laccase_{ox}$ — $Substrate_{red}$

O_2 — $Laccase_{red}$ ← → $Substrate_{ox}$

Figure 17.14 *General reaction scheme of laccase mediated oxidation*

If laccases are to be used in targeted coloration processes, the reactions have to be restricted to the surface or to a certain depth beneath the surface. The reaction conditions need to be carefully optimized in order to preserve the mechanical performance of the material. A variety of processes can be designed for laccase-assisted dyeing of wood-based material. While strictly superficial application of laccase/phenol can be accomplished by painting (using brushes or roller coaters), by spraying (using nozzles of various geometries) or by pouring the formulation on to the wood material surface, more in-depth treatment can be achieved by applying impregnation procedures in different degrees of severity (as described before).

Although the detailed treatment conditions will vary slightly depending on the type of phenols and enzyme used, a basic formulation for direct surface coloration may contain the phenolic coupling substance and the enzyme with or without a mediator present in cases where the phenol is not reactive towards laccase. Here, the laccase will generate in situ radicals on the lignin of the wood substrate and the grafting of the phenol will take place in a subsequent step. Alternatively, a two-step treatment is possible where in a first step laccase is applied to the surface for radical generation and a reactive phenol is subsequently contacted to the surface. When peroxidases are used, the mixture of dye precursor and enzyme may be applied in a first step and subsequently the surface may be subjected to treatment with hydrogen peroxide. However, these reactions have not yet been studied in detail and only preliminary results have been obtained by the authors (Pöckl, 2007). The most promising treatment conditions were described by Kandelbauer *et al.* (2007b) in a filed patent.

17.4.7 Radiation

17.4.7.1 Basics

Penetration of electromagnetic radiation into wood had, for example, been described by Browne and Simonson (1957). They showed that ultraviolet (UV) radiation has a limited penetration depth (less than the thickness of even thin veneers) and the transmission of visible light (VIS) depends on color depth. For example, VIS is transmitted better in sapwood than in heartwood. Furthermore, they showed that photo-induced color changes can be found well beyond the layers that were penetrated by UV radiation. For example, discoloration could be found up to a depth of about 0.25 mm. Therefore they posited that parts of the VIS might also be involved in the process of photodegradation. Kataoka *et al.* (2007) concluded that violet light is the part of the visible spectrum that extends photodegradation beyond the zone affected by UV radiation. However, Browne and Simonson (1957) also mentioned that transportation of degraded substances with water flow (e.g. due to precipitation) might have a certain secondary effect on wood altering due to radiation.

Analyses from Kataoka *et al.* (2005) showed significant effects of photodegradation up to a depth of 600 μm. Furthermore, they found a negative relation between photodegradation and wood density. Hon and Ifju (1978) showed for some coniferous species that UV radiation penetrates up to 75 μm and VIS up to 200 μm and that radiation-induced color changes in deeper layers must have secondary reasons. They suggested that a reduction of free radicals might help to increase color stability (which is, for example, nowadays done in coating systems). Accordingly, reactions of hydrogen peroxide and the contribution of generated or quenched singlet oxygen were described by Hon *et al.* (1982). Production of hydrogen peroxide also seems to be attributed to longer wavelengths within the UV spectrum according to Hon and Feist (1992).

17.4.7.2 Color

Kataoka and Kiguchi (2001) mentioned that among all possible environmental influences on wood color solar radiation is the most important one. As radiation energy is indirectly proportional to its wavelength, UV radiation is more photochemically active than VIS. Andray *et al.* (1998) discussed the influence of increasing UV-B radiation due to depletion of ozone on a product's service lifetime. As photo-induced color changes originate from chemical decomposition, such processes can lead to massive loss of mechanical properties.

However, Kataoka *et al.* (2007) showed that VIS also participates in the photodegradation of wood. They found a positive correlation between the penetration of light into wood and the wavelength between 246 and 496 nm. Also increasing the depth of photodegradation up to the violet region (403 nm) could be found. Furthermore, even the blue part of the VIS was found to be responsible for bleaching reactions, but was not involved in lignin degradation.

Changes in color due to radiation is usually unwanted and often the reason for customer complaints. Due to the influence of radiation, wood color can turn grey, yellow, red-orange or brown (Sandermann and Schlumbom, 1962). Color stability is defined as the resistance of color against radiation. Reactions of UV radiation with wood take place at several functional groups and were described by Fengel and Wegener (1984) for lignin and cellulose.

According to Turkulin and Sell (2002), photo-induced degradation coupled with a loss of tensile properties during the weathering process follows a characteristic three-phase pattern. A slight initial increase is followed by two phases of pronounced decreases. These phases could be attributed to radio-induced crosslinking followed by degradation of lignin and cellulose. Anatomical changes are reported for the breakdown of compound middle lamella, thinning of cell walls and increasing brittleness of secondary cell walls.

Yellowing (pronounced positive values for Δb^*) of several species due to UV radiation is due to degradation of lignin towards oligomeric chromophores (Tolvaj and Faix, 1995). Degradation of aromatic structures towards nonconjugated carbonyl and carboxyl groups as well as dehydrations, oxidations to carboxyl groups and formation of ester are reported. Müller *et al.* (2003) reported a 20 % loss of lignin at the surface of European spruce (*Picea excelsa* L.) due to UV degradation leading to quinine formation. Opening reactions of glucopyranose rings due to UV radiation were reported by Papp *et al.* (2004). However, Tolvaj and Faix (1995) mentioned that maybe not all oxidation products contribute to the color change.

Oltean *et al.* (2008) compared 16 of the most traded European wood species concerning their indoor color stability. For example, they showed that color change due to radiation is most expressed at the beginning and fades out over time. However, comparing the results after 120 and 600 h of radiation, ranking of color stability of the investigated species was shown to be partly different.

An overview on color changes (lightness and color) due to UV radiation of 20 central European wood species was given in Weigl *et al.* (2007). For example, it was found that ammoniation leads to a significant lightness stabilization for plum (*Prunus domestica* L.), black locust (*Robinia pseudacacia* L.), Norway spruce (*Picea abies* (L.) Karst.), maple (*Acer* sp.), silver fir (*Abies alba* Mill.), birch (*Betula pendula* Roth), hornbeam (*Carpinus betulus* L.), European beech (*Fagus sylvatica* L.), black alder (*Alnus glutinosa* Gaertn.), cherry (*Prunus avium* L.), oak (*Quercus* sp.) and pear (*Pyrus* sp.). Concerning the overall color stabilization, a positive effect could be found for almost the same species: plum, black locust, Norway spruce, maple, silver fir and oak. Also other color modification methods show improvements in color resistance. For example, the same samples of black alder (*Alnus glutinosa* Gaertn.) as actually shown for steaming (Figure 17.8) had been UV radiated for 24 h (Figure 17.15). Obviously color stabilization due to steaming increases with steaming time.

Figure 17.15 *Color impression of black alder (Alnus glutinosa Gaertn.) steamed for 0, 1, 2, 3, 4, 8 (first row) and 10, 12, 14, 16, 18 and 30 days (second row) after 24 h UV radiation (See Colour Plate 13)*

This hypothesis can also be supported by color measurements (Figure 17.16). Interestingly, fadeout of the stabilization due to steaming for the four different color change parameters (i.e. ΔL^*, Δa^*, Δb^* and ΔE^*_{ab}) appeared after different steaming times. Stabilization of the red-green component (Δa^*) only appeared for a one-day steaming process. Further steaming leads to a change of sign and a fadeout after approximately two days of steaming at almost the same absolute level as that for the nonsteamed sample. Also for ΔL^* the optimum steaming time would be one day and further steaming led to a change in sign. In this case the fadeout appeared after approximately five days of steaming at about half of the original absolute level. A change in lightness due to UV radiation is in general also more obverse than change in the red-green component.

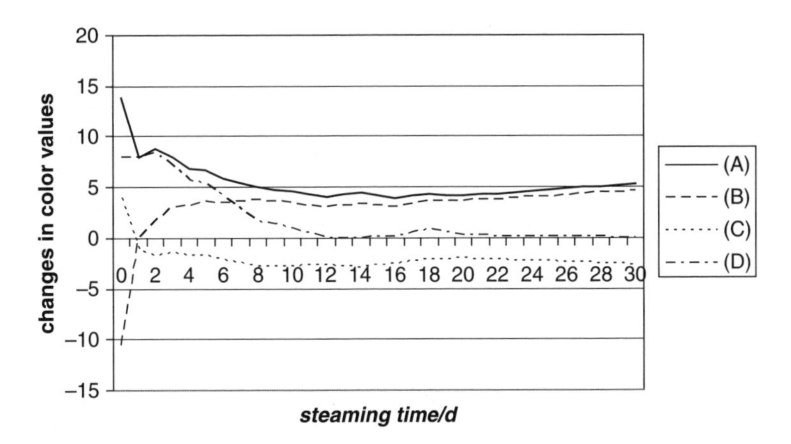

Figure 17.16 Influence of steaming time on the UV stabilization of color values for black alder (*Alnus glutinosa* Gaertn.): (A) $\Delta E_{ab}*$, (B) $\Delta L*$, (C) $\Delta a*$ and (D) $\Delta b*$

Major effects due to steaming in order to prevent UV-based color changes can be found for the total color change (ΔE^*_{ab}) and the change in the yellow-blue component (Δb^*). The strong association between these two values does not surprise as UV-radiated wood showing a significant color change almost always turns yellowish. This yellowing effect can be totally eliminated by steaming for at least 12 days or more. Therefore in this particular case the overall color change (ΔE^*_{ab}) due to UV radiation is mainly influenced by lightness changes (ΔL^*) and for short-time steaming also by the change in the yellow-blue component (Δb^*).

Comparing the results of ΔE^*_{ab} and ΔL^* with the literature values shows that steamed wood can, for example, compete with tropical wood species such as jatobá (*Hymenaea courbaril* L. var. *courbaril*), garapeira (*Apuleia molaris* Spruce ex Benth.), marupá (*Simarouba amara* Aubl.) or angelim vermelho (*Dinizia excelsa* Ducke) (Pastore *et al.*, 2004).

17.4.7.3 Technology

Radiation coupled with all these material property changes can be seen as a limiting factor for the service lifetime of products or as a color modification method. According to the first case, selection of capable protection methods is required. Such methods can be constructive solutions (e.g. shading of sensible products or filtering of radiation) as well as protection by modification or surface treatment. The latter has actually been partly discussed in Section 17.2 about coatings.

Radiation as a technological method of color modification can be used as a kind of pre-ageing of wooden surfaces in order to minimize color changes of the final product. However, it can only be used to a certain extent due to the limiting factors concerning mechanical stability described above.

Seeling *et al.* (2007) proved that UV radiation can partly be used to align color of the discolored red heartwood and nondiscolored regions within European beech wood (*Fagus sylvatica* L.). An advanced treatment is the combination of radiation with another color modification technique. Mitsui *et al.* (2001) described the benefits of the increased color change of wood that had been irradiated previously to a heat treatment. A reduction in lightness as well as an increase in red and yellow saturation were more pronounced due to the preliminary radiation of wood compared to nonradiated references. The observed effects increased with time, temperature and humidity. In general, coniferous species are more sensitive to this treatment than broad-leaved species, which is considered to be related to different lignin contents (Mitsui, 2004). However, chemical reactions in both treatment steps differ from each other and give an interesting synergetic effect.

17.5 Outlook

The color of wood and wood-based products has brought natural variability that can even become intensified due to several methods. Creating new and especially exquisite colors gives the potential to react to market changes and even change market position.

In times of ecological and/or environmental consciousness, questions for renewable resources are always loud. Wood fulfills this criterion, but many applications such as glueing systems or coatings do not. Modifying wood color by changing it using energy-saving methods or dyeing it with renewable resources might be a further step towards an increased positive life-cycle assessment of wood-based products.

Some of the methods presented here are state-of-the-art techniques carried out for centuries and others are potential future applications. Adaptation of existing color modification and dyeing techniques and the creation of new methods such as the use of herbal plant dyes are of industrial and scientific interest. These new processes will probably help wood and wood-based products to a further positive life-cycle assessment and a further positive image.

References

Aaslyng, D., Sorensen, N. H. and Rorbaek, K. (1999) Laccases with improved dyeing properties, United States Patent, US 5,948,121.

Alexiou, P. N., Marchant, J. F. and Groves, K. W. (1990) Effect of pre-steaming on moisture gradients, drying stresses and sets, and face chacking in regrowth *Eucalyptus pilularis* Sm., *Wood Science and Technology*, **24**, 201–209.

Amburgey, T. L. and Johnson, B. R. (1979) Comparing ammoniation and water-repellent preservative treatments for protecting wood window units from decay, *Forest Products Journal*, **29**(2), 23–28
Andray, A. L., Hamid, S. H., Hu, X. and Torikai A. (1998) Effects of increased solar ultraviolet radiation on materials, *Journal of Photochemistry and Photobiology B: Biology*, **46**, 96–103.
Anonymous (1924) *Handbuch der Mikroskopisch=Bakteriologischen Farbstoffe und Farblösungen*, Dr G. Grübler & Co., Leipzig.
Ayadi, N., Lejeune, F., Charrier, F., Charrier, B. and Merlin, A. (2003) Color stability of heat-treated wood during artificial weathering, *Holz als Roh- und Werkstoff*, **61**(3), 221–226.
Bariska, M. (1969) Plastifizierung des Holzes mit Ammoniak in Theorie und Praxis, *Holz-Zentralblatt*, **95**(84), 1309–1311.
Bariska, M. (1974) *Physikalische und physikalisch chemische Änderungen im Holz während und nach NH3-Behandlung*, Habilitation at the ETH, Zurich.
Berzin'š, G., Rocens, K. and Rumba, A. (1970) Über die Festigkeit und Elastizität des mit Ammoniak behandelten und verdichteten Holzes, *Holztechnologie*, **11**(1), 48–52.
Bieleman, J. (1998) *Lackadditive*, Wiley–VCH, Weinheim.
Brauner, A. B. and Loos, W. E. (1968) Color changes in black walnut as a function of temperature, time, and two moisture conditions, *Forest Products Journal*, **18**(5), 29–34.
Browne, F. L. and Simonson, H. C. (1957) The penetration of light into wood, *Forest Products Journal*, **7**(10), 308–314.
Brunner, R. (1987) *Die Schnittholztrocknung*, Buchdruckwerkstätten Hannover GmbH, Hannover.
Buchholzer, P. (2005) Durchgefärbte MDF im Markt angekommen, *Holz-Zentralblatt*, **62**, 808–809.
Burtin, P., Jay-Allemand, C., Charpentier, J. P. and Janin, G. (2000) Modifications of hybrid walnut (*Juglans nigra* 23 × *Juglans regia*) wood colour and phenolic composition under various steaming conditions, *Holzforschung*, **54**, 33–38.
Charrier, B., Haluk, J. P. and Janin, G. (1992) Prevention of brown discoloration in European oakwood occurring during kiln drying by a vacuum process: colorimetric comparative study with a traditional process, *Holz als Roh- und Werkstoff*, **50**, 433–437.
Charrier, B., Haluk, J. P. and Metche, M. (1995) Characterization of European oakwood constituents acting in the brown discolouration during kiln drying, *Holzforschung*, **49**, 168–172.
Choong, E. T., Shupe, T. F. and Chen, Y. (1999) Effect of steaming and hot-water soaking on extractive distribution and moisture diffusivity in southern pine during drying, *Wood and Fiber Science*, **31**(2), 143–150.
Davidson, R. W. and Baumgardt, W. G. (1970) Plasticizing wood with ammonia – a progress report, *Forest Products Journal*, **20**(3), 19–25.
Dawson-Andoh, B. E., Wiemann, M., Matuana, L. and Baumgras, J. (2004) Infrared and colorimetric characterization of discoloured kiln-dried hard maple lumber, *Forest Products Journal*, **54** (1), 53–57.
DIN 6174 (1979) *Farbmetrische Bestimmung von Farbabständen bei Körperfarben nach der CILAB-Formel,* DIN-Deutsches Institut für Normung e.V.
DIN 55945:2005-08 (2007) *Beschichtungsstoffe und Beschichtungen – Ergänzende Begriffe zu DIN*, EN ISO 4618 DIN-Deutsches Institut für Normung e.V.
Douek, M. and Goring, D. A. I. (1976) Microscopical studies on the peroxide bleaching of Douglas fir wood, *Wood Science and Technology*, **10**, 29–38.
Ejechi, B. O., Obuekwe, C. O. and Ogbimi, A. O. (1996) Microchemical studies of wood degradation by brown rot and white rot fungi in two tropical timbers, *International Biodeterioration and Biodegradation*, **38**(2), 119–122.
Fabbrini, M., Galli, C., Gentili, P. and Macchitella, D. (2001) An oxidation of alcohols by oxygen with the enzyme laccase and mediation by TEMPO, *Tetrahedron Letters*, **42**(43), 7551–7553.
Fabbrini, M., Galli, C. and Gentili, P. (2002a) Radical or electron-transfer mechanism of oxidation with some laccase/mediator systems, *Journal of Molecular Catalysis B: Enzymatic*, **18**(1–3), 169–171.
Fabbrini, M., Galli, C. and Gentili, P. (2002b) Comparing the catalytic efficiency of some mediators of laccase, *Journal of Molecular Catalysis B: Enzymatic*, **16**(5–6), 231–240.
Faber, K. (1992) *Biotransformations in Organic Chemistry*, Springer, Berlin.

Fackler, K., Gradinger, C., Hinterstoisser, B., Messner, K. and Schwanninger, M. (2006) Lignin degradation by white rot fungi on spruce wood shavings during short-time solid-state fermentations monitored by near infrared spectroscopy, *Enzyme and Microbial Technology*, **39**(7), 1476–1483.
Fengel, D. and Wegener, G. (1984) *Wood Chemistry, Ultrastructure, Reactions*, Walter de Gruyter & Co., Berlin.
Flocken, J., Walking, H., Buhrmester, E. and Laudage, G. (1975) *Lehrbuch für Tischler*, Vol. 2, Schroedel Schulbuch Verlag, Hannover.
Gerlach, D. (1984) *Botanische Mikrotechnik, Eine Einführung*, Georg Thieme Verlag, Stuttgart.
Graf, G., Koch, H., Schiene, R. and Schuster, E. (1971) Herstellung und Fastigkeit von Platten aus ammoniakplastifizierten Fasern (I). *Holztechnologie*, **12**(4), 235–238.
Graf, G., Koch, H., Schiene, R. and Schuster, E. (1972) Herstellung und Fastigkeit von Platten aus ammoniakplastifizierten Fasern (II). *Holztechnologie*, **13**(3), 152–155.
Grekin, M. (2007) Color and color uniformity variation of Scots pine wood in the air-dry condition, *Wood and Fiber Science*, **39**(2), 279–290.
Hansmann, C. (2002) *Einflüsse von Hochfrequenz-Vakuum-Trocknung auf Trocknungsspannungen und Holzfarbe bei Fagus sylvatica L.*, Master's Thesis, University of Natural Resources and Applied Life Sciences, Vienna.
Hansmann, C. (2007) Homogenising the colour of beech wood by drying – with special attention to the red heart, in Workshop on *Steaming of Beech Wood*, University of West Hungary, Faculty of Wood Sciences, Sopron, 3 March 2007.
Harris, R. A., Schroeder, J. G. and Addis, S. A. (1989) Steaming of red oak prior to kiln drying: effects on moisture movement, *Forest Products Journal*, **39**(11/12), 70–72.
Hesse, M., Meier, H. and Zeeh, B. (2005) *Spektroskopische Methoden in der organischen Chemie*, 7th edition, Thieme, Germany.
Hildebrand, R. (1979) *Schnittholztrocknung*, NT-Raidwangen.
Hill, C. (2006) *Wood Modification – Chemical, Thermal and Other Processes*, John Wiley & Sons, Ltd, Chichester, West Sussex.
Hoadley, R. B. (1990) *Identifying Wood – Accurate Results with Simple Tools*, The Taunton, Inc.
Hon, D. N. S. and Feist, W. C. (1992) Hydroperoxidation in photoirradiated wood surfaces, *Wood and Fiber Science*, **24**(4), 448–455.
Hon, D. N. S. and Ifju, G. (1978) Measuring penetration of light into wood by detection of photo-induced free radicals, *Wood Science*, **11**(2), 118–127.
Hon, D. N. S. and Minemura, N. (2001) Colour and discoloration, in D. N. S. Hon and N. Shiraishi (eds), *Wood and Cellulosic Chemistry*, Marcel Dekker, Inc., New York.
Hon, D. N. S., Chang, S. T. and Feist, W. C. (1982) Participation of singlet oxygen in the photo degradation of wood surfaces, *Wood Science and Technology*, **16**, 193–201.
Kalnin'š, A. J., Darzin'š, T. A., Jukna, A. D. and Berzin'š, G. V. (1967) Physikalisch-mechanische Eigenschaften mit Ammoniak chemisch plastifizierten Holzes, *Holztechnologie*, **8**(1), 23–28.
Kalnin'š, A., Berzin'š, G., Skrupskins, W. and Rumba, A. (1969) Chemisch plastifiziertes Holz anstelle von Buntmetallen und importiertem Holz, *Holztechnologie*, **10**(1), 17–23.
Kandelbauer, A., Cavaco-Paulo, A. and Gübitz, G. M. (2007a) Biotechnological treatment of textile dye effluent, in R. M. Christie (ed.), *Environmental Aspects of Textile Dyeing, CRC Press*, Woodhead Publishing, Cambridge, pp. 212–232.
Kandelbauer, A., Paar, A., Weigl, M., Grabner, M. and Pöckl, J. (2007b) *Verfahren zum Verfärben von Holz*, filed Austrian Patent.
Kataoka, Y. and Kiguchi, M. (2001) Depth profiling of photo-induced degradation in wood by FT-IR microspectroscopy, *Journal of Wood Science*, **47**, 325–327.
Kataoka, Y., Kiguchi, M., Fujiwara, T. and Evans, P. D. (2005) The effects of within-species and between-species variation in wood density on the photo degradation depth of sugi (*Cryptomeria japonica*) and hinoki (*Chamaecyparis obtuse*), *Journal of Wood Science*, **51**, 531–536.
Kataoka, Y., Kiguchi, M., Williams, R. S. and Evans, P. D. (2007) Violet light causes photo degradation of wood beyond the zone affected by ultraviolet radiation, *Holzforschung*, **61**, 23–27.
Kollmann, F. (1955) *Technologie des Holzes und der Holzwerkstoffe*, Zweite Auflage, Zweiter Band, Springer-Verlag, Berlin.

Kollmann, F. F. P. and Côté, W. A. (1968) *Principles of Wood Science and Technology, I, Solid Wood*, Springer-Verlag, Berlin.
Kürschner, K. and Melcerová, A. (1965a) Über die chemischen Veränderungen des Buchenholzes bei thermischer Behandlung – Teil I. Chemische Veränderungen von Sägespänen bei 1-28 tägiger Erhitzung auf 80–160 °C, *Holzforschung*, **19**(6), 161–171.
Kürschner, K. and Melcerová, A. (1965b) Über die chemischen Veränderungen des Buchenholzes bei thermischer Behandlung – Teil II. Chemische Veränderungen von Buchenholz-Kanteln bei 1-2 tägiger Erhitzung auf 80–130 °C unter besonderer Berücksichtigung der UV-Absorptionsspektren, *Holzforschung*, **19**(6), 171–178.
Lang, G. and Cotteret, J. (2002) Dyeing composition containing a laccase and keratinous fiber dyeing methods using same, US Patent US 6,471,730 B1.
Langendorf, G. and Eichler, H. (1982) *Holzvergütung*, VEB Fachbuchverlag, Leipzig.
Loader, N. J., Robertson, I., Barker, A. C., Switsur, V. R. and Waterhouse, J. S. (1997) An improved technique for the batch processing of small wholewood samples to α-cellulose, *Chemical Geology*, **136**, 313–317.
Luostarinen, K. and Luostarinen, J. (2001) Discolouration and deformations of birch parquet boards during conventional drying, *Wood Science and Technology*, **35**, 517–528.
Luostarinen, K., Möttönen, V., Asikainen, A. and Luostarinen, J. (2002) Birch (*Betula pendula*) wood discolouration during drying. Effect of environmental factors and wood location in the trunk, *Holzforschung*, **56**, 348–354.
MacLeod, I. T., Scully, A. D., Ghiggino, K. P., Ritchie, P. J. A., Paravagna, O. M. and Leary, B. (1995) Photo degradation at the wood–clearcoat interface, *Wood Science and Technology*, **29**, 183–189.
Militz, H. (2002) Heat treatment technologies in Europe – scientific background and technological state-of-art, in Proceedings of the Conference on *Enhancing the Durability of Lumber and Engineered Wood Products*, Kissimmee, 11–13 February 2002.
Mitsui, K. (2004) Changes in the properties of light-irradiated wood with heat treatment, Part 2. Effect of light-irradiation time and wavelength, *Holz als Roh- und Werkstoff*, **62**, 23–30.
Mitsui, K., Takada, H., Sugiyama, M. and Hasegawa, R. (2001) Changes in the properties of light-irradiated wood with heat treatment, Part 1. Effect of treatment conditions on the change in color, *Holzforschung*, **55**(6), 601–605.
Müller, U., Rätzsch, M., Schwanninger, M., Steiner, M. and Zöbl, H. (2003) Yellowing and IR-changes of spruce wood as result of UV-irradiation, *Journal of Photochemistry and Photobiology B: Biology*, **69**, 97–105.
Nguila Inari, G., Petrissans, M. and Gerardin, P. (2007) Chemical reactivity of heat-treated wood, *Wood Science and Technology*, **41**, 157–168.
Oliveira, R. L., Colodette, J. L., Eiras, K. M. and Ventorim, G. (2006) The effect of wood supply and bleaching process on pulp brightness stability, *R. Árvore, Vicosa-MG*, **30**, 439–450.
Oltean, L., Teischinger, A. and Hansmann, C. (2008) Wood surface discolouration due to simulated indoor sunlight exposure, *Holz als Roh- und Werkstoff*, **66**(1), 51–56.
Oniśko, W. and Matejak, M. (1971) Einfluß 25%iger Ammoniaklösung auf die physikalischen und mechanischen Eigenschaften des Holzes, *Holztechnologie*, **12**(1), 45–54.
Onuki, T., Noguchi, M. and Mitamura, J. (1999) Hair dye compositions, European Patent Application EP 1 142 561 A1.
Papp, G., Preklet, E., Kosiková, B., Barta, E., Tolvaj, L., Bohus, J., Szatmari, S. and Berkesi, O. (2004) Effect of UV laser radiation with different wavelengths on the spectrum of lignin extracted from hard wood materials, *Journal of Photochemistry and Photobiology A: Chemistry*, **163**, 187–192.
Parham, R. A., Davidson, R. W. and de Zeeuw, C. H. (1971) Radial–tangential shrinkage of ammonia-treated loblolly pine wood, *Wood Science*, **4**(3), 129–136.
Pastore, T. C. M., Santos, K. O. and Rubim, J. C. (2004) A spectrocolorimetric study on the effect of ultraviolet irradiation of four tropical hardwoods, *Bioresource Technology*, **93**, 37–42.
Pöckl, J. (2007) *Neue Methoden der Laubholzvergütung*, PhD Thesis, University of Natural Resources and Applied Life Sciences, Vienna.
Roffael, E. and Schaller, K. (1971) Einfluß thermischer Behandlung auf Cellulose, *Holz als Roh- und Werkstoff*, **29**(7), 275–278.

Rothkamm, M., Hansemann, W. and Böttcher, P. (2003) *Lack Handbuch Holz*, DRW-Verlag, Leinfelden-Echterdingen.

Ruzin, S. E. (1999) *Plant Microtechnique and Microscopy*, Oxford University Press, Inc., New York.

Sandermann, W. and Augustin, H. (1963a) Chemische Untersuchungen über die thermische Zersetzung von Holz. Erste Mitteilung: Stand der Forschung, *Holz als Roh- und Werkstoff*, **21**, 256–265.

Sandermann, W. and Augustin, H. (1963b) Chemische Untersuchungen über die thermische Zersetzung von Holz. Zweite Mitteilung: Untersuchungen mit Hilfe der Differential-Thermoanalyse, *Holz als Roh- und Werkstoff*, **21**, 305–315.

Sandermann W., and Schlumbom, F. (1962) Über die Wirkung gefilterten ultravioletten Lichtes auf Holz – Zweite Mitteilung: Änderung von Farbwert und Farbempfindung an Holzoberflächen, *Holz als Roh- und Werkstoff*, **20**(8), 286–291.

Schleusener, J. (2003) *Oberflächenlexikon – Holz, Holzwerkstoffe, Innenausbau, Möbel*, 2nd edition, DRW-Verlag, Leinfelden-Echterdingen.

Schmidt, K. (1986) Untersuchungen über die Ursachen der Verfärbungen von Eichenholz bei der technischen Trocknung, *Holzforschung und Holzverwertung*, **38**, 25–36.

Schroeder, M., Pereira, L., Rodríguez Couto, S., Erlacher, A., Schoening, K.-U., Cavaco-Paulo, A. and Guebitz, G. M. (2007) Enzymatic synthesis of tinuvin, *Enzyme and Microbial Technology*, **40**(7), 1748–1752.

Schuerch, C. (1964) Wood plasticization – principles and potential, *Forest Products Journal*, **14**(9), 377–381.

Seeling, U., Ohnesorge, D., Helzle, C., Burgbacher, C., Németh, R., Tolvaj, L., Teischinger, A., Hansmann, C., Mitteramskogler, H., Huber, H., Oliver, J. V., Abián, M. A., Pons, L. and Custodio, R. (2007) *Red Heartwood Handbook*, University of Freiburg, Institute for Forest Utilisation and Work Science, Freiburg.

Simpson, W. T. (1975) Effect of steaming on the drying rate of several species of wood, *Wood Science*, **7**(3), 247–255.

Sundqvist, B. and Morén, T. (2002) The influence of wood polymers and extractives on wood color induced by hydrothermal treatment, *Holz als Roh- und Werkstoff*, **60**, 375–376.

Ters, T., Grabner, M., Willför, S. and Hinterstoisser, B. (2006) *Pinus sylvestris* and *Pinus nigra* – differences in extractives content and composition, in Proceedings of the 9th European Workshop on *Lignocellulosics and Pulp*, Vienna, 27–30 August 2006.

Thurston, C. F. (1994) The structure and function of fungal laccases, *Microbiology*, **140**, 19–26.

Tinkler, C. K. (1921) 'Fumed' oak and natural brown oak, *Biochemical Journal*, **15**, 477–486.

Tolvaj, L. and Faix, O. (1995) Artificial ageing of wood monitored by DRIFT spectroscopy and CIE L*a*b* color measurements, 1. Effect of UV light, *Holzforschung*, **49**(5), 397–404.

Tolvaj, L., Horvath-Szovati, E. and Safar, C. (2000) Color modification of black locust by steaming, *Drevasky Vyskum*, **45**(2), 25–32.

Torgovnikov, G. I. (1993) *Dielectric Properties of Wood and Wood-Based Materials*, Springer-Verlag, Berlin.

Trübswetter, T. (2006) Holztrocknung, Verfahren zur Trocknung von Schnittholz – Planung von Trocknungsanlagen, Carl Hasner Verlag, München.

Turkulin, H. and Sell, J. (2002) Investigations into the photo degradation of wood using microtensile testing, Part 4: Tensile properties and fractography of weathered wood, *Holz als Roh- und Werkstoff*, **60**, 96–105.

Uysal, B., Atar, M. and Özcifci, A. (1999a) The effects of chemicals for using the bleaching of wood surfaces on the layer hardness of varnish, *Tr. Journal of Agriculture and Forestry*, **23**, 443–450.

Uysal, B., Atar, M. and Özcifci, A. (1999b) The effects of wood bleaching chemicals on the bending strength of wood, *Tr. Journal of Agriculture and Forestry*, **23**, 615–619.

Vanek, M. (1986) Trocknungsspannungen: Spannungsermittlung bei einer Buchentrocknung mittels Dehnungsmeßstreifen, *Holzforschung und Holzverwendung*, **38**, 36–42.

Vanek, M. (2001) Holztrocknung am Institut für Holzforschung, in A. Teischinger and R. Stingl (eds),. *Lignovisionen, Eine Holzzeitgeschichte – Konturen der Forschung und Lehre in Österreich*, Vol. **1**, Schriftenreihe des Institutes für Holzforschung an der Universität für Bodenkultur Wien, Vienna.

Van Groenou, H. B., Rischen, H. W. L. and van den Berge, J. (1952) *Wood Preservation during the Last 50 Years*, A. W. Sijthoff's Uitgeversmaatschappij N.V., Leiden.
Voigt, J. and Lentner, A. (2007) Direct multi-color printing on wood and wood composite panels, in Proceedings of the Interior Surface Conference, Amsterdam, 28–29 March 2007.
Wagenführ, R. and Scheiber, C. (1985) *Holzatlas*, VEB Fachbuchverlag, Leipzig.
Ward, J. C. and Simpson, W. T. (1991) *Dry Kiln Operator's Manual*, US Department of Agriculture, Forest Service, Forest Products Laboratory, Madison, Montana.
Weigl, M., Pöckl, J., Müller, U., Pretzl, H. and Grabner, M. (2007) UV-resistance of ammonia treated wood, in C. A. S. Hill, D. Jones, H. Militz and G. A. Ormondroyd (eds), Proceedings of the 3rd European Conference on Wood Modification, Cardiff, 15–16 October 2007.
Wiberg, P. (1996) Colour changes of Scots pine and Norway spruce – a comparison between different treatments, *Holz als Roh- und Werkstoff*, **54**, 349–354.
Wicks, Z. W., Jones, F. N. and Pappas, S. P. (1999) *Organic Coatings – Science and Technology*, 2nd edition, Wiley Interscience, New York.
Willför, S., Hemming, J., Reunanen, M. and Holmbom, B. (2003) Phenolic and lipophilic extractives in Scots pine knots and stemwood, *Holzforschung*, **57**, 359–372.
Wimmer, R., Dowens, G. M., Evans, R., Rasmussen, G. and French, J. (2002) Direct effects of wood characteristics on pulp and handsheet properties of *Eucalyptus globulus*, *Holzforschung*, **56**, 244–252.

18

Natural Colorants in Textile Dyeing

Rita A. M. Mussak and Thomas Bechtold

18.1 Introduction

Natural dyes and pigments have been used rigorously for millennia, up to the middle of the 19th century. The invention of the first synthetic dyes changed the situation and were substituted for natural colorants almost completely. However, in some niche segments or specialty segments of the market natural dyes survived.

Recently the awareness of environment as well as increasing disputes about the risks of synthetic dyes resulted in growing interest in natural resources, environmentally friendly products and new strategies. The application of natural sources – particularly plants – as the origin for colorants seems to be a promising approach.

Due to the fact that historical aspects as well as regional specialties are already discussed in specific chapters of this book, this chapter focuses on the present situation and new developments in the field of natural dyeing. Outstanding highlights are discussed, representative examples are given and a general discussion of the main important factors is included. Direct comparison of advantages and disadvantages as well as reasons for and against natural coloration will lead to individual decisions and may result in unique strategies to implement natural colorants into technical processes and products.

Since plant dyeing does not exist on a commercial scale in the modern textile industry currently, most of the studies are based on laboratory results or model processes. Recently, the Austrian government started a new funding program to support environmental motivated research. Projects in different areas like industrial innovation, energy management and home building construction are summarized in their aim of

Handbook of Natural Colorants Edited by Thomas Bechtold and Rita Mussak

'Technologies for Sustainable Development' (detailed information is available on: www.fabrikderzukunft.at). A series of projects were performed with a focus on textiles:

- available resources (Hartl and Vogl, 2001a, 2001b);
- the potential of renewable resources for textile dyeing (Ganglberger and Geissler, 2001);
- analysis of the fundamental structure and consideration of the essential requirements (Ganglberger and Geissler, 2003);
- realization and implementation in a state-of-the-art dyehouse (Rappl, 2005);
- product management and marketing (Grill *et al.*, 2004);
- minimizing risks for a virtual pilot company (Pladerer and Wenisch, 2007).

The research group at the Research Institute for Textile Chemistry and Textile Physics (University of Innsbruck) is one of the active partners in these projects focusing on plant colorants. Most of the results presented in this chapter will be extracted from research results of this working group.

18.2 Reasons for Natural Coloration

The superior success of synthetic colorants since their introduction in the 19th century has been described by Brian Glover as follows: 'In summary, they allowed people to make more money' (Glover, 1998). Actually, synthetic colorants showed enormous advantages: brilliant colors, more variations in color shade and color depth, independence from agriculture and/or farming, better fastness properties, increased repeatability, better standardization features, easier handling and enlarged application fields (substrates, machines) (Rath, 1972; Trotman, 1984; Zollinger, 2003; Mussak, 2008). In the meantime, however, the circumstances changed significantly.

Several reasons for the new, increasing interest in natural dyes can be identified. Some of them are summarized in Table 18.1. To obtain a more detailed and structured impression four categories, named innovation, economics, personal and ethic reasons, have been established. Therefore the categories defined the reasons and were classified in a subjective manner. If more than one category seemed to be suitable all those categories have been marked.

The reasons for innovation and economics are the most important factors for company decisions. Market research, governmental regulations and cost analysis are of huge importance in decision-making processes in industry. Established knowledge of consumer attitudes towards a product and good estimations of future trends may lead to economical success, the surviving of an enterprise or nonprofitable investigation followed by increased economical pressure (Geissler *et al.*, 2006).

The successful introduction of a new product is not only dependent on the product's properties but also to a large extent on the marketing strategy used. The increasing awareness of health and environmental aspects seems to work as a catalyst when innovative products are presented. Highlighting the luxury or additional benefits of a product also serves as a driving force to attract consumer attention (Grill *et al.*, 2004).

Table 18.1 Reasons for natural dyes: a set of reasons for the increasing interest in natural dyes has been listed and categorized in four different classes

	Categories			
Reasons	Innovation	Economics	Personal reasons	Ethics
New products	X			
Better range of price	X	X		
Luxury product	X	X		
Health aspects	X		X	
Medicinal benefit	X		X	
New sources of income		X		
Specific costumer group		X		
Increasing oil price		X		
Governmental regulations		X		
Waste and waste water treatment		X		
Limited resources		X		X
Fashion		X	X	
Change in mind of society		X		X
Change in mind of costumers			X	X
Protect environment			X	X
Resources for future generations		X	X	X

As far as economic reasons are concerned two types should be distinguished: on the one hand, better levels of prices and products with a higher value can increase the company's value; on the other hand, new legal limits and governmental rules or changed consumer demands can increase the investments required. One well-known example in the textile industry is the increasing oil price. Oil is needed twice: first of all, synthetic dyestuff is based on raw oil fractions; secondly, the dyeing and finishing processes need high amounts of energy. Both aspects are strongly dependent on the economical background of a product: in terms of resources and in terms of production costs (Ganglberger and Geissler, 2003; Rappl, 2005).

One other example should clarify the situation. A new legal limit for the release of copper in the waste water can cause additional costs due to adaptations in waste water treatment, research of alternative products, substitution of chemicals, modification of the process or in rare cases financial punishment by involved government authorities.

Analysis of personal decision-making processes and reasons for customers to purchase a product are very difficult to assess. These are mostly very subjective and might change regularly. One tool of investigation is marketing studies and trend analysis (Grill *et al.*, 2004; Geissler *et al.*, 2006). As far as natural dyes are concerned, the most challenging aspect as well as the reason that offers the highest potential might be the factor of fashion.

18.3 Analysis of a Dyeing Process

Figure 18.1 illustrates a general dyeing process. Starting form an undyed/raw substrate the main components of the coloration process are dyestuff, water and energy. Additionally different chemicals like mordants, auxiliaries and/or detergents are necessary depending

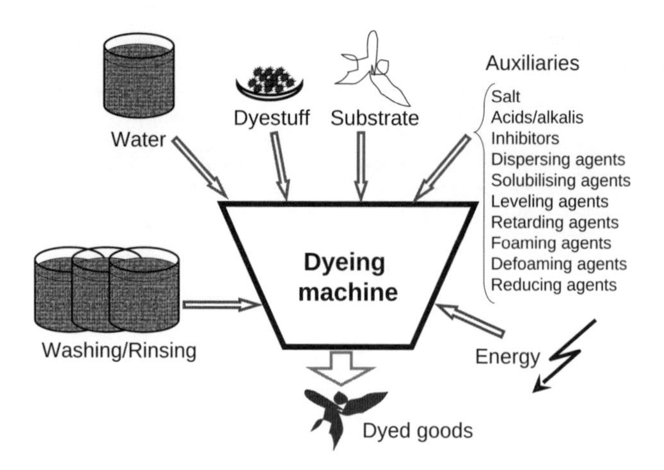

Figure 18.1 *Dyeing process: requirements for a general dyeing process. (See Colour Plate 14)*

on the properties of the dyestuff–substrate interaction. All components are applied in a dyeing machine, resulting in a colored textile product.

Improving a system requires a detailed analysis of the single components. Therefore a discussion of the input factors is executed below.

18.3.1 Water

The textile industry is one of the biggest consumers of high-quality water (Bechtold *et al.*, 2004a). Water is used in nearly all stages of the coloration process either in a direct or in an indirect way. While the former means the use of water for the preparation of dye baths, washing and rinsing liquors or various pre/after-treatments, indirect water consumption embraces thermal needs like heating, cooling, steaming and occasionally drying steps (Mussak and Bechtold, 2008). Generally, the dyeing of 1 kg of textile requires 100–200 L of fresh water (Bechtold *et al.*, 2004a). An increased number of washing/rinsing steps or after/pre-treatment processes leads to a higher water consumption. Besides the dyestuff, the dyeing machine and the corresponding dyeing technology influence the total water input. While continuous coloration techniques deal with rather small volumes (concentrates), for batchwise techniques the demand of water increases up to 40 L per kilogram of textile and treatment step (e.g. in a paddle dyeing machine).

Some examples based on literature data (Peter and Rouette, 1989; Wulfhorst, 1998) are listed in Table 18.2.

18.3.2 Energy

Energy and indirect water consumption usually go together with direct water consumption of a certain treatment step. Heating and cooling of high volumes in batchwise dyeing technology especially need energy (Mussak and Bechtold, 2008). Therefore minimization of the volume of the liquor baths, optimized processes including washing, rinsing as well as

Table 18.2 Comparison of dyeing technologies: for different standard dyeing technologies the liquor ratio used varies (liquor ratio = mass of textile in kg : volume of treatment bath in liters)

Dyeing technology	Dyeing machine / unit	Liquor ratio (average)
Continous	Foulard	0.7–0.8
Batchwise	Jet (airflow)/jig	1:3–1:5
	Jet (overflow)/winch	1:10–1:20
	Beaker/cooker/paddle dyeing machine	1:20–1:40

pre- and after-treatment and reduction of the number of drying steps to a minimum is required, although any negative influence on the final quality of the dyeings has to be avoided.

18.3.3 Dyestuff and Chemicals (Mordants and Auxiliaries)

Starting from the first synthetic dyes, a huge number of new colorants, for different applications and substrates, offering a wide range of coloration properties, have been invented (Zollinger, 1987; Mueller, 2003). In the following decades some of these turned out to have potential risks for human health or the environment. Nowadays strict regulations limit the application of colorants and prohibit use of harmful dyes (Mueller, 2003).

Dyestuff is mainly produced from refinery precursors. There is no general synthesis due to the high number of different colorant types and fields of application. Therefore generalization of specific production features for selected synthetic colorants is questionable. Individual elected examples will not reflect the reality and generalization is hardly possible. Furthermore, there is a wide range of variations in coloration technology. Low concentrations are used in exhaust processes with a rather high liquor ratio, while higher concentrations of chemicals and dyes are applied in continuous processes, e.g. pad batch dyeing or pad steam dyeing. Essential parameters that indicate the coloration properties of colorants are mentioned below:

- color strength and color strength per mass of dye;
- available shade and color depth;
- dye-ability of substrates;
- affinity of colorant to substrate;
- degree of exhaustion and fixation;
- commercial product type;
- solubility in dye bath (water);
- ionic charge (ionic properties);
- machinery requirements;
- physical, chemical and toxicological properties.

Colorants are divided into dyes and pigments according to their solubility in water. While dyes are dissolved in water and aqueous solutions, pigments are applied as insoluble materials, e.g. from dispersions. While binding of dyes occurs via ionic attraction, hydrogen bonding and van der Waals forces, binding of pigments on textile substrates is usually achieved by binder polymers, which fix the pigment on the surface of the fibers. Alternatively, pigments can be added directly to the synthetic fibers (e.g. polyester, PP, etc.), spinning dope or molten polymer and thereby can be incorporated into the fiber structure.

Depending on the affinity of the colorant molecule to the fiber, the required amount of dyestuff for a desired color depth will vary. The fixation degree of dyes can be calculated by quantification of the dyestuff concentration in the dye bath before and after dyeing (Zollinger, 2003).

Unfixed dyestuff is of growing interest for two reasons: firstly, this percentage of dyestuff is not available for the coloration process while increasing costs and, secondly, the dye is present in the waste water and has to be removed in order to avoid colored effluents and fulfill legal limits for textile effluents (Bechtold *et al.*, 2004a). Both aspects represent key factors in cost analysis.

In natural dyeing mordants increase the fixation of the dyestuff in various ways. In many cases mordants are metal salts that can form a metal complex with the natural colorant, which exhibits increased affinity to the substrate. Depending on the metal character the complex formation does not only strengthen dyestuff fixation on the substrate but also changes the color of the dyeing. In some cases the resulting change in shade can be seen as an opportunity to steer color in a wider range.

Depending on the substrate–colorant combination a considerable amount of auxiliaries is needed. Table 18.3 highlights some examples.

Table 18.3 *Chemical requirements for selected dyestuff classes: due to very specific physical and chemical properties of the colorant molecules different dyestuff classes require certain chemicals and auxiliaries. Column 3 shows the main function of the chemicals added*

Dyestuff class	*Required chemicals/ additives*	*Consequence*
Direct dyes	Salt	Increase affinity to substrate
Metal–complex dyes	Metal salts	Complex formation
Reactive dyes	Alkali, salt	Chemical reaction in alkaline solution, salt for increased affinity
Vat/sulfur dyes	Alkali, reducing agent, dispersing agent, peroxide	Reduction of dyestuff, reduced form is soluble in solvent, reoxidation of leuco form, dispersion of vat pigments
Mordant dyes	Mordant	Fixation
Natural dyes	Mordant optional	Fixation, shift in color shade, increasing fastness properties

In many cases the addition of auxiliaries is imperative; for others it leads to improvement in handling or exhaustion of dyestuff, e.g. increased affinity, increased wetting ability, adjustment of dyeing conditions. Chemicals and auxiliaries also can be part of the final colorant formulation.

While the amount of colorant is determined by the color to be achieved on the substrate, the amount of auxiliaries and mordants might be reconsidered. Generally, optimization towards the required minimum will lead to dwindled costs and advantages in waste water treatment (Bechtold *et al.*, 2003).

18.3.4 Machinery

A careful selection of the machinery according to a specific dyeing technology, suitable handling of the substrate, the available equipment and knowledge of the advantages and disadvantages of the dyeing concept is essential for successful implementation. Every piece of dyeing machinery, e.g. pad batch, jig, winch or jet, brings specific difficulties that have to be overcome. For example, lighter side/middle variation in color depth in continuous dyeing processes, stripes in the case of jet dyeing and unequal dyeing results in terms of cone dyeing due to filtration of insoluble material are typical challenges for a dyer (Wulfhorst, 1998).

18.4 Basics of Natural Dyeings

Using natural dyes for textile dyeing adds further aspects of technology and product quality to the existing processing, which have to be considered with care. Challenges that have to be overcome focus on the stability of the dyestuff through the isolation step (extraction), the application process (dyeing) and the final/customer use. Technical questions have already been targeted and application strategies are available. Actually in many cases the problems to be solved are similar to those already existing for synthetic colorants. Currently, the availability of raw material, the handling, product properties and standardization are under investigation (Hartl and Vogl, 2001a, 2001b; Ganglberger and Geissler, 2001, 2003; Bechtold *et al.*, 2003, 2006, 2007a; Rappl, 2005; Mussak and Bechtold, 2006, 2008; Pladerer and Wenisch, 2007; Mussak, 2008).

18.4.1 Requirements of the Dyestuff

18.4.1.1 Gamut and Color Shade

Although in nature thousands of different colors and shades exist, it is difficult to obtain the pure dyestuff. As a rule a basic set of dyes – namely yellow, red, blue, green and black – is absolutely necessary for a complete color gamut. The resulting range of colors can be enlarged by dyestuff mixtures and repeated coloration steps (Bechtold *et al.*, 2003; Ganglberger and Geissler, 2003; Mussak, 2008). Table 18.4 represents a list of plant sources that can be used for dyestuff isolation/extraction.

18.4.2 Dye-ability of Substrates

There are many types of fibers and substrates used in the textile industry. Independent of the dyestuff sources (natural or synthetic), the combination of colorant and substrate is essential. Nevertheless, from a dyer's point of view *one* colorant for *many* substrates would be desirable. The dye-ability of a substrate varies according to the material properties, e.g. surface properties, ionic character, roughness, structure, penetration and swelling properties (Hihara *et al.*, 2002).

Table 18.4 Plant material and its color: plant material and possible resulting colors are shown. Additionally the chemical dyestuff group and references for additional information are given (Sequin, 1981; Taylor, 1985, Sewekow, 1988; Knaggs, 1992; Schweppe, 1993; Hill, 1997; Ganglberger and Geissler, 2001, 2003; Bechtold et al., 2003, 2004b, 2006, 2007a; Kannan, 2005; Rappl, 2005; Rappl et al., 2005; Guimot et al., 2006).

Dominant color	Chemical dyestuff group	Plant	Notice	Reference
Red, violet,	Anthocyanin	Grapes, black alder, black currant, blackberries	Sensitive to pH and heat, applicable with tannin mordant	Bechtold et al., 2007b
Rose, red	Anthrachinon	Madder	Dyestuff in the roots, difficult for isolation	Grosjean et al., 1987
Yellow, brown	Flavonoid	Canadian golden rod, reseda, dyer's chamomile	Light sensitive	Bechtold et al., 2007c; Angelini et al., 2003
Blue	Indigo	Indigofera (Indian indigo), isatis (woad), dyer's knotweed	Different dyestuff group = vat dye, different dyeing procedure	Bechtold et al., 2002
		Privet berries	Light blue shades, berries are slightly poisonous	Ganglberger and Geissler, 2003; Taylor, 1985
Yellow, orange, red	Carotenoids	Carrots, pumpkin, saffron, annatto seeds	Instable against light and heat, insoluble in water, soluble in oil (lipophil)	Chandrika et al., 2003
Green	Chlorophylls	Spinach, grass, parsley, broccoli, nettle	Soluble in water (lipophil), sensitive to light and heat	Schweppe, 1993

Table 18.5 Substrate depending CIE-Lab-coordinates: aqueous extracts of green nut shell and onion skin were used for dyeing different substrates without mordants. Hunter CIE-Lab coordinates are given. In all dyeings the liquor ratio for extraction and dyeing was 1:20

Plant	Substrate	L^*	a^*	b^*
Green nut	Wool	41.28	+9.88	+19.27
	Linen	37.81	+6.20	+18.57
	Cotton	65.35	+0.79	+15.35
	Polyamide	53.20	+2.26	+13.71
Onion	Wool	49.53	+19.29	+33.06
	Linen	68.02	+8.14	+22.73
	Polyamide	48.67	+13.91	+33.64

As an example, the behavior of different plant extracts has been investigated by using the same dyestuff solution for dyeing wool, cotton, flax and polyamide. Results are given in Table 18.5.

18.4.2.1 Fastness Criteria

The quality of dyeing is described in terms of fastness properties. To assess the fastness properties of dyeing a set of specific test methods (DIN, 1985) has been established.

According to this norm test methods are preformed and fastness behavior is described with marks. The behavior of dyeings to various treatments such as chlorine bleach, chemical cleaning and steam ironing is also tested in specialized procedures.

Due to different treatments during the production and in final use a high number of properties that are evaluated exist. The following scheme in Figure 18.2 shows some of the most important categories of textile test methods.

identification / analysis:	**dyeing properties:**	**physical properties:**
alkali content chelating agents extractable contents fiber analysis finishes fluidity ash content formaldehyde release hydrogen peroxide mercerization transmittance or blocking of UV Radiation whiteness **resistance tests**: stain resistance insect resistance fire resistance abrasion resistance weather resistance	chelating agents dispersibility dusting migration / transfer foaming propensity color strength of dye solution speckiness thermal fixation **efficiency tests**: wetting agents rewetting agents mercerization agents **water repellency tests**: spray test rain test penetration test absorption test hydrostatic pressure test	abrasion resistance absorbency ageing appearance chlorine retained tensile loss cleaning dimensional changes dry cleaning electrical resistivity oil repellency wrinkle recovery retention of crease **biological properties**: antifungal activity antibacterial activity antibacterial finishes antimicrobial activity amylase enzymes insect resistance

Figure 18.2 *Overview textile test methods: depending on different parameters that are essential in the production or during the lifetime of a product different properties are studied according to international norm test methods*

Fastness tests and the resulting marks are important tools for assessing the quality and the stability of dyeing. In many cases the fastness properties are strongly related to the substrate type and mordant used for dyestuff fixation. Besides the dyestuff itself there are many factors that influence the fastness, such as the substrate, the surrounding conditions (water, solvents, chemicals, temperature, humidity, light intensity and light source, etc.) pre- and after-treatments, as well as dyestuff distribution in the fibers/textile and also the amount of dyestuff fixed on the goods (Rys and Zollinger, 1982; Grosjean *et al.*, 1987; Zollinger, 1987; Freeman *et al.*, 1996; Nguyen *et al.*, 2000; Hihara *et al.*, 2001; Batchelor *et al.*, 2003; Novotna *et al.*, 2003; Podsiadly *et al.*, 2003; Yoshizumi and Crewes, 2003).

In natural dyeing color fastness of the natural dyes requires considerable attention and careful selection of materials and processes. The color fastness quantifies the color change on a dyed material under specific conditions and also the transfer of dyestuff to uncolored adjacent material (bleeding). In Table 18.6 the main test methods used to study color fastness have been summarized and international norm methods are listed.

Table 18.6 *Examples of color fastness properties: a list of color fastness test methods and the specific national norm*

Color fastness to	DIN	ISO	AATCC
Light	54 003	105–B01	16–1998
	54 004	105–B02	
Water	54 005	105–E01	107–1997
	54 006		
Sea water/chlorinated water	54 007	105–E02	106–1997
	54 019	105–E03	162–1997
Washing/laundry	54 010–54 014	105–C01/C06	61–1996
Perspiration	54 020	105–E04	15–1997
Acids and alkalis	54 028	105–E05	6–1994
	54 030	105–E06	

Absolute central properties used to characterize dyed textile materials are the color fastness to light (light fastness), the color fastness to washing (household wash/laundry) and the fastness to water (wet fastness). These three parameters form the basic evaluation tool to assess color stability of dyeings in garments.

Natural dyes have been used up to the end of the 19th century. At that time such dyeings had reached quite a good level of fastness. Unfortunately, focused research which had started in the 17th century (Friedlaender, 1909; Wescher, 1938; Rys and Zollinger, 1982; Zollinger, 1987; Smith and Wagner, 1991; Schweppe, 1993; Sewekow, 1995; Freeman *et al.*, 1996; Schmidt, 1997; Mueller, 2003; Guinot *et al.*, 2006) stopped immediately after the introduction of synthetic colorants and even the existing knowledge about natural dyes of that time was lost. However, in the last few years there has been renewed interest. Starting from accessible literature and available plant sources, dyeing experiments have been performed (besides references cited in this chapter also see other chapters in this book). As a result a list of potential colorants derived from plant sources has been created. Some of the most promising candidates are given in Table 18.7.

In an extensive study plant material available in Austria was tested for a possible use as source for natural colorants to be applied in textile dyeing. Two main sources were identified: (a) primary products from agriculture and (b) plant material like waste and by-products from industry. Especially side- and by-products of high quality from the timber industry and food production companies could be identified as promising raw material for natural colorants released from suppliers. The results in Table 18.7 are some of the highlights from research of Bechtold and co-workers (Ganglberger and Geissler, 2001, 2003; Bechtold *et al.*, 2003, 2004b, 2006; Geissler *et al.*, 2005, 2006; Rappl, 2005; Rappl *et al.*, 2005) on natural dyes in the last few years. The second and third columns show color fastness marks (light fastness and wet fastness) that have been reached during the projects. While for light fastness marks from 1 to 8 are given in the case of wet fastness only marks from 1 to 5 are possible. For both tests the higher resistance is indicated by the higher number. The range of fastness marks given in the table demonstrates the influence of the substrate type on the achievable fastness level. Although there are dyeings that exhibit poor color fastness (marks 1 or 2) some of the results are encouraging and scale-up is in progress.

Table 18.7 Selected screening results: a list of promising screening results from the Austria research group (Ganglberger and Geissler, 2001, 2003; Rappl, 2005; Grill et al., 2004; Pladerer and Wenisch, 2007)

Plant material	Color fastness		Availability in Austria	Screening results
	Light	Wet		
Canadian golden rod	1–4	3–5	Collection/ agriculture	Flax, wool
Onion peels	1–4	4–5	Food industry	Wool, PA
Nut shell	1–4	2–5	Agriculture	Except PA!
Bark	3–4	2–5	Timber industry	Ash tree Sticky alder tree
Grape pomace	1–4	2–5	Wine production	Light fastness! Tannin
Black elder	1–3	3–5	Juice production	Light fastness!
Tea	3–4	4–5	Iced tea production	Wool

The biggest challenge as far as color fastness is concerned might be the fulfillment of light fastness requirements. The stability of colorations can be improved in various ways. Generally the addition of mordants is suitable to improve the light stability. However, some of the mordants – especially iron salts – lead to a shift in the resulting color.

Textile auxiliaries are also in use to improve the quality of dyeing. Some of them were also applied in natural dyeing and showed promising results. While Lee recommended an UV absorber on protein fibers (Lee *et al.*, 2001), Oda suggests singlet oxygen quenchers to increase the light stability (Oda, 2001a, 2001b). Other groups focus on the coloration technology and the machinery used to achieve further improvements.

Generally, in order to discuss light-induced degradation processes more information about photocatalytic reactions is needed. Based on the literature (Timpe and Neuenfeld, 1990; Hihara *et al.*, 2002; Batchelor *et al.*, 2003; Oakes, 2003; Yang *et al.*, 2005), Mussak focused on the light-induced photodegradation process of natural dyes. Berberine, a yellow natural dye, served as a model system for the dyestuff degradation process. Dyed films were irradiated by an artificial light source while the degradation reaction was observed spectro-photometrically. Subsequent analyses of the kinetics were preformed and a model system for the degradation was generated. As a result, for berberine, which is a rather light-sensitive dye, a two-step degradation mechanism has been suggested (Mussak, 2008).

18.4.2.2 Costs

In the past natural dyes were mainly derived from agriculture (Schweppe, 1993; Mueller, 2003). Up to the end of the 19th century, both animals and plants were used as sources for dyestuff isolation. Nowadays, primarily natural colorants isolated from vegetable material are used as the potential dyestuff source. For this purpose traditional dye plants have been cultivated. More recently other resources from industrial sectors have been evaluated.

Particularly raw material from the food and timber industries was found to be suitable for dyestuff extraction and furthermore is available in considerably high amounts from local sources (Bechtold *et al.*, 2006). As a matter of fact these side- and by-products (waste) lower the average price, which makes natural colorants even more attractive.

A mixed calculation of raw material from at least three main sources in farming and secondary products from industry (timber and food) results in an average price of €2 per kilogram of dyed textile (Bechtold *et al.*, 2003, 2004b). As an estimate 1 kg of plant material will be required to dye 1 kg of textile goods, this indicates that the use of plant dyes available from different sources including secondary products will be competitive in cost to that of synthetic dyes. Detailed information is given in Chapter 21, which focuses on 'Economic Aspects of Natural Dyes'.

18.4.2.3 Handling

The handling of dyestuff is one of the critical points in terms of natural colorant application. Although extraction of plant material immediately *before* dyeing in the dye house would be advantageous, the handling of the colorants compared to synthetic dyestuff will change. One suggested strategy is the distribution of dried, standardized plant material in permeable bags. At the time these bags are extracted and used for coloration no further purification is required besides a filtration step (Ganglberger and Geissler, 2001, 2003). One of the Austrian projects focusing on the acceptance of different dyestuff handling showed that this might hinder but not block the introduction of natural dyes (Rappl, 2005).

Generally, application methods and commercial products similar to the present use of synthetic dyes are preferable. Therefore the production of concentrates (solid or liquid) should be considered carefully. As a matter of fact concentrates would be favorable in dyeing techniques where high dyestuff content and respectively low liquor ratios are used, e.g. continuous dyeing by a Foulard unit or batchwise dyeing with a jet or jig apparatus (Mussak and Bechtold, 2006; Mussak 2008; Mussak and Bechtold, 2008).

18.4.3 Standardization of the Dyestuff

Natural raw material shows variations of quality, dyestuff content and composition of the plant due to the effects of weather/climate, soil, fertilization, location, etc. Such materials can be used for handicraft dyeing where the uniqueness of a product is well advertised.

In other words, the dyeing result alters depending on the crop as long as no product standardization has been performed to adjust plant materials to a fixed standard. For enlarged application of natural dyes in the textile industry a certain level of repeatability is required. Basically the standardization of agricultural material is more important than that of a secondary product, e.g. from the food industry. Especially in the food sector strict quality assurance of the final product (e.g. fruit juice) already results in constant constitution of the waste. Although this may facilitate the standardization an assessment of the dyeing properties and respectively dyestuff content is necessary. In general, information on the quality of plant material and related coloration features may be obtained from dyeing experiments. Unfortunately, extraction of natural dyestuffs and subsequent coloration requires a minimum of two hours of process time. Therefore, indicating parameters that will allow prediction of the later dyeing result would be essential (Ganglberger and Geissler, 2001, 2003; Rappl, 2005; Pladerer and Wenisch, 2007).

Two representative examples for plant material analysis have been demonstrated in the literature (Bechtold, 2007b, 2007c): on the one hand the ACY (total anthocyanin concentration) for anthocyanins according to the pH differential method (Waterhouse, 2001; Giusti and Wrolstad, 2003) and on the other hand the TPH (total phenol content) according to Slinkard and Singleton as well as the iron(II)–complex concentration in the case of flavonoids (Slinkard and Singleton, 1977; Singleton *et al.*, 1999).

18.4.3.1 Anthocyanins

Different batches of pomace – the grape's residue of wine production – were analyzed and the total anthocyanin concentration (ACY) in terms of cyanidin-3-glycosides was determined (Wrolstad, 2000; Howard *et al.*, 2003; Mussak and Bechtold, 2006). Simultaneously, cotton fabrics were dyed and the resulting color was evaluated. Characterization of the dyed samples was performed by measurement of diffuse reflectance and calculation of K/S values according to the theory of Kubelka-Munk. The results show a direct correlation of the K/S value with the concentration of extracted anthocyanin dyes in solution, though variation of the dyestuff composition in the extract was observed in the transmission spectra (Bechtold *et al.*, 2007b).

18.4.3.2 Flavonoids

Different batches of Canadian golden rod material were used to investigate variations in dyestuff content due to different years of harvest and time of storage. The total soluble phenol content (TPH) and the amount of metal complex (iron(II)–complex) formed after additions of iron salts were determined. Wool yarns and cotton fabrics were dyed with the extracts obtained using identical conditions in all batches. The photometrically determined parameters (TPH and Fe(II)–complex) were related to the final dyeing result. In this case neither the total phenolic compound nor the metal–complex concentration was correlated to the dyeing result (Bechtold *et al.*, 2007c).

These two examples show the relevance and difficulties of standardization. While for anthocyanins a suitable parameter could be identified, both evaluated procedures tested for flavonoids failed.

Besides the isolation of the coloring compounds from plant material, the individual affinity of the extracted dyes towards the substrate is of importance. Even if a considerably high amount of colored components are present in the extract only the share of molecules that sorbs on the substrate will be of relevance for the dyeing (Mussak, 2008).

18.4.4 Ecological Aspects

The application of plant-derived colorants for textile dyeing does not lead automatically to 'ecological' products. A critical investigation of all steps of the process, including the dyestuff isolation, the dyeing procedure and subsequent emissions is necessary in terms of the eco-potential of a product. A detailed discussion about the ecological aspects is available in the literature (Geissler *et al.*, 2005, 2006; Bechtold *et al.*, 2007a; Mussak, 2008). Essential restrictions for an environmentally friendly dyeing strategy are:

- limitations of chemicals and solvents to avoid hazardous effluents;
- substitutions of primary agricultural products by waste, side- and by-products if available or collection from the wild;
- careful selection of mordants (heavy metals must not be used);
- minimization of inputs (water, energy).

18.4.5 Aspects of Application

Generally a natural dyestuff requires the same handling procedure as a synthetic one. Depending on the commercial type of natural dye only the preparation of the dye bath is different, compared to synthetic dyes. In the case of using dried standardized plant material the isolation of the coloring matters in terms of extraction is necessary as an additional step.

18.4.5.1 Extraction

Due to ecological and economical aspects the extraction of natural dyes from plant material is limited to water as solvent. The explicit use of water causes lower costs and simplifies the waste water treatment requirements (Dalby, 1993; Mussak, 2008). If other solvents are used and/or chemicals are added, considerable amounts of inputs are lost in the plant material (Bechtold *et al.*, 2008). This may lead to complicated aftertreatments of extracted plant material, to additional costs from solvent/chemical consumption and expensive deposition of contaminated wastes. Important factors to be considered for the dye extraction are the extraction time and temperature, as shown in Figure 18.3.

Figure 18.3 shows the extraction degree versus the extraction time for three different types of plant material. Canadian golden rod is a yellow flower, possessing

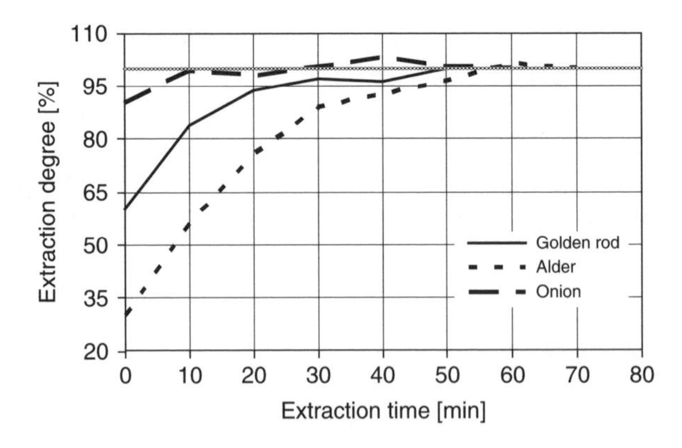

Figure 18.3 *Extraction properties of different plants: the extraction degree versus the extraction time for three different parts of plants (Canadian golden rod, black alder bark and onion skins) is shown. All extractions were done with water at boiling temperature at a liquor ratio of 1:20*

many very small blossoms, which are extracted with water in a similar way to preparing tea. In the case of black alder bark, the dried bark is grind to powder and filled into bags that are boiled in water. To isolate the dyestuff from onion skin the papery dry outer skins of, preferably, red onions are extracted in boiling water. Due to differences in the specific surface, the particle diameters and the accessibility of the plant material vary according to the extraction degree measured after a defined time. While the maximum extractable dyestuff content for onion skins is reached after about 10 minutes it takes about one hour to reach the upper level of exhaustion in the case of black alder bark. The extraction properties of Canadian golden rod material is somewhere in between, with the maximum obtained after about 20 minutes. These data are of relevance if mixtures of plant material are used in the same batch and if a standard procedure for extraction and quality determination is defined.

In some cases further adjustment of the pH value may increase the amount of dyestuff extracted, though it needs careful consideration of the overall chemical input due to the economical and ecological aspects discussed above.

18.4.6 Dyeing Technology

The general dyeing technology is hardly modified. By starting with the plant extract as the dyeing solution a further preparation step is no longer necessary. However, the low-temperature stability of some dyes (e.g. anthocyanines) has to be respected in order to avoid dye degradation.

In cooperation with well-known Austrian textile companies the Research Group of Natural Dyes at the Research Institute for Textile Chemistry and Textile Physics, University of Innsbruck, developed an innovative textile dyeing process based on natural dyes. One of the essential parameters was to limit energy requirements strictly. Therefore a strategy of combined dyestuff extraction and subsequent dyeing was suggested. If local as well as temporal gaps are minimized (in the ideal case it is skipped) and the hot dye liquor is used immediately for coloration purposes, up to 60 % of the energy might be saved (Ganglberger and Geissler, 2001, 2003; Bechtold *et al.*, 2007d; Mussak, 2008). Figure 18.4 illustrates the temperature profile of a combined extraction and dyeing process.

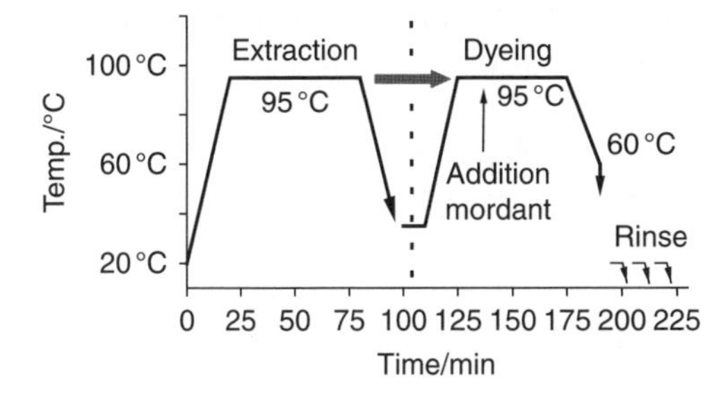

Figure 18.4 Temperature profile of the extraction/dyeing process: the temperature profile of a combined extraction/dyeing process is shown. The hot extract is directly used to dye the substrate

18.4.7 Mordanting

Mordants are used for fixation of the dyestuff, increasing the fastness properties or variation in the color appearance. This last effect can be a desired effect as well as an unwanted phenomenon. Especially the use of iron mordants leads to substantial color differences and primarily results in a color shift towards dark shades. On the contrary, alum salts as mordants intensify the obtained color brilliance but do not influence the color shade to the same extent. Selected representative examples are given in Table 18.8.

Table 18.8 *Color shift due to mordants: aqueous plant extracts of dyer's chamomile, madder and rhubarb are used for dyeing wool or cotton. One trial was done without a mordant. For the other two dyeings with iron and alum mordant was used at a concentration of 5 g/L. The liquor ratio of extraction and dyeing was 1:20*

Plant material	Substrate	Mordant	CIE-Lab values		
			L^*	a^*	b^*
Dyer's chamomile	Wool	—	84.82	−7.37	+29.15
		Iron	33.93	−2.00	−8.50
		Alum	83.26	−5.05	+73.81
Madder	Cotton	—	71.23	+11.97	+18.04
		Iron	57.18	+5.26	+4.71
		Alum	64.17	+23.07	+20.55
Rhubarb	Wool	—	49.05	+9.17	+50.04
		Iron	26.16	−0.19	+14.77
		Alum	52.22	+6.78	+52.60

One of the big challenges for natural dyestuff application is that only very few sources for red and blue colors are available. One outstanding possibility to fill this gap is the addition of tannin or tannin agents. Generally they can be used for color shifts toward darker shades. However, in combination with anthocyanins all kinds of variations from blue via violet to red are observed (Bechtold *et al.*, 2007b).

Basically, three different types of mordanting strategies can be distinguished: pre-, after- and meta-mordanting. The processes mainly vary in the time of mordant addition. While pre- and after-mordanting require an additional treatment step in a separate bath (a mordant solution with defined concentration of the metal salt), the simple addition of a concentrated salt solution directly to the dye bath is used in the so-called meta-mordanting process. Figure 18.5 illustrates the three different strategies schematically.

Pre- and after-mordanting are very similar besides the time of mordant addition. Both require two fillings (treatment baths), which doubles the water input. However, repeated cycles of mordanting/dyeing and re-use of the dyeing solution and the mordant baths is possible if the solutions can be stabilized and contamination is avoided (Kumar and Bharati, 1998; Mussak, 2008). On the contrary meta-mordanting procedures, where the mordant is added after some minutes, show an advantage in handling because only one dye bath exists and the water consumption is minimized. Although this option seems to be favorable one disadvantage is the hindrance of repeated dyeing cycles because the dyeing

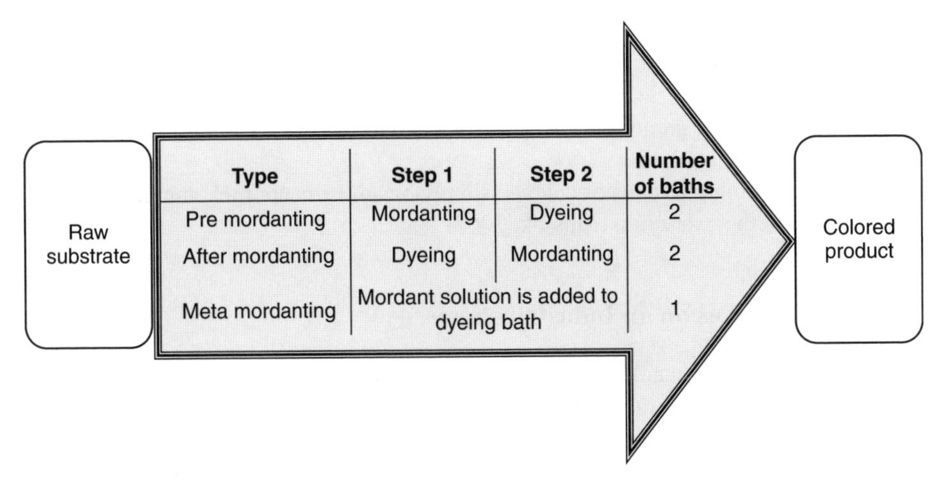

Figure 18.5 *Types of mordant techniques: three different mordant techniques are shown. They differ in the time of mordant addition and the number of baths required*

liquor is contaminated with mordant and re-use is therefore not possible for a second dyeing.

Generally small variations in the color depth are found if the three mordanting types are compared. In most cases pre-mordanting leads to darker shades compared to meta-mordanting, as shown in the literature (Kumar and Bharati, 1998; Mussak, 2008).

18.4.8 Standardization of the Coloration Process

Standardization of the process generally includes the transfer of a dyeing procedure to the specific machinery used. Every dyeing apparatus has its own properties; therefore a careful adaptation of the processing parameters with respect to substrate, mordant, auxiliaries and also the machinery is required. The first step is to set up a model system. This can be a laboratory dyeing machine or a semi-scale industrial machine suited to generate basic data about dyeing conditions. The transfer of the technology to full-scale operation and fine adjustment of the dyeing conditions are performed in full-scale tests on the industrial equipment.

18.4.9 Mixtures of Plant Material

Bechtold *et al.* suggested a natural dyeing procedure where the dried standardized plant material is packed in bags. These bags contain defined amounts of plant dyestuff which permit easier handling of the raw material for dyestuff extraction. The bags are permeable for solvent (water) and therefore allow a dyestuff extraction procedure that is comparable to tea preparation. Due to the bag's material (preferably paper or even textile fabric) the bags are biodegradable and so can easily be disposed by composting or combustion. Further advantages of such a concept are as follows:

(a) The available color gamut can easily be widened by joint extraction of bags filled with different plant materials.
(b) Different color depths can be achieved simply by adjustment of the number and size of bags used in the extraction (Mussak, 2008).

18.5 Natural Dyes on an Industrial Scale

While natural dyes in handicraft, food production and cosmetics are well known, the application for textile coloration on an industrial scale is quite innovative. The Austrian Project Group on Natural Dyes combines expertise in different fields: textile chemistry, ecology, agriculture, sustainable management, marketing and textile production. The research activities also include the transfer of natural dyeing processes on industrial dyeing equipment. Therefore the first trials on enlarged scales, up to 50 kg of textile goods per dyeing, were preformed and first prototypes were produced. Selected examples are summarized here.

18.5.1 Hank Dyeing of Woolen Yarn and Production of Woolen Caps

Woolen yarns were dyed with aqueous extracts of black alder bark, dyer's chamomile and onion skin. To enlarge the range of resulting colors, dyeings with and without mordants were preformed. Finally six different colors were dyed. The colored yarns were then used by a knitwear manufacturer for the production of stylish sports caps. Two examples are shown in Figure 18.6.

Figure 18.6 Woolen caps dyed with plant extracts: two prototypes of caps produced from woolen yarn that was dyed with plant extracts (onion skin with alum mordant or without mordant; black alder bark extract with alum or iron mordant; dyer's chamomile extract with alum or iron mordant)

Generally the levelness of the hank dyed wool yarn is limited and is directly related to a continuous movement of the woolen hanks in the dyeing liquor. The dyed material could be processed to knitwear without problems. No restriction in the knitting and sewing of the caps that were due to the natural dye was observed.

18.5.2 Dyeing of Cones in a Yarn Dyeing Machine

Cones with woolen yarn were dyed with an aqueous extract of dyer's chamomile. In a first series of experiments the use of the alum mordant was optimized. As a result the input concentration of alum could be reduced from 5 to 1 g/L, which however was accompanied by a slight reduction in color brilliance.

A key factor in terms of yarn dyeing on cones is the evenness of the dyeing through the cone. In some cases slight shade variations between the outer and inner layers of the cones were observed. Remarkable improvement in the levelness could be achieved by adjustment of the cycle times for changing the flow direction through the cones (inside/out, outside/in). Filtration effects were of minor relevance but should be considered. Careful pre-filtration, e.g. through a filtration cloth, before addition of the extract to the dyeing apparatus is required in order to avoid filtration effects.

18.5.3 Dyeing of Cotton Fabric on a Jet Dyeing Machine

Cotton fabric (50 meters) was dyed on a jet dyeing machine with an extract of dyer's chamomile with a meta-mordation procedure. A brilliant yellow color shade was achieved, but light/dark strips were observed on the fabric after drying. The average color difference $\Delta E = 1.4$ led to a somewhat uneven and flickering impression of the fabric. However, this might be highlighted in terms of marketing of natural dyed products and in advertisements of a 'new' product. Critical points of this technique are the filling of the machine with the extract. Up to a certain minimum level of liquor/extract in the jet the goods cannot be moved and the risk of stripe formation is quite high. Further, addition of the mordant has to be done with care, e.g. as a dosage procedure, similar to the alkali dosage in the reactive dyeing of cotton.

18.5.4 Dyeing of Cotton Fabric on a Jig Dyeing Machine

Different plant extracts were used for dyeing cotton and linen fabric. Repeated use of one extract was also studied. A comparison of the dyeing results showed that there are only slight differences in shade in the dyeings, which is due to the low affinity of the dyestuff. Analysis of the 5 m fabrics showed that the variations in terms of color differences reached $\Delta E = 0.6$ on average and 1.8 maximum. Such rather small deviations prove the potential of natural dyes being dyed on textiles with an acceptable degree of levelness.

18.5.5 Fabric Dyeing on a Garment Dyeing Machine

Cotton and linen fabric were dyed with an extract of red onion peels and iron mordant. For this experiment a meta-mordanting technique was applied. The dyestuff shows a higher affinity toward the linen substrate. In both cases rather uneven dyeing results were found, which was explained by wrinkle forming in the textile goods during the coloration process. Precipitation of iron components was not observed.

18.5.6 Dyeing of Polyamide Tights in a Paddle Dyeing Machine

From the pre-screening experiments on a laboratory scale four suitable plant sources could be identified for dyeing of polyamide. A ready-to-use concept for application of natural dyes for polyamide tights dyeing was elaborated. After evaluation of fastness properties, four natural colorants were found to be suitable for large-scale dyeing.

Paddle dyeing techniques used for exhaust dyeing of polyamide tights require rather high volumes of dyeing liquor, which increases the water and energy consumption of a dyeing process. In natural dyeing large volumes are already present from the extraction step due to the limited extraction capacity of water and the rather low concentration of dyestuff in the plant material. The combination of the extraction/dyeing units leads to an optimized process with regard to the overall energy balance.

18.6 Conclusion

Natural colorants have been used for textile dyeing for millennia, and they still have the potential to color textile materials. However, the circumstances in textile coloration changed and in modern textile production not only is the color important but also a high number of other aspects have to be considered. Recently, 'ecological' aspects besides the economical factors became more important for consumer decisions to purchase a certain product.

Colorants from plants can be extracted and used for textile dyeing in the laboratory as well as in industry. At the moment textile companies all over the world show a growing interest in plant dyeing and start their individual trials to re-establish the technology in their production line. However, in the end the final decision will be made by the consumer. In case of an increasing public request for natural dyed products – from our present point of view – the industry will be ready to supply what is required by the market.

Acknowledgment

The authors gratefully acknowledge the State Government of Vorarlberg for financial support and the Versuchsanstalt and HTL Dornbirn for the use of their equipment. They also want to acknowledge all companies that supplied them with material: Lenzing AG, Leinenweberei Vieböck, Getzner Textil, Schoeller Wolle and Mr Prischink. They also thank the members of the Institute for Applied Ecology for their cooperation and want to acknowledge all partners of the Project Group (Programmlinie Fabrik der Zukunft). A part of this work was supported by the Austrian Federal Ministry for Transport, Innovation and Technology.

References

Angelini, L. G., Bertoli, A., Rolandelli, S. and Pistelli, L. (2003) Agronomic potential of *Reseda luteola* L. as new crop for natural dyes in textile production, *Indust. Crops Prod.*, **17**, 199–207.

Batchelor, S. N., Darr, D., Colemann, C. E., Fairclough, L. and Jarvis, A. (2003) The photofading mechanism of commercial reactive dyes on cotton, *Dyes and Pigments*, **59**, 269–275.

Bechtold, T., Turcanu, A., Geissler, S. and Ganglberger, E. (2002) Process balance and product quality in the production of natural indigo from *Polygonum tinctorium* Ait. applying low technology methods, *Bioresource Technol.*, **81**, 171–177.

Bechtold, T., Turcanu, A., Ganglberger, E. and Geissler, S. (2003) Natural dyes in modern textile dyehouses – how to combine experiences of two centuries to meet the demands of the future?, *J. Clean. Prod.*, **11**, 499–509.

Bechtold, T., Burtscher, E. and Hung, Y. (2004a) Treatment of textile wastes, in L. K. Wang, Y.-T. Hung, H. H. Lo and C. Yapijakis (eds), *Handbook of Industrial and Hazardous Wastes Treatment*, 2nd edition, Marcel Dekker, Inc., New York, ISBN: 0-8247-411-5.

Bechtold, T., Kaulfuss, P., Mahmut, A., Geissler, S. and Ganglberger, E. (2004b) Naturfarbstoffe in Mitteleuropa – Rohstoffquellen und färberische Qualität, *Melliand Textilberichte*, **85**(4), 268–272.

Bechtold, T., Mussak, R., Mahmud-Ali, A., Ganglberger, E. and Geissler, S. (2006) Extraction of natural dyes for textile dyeing from coloured plant wastes released from food and beverage industry, *J. Sci. Food Agric.*, **86**, 233–242.

Bechtold, T., Mahmud-Ali, A. and Mussak, R. (2007a) Natural dyes from food processing wastes, Chapter 19, in K. Waldron (ed.), *Handbook of Waste Management and Co-product Recovery in Food Processing*, Woodhead Publishing Ltd, Cambridge.

Bechtold, T., Mahmud-Ali, A. and Mussak, R. (2007b) Anthocyanin dyes extracted from grape pomace for the purpose of textile dyeing, *J. Sci. Food Agric.*, **87**, 2589–2595.

Bechtold, T., Mussak, R. and Mahmud-Ali, A. (2007c) Natural dyes for textile dyeing – comparison of methods to assess quality of Canadian golden rod plant material, *Dyes and Pigments*, **75**, 287–293.

Bechtold, T., Mahmud-Ali, A. and Mussak, R. (2007d) Reuse of ash-tree bark as natural dye for textile dyeing – process conditions and process stability, *Coloration Technology*, **123**, 271–270.

Bechtold, T., Mahmud-Ali, A., Ganglberger, E. and Geissler, S. (2008) Efficient processing of raw material defines the ecological position of natural dyes in textile production, *International Journal of Environment and Waste Management*, **2**(3), 215–232.

Chandrika, U. G., Jansz, E. R., Wickramasinghe, S. M. D. N. and Warnasuriya, N. D. (2003) Carotenoids in yellow- and red-fleshed papaya (*Carica papaya* L.), *J. Sci. Food Agric.*, **83**, 1279–1282.

Dalby, G. (1993) Greener mordants for natural coloration, *J. Soc. Dyers Color.*, **109**, 8–9.

Deutsches Institut für Normung (DIN) (1985) *DIN Taschenbuch 16, Textilprüfung 3, Farbechtheit, Normen*, Beuth Verlag, Berlin and Köln.

Freeman, H. S., Hinks, D. and Esancy, J. (1996) Genotoxicity of azo dyes: bases and implications, Chapter 7, in A. T. Peters and H. S. Freeman (eds), *Physico-Chemical Principles of Color Chemistry*, Vol. 4, Chapmann & Hall, Glasgow.

Friedlaender, P. (1909) Ueber den Farbstoff des antiken Purpurs aus *murex brandaris*, *Ber. Dtsch. Chem. Ges.*, **42**, 765–770.

Ganglberger, E. and Geissler, S. (2001) *Potential an nachwachsenden Rohstoffen unter Aspekten der Nachhaltigkeit: Produktion von farbstoffliefernden Pflanzen in Österreich und ihre Nutzung in der Textilindustrie*, Projektendbericht, Bundesministerium für Verkehr, Innovation und Technik, Wien.

Ganglberger, E. and Geissler, S. (2003) *Farb&Stoff – Sustainable Development durch neue Kooperationen und Prozesse*, Projektendbericht, Bundesministerium für Verkehr, Innovation und Technik, Wien.

Geissler, S., Klug, S., Bechtold, T. and Mussak, R. (2005) Marketing products from renewable resources: the example of natural dyes textiles produced by industry, *Greentech Newsletter*, **8**(1), 2–3.

Geissler, S., Mussak, R., Bechtold, T., Klug, S. and Vogel-Lahner, T. (2006) Nachwachsend = nachhaltig? Eine Analyse am Beispiel pflanzlicher Textilfärbung, *GAIA*, **15**(1), 44–53.

Giusti, M. M. and Wrolstad, R. E. (2001) Characterisation and measurement of anthocyanins by UV–visible spectroscopy, in R. E. Wrolstad (ed.), *Current Protocols in Food Analytical Chemistry*, John Wiley & Sons Inc., New York, pp. F1.2.1–F1.2.13.

Glover, B. (1998) Doing what comes naturally in the dyehouse, *J. Soc. Dyers Color.*, **114**, 4–7.

Grill, E., Fux, B., Pfeiffer, E. and Rohrschacher, E. (2004) *Trademark*, Projektendbericht, Teil 1–3, FH Wiener Neustadt, Diplomstudiengang Wieselburg für Produkt- und Projektmanagement, Wieselburg.

Grosjean, D., Whitmore, P. M., De Moor, C. P. and Cass, G. R. (1987) Fading of alizarin and related artist's pigments by atmospheric ozone: reaction products and mechanisms, *Environ. Sci. Technol.*, **21**, 635–643.

Guinot, P., Rogé, A., Garagadennec, A., Garcia, M., Dupont, D., Lecoeur, E., Candelier, L. and Andary, C. (2006) Dyeing plants screening: an approach to combine past heritage and present developement, *Color. Technol.*, **122**, 93–101.

Hartl, A. and Vogl, C. R. (2001a) *Faser- und Färbepflanzen aus biologischem Landbau: Erzeugung, Verarbeitung und Vermarktung*, Projektendbericht 1a, Bundesministerium für Verkehr, Innovation und Technik, Wien.

Hartl, A. and Vogl, C. R. (2001b) *Faser- und Färbepflanzen aus ökologischem Landbau: Anbauversuche Fasernessel*, Projektendbericht 1b, Bundesministerium für Verkehr, Innovation und Technik, Wien.

Hihara, T., Okada, Y. and Morita, Z. (2001) Photofading, photosensitization and the effect of aggregation on the fading of triphenodioxazine and copper phthalocyanine dyes on cellulosic film, *Dyes and Pigments*, **50**, 185–201.

Hihara, T., Okada, Y.and Morito, Z. (2002) Photo-oxidation and -reduction of vat dyes on water swollen cellulose and their lightfastness on dry cellulose, *Dyes and Pigments*, **53**, 153–177.

Hill, D. J. (1997) Is there a future for natural dyes? *Rev. Prog. Color.*, **2**, 18–25.

Howard, L. R., Clark, J. R. and Brownmiller, C. (2003) Antioxidant capacity and phenolic content in blueberries as affected by genotype and growing season, *J. Sci. Food Agric.*, **83**, 1238–1247.

Kannan, S. S. and Geethamalini, R. (2005) Natural dyes, a brief glimpse, *Textile Asia*, **2**, 32–39.

Knaggs, N. S. (1992) Dyestuff of the Ancients, *Am. Dyestuff. Rep.*, **81**, 109–111.

Kumar, V. and Bharati, N. V. (1998) Studies on natural dyes: *Mangifera indica bark, Am. Dyest. Rep.*, **87**(9), 18–22.

Lee, J. J., Lee, H. H., Eom, S. I. and Kim, J. P. (2001) UV absorber aftertreatment to improve lightfastness of natural dyes on protein fibres, *Color. Technol.*, **117**, 134–138.

Mueller, W. (2003) *Handbuch der Farbenchemie: Grundlagen, Technik, Anwendungen*, ecomed-Verlagsgesellschaft, Weinheim.

Mussak, R. (2008) *Naturfarbstoffe in der modernen Textilindustrie*, VDM Verlag Dr Mueller, Saarbruecken, ISBN: 978-3-8364-9707-7.

Mussak, R. and Bechtold, T. (2006) The position of natural dyes in the Austrian textile industry, in Proceedings of the International Symposium in Kyoto on *Dyeing and Finishing of Textiles*, co-work of the 120th Committee, JSPS and 21st Century COE Program at Shinshu University, Kyoto, pp. 21–28.

Mussak, R. and Bechtold, T. (2008) Renewable resources for textile dyeing – technology, quality, and environmental aspects, in Proceedings of the IFATCC International Congress, Barcelona.

Nguyen, T., Martin, J. W., Byrd, E. and Embree, N. (2000) *Effects of Relative Humidity on Photodegradation of Acrylic Melamine Coatings: A Quantitative Study*, Vol. 83, American Chemical Society, PSME, Washington, DC.

Novotna, P., Boon, J. J., van der Horst, J. and Pacakova, V. (2003) Photodegradation of indigo in dichloromethane solution, *Color. Technol.*, **119**, 121–127.

Oakes, J. (2003) Principles of colour loss. Part 2: degradation of azo-dyes by electron transfer, catalysis and radical routes, *Rev. Prog. Color.*, **33**, 72–84.

Oda, H. (2001a) Improvement of the light fastness of natural dyes: the action of singlet oxygen quenchers on the photofading of red carthamin, *Color. Technol.*, **117**, 204–208.

Oda, H. (2001b) Improvement of light fastness of natural dyes. Part 2: effect of functional phenyl esters on the photofading of carthamin in polymeric substrates, *Color. Technol.*, **117**, 257–261.

Peter, M. and Rouette, H. K. (1989) *Grundlagen der Textilveredlung: Handbuch der Technologie, Verfahren und Maschinen*, 13th edition, Deutscher Fachverlag, Frankfurt and Main.

Pladerer, C. and Wenisch, A. (2007) *Riskmin: Risikominimierung entlang der Wertschöpfungskette*, Projektendbericht, Bundesministerium für Verkehr, Innovation und Technik, Wien.

Podsiadly, R., Sokolowska, J., Marcinek, A., Zielonka, J., Chrzescijanska, E. and Socha, A. (2003) Electrochemical and photochemical reduction of a series of azobenzene dyes in protic and aprotic solvents, *Color. Technol.*, **119**, 269–274.

Rappl, B. (2005) *Trademark Farb&Stoff – Von der Idee zum marktfähigen Handelsprodukt: Pflanzenfarben für die Textilindustrie*, Projektendbericht, Bundesministerium für Verkehr, Innovation und Technik, Wien.

Rappl, B., Bechtold, T. and Mussak, R. (2005) Zwiebeln auf Wolle – Pflanzenfarben für die Textilindustrie, *Textilveredlung*, **40**, 4–6.

Rath, H. (1972) *Lehrbuch der Textilchemie*, Springer-Verlag, Berlin, ISBN: 3-540-05587-8.

Rys, P. and Zollinger, H. (1982) *Farbstoffchemie – Ein Leitfaden*, 3rd edition, Verlag Chemie.

Schmidt, H. (1997) Indigo – der König der Farbstoffe wird 100, *Melliand Textilberichte*, **6**, 418–421.

Schweppe, H. (1993) *Handbuch der Naturfarbstoffe: Vorkommen, Verwendung, Nachweis*, ecomed Verlagsgesellschaft, Landsberg am Lech.

Sequin-Frey, M. (1981) The chemistry of plant and animal dyes, *J. Chem. Edu.*, **58**, 301–305.

Sewekow, U. (1988) Naturfarbstoffe – eine Alternative zu synthetischen Farbstoffen?, *Melliand Textilberichte*, **4**, 271–276.

Sewekow, U. (1995) Naturfarbstoffe in der heutigen Textilfärberei, *Melliand Textilberichte*, **5**, 330–336.

Singleton, V. L., Orthofer, R. and Lamuela-Raventos, R. M. (1999) Analysis of total phenols and other oxidation substrates and antioxidants by means of folin-ciocalteu reagent, *Methods in Enzymology*, **299**, 152–178.

Slinkard, K. and Singleton, V. L. (1977) Total phenol analysis: automation and comparison with manual methods, *Am. J. Enol. Vitic.*, **28**, 49–55.

Smith, R. and Wagner, S. (1991) Dyes and the environment: is natural better?, *Am. Dyest. Rep.*, **80**, 32–34.

Taylor, G. W. (1985) Natural dyes in textile applications, *Rev. Prog. Coloration*, **16**, 53–61.

Timpe, H. J. and Neuenfeld, S. (1990) Dyes in photoinitiator systems, *Kontakte (Darmstadt)*, **2**, 28–35.

Trotman, E. R. (1984) *Dyeing and Chemical Technology of Textile Fibres*, 6th edition, Charles Griffin Company Ltd, Bucks, England, p. 60, ISBN: 8 852642679.

Waterhouse, A. L. (2001) Determination of total phenolics, in R. E. Wrolstad (ed.), *Current Protocols in Food Analytic Chemistry*, John Wiley & Sons Inc., New York, pp. I1.1.1–I1.1.8.

Wescher, H. (1938) Große Lehrer der Färbekunst im Frankreich des 18. Jahrhunderts, *Ciba-Rundschau*, **22**, 783–799.

Wrolstad, R. E. (2000) Natural food colorants, *IFT Basic Symposium Series*, **14**, 237–252.

Wulfhorst, B. (1998) *Textile Fertigungsverfahren: Eine Einführung*, Carl Hanser Verlag, München and Wien.

Yang, J., Chen, C., Ji, H., Ma, W. and Zhao, J. (2005) Mechanism of TiO_2-assisted photocatalytic degradation of dyes under visible irradiation: photoelectrocatalytic study by TiO_2-film electrodes, *J. Phys. Chem. B*, **109**, 21900–21907.

Yoshizumi, K. and Crewes, P. C. (2003) Characteristics of fading of wool cloth dyes with selected natural dyestuffs on the basis of solar radiant energy, *Dyes and Pigments*, **58**, 197–204.

Zollinger, H. (1987) *Color Chemistry: Syntheses, Properties and Applications of Organic Dyes and Pigments*, VCH Verlagsgesellschaft, Weinheim, Germany, 2003

Zollinger, H. (2003) *Color Chemistry, Syntheses, Properties, and Applications of Organic Dyes and Pigments*, 3rd edition, Wiley–VCH, Weinheim, Germany, Chapters 8, 11 and 12, ISBN: 3-906390-23-3.

19

Natural Colorants in Hair Dyeing

Thomas Bechtold

19.1 Introduction

The desire to influence the individual appearance is as old as mankind itself. Thus coloration of hair and skin is found almost everywhere and numerous cultural expressions of such a habit have been reported.

In today's modern societies a considerable share of women and men apply colorants to change the colour of their hair; for example, 75 % of all American women are estimated to dye their hair [1]. In Europe more than 60 % of women and 5–10 % of men colour their hair, with a mean frequency of 6–8 times per year. The market for hair dyes in the EU has been estimated in 2004 at €2.6 billion [2]. Examples for relevant motivations for hair coloration are:

- demonstration of individuality by hair colour;
- fashion trends and cosmetic aspects;
- desire to cover white hair to maintain a youthful appearance.

Cosmetologists distinguish hair colorants with regard to the achieved permanence of the dyeing between temporary hair dyes, semi-permanent or permanent hair dyes, which will depend both on the type of dyestuff used and the conditions applied. The main difference is in the number of shampoos the dyeing will last, which ranges from 6 to 12 for semi-permanent colours to permanent dyeings which cannot be washed out.

While some traditional coloration of skin and hair is based on the application of pigments from, for example, clay or carbon, the focus of this chapter will be directed on the application of natural dyes to human hair. The colorants will be sorted with regard to their chemical nature and source.

Handbook of Natural Colorants Edited by Thomas Bechtold and Rita Mussak

19.2 Human Hair

As an average, 80 000 to 150 000 hairs grow on the human scalp. The close outer layer of a hair is called the cuticle [3]. This shed layer forms a roofing tile-like outer layer, which has to be opened to enable the colorants to access the core of the hair, the so-called cortex. The cortex is built up from spindle cells, which in the case of wool consists of an orto- and a para-cortex.

Chemically hair consists of α-keratin, a fibrous insoluble protein, which is also the basis for other biological derivatives of ectoderm, such as wool, feathers, nails and horns [3]. While cystein-rich α-keratins contain up to 22 % of cystine, the more flexible keratines of skin, hair and wool contain up to 10–14 % of cystine [4]. Further important amino acids used to build α-keratin are leucine, glutamine, arginine and serine [5]. In wool hair the cortex is formed by two types of corticular cells: regular smaller cells forming the chemically more reactive orto-cortex and larger irregular more crosslinked para-cortex cells [6]. However, in human hair the cortex is mainly built up from ortho-cortex, but transition forms to the para-cortex are also found [3].

The natural colour of hair is formed by granules of melanin, which are deposited in the stem of the hair during growth. Oxidation of tyrosine leads to the formation of black and brown eumelanin, found in dark hair colours, while in yellowish-blond, ginger and red coloured hair sulfur-containing pheomelanins are present. Such compounds are complex polymeric mixtures based on benzene-1,2-diol/1,2-benzoquinone fused to pyrrole units [7]. The observed colour of these insoluble pigments is based mainly on light-scattering effects.

The similarities in physical structure and chemical behaviour between wool fibres and human hair would allow a dyestuff chemist to expect that the dyeing behaviour of human hair will be similar to that of wool. In fact, similar chemical concepts can be applied, but wool dyeing is usually performed from aqueous solution at temperatures near to boiling while human hair dyeing is executed directly onto the scalp of the customer. Therefore completely different strategies have to be followed [8].

19.3 General Requirements on Hair Dyeing Concepts

Technical wool dyeing processes can focus on the technical quality of the dyeing without the restrictions that have to be observed in hair dyeing as a cosmetic procedure.

Aspects to be considered in the development and application of chemical concepts (including natural colorants) used for dyeing of human hair are complex. Requirements and limitations derive from:

- aspects of practical application;
- product safety and health risks;
- durability and quality of the dyeing.

Relevant aspects and parameters that will define restrictions and limits during application of dyes for hair dyeing and some estimates for limits are summarized in Table 19.1. These values should be understood as representative examples but cannot be taken as general limits. Dependent on the formulation used and the chemicals applied in a dyeing recipe, the applicable range of a parameter, e.g. pH or temperature, can vary.

Table 19.1 Aspects and parameters that can define application limits and representative examples for critical chemicals and limits

Aspect	Parameter	Chemical/limit
Practical application		
	Formulation (paste, liquid spray)	–
	Storage	–
	Simple handling	–
	Odour	Sulfide, ammonia
	Hair damage	Peroxide bleach on mordant dyeing Alkali and reducing agents
	Low staining of skin	–
Product safety/health risks		
	Temperature	20–40 °C (–60 °C)
	Time	5–50 min
	pH	4–9
	Toxicity, carcinogenic potential, sensibilization, allergic potential, low resorption through skin	Pb, Ni salt, pyrogallol, *p*-phenylenediamine
Quality of dyeing		
	Range of colours available	
	Fastness properties	Fastness to rubbing, light, water, sea-water, sweat,
	Stability against other hair treatments	e.g. conditions of permanent wave formation, colour bleaching
	Even dyeing from roots to end of hair	
	Reproducible shade	
	Low influence on handling ability and lustre of hair	

Reproducibility of shade is a significant problem in hair dyeing. In textile dyeing operations, standardization of pretreatment and dyeing processes can be achieved to a high degree and variability in wool quality can be kept relatively low. Textile dyeing operations are usually performed only once on the same substrate; a re-dyeing following a discharge of the first dyeing is done only in the case of failure of the first dyeing.

The result of a hair dyeing is dependent on the individual hair quality, which also includes the individual chemical history of the hair to be dyed. Repetitive dyeings and overdyeing are thus a practical reality in hair dyeing. Due to the varying individual hair properties standardization of hair quality is difficult and results of dyeing procedures will vary.

19.4 Chemical Principles of Dyestuff Binding

Due to the chemical similarities between the protein fibres of hair, wool and silk the concepts of dyestuff binding, respectively formation and fixation, are similar [7]. The main binding mechanisms are based on van der Waals forces, hydrogen bonds and also metal–complex formation. Due to the less drastic chemical conditions (pH, temperature and also time) low molecular weight dyes that easily diffuse into the hair are preferable. A wide variety of natural dyes that are able to fulfil these requirements has been used from ancient time to the present [9, 10].

The limited degree of dyestuff fixation also permits a certain mobility of these substances. As a result, in many cases natural dyes also include additional properties, e.g. astringent, anti-inflammatory and antibacterial activity [11].

19.5 Relevant Natural Dyes for Hair Dyeing

19.5.1 Naphthoquinone Dyes – Henna and Walnut

19.5.1.1 Henna, CI (Colour Index) Natural Orange 6

Henna is probably one of the oldest sources of natural dyes, known to many cultures for thousands of years [12]. For example, in 3000 BC henna was used for dyeing hair and fingernails and for ceremonial body art. An excellent overview of the historical and geographical aspects of henna is given by Cartwright-Jones [13].

Henna, e.g. from *Lawsonia inermis* L., contains 2-hydroxy-1,4-napthoquinone (CI 75480, lawson, Figure 19.1) as a dyestuff. In the plant lawson is supposed to be present as the glucosides hennosid A, B and C, from which lawson is released by enzymatic hydrolysis. A typical content of the leaves is 1 % w/w of lawson. As a minor component luteolin (3′,4′,5,7-tetrahydroxy-flavone) is extracted [14]. The usual application on hair is done as a kataplasma dyeing, in which a warm paste of powdered henna leaves is directly applied on the hair.

To prepare the paste henna is acidified with weak organic acids, e.g. from lemon juice or orange juice, and mixed to form a paste and rested for some hours. The small lawson molecule is able to penetrate skin and diffuse into hair. Mordants or fixatives are not required. A reaction with the keratin is supposed to occur, which yields the final brick-coloured dyestuff. During the fixation period of 1–3 hours the hair is covered to keep warm and avoid drying. Depending on the conditions used reddish to brown shades are obtained.

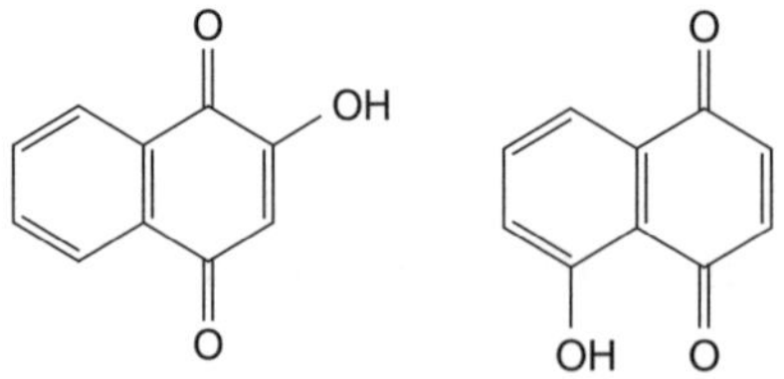

Figure 19.1 *Chemical structure of lawson (2-hydroxy-1,4-naphthoquinone) and juglon (5-hydroxy-1, 4-naphthoquinone)*

In practice, besides lawson further components are extracted from henna, which influence the shade of the final colour. Thus besides dyeing conditions time, pH and temperature also the rate of extraction of lawson determines the final outcome of the dyeing [15]. Due to the low toxicity potential of henna growing interest is observed in the cosmetic industry to reintroduce henna as a dyestuff in commercial hair dyeing products [16–23].

The application of henna in a mixture with other dyes is discussed in more detail in Chapter 6, which considers the natural dye. Besides the addition of oxidants [24, 25] and reducing agents [26, 27] polymers like collagen have also been proposed to improve the dyeing behaviour [28]. After-treatment of hair dyed with natural dyes with cationic dye compositions has been shown to improve colour stability in shampooing [29].

Reng is a traditional mixture of lawson with indigo. The addition of a blue indigo shade to the henna shifts the coppery colour of henna to brunette and dark brown. In the paste, part of the indigo is assumed to be present in reduced form, most probably released from the hydrolysis of the glycosidic precursor indican. The primarily formed indoxyl then couples to another indoxyl to form leuco-indigo, the reduced form of indigo, which is oxidized at the end of the dyeing step during the rinse. Thus the fresh dyeings with reng can exhibit a greenish colour from the reduced indigo, which disappears during the first days after dyeing. Consecutive dyeing first with henna and then indigo leaves will yield dark colours up to black.

19.5.1.2 Juglon, CI Natural Brown 7

The dyeing principle in extracts from green walnut shells and walnut leaves (*Juglans regia* L.) is 5-hydroxy-1,4-naphthoquinone (juglon, CI 75500). The thermodynamics of juglon to dye natural fibres including hair and synthetic fibres have been studied in detail. At high dyestuff concentrations dark shades are obtained, which is explained by aggregation of the small molecule inside the fibre [30]. Juglon is chemically rather unstable and tends to polymerize to brown pigments. Juglon can be extracted from young leaves, where it is present as 5-hydroxy-naphthohydrochinone-4-β-D-glycosid and from green walnut shells, which contain hydrojuglonglucoside, hydrojuglon, juglon and also tannins.

Juglon has been used for hair dyeing for a long time and there is growing interest in the use of natural dyes for hair coloration. The use of juglon has also been described in more recent literature [9, 15, 29, 31–33]. Usually the application of juglon is made as a direct dye,;thus no mordants or other fixatives are required. Juglon is a quite bioactive molecule which is widely used in pharmaceutical applications, but is also known to cause skin irritation and toxic effects [34–36].

19.5.2 Indigo

Indigo is a blue vat dye (CI Vat Blue 1) that can be obtained from natural sources or synthesis. The farming, extraction, analysis and application aspects are described in Chapters 7 and 8 of this book devoted to indigoid dyes. While in textile dyeing the blue shade is a desirable colour, in hair dyeing indigo is used to shift shades, e.g. from the coppery colour of henna to dark brown or black. For hair dyeing an application of the oxidized pigment form leads to deposition of pigment on the hair and dyeings with poor durability are obtained.

In the major part of the processes dyeing proceeds from the leuco-indigo form, which can be assumed to be present in precipitated protonated form as indigo white. For hair dyeing the pH of the reduced indigo bath has to be kept much lower than pH 11.5–12.5, which is mainly used in textile dye baths. Due to the lower pH in the hair dyeing paste, the protonated leuco-indigo will be present mainly in the precipitated dispersed form [37]. The protonation of leuco indigo as a function of pH and the species present in the solution, respectively precipitating from the liquid formulation, is given in Figure 19.2.

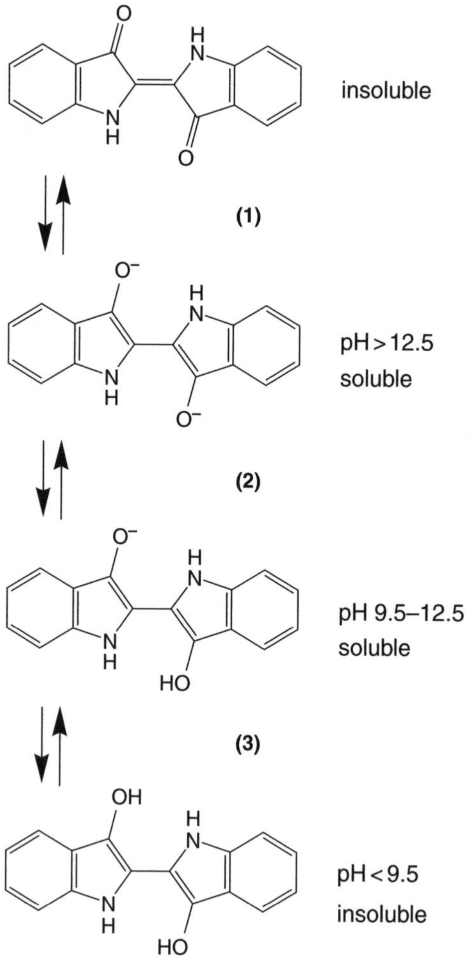

insoluble

(1)

pH > 12.5
soluble

(2)

pH 9.5–12.5
soluble

(3)

pH < 9.5
insoluble

Figure 19.2 *Reduction of indigo in alkaline solution (1) and protonation steps (2) and (3) of leuco-indigo as a function of pH [38]*

The reduced indigo can be obtained in different ways:

- reduction of indigo during the preparation of the paste used for hair dyeing, e.g. by addition of reducing agents [39];
- reduction of indigo by reactions with the plant material in the paste, e.g. observed in indigo/henna pastes (reng);
- use of indigo precursors, e.g. indican, which are hydrolysed in situ to release indoxyl, which couples to leuco-indigo [40].

Numerous examples for the application of indigo/indigo precursor-containing leaves are given in the literature [41–43]. Which reaction is actually the predominant pathway will be dependent on the chemical form in which the dye is present in the plant material used, e.g. as an indigo precursor or oxidized indigo dye.

Natural indigo in the form of ground indigo leaves can also be used in combinations with synthetic direct dyes [31, 44, 45].

19.5.3 Metal Complexes

An important procedure widely used in natural dyeing is based on the formation of metal complexes between the extracted natural dye and a mordant. In this case metal salts are used as mordants.

Metal complexes are quite common as synthetic dyes in wool and polyamide dyeing. However, for the dyeing of hair the selection of metal complexes has to be performed taking certain precautions. Contrary to textile dyeing, the hair can be bleached after a certain time to lighten the colour or to overdye using a different shade. The main reagent used for bleaching is hydrogen peroxide, which can be decomposed by catalytic effects due to the presence of Fe, Cu or Mn salts, a reaction that can cause serious damage to the hair [46].

Important representatives for natural dyes that can form metal complexes are shown in Figure 19.3. Haematoxylum (CI Natural Black 1) can be extracted from the wood of the logwood tree (*Haematoxylum campechianum* L) by hot water extraction [47]. Most

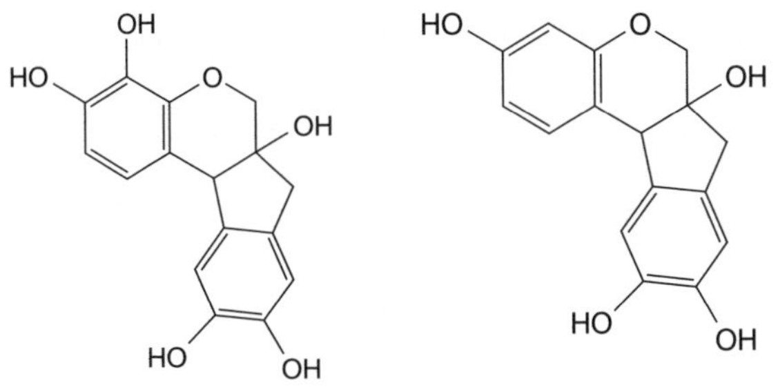

Figure 19.3 Structure of haematoxylum (3,4′,5′,7,8-pentahydroxy-2′3-methylen-neoflavan) and brasilin (3,4′,5′,7-tetrahydroxy-2′,3-methylen-neoflavan)

probably the complexation with the mordant occurs at the neighbouring phenolic hydroxyl groups. Brasilin (CI Natural Red 24) can be extracted from sappan wood, Brazilwood and other *Caesalpinia* species [47].

Hematoxylin or brasilin can be applied to dye hair in a sequential process that first applies a Cu-mordant on the hair and then forms the metal complex by treating the pre-mordanted hair with extracts from Brazilwood, campeche wood, sappan wood or pernambuk wood [48]. Fixation of the extracted haematoxylum or brasilin can be increased by pre-mordanting [49]. Further patents claim that the combination of henna with haematoxylum, haematin or brasilin alters the reddish shade of henna to dark brown [20, 50].

Another class of metal complex based hair dyes uses the formation of iron complexes with gallo-tannins and tannins. While the dyes exhibit only low substantivity for hair, the use of Fe or Cu salt mordants will increase their affinity by formation of complexes.

A two-step procedure for hair dyeing with catechu (CI Natural Brown 1) as the first step and dyestuff fixation by addition of Fe(II) salt has also been proposed in the literature [51]. A tannin–Fe-salt-based recipe has been proposed in Reference [52]. An example for gallotannin-based hair dyeings is given in Reference [53]. Also gallotanins from *Punica granatum* pericarp have been proposed in a mixture with wood from *Caesalpinia sappan* and Fe(II) or Fe(III) salt mordants [54].

The extraction of tannins from different barks and leaves, e.g. from *Juglans regia* or *Punica granatum*, has also been proposed for hair dyeing [55–58]. Tea has been used for hair dye compositions as a natural source for extraction of tannic substances [59].

Rastik is a traditional combination of metal-reaction dyes and plant dyes used in the Orient. The mixture containts pyrogallol from roasted gallnuts and iron and copper salts, and henna and cobalt salts can also be part of the mixture [60]. Such mixtures can be used to dye hair black, but as pyrogallol is forbidden in cosmetics, the use of such mixtures has to be considered critically.

The formation of metal complexes of natural dyes on iron, chromium or manganese containing hair has also been described in the literature [60, 61].

19.5.4 Metal Reaction Dyes

This technique could be considered to be in between the synthetic dyes and natural dyes. The general principle is to form insoluble metal salts, e.g. Ag, Cu and Ni sulfides, on the hair. In the case of silver the reduction yields the finely divided elemental form, which reacts with the cystine to form the corresponding sulfide [62]. As an example, the use of silver nitrate in formulations for dyeing of eyebrows and eyelashes is permitted with a concentration of up to 4 % [63]. The metal salt is mixed with a developer solution, e.g. containing pyrogallol, which then starts the reduction reaction. An example for the application of Ag-salt-based dyeing in combination with plant dyes such as henna is given in Reference [64].

Considerable physiological risks exist when a formulation contains toxic heavy metals, e.g. Ni, Bi or Pb. Thus such formulations have been banned in many countries [3].

19.5.5 Anthraquinoid Dyes

Extracts of the roots of madder (*Rubia tinctorum* L.) contain alizarin (1,2-dihydroxy-anthraquinone) as the main dyeing components. Other anthraquinoid components obtained from the roots are pseudopurpurin (1,2,4-trihydroxy-3-carboxy-anthraquinone), rubiadin (1,3-dihydroxy-2-methylanthraquinone) and lucidin (1,3-dihydroxy-2-hydroxymethyl-anthraquinone). As lucidine has been found to exhibit mutagenic properties the removal of lucidin and also rubiadin has to be considered before application [65, 66].

The use of madder extract has also been proposed as a natural dye in hair dyeing [67] and the use in mixture with other plant dyes has been described [23]. However, compared to henna or indigo the application of madder extracts for hair dyeing is less common.

19.6 Specialities

The use of a Cu–chlorophyll complex for dyeing of hair has been proposed in the literature [68]. The exchange of the centre ion magnesium by copper improves stability considerably, but could be interpreted as synthetic step. Another example for a semi-synthetic procedure is the oxidation of 3,4-dihydroxy-phenylalanine (DOPA) to yield 5,6-dihydroxyindole. The indole is a precursor for the formation of black eumelanins, which are known as natural hair pigments. Incorporation of sulfur-containing components, e.g. cystein, to modifiy the colour of the pigments has also been described [69].

19.7 Regulations

In the EU hair dye products are cosmetic products regulated by Council Directive 76/768/EEC of 27 July 1976 on the approximation of the laws of the Member States relating to cosmetic products (Cosmetic Directive 76/768/EEC). The substances that are used in formulations of hair dyes are regulated within the EU framework of the Cosmetics Directive. As a general strategy a positive list of hair dye substances that are considered safe for human health and allowed for use by the cosmetics industry will be established [70]. Natural dyes from plant material will thus have to be considered in order to be included in that regulatory structure before they can be used for hair dyeing in the future.

References

1. http://science.howstuffworks.com/hair-coloring.htm.
2. http://ec.europa.eu/enterprise/cosmetics/html/som_hairdyes.htm.
3. O. A. Neumüller, *Römpps Chemie-Lexikon*, Vol. 3, 8th edition, Franckh'sche Verlagshandlung, W. Keller & Co., Stuttgart, ISBN: 3-440-04513-7, 1983.
4. L. A. Lehninger, *Biochemistry*, Worth Publishers, Inc., New York, ISBN: 0-87901-047-9, 1976, p. 126.
5. H. Rath, *Lehrbuch der Textilchemie*, Springer-Verlag, Berlin, Germany, ISBN: 3-540-05587-8, 1972.
6. E. R. Trotman, *Dyeing and Chemical Technology of Textile Fibres*, 6th edition, Charles Griffin Company Ltd, ISBN 8 852642679, Bucks, England, 1984, p. 60.

7. H. Zollinger, *Color Chemistry, Syntheses, Properties, and Applications of Organic Dyes and Pigments*, 3rd edition, Chapters 8, 11 and 12, Wiley–VCH, Weinheim, Germany, ISBN 3-906390-23-3, 2003.
8. J. S. Anderson, The chemistry of hair colorants, *Journal of the Society of Dyers and Colorists*, **116**, 193–196 (2000).
9. M. Akram and W. Wolff, Granulation of powdered direct hair dyes, German Patent 4233874, 19940414, Priority DE 1992-4233874, 19921008 (1992).
10. U. Lenz, Natural dyestuffs in hair cosmetics, in Forum *Färberpflanzen*, Dornburg, Germany, 4–5 June 1997, Fachagentur Nachwachsende Rohstoffe, Guelzow, Germany, 1997, pp. 145–151.
11. A. C. Dweck, Natural ingredients for colouring and styling, *International Journal of Cosmetic Science*, **24**, 1–16 (2002).
12. C. Cartwright-Jones, *Developing Guidelines on Henna: A Geographical Approach*, Essay for Master's Degree, Kent State University, Kent, Ohio, www.hennapage.com, August 2006.
13. C. Cartwright-Jones, *Henna for Hair 'How-To' Henna*, TapDancing Lizard™ LLC, Stow, Ohio, 2006.
14. J. E. Park and K. W. Oh, Characterization of wool dyeing with henna, *Journal of the Korean Fiber Society*, **41**(5), 322–327 (2004).
15. K. C. James, S. P. Spanoudi and T. D. Turner, The absorption of lawsone and henna by bleached wool felt, *Journal of the Society of Cosmetic Chemists*, **37**(5), 359–367 (1986).
16. H. Hoeffkes and B. Bergmann, Low-dust powdered natural hair dyes, DE 19600225, 19970710, Application: DE 96-19600225, 19960105 (1996).
17. M. Akram, W. Wolff, S. Schlagenhoff, S. Schwartz and A. Kleen, Hair colorant and hair colorant slurry for dyeing human hair, European Patent Application 1997, 782845, 19970709, Application: EP 96-102822, 19960226 (1996).
18. K. Kirchmayr and H. Hoeffkes, Two-component natural hair dye containing plant pigment and thickener, PCT International Application 1996, WO 9618376, 19960620, Application: WO 95-EP4766, 19951204 (1995).
19. R. Bauer, Two-step hair dyeing, involving nonoxidative and natural dyes. German Patent DE 3829102, 19890810, Application: DE 88-3829102, 19880827 (1988).
20. G. Rosenbaum, J. Cotteret and J. F. Grollier, Hair dye composition containing a mixture of non-exhausted vegetable powder, a direct dye of a natural origin, and a diluent, US Patent US 5447538, 1995-09-05, Priority: FR 81-3946, 19810227 (1981).
21. J. K. Kim and I. H. Lee, Hair tint composition comprising hair dyeing agents which show excellent coloring properties under both acid and basic conditions, respectively, Korean Patent KR 2005015146, 20050221, Application: KR 2003-53830, 20030804 (2003).
22. N. G. Harishchandra and B. A. Singh, Hair colouring conditions composition method of preparation thereof, Indian Patent Application: IN 2005MU01267, 20070824, Application: IN 2005-MU1267, 20051007 (2005).
23. H. J. Kim, Composition for hair dyeing by using natural dyes which improves safety to human and method for dyeing, Korean Patent KR 2005117117, 20051214, Application: KR 2004-42326, 20040609 (2004).
24. T. Amano, Hair dye compositions containing natural dye, and method and kit for hair dyeing, Japan Patent JP 2006045180, 20060216, Application: JP 2004-256292, 20040806 (2004).
25. D. I. Kim and Y. S. Na, Hair dyes and bleaches containing henna and aloe, Korean Patent KR 2006129874, 20061218, Priority: KR 2005-50583, 20050613 (2005).
26. J. K. Kim and I. H. Lee, Hair tint composition comprising hair dyeing agents which show excellent coloring properties under both acid and basic conditions, respectively, Korean Patent KR 2005015146, 20050221, Application: KR 2003-53830, 20030804 (2003).
27. A. Togane, Hair dyeing mousses containing reductive liquid containing henna powder or extract, Japan Patent JP 2002348215, 20021204, Priority: JP 2001-193719, 20010523 (2001).
28. J. Kimura, Hair dyes containing lawson and collagen, hair dyeing using them, and *hair* washing after it, Japan Patent JP 2003261429, 20030916, Application: JP 2002-64027, 20020308 (2002).
29. A. Kleen and J. Terrier, Hair dye composition with plant dyes, European Patent Application: EP 1800652, 20070627, Application: EP 2006-24172, 20061122 (2006).

30. D. B. Gupta and M. L. Gulrajani, Studies on dyeing with natural dye juglone (5-hydroxy-1, 4-naphthoquinone), *Indian Journal of Fibre and Textile Research*, **18**(4), 202–206 (1993).
31. M. Akram, W. Wolff, S. Schlangehoff, S. Schwartz and A. Kleen, Hair colorant and hair colorant slurry for dyeing human hair, European Patent Application 1997: EP 782845, 19970709, Priority: DE 96-19600221, 19960105 (1996).
32. R. Bauer, Two-step hair dyeing, involving nonoxidative and natural dyes, German Patent DE 3829102, 19890810, Application: DE 88-3829102, 19880827 (1988).
33. A. Maric, Extraction of active dyes from juglans regia peel, Chinese Patent, CN 1912011, 20070214, Priority: CN2005-10044243, 20050811 (2005).
34. I. Neri, F. Bianchi, F. Giacomini and A. Patrizi, Acute irritant contact dermatitis due to Juglans regia, Contact Dermatitis, **5571**, 62–63 (2006).
35. T. K. Auyong, B. A. Westfall and R. I. Russel, Pharmacological aspects of juglone, *Toxicon*, **1**, 235–239 (1963).
36. J. J. Inbaraj and C. F. Chiqnell, Cytotoxic action of juglone and plumbagin: a mechanistic study using HaCaT keratinocytes, *Chemical Research in Toxicology*, **17/1**, 55–62 (2004).
37. J. N. Etters, Indigo dyeing of cotton denim yarn: correlating theory with practice. *Journal of the Society of Dyers and Colourists*, **109**, 251–255 (1993).
38. Southeastern Section, AATCC, Effect of dyebath pH on color yield in indigo dyeing of cotton denim yarn, *Textile Chemist and Colorist*, **12**, 25–31 (1989).
39. T. Kripp, B. Grasser and U. Lenz, Hair dyes containing herbal dyes and reducing agents and method for use, German Patent DE 10360204, 20050728, Application: DE 2003-10360204, 20031220 (2003).
40. Y. Yamoaka, Hair dye composed of indigo plant preparation containing indican and another plant preparation containing β-glucosidase, German Patent (2003) DE 10131385, Application: DE 2001-10131385, 20010625 (2001).
41. G. S. Jin, Hair dyes containing henna and indigo, Korean Patent KR 2006042467, 20060515, Application: KR 2004-91051, 20041109 (2004).
42. T. H. Gwakand H. Y. Lee, Natural plant hair-dye composition which overcomes color limitation of henna dye, maintains hair health, and does not stimulate skin of head and hair-dyeing method using same, Korean Patent KR 2005026759, 20050316, Application: KR 2003-62382, 20030906 (2003).
43. X. Cai, Manufacture of traditional chinese medicine hair dye, Chinese patent CN 101062004, 20071031, Application: CN 2007-10028148, 20070524 (2007).
44. A. Sallwey, M. Schmitt and U. Lenz, Indigo-based compositions for coloring human hair, German Patent DE 19736553, 19981210, Application: DE 97-19736553, 19970822 (1997).
45. M. Akram, W. Wolff, S. Schlagenhoff, S. Schwartz and A. Kleen, Hair colorant and hair colorant slurry for dyeing human hair, European Patent Application: EP 782845, 19970709, Application: EP 96-102822, 19960226 (1996).
46. H. Rosenberger, K. H. Riedel and H. Lasen, Das Neue Friseurhandbuch, Verlag Dr Max Gehlen, Bad Homburg vor der Höhe, Germany, ISBN: 3-85207-403-7, 1981.
47. H. Schweppe, *Handbuch der Naturfarbstoffe*, ecomed Verlagsgesellschaft, Landsberg and Lech, Germany, ISBN: 3-933203-46-5, 1993.
48. J. F. Grollier, Hair dyes containing copper compound and brazilin or hematoxylin, German Patent DE 3715226, 19871112, Application: DE 87-3715226, 19870507 (1987).
49. Y. Murakami, Hair-dyeing/styling compositions containing natural dyes and inorganic white pigments, Japanese Patent JP 3309221, 20020729, Application: JP 2001-130679, 20010427 (2001).
50. G. Machida, Natural products for hair dyes, Japan Patent JP 2005060293, 20050310, Application: JP 2003-291941, 20030812 (2003).
51. X. Hou, H. Gao, Q. Liu and H. Wu, Method for preparing natural hair dyes from plant pigment catechu and its application method, Chinese Patent CN 101164528, 200880423, Patent Application: CN 1013-3373, 20071019 (2007).
52. B. Wan, Hair dyes containing tannings from *Eclipta* and iron components in buckwheat flour, Chinese Patent CN 101099717, 20080109, Patent Application: CN 1005-2242, 20070522 (2007).
53. Z. Zhou, R. Shi and L. Zhang, Black hair dye containing *Galla chinensis* extract and its preparation method, Chinese Patent CN 1989940, 20070704, Patent Application: CN 1011-2379, 20051230 (2005).

54. R. Shi, Z. Zhou, L. Zhang and G. Song, Hair dyes containing extracts from natural plants of *Punica* and *Caesalpinia*, Chinese Patent CN 1923164, 20070307, Patent Application: CN 1002-9176, 20050829 (2005).
55. A. Maric, Extraction of active dyes from *Rumex madaio* root and leaf, Chinese Patent CN 1912010, 20070214, Patent Application: CN 1004-4242, 20050811 (2005).
56. A. Maric, Extraction of active dyes from *Juglans regia* peel, Chinese Patent CN1912011, 20070214, Patent Application CN 1004-4243, 20050811 (2005).
57. A. Maric, Extraction of active dyes from *Punica granatum* peel and leaf, Chinese Patent CN 1912012, 20070214, Patent Application CN 1004-4244, 20050811 (2005).
58. A. Maric, Extraction of active dyes from *Platanus acerifolia* bark, Chinese Patent CN 1912009, 20070214, Patent Application CN1004-4241, 20050811 (2005).
59. H. Hoeffkes and B. Bergmann, Plant-derived hair dye compositions containing natural emulsifiers, German Patent DE 19548291, 19970626, Application: DE 95-19548291, 19951222 (1995).
60. K. Kirchmayr and H. Hoeffkes, Two-component natural hair dye containing plant pigment and thickener, PCT International Application: WO 9618376, Application: WO 95-EP4766, 19951204 (1995).
61. F. Du and S. F. Yang, Research and development on nontoxic natural hair colorant,*Ranliao Yu Ranse*, **41**(2), 91–92 (2004).
62. *Ullmanns Enzyklopädie der Technischen Chemie*, 3rd edition, Vol. 10, Urban und Schwarzenberg, München-Berlin, Germany, 1958, p. 736.
63. European Comission, Council Directive (76/768/EEC) of 27 July 1976, on the approximation of the laws of the Member States relating to cosmetic products, 1976L0768–19.09.2007, http://ec.europa.eu/enterprise/cosmetics/html/consolidated_dir.htm.
64. K. Shiraki and D. Setoshima, Hair-dyeing cosmetic compositions containing silver salts, organic acid salts, and plant dyes. Japanese Patent JP2006176486, 20060706, Application: JP 2004-382789, 20041220 (2004).
65. G. C. H. Derksen, *Red, Redder, Madder – Analysis and Isolation of Anthraquinones from Madder Roots* (*Rubia tinctorum*), Dissertation, Wangeningen University, The Netherlands, ISBN: 90-5808-462-0, 2001.
66. Y. Kawasaki, Y. Goda and K. Yoshihira. The mutagenic constitutents of *Rubia tinctorum*, *Chemical and Pharmaceutical Bulletin (Tokyo)*, **40**, 1504–1509 (1992).
67. R. Belle, Hair dye composition prepared from *Rubiaceae* extracts, French Patent FR 2483226, 19811204, Application: FR 80-11880, 19800528 (1980).
68. K. Mizumaki, Hair dyes containing natural pigments, Japanese JP 02160716, 19900620, Application: JP 88-314704, 19881213 (1988).
69. K. C. Brown, E. Marlowe, G. Prota and G. Wenke, A novel natural-based hair coloring process, *Journal of the Society of Cosmetic Chemists*, **48**(3), 133–140 (1997).
70. http://ec.europe.eu/enterprise/cosmetics/html/cosm_hairdyes_strategy.htm.

Part V
Environmental

20

Environmental Aspects and Sustainability

Erika Ganglberger

20.1 Introduction

Over the past few years natural textiles have been developed out of a growing awareness of the environmental, health-related and social problems caused by the conventional production of textiles. Meanwhile there are a great many textiles, especially clothes, that are advertised using terms like 'natural', 'eco' or 'bio', but the labelling refers to rather different quality standards. Products may be labelled as 'sustainable' on condition that the raw materials come from organic farming and that manufacturing processes up to the finished garment comply with ecologically and socially acceptable production methods. Until now the composition of dyestuff used for colouring is not part of any labelling.

Admittedly nowadays natural colorants are hardly used in the textile sector. Although for most of the thousands of years in which dyeing has been used by humans to decorate clothing or fabrics for other uses, the primary source of dye has been natural, with the dyes extracted from animals or plants. However, in the last 150 years, man has produced artificial dyes to achieve a broader range of colours and to render the dyes more stable to washing and general use. Nowadays natural dyes are only used on a small scale, mostly applied in handicrafts, and retailed in market niches.

As long as natural dyes reach only a low production volume, environmental aspects were neglected. For a revival of natural colorants the environmental benefits have to exceed the present situation defined by synthetic materials. From the sustainability point of view it is a desirable development to use natural colorants to a greater extent. An intensified use of renewable raw materials represents a substantial contribution to sustainable development

Handbook of Natural Colorants Edited by Thomas Bechtold and Rita Mussak

as fossil resources are conserved, the environmental impact is reduced throughout the whole life cycle and agricultural land is preserved. Nevertheless, mitigation of environmental impacts along the whole product life cycle has to be considered to gain environmental sustainability: The total expenditure concerning renewable resources, energy consumption, waste and waste water must be considered as important aspects of the process.

In Austria several industry-oriented studies in series [1–4] worked on dye plants, aiming to establish the fundamentals of dye plant use in the textile industry. The following text is based on the findings from these studies.

The starting point of the project was to gather all sectoral requirements for supply and application of the dye plants: criteria for the supply of raw material and requirements of the trades and industries that process the material. Due to the diversity and heterogeneity of natural colorants, it is difficult to make universal statements. In fact, it is of great importance to set limits concerning the plant dyes observed and the dyeing procedure applied. In the following, criteria and requirements within the diverse stages of the life cycle and some nonallocated scopes are mapped using the main focus of ecology and sustainability. Finally it is a challenge to know how to deal with these contradicting interests and how to find a feasible answer.

20.2 Supply of Plant Material

20.2.1 Cultivation of Dye Plants

Inherently cultivation of dye plants contributes to the conservation of agricultural land and to the creation and safeguarding of jobs with regional value added. However, this presumption implies two important prerequisites: sustainable farming of dyeing plants that are indigenous or adequate for local conditions according soil and climate.

20.2.1.1 Organic Farming

Advances in biochemistry (nitrogen fertilizer) and engineering (the internal combustion engine) in the early 20th century led to profound changes in farming. Meanwhile conventional agriculture is criticized for depleting natural resources, particularly fossil fuels and fresh water, and seriously polluting the soil, water and air [5, 6]. Organic farming is a form of agriculture that avoids or largely excludes the use of synthetic fertilizers and pesticides, plant growth regulators and livestock feed additives. As far as possible, organic farmers rely on crop rotation, crop residues, animal manures and mechanical cultivation to maintain soil productivity and tilth to supply plant nutrients and to control weeds, insects and other pests.

According to the international organic farming organization IFOAM [7]: 'The role of organic agriculture, whether in farming, processing, distribution, or consumption, is to sustain and enhance the health of ecosystems and organisms from the smallest in the soil to human beings.'

20.2.1.2 Sustainable Farming

Although it is common to equate organic farming with sustainable agriculture, the two are not synonymous. Sustainability in agriculture is a broader concept, with considerations on many levels, such as environmental health, economic profitability, and social and economic equity. With regard to organic farming methods, one goal of sustainability would be to approach as closely as possible a balance between what is taken out of the soil with what is returned to it, without relying on outside inputs. Organic farming today is a small part of the agricultural landscape, with a relatively minor impact on the environment. As the size of organic farms continues to increase, a new set of large-scale considerations will eventually have to be tackled. Large organic farms that rely on machinery and automation, and purchased inputs, will have similar sustainability issues to those that large conventional farms have today [8].

20.2.1.3 Native Species

In Europe cultivation of dye plants shows a fine tradition. Since the Middle Ages the cultivation of dye plants and the further processing and dyeing had been an important economic factor. Depending on the climate, various plants served as sources for natural dyes. However, with the invention of synthetic dyes at the end of the 19th century, natural dyes lost their economic significance. Presently cultivation of dye plants is mainly restricted to botanical gardens and museums for experiments and demonstration.

As there are about 1100 plant species [9] that can be used for dyeing, it is important to screen and select species that are fit for modern sustainable cultivation techniques (see Table 20.1). In Germany (Thüringer Landesanstalt für Landwirtschaft), 108 dye plant species were assessed on their suitability for modern cultivation systems, on yields and on dyeing quality. Madder (*Rubia tinctorum*), weld (*Reseda luteola*), Canadian golden rod (*Solidago canadensis*), dyer's chamomile (*Anthemis tinctoria*) and dyer's knotwed (*Polygonum tinctorium*) were considered to play a decisive role in future dye plant cultivation and processing [10].

Table 20.1 *Agricultural criteria for sustainable cultivation of dye plants*

- Demands according to the site, including parameters of soil and climate
- Requirements of organic farming (no or low need for fertilizer, pesticides, etc.)
- Available and verified cultivation instructions for conventional farming methods with the possibility to adopt the recommended techniques necessary for organic farming
- Cropping can be done with 'common' equipment
- Dye plants with a high content of dyestuff

20.2.2 Residual Materials and By-products

In the Austrian project series, implementation of natural dyes in the textile industry was the main goal. Therefore the requirements stated by the modern dyehouse and textile industry were mapped to overcome the lack of linkage between the supply side (mainly agriculture) and the demand side (textile industry) (see Table 20.2).

Table 20.2 Requirements stated by the modern dyehouse and textile industry

- High-quality shades of red, blue and yellow
- Composition of a single dyestuff class that facilitates the possibility for varying colour depth and shade by dyestuff mixtures and mixed mordants
- Existing machinery is also used for dyeing with natural dyestuff
- Good fastness properties on diverse substrates
- Economic benefit

As modern dyehouses demand a single dyestuff class including red, yellow and blue, a good set of local dye plants was indispensable. Therefore it was important to search for further resources providing colouring matter. Screening relevant literature, tea, wine grapes, onions, nutshells, bark and other 'unusual' sources were referred to as dyestuff. Thinking about the origins of these materials, by-products and residual material from food and wood processing seemed promising.

Shifting industrial processes from linear (open-loop) systems, in which resource (and capital) investments move through the system to become waste, to a closed-loop system, where waste becomes input for new processes, is one of the fundamental ideas of modern industrial ecology. Using by-products and residual material for dyeing textiles supports this idea as material life cycles become extended with an additional life stage.

Additionally by-products and residual material are very welcome according to economical aspects, as these secondary or incidental products must not be produced primarily. Therefore the supply of such plant material should be cheaper. Due to the fact that a basic condition for large-scale use of natural dyes is an acceptable cost of the dyestuff, a mixture of agricultural cultivated dye plants and dyeing material derived from residues and by-products might offer acceptable costs for the entire colour portfolio.

20.2.2.1 Criteria for Residual Material

Whereas for a sustainable supply of dye plants organic farming of indigenous plants is a prerequisite, these criteria are not appropriate for residual material and by-products. Accumulation of the material takes place independently from the use as dyeing matter. Using this material for dyeing includes an ecological benefit in any case, namely that the local 'waste' is not a burden. Nevertheless, it is important to set clear limits according to the ecology. For the supply of plant material derived from by-products and residues the most important criterion is the local availability of the material to avoid long transportation.

20.2.3 Selection Process for a Sustainable Supply of Plant Material

In the Austrian project series implementation of the presented criteria and the provision of the requirements stated by the demand side led to the selection process shown in Figure 20.1

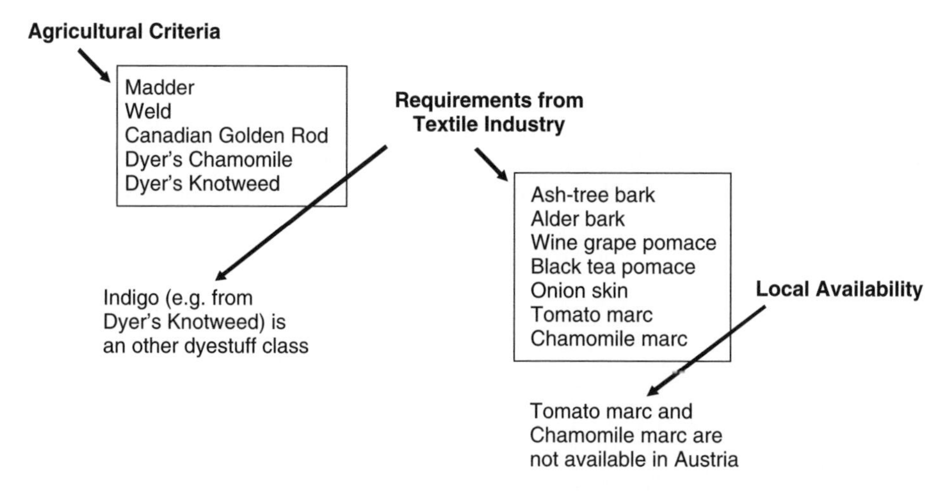

Figure 20.1 Selection of plant material derived from agricultural cultivation and food and wood processing

From all known dye plants the species that achieved the agricultural criteria for sustainable cultivation in Austria were chosen to prove if they fitted the requirements stated by the modern dyehouse and textile industry. It was found that madder, weld, Canadian golden rod and dyer's chamomile met the demands. As indigo belongs to another dyestuff class, dye plants delivering indigo (e.g. dyer's knotweed) were excluded due to the requirement for a single dyestuff class.

The main criteria for inclusion of residual material and by-products containing colouring matter was whether they were available locally. As tomato marc and chamomile marc are not available in Austria, these by-products were excluded.

20.3 Processing to Dyestuff

A prerequisite for an industrial use of plant dyes is the all-season availability of the dyestuff. However, dye plants vary in their time of harvest. Also, by-products and residual material do not necessarily accumulate all year; especially in food and wood processing there are seasonal varieties. In addition, the dyestuff-containing plant material ranges from perishable to durable. Therefore it is essential to define a trade product that is seasonally independent, available all the year, durable and standardized.

In principle there are two possibilities for processing the plant material:

- extraction of the plant material and creation of a colour batch or
- dried and shredded dye plant.

In both cases the dyestuff originates from plant material and must be extracted. The difference in processing concerns the time of extraction and the trade product. In the first case the trade product is a liquid or solid concentrate of extracted dyestuff. Extraction and concentration of the natural dye from the plant material is done within the production of the trade product. In the other case the trade product is stabilized plant material. Thereby extraction takes place at the dyehouse. An important parameter of the trade product is a standardized amount of colouring matter within the given unit of plant material. Depending on the kind of preservation, different procedural steps are required (see Table 20.3).

Table 20.3 *Procedural steps towards the trade product* [3]

Colour batch	*Plant powder bag*
• Extraction of plant material	• Stabilization/drying
• Filtration	• Shredder
• Evaporation	
• Standardization	
• Packaging	• Standardization
	• Packaging
• Dilution (production of dye bath)	• Extraction (extract = dye bath)
	• Filtration

From an ecological point of view there are two crucial points:

- procedure-added energy consumption and
- choice of the extraction solvent.

20.3.1 Energy Consumption

In the Austrian project series the extraction of plant material was limited to a simple hot water extraction. According to the energy consumption a rough estimation based on the following assumptions was carried out:

- Liquor ratio of the plant extraction 1:10 (1 kg of plant material:10 L liquor)
- Production of 100 kg of dyed substrate requires 100 kg of colouring plant material
- Liquor ratio of the dye bath 1:10 (1 kg of substrate:10 L liquor)
- Evaporation of 1 kg of water consumes about 0.3 kW h
- Specific heat of water: 4.18 kJ / kg °C
- Energy content of fuel: 40 000 kJ / kg

The estimation of the energy assumption shown in Table 20.4 clearly points out that it is very advantageous to use plant powder bags. Thereby the aqueous extraction is performed near to the dyeing apparatus, directly followed by the dyeing process. The hot extract can be used as a dye bath, thus saving energy input for evaporating and up-heating.

Table 20.4 *Rough estimation of the procedure-added energy consumption [11]*

Colour batch	*Plant powder bag*
Extraction of plant material: 100 kg plant material 1000 L water Extraction at 95 °C Heat quantity: **314 MJ**	Stabilization/drying: Energy consumption depending on the humidity ratio of the material Potentially energy-intensive process Energy input: **180 MJ**
Evaporation: Membrane technique (0.3 kW h/kg) Volume reduction to 10 kg Energy input: **961 MJ**	Comminution: Energy consumption depending on the character of the material Consumption of electricity for mechanical cutting
Transport: About 50 kg fuel is necessary to transport 30 ton over a distance of 100 km. That corresponds to an energy input of 2000 MJ Proportion for 10 kg: **0.7 MJ**	Transport: About 50 kg fuel is necessary to transport 30 ton over a distance of 100 km. That corresponds to an energy input of 2000 MJ Proportion for 50 kg: **3.5 MJ**
	Extraction of plant material on site: 100 kg plant material 1000 L water Extraction at 95 °C Heat quantity: **314 MJ**
Production of dye bath: Dilution of batch 1:10 Heating to 95 °C Heat quantity for 1000 L: **314 MJ**	Extract = dye bath: If the hot extract is promptly used, no additional energy input is required
Total: **1589.7 MJ**	Total: **497.5 MJ**

Note. The energy use within the dyeing procedure to keep the temperature at 95 °C was completely neglected.

In the case of the colour batch, the substantial amount of energy required to raise the temperature of a large quantity of water is needed twice – while extracting the plant material and to heat the dye bath. Moreover, concentration via evaporation is energy consuming. The main advantage of using colour batches is that it has the same handling as synthetic dyestuff. Moreover, colour batches show increased concentration accompanying a reduction in volume and weight. Especially for increasing transportation distances, this fact might be advantageous, but the comparison shows that transportation makes a quite small contribution to the overall situation.

If the plant material, e.g. pomace and marc, includes a very high water content, a distinction could be drawn due to the requirement of an energy-intensive stabilization. A small dyestuff content might also shift the situation.

20.3.2 Water Consumption

Within the Austrian project series hot water was chosen as the environmentally harmless extraction solvent. According to the water consumption, plant powder bags are also advantageous as the water used for extraction is directly used for the dyeing procedure.

Table 20.5 Water consumption within the procedure

Colour batch	Plant powder bag
Extraction of plant material: 1000 L water	Extraction of plant material on site: 1000 L water
Production of dye bath: 1000 L water	Extract = dye bath: No additional water input required
Total: **2000 L water**	Total: **1000 L water**

Note. The estimation is limited to the dyeing step. Water consumption for the subsequent washing procedure was completely neglected.

20.4 Application of Colouring Matter

The predominance of synthetic dyes hindered the continuous development and adaption of natural dyeing to the changing requirements of modern dyehouses. Therefore nowadays there is a considerable gap between knowledge about natural dyes and the demands of commercial dyeing processes. Over the past few years, several research activities on plant dyes used in textile dyeing were performed and encouraging results regarding colour depth, shade and fastness properties were described [12–15]. However, the applied dyeing procedures are mainly two-bath dyeings including a separated mordanting step to improve colour fastness properties. The numerous variations of plant sources and dyeing processes proposed in the literature are rather difficult to handle in modern dyehouses as rapid changes in trends and fashion demand a basic database describing practicable applications of natural dyes on different substrates.

Bechtold *et al.* investigated an environmentally sound one-bath dyeing process where the mordant was added directly into the dye bath [16]. The process is shown in Figure 20.2. Easy handling, acceptable time consumption and good reproducibility are thereby met. From an ecological point of view, one-bath dyeing is also desirable due to water consumption and the lack of energy consumption for intermediate drying steps that might be necessary. To lower the ecological impact, it is possible to use mordant baths and dye baths more than once, but storing, stability and all kinds of contamination must be checked regularly.

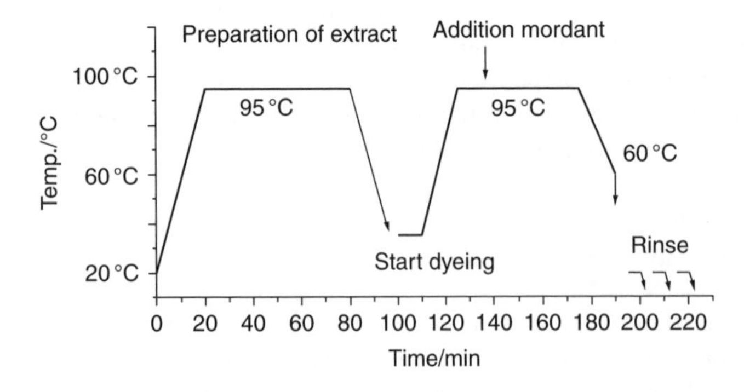

Figure 20.2 Temperature–time diagram of the one-bath dyeing process with extracted natural dyes

Furthermore, Bechtold limited the use of mordants to iron or alum mordants as Co, Sn or Cr salts cause problems with the effluents released from the dyeing process. Thus waste water limits defined for heavy metal concentrations may be affected.

20.4.1 Dyeing Procedure

The dye lot is prepared by extraction of plant material with boiling water. After separation of the insoluble residue, the textile fibres are immersed into the dyestuff containing the dye lot. In some cases a mordant (iron sulfate/chloride or alum) is added after 15 minutes at 95 °C. The dyeing temperature is held at 95 °C for a further 35 min. After cooling the dye bath to 60 °C, the unfixed dyestuff is removed by rinsing the textile with cold water. The resulting dyeing is characterized by colour depth, shade and fastness properties.

20.5 Considerations Concerning the Life Cycle

Industrial waste management policies have been changed from 'end-of-pipe' approaches, where pollution control demands further treatment of process wastes, to 'cleaner production'. In 'cleaner production', modifications based on input–output analysis are performed to minimize process emissions [17]. Reduction, recycle and re-use of waste are the main focuses for consideration. The 'zero emission' approach produces a further step towards efficiency: waste products of one industry/sector become value-added inputs for another. The term 'zero emission' refers to an idealized goal without measurable impact on the environment.

A dyeing process with a natural dyestuff based on locally available resources fits the 'zero emission' approach. Especially using by-products and residual material from food and wood processing for dyeing purposes contributes to this idea. If the overall process meets the intentions of 'zero emission', a careful check is necessary because within the life cycle diverse resources are consumed and several emissions are released.

In the following, the succession of life cycle stages including the most important inputs and outputs are outlined. An overview about all inputs and outputs within the dyeing procedure is shown in Figure 20.3.

20.5.1 Raw Material

Depending on the source of raw material there are essential differences with regard to the ecological impact. Using by-products and residual material from food and wood processing corresponds to the zero emission idea and so the ecological impact can be neglected.

In contrast, cultivation of dye plants includes a high ecological impact. As some bright colours are not available from residual material, it is necessary to complete the essential colour range with dye plants. Cultivation of dye plants may contribute to biodiversity and agricultural value-added. However, this aspect is controversial because agricultural areas used for food production are used and increasing competition can mark up the price for edible products [18].

Anyway, agricultural production includes crop growing, harvest, irrigation and manuring. Thus energy, water and chemicals (fertilizer, pesticides) are the main inputs. In the case of organic or sustainable farming, the use of chemicals is largely excluded.

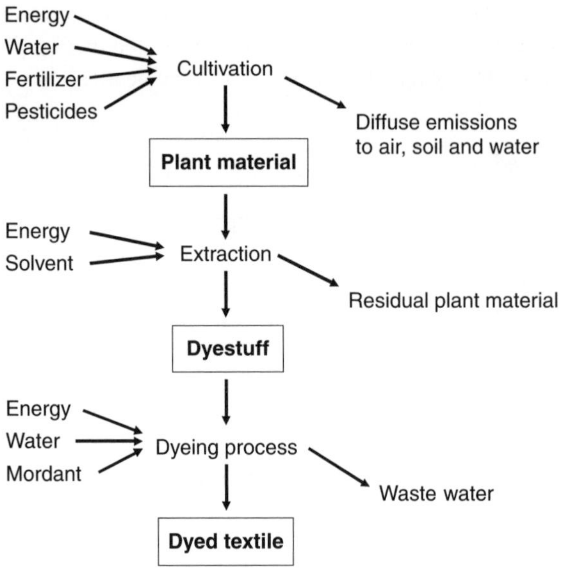

Figure 20.3 *Inputs and outputs within the life cycle*

Cultivation of dye plants is also connected with undesired outputs: mainly diffuse emissions like exhaust emissions of the machinery, and emissions to soil, water and air depending on the use of fertilizers and pesticides.

20.5.2 Processing of Raw Material

Independent of the plant material's origin, subsequently the material must be processed to form a stable and standardized colouring product. As mentioned before, there are two possibilities: creation of a colour batch or fabrication of plant powder bags. In both cases the formation of the trade product is linked with a considerable energy input, although the plant powder bags perform better.

20.5.3 Extraction of Dyestuff

Regardless of the chosen trade product, the dyestuff must be extracted from the plant material anyway. The crucial ecological impact at this stage is the selection of solvent for the extraction procedure.

If organic solvents are chosen, potentially an increased dyestuff yield can be achieved but at the same time there are several problems in series: organic solvents cause substantial additional costs for purchasing and significantly higher expenses for waste water treatment and waste processing. Of course, it is possible to recycle the solvent, but depending on the corresponding pick-up value of the plant material there is a considerable amount of solvent remaining in the waste [19].

In the case of hot water extraction, the waste water load is limited to the biological content degraded in conventional aerobic processes and the insoluble residue is biodegradable and might be used for animal feed or soil conditioning.

20.5.4 Dyeing Procedure

Within the application of the dyes there are two environmental remarkable variables: decision on one-bath or two-bath dyeing and selection of the mordant.

Regarding one-bath or two-bath dyeing, one-bath dyeing is clearly advantageous for water consumption and the amount of waste water. In addition, the energy input might be significantly higher due to additional intermediate drying steps.

A mordant is a substance used to fix dyes on fabrics. Using a mordant may have different effects on the shade obtained after dyeing and also on the fastness properties. Therefore mordants are used to broaden the range of shades and to improve colour fastness properties.

Mordants such as tannin, alum and certain salts of aluminium, chromium, copper, iron and tin should be considered very carefully due to the health aspect as well as the legal waste water limits. Depending on the ions involved there are different limit values for the concentration and the load [11].

Especially heavy metals are often detrimental to organisms by forming poisonous soluble compounds. Toxicity evolves when metals imitate the action of an essential element in the body, interfering with the metabolic process to cause illness. Metals in an oxidation state abnormal to the body may also become toxic: chromium(III) is an essential trace element, but chromium(VI) is a carcinogen. Toxicity is always a function of solubility. Insoluble compounds as well as the metallic forms often exhibit negligible toxicity. A common characteristic of toxic metals is the chronic nature of their toxicity. Toxic metals can bioaccumulate in the body and in the food chain. Decontamination for toxic metals is rather difficult as they cannot be destroyed. Toxic metals must be made insoluble, collected or avoided [10].

20.5.5 Transport

Transport is a major use of energy and transport burns most of the world's petroleum. Transport is responsible for air pollution as particulates (soot), carbon monoxide, nitrous oxides and greenhouse gas emissions are created. Other environmental impacts of transport systems include traffic congestion, and toxic runoff from roads and parking lots can also pollute water supplies and aquatic ecosystems. Therefore a sustainable economy implies short haulage distances and closed regional material cycles [20].

For natural dyeing closed regional material cycles are realized by use of traditional dye plants and locally available colouring by-products and residual material.

Another important factor is the transport of the colouring matter. Depending on the processing of the plant material a colour batch or dry plant powder is the colouring trade product. As the dyestuff content of plant material is quite low, the weight of dry plant powder is relatively high as measured by the colouring potency. In the case of long transportation distances it might be ecologically reasonable to favour a concentrated colour batch in spite of the high energy input within production. Increased dyestuff concentration and reduction in volume and weight are definitely advantageous with regard to transportation.

In any case, it is important to give attention to transport distances. Starting with the selection of the plant material, distances must be considered for all transportations up to application of the dyestuff in the dyehouse.

20.6 Conclusion

To gain a real sustainable effect application must be practically, applied otherwise it won't be implemented and there is no effect at all. Therefore, implementation in the industrial procedure is very important in order to cause a sensible sustainable effect.

Due to the diversity and heterogeneity of dye plants it is not possible to make universal statements about questions that must be answered before starting implementation on a large scale. To obtain tangible outcomes, it is of great importance to set limits concerning the plant dyes observed and the dyeing procedure applied.

In this chapter, criteria used to achieve an environmentally oriented implementation of natural dyes were explicated and the supply of raw material, the processing to dyestuff and the dyeing process per se were defined.

20.6.1 Dealing with Sustainability

To adapt the use of vegetable dyeing materials in textile colouring according to the technical, economic and ecological requirements of the 21st century is always a challenge in decision making. In modern ecological economics the problem of conflicting values are well known, as efficiency, equity and sustainability must be considered at the same time. Multicriteria methods [21] were found to be useful tools to support decision making about complex situations such as those concerned with sustainable development issues. These methods deal with conflicts in a structured and transparent way.

In the Austrian project series a structured and consistent approach took place (see Figure 20.4). In the following a short review on the proceedings is given.

Regarding the use of plant dyes in the textile industry the goal of achieving sustainable development was split into subgoals of occasionally contradicting fulfilment. Thus, approaching one goal can sometimes create a deviation from another goal. The rather complex problem context was addressed by defining a superordinate goal and by means of discussion and dialogue. A continuous decision-making process accompanied the entire project series in the form of an elimination procedure covering three subsequent categories:

- Technical feasibility
- Operational feasibility
- Economical feasibility

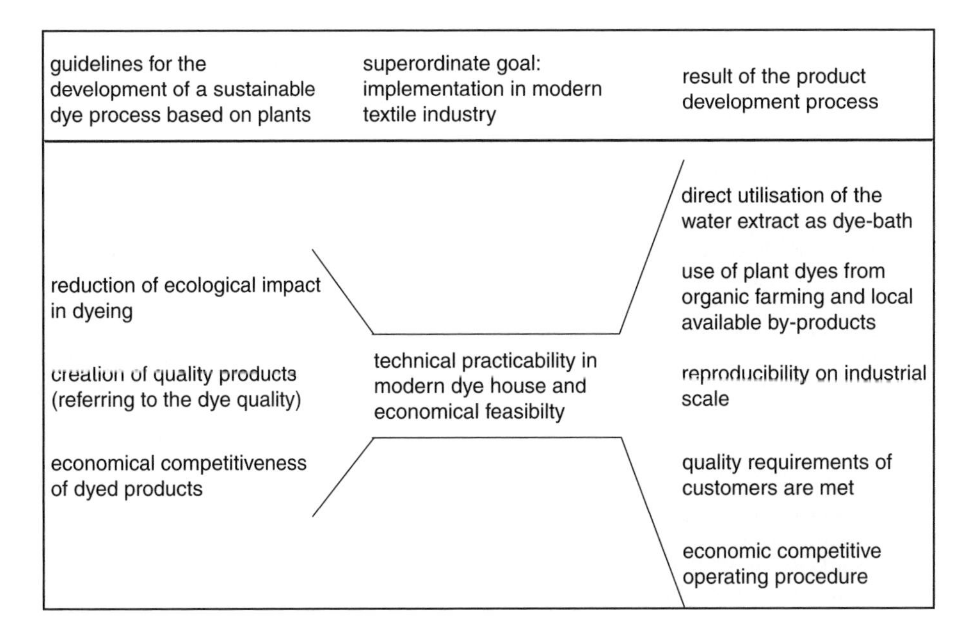

Figure 20.4 *'Austrian' approach to design a sustainable dyeing procedure with natural dyes*

Sustainability is a characteristic of a process or state that can be maintained at a certain level indefinitely. For planet Earth, it is thus to provide the best outcomes for the human and natural environments, both now and into the indefinite future. One of the most often cited definitions of sustainability is the one created by the Brundtland Commission. The Commission defined sustainable development as development that 'meets the needs of the present without compromising the ability of future generations to meet their own needs [22]'. Due to the complexity of ecological, economical and social interests, sustainable design of practicable processes implies a challenge.

References

1. S. Geissler, E. Ganglberger, T. Bechtold, S. Sandberg, O. Schütz, A. Hartl and R. Reiterer, *Produktion von Farbstoffliefernden Pflanzen in Österreich und ihre Nutzung in der Textilindustrie*, Berichte aus Energie- und Umweltforschung 2/2001, Bundesministerium für Verkehr, Innovation und Technologie, Wien, 2001.
2. S. Geissler, E. Ganglberger, T. Bechtold, A. Mahmut, A. Hartl and O. Schütz, *FARB und STOFF – Sustainable Development durch neue Kooperationen und Prozesse*, Berichte aus Energie- und Umweltforschung 25/2003, Bundesministerium für Verkehr, Innovation und Technologie, Wien, 2003.
3. B. Rappl, Ch. Pladerer, M. Meissner, N. Prauhart, G. Roiszer-Bezan, B. Friedrich, E. Egger-Rollig, E. Ganglberger, S. Geissler, T. Bechtold, R. Mussak, A. Mahmud-Ali, A. Grimm, G. Jasch and E. Freudenthaler, *TRADEMARKFarb&Stoff. Von der Idee zum marktfähigen Handelsprodukt: Pflanzenfarben für die Textilindustrie*, Berichte aus Energie- und Umweltforschung 21/2006, Bundesministerium für Verkehr, Innovation und Technologie, Wien, 2006.

4. A. Wenisch and C. Pladerer, *Risikominimierung entlang der Wertschöpfungskette vom pflanzlichen Rohstoff bis zum Farbstoff*, Berichte aus Energie- und Umweltforschung 08/2007, Bundesministerium für Verkehr, Innovation und Technologie, Wien, 2007.
5. C. A. Edwards, The importance of integration in sustainable agricultural systems, *Agriculture, Ecosystems and Environment*, **27**, 25–35 (1989).
6. P. Mäder, A. Fließbach, D. Dubois, L. Gunst, P. Fried and U. Niggli, Soil fertility and biodiversity in organic farming, *Science*, **296**, 1694–1697 (2002).
7. http://www.ifoam.org/about_ifoam/principles/index.html.
8. A. Trewavas, A critical assessment of organic farming-and-food assertions with particular respect to the UK and the potential environmental benefits of no-till agriculture, *Crop Protection*, **23**, 757–781 (2005).
9. R. Hofmann, *Färbepflanzen*, Dissertation at the University of Vienna, Austria, 1989.
10. A. Biertümpel, A. Wurl, A. Vetter and R. Bochmann, Anbau von Färberpflanzen zur Gewinnung von Farbstoffextrakten für die Applikation auf Textilmaterial, *Berichte über Landwirtschaft*, **78**, 408–420 (2000).
11. S. Geissler, R. Mussak, T. Bechtold, S. Klug and T. Vogel-Lahner, Nachwachsend = nachhaltig? Eine Analyse am Beispiel pflanzlicher Textilfärbung, *GAIA*, **15**, 44–53 (2006).
12. H. T. Deo and B. K. Desai, Dyeing cotton and jute with tea as a natural dye, *Journal of the Society of Dyers and Colourists*, **115**, 224–227 (1999).
13. N. Bhattacharya, B.A. Doshi and A.S. Sahasrabudhe, Dyeing jute with natural dyes, *American Dyestuff Reporter*, **87**, 26–29 (1998).
14. K. Nishida and K. Kobayashi, Dyeing properties of natural dyes from vegetable sources (Part II), *American Dyestuff Reporter*, **81**, 26 (1992).
15. U. Brückner, S. Struckmeier, J. H. Dittrich and R. D. Reumann, Zur Echtheit von Färbungen mit ausgewählten Naturfarbstoffen auf Synthesefasergeweben, *Textilveredelung* **32**, 112–115 (1997).
16. T. Bechtold, A. Turcanu, E. Ganglberger and S. Geissler, Natural dyes in modern textile dyehouses – how to combine experiences of two centuries to meet the demands of the future? *Journal of Cleaner Production*, **11**, 499–509 (2003).
17. A. Yacooub and J. Fresner, *Half is Enough – An Introduction to Cleaner Production*, LCPC Press, Beirut, Lebanon, 2006.
18. K. Buchgraber, Fachzeitschrift des Bundesministeriums für Land- und Forstwirtschaft, Umwelt und Wasserwirtschaft, Landlicher Raum, Jahrgang 2007, http://www.laendlicher-raum.at/article/articleview/53550/1/10406.
19. T. Bechtold, A. Mahmud-Ali, E. Ganglberger and S. Geissler, Efficient processing of raw material defines the ecological position of natural dyes in textile production, *International Journal of Environment and Waste Management*, **2,** 215–232 (2008).
20. J. Fuglestvedt, T. Berntsen, G. Myhre, K. Rypdal and R. BieldtvedtSkeie, Climate forcing from the transport sectors, *PNAS*, **105**, 454–458 (2008).
21. G. Munda, *Multicriteria Evaluation in a Fuzz Environment. Theory and Application in Ecological Economics*, Physical Verlag, Heidelberg, 1995.
22. United Nations, Report of the World Commission on Environment and Development, General Assembly Resolution 42/187, 11 December 1987.

21

Economic Aspects of Natural Dyes

Susanne Geissler

21.1 Introduction

Natural colorants from plants and minerals were used before the development of oil-based dyes. After a long break, natural dyes have regained ground. Today, natural colorants are used in various products such as coatings in the building sector, in children's handicrafts, leather products, and textiles [1].

In sales, customers are typically addressed through ecologically oriented marketing, the message being: buy green, pay more and help the environment. The environment benefits from the production of green products, and so does people's health, thanks to clean air, water and soil.

However, is it always necessary to market green products as environmentally friendly and healthy or is it possible to apply environmentally friendly production procedures as part of the corporate quality standards? Depending on the implementation concept, natural colorants have the potential to create local jobs, to reduce the environmental impact and, as a result, to give companies a competitive edge. Therefore, from the sustainability point of view, it might also be a desirable development to use natural colorants beyond the so-called green markets. However, in order to achieve this, economic success is a basic condition, requiring a competitive cost structure as the fundamental prerequisite. In addition, a detailed analysis of consumer expectations and marketing positions helps to identify the potential of natural colorants.

In this chapter, the textile sector is subject to analysis. An Austrian project series on the development of natural colorants for the textile industry has demonstrated that the first steps have been made to lift natural dyes from the 'green', and so-called 'handicraft' or 'home-made' niche, to a more industrial scale. The following text is based on the scientific studies carried out as part of the Austrian project series [2–5].

Handbook of Natural Colorants Edited by Thomas Bechtold and Rita Mussak

What are the requirements for using natural colorants successfully, not only in green markets, but also beyond? To answer this question, the following issues are tackled in this chapter:

- Basic conditions for the industrial use of natural colorants – demands that must be met
- Challenges for the industrial use of natural colorants – the conditions set by industry
- Consumer expectations – what do consumers want and how to find out
- Production costs of natural colorants products – primary production versus residue utilization
- Closed-loop economy: towards the zero-emission and zero-waste society – natural colorants and their potential to link up with regional closed-loop economies
- Conclusions: aspects influencing market development for natural colorants

Cost data presented in this paper were generated in exemplary studies dealing with the natural dyeing of textiles, with the aim of providing basic information about economic aspects. This information should serve as the basis for more detailed economic analyses to come, for example, in the course of preparing the implementation of a natural colorant production plant. Although the following analysis is based on a case study that cannot be generalized, results provide valuable information and orientation concerning the economics of natural dyes for textiles.

21.2 Basic Requirements for the Industrial Use of Natural Colorants

Companies interested in applying natural colorants need to ascertain the realization of the following preconditions before giving the green light to go ahead to designers, marketers and producers:

- compliance with requested colour fastness properties;
- feasible cost of process engineering;
- acceptable cost of natural colorants and dyeing processes;
- competitive cost of naturally dyed products;
- security of colorant supply.

Precondition 1: compliance with requested colour fastness properties. When purchasing textiles, customers expect a certain level of quality in terms of colour change through light exposure, washing, sweating and abrasion. Therefore it is not possible to go below these essential colour fastness properties.

Precondition 2: feasible cost of process engineering. Application of natural dyes might require adaptations in process engineering. The dyeing process should be analysed to find out whether it is possible to make use of the existing equipment or whether the installation of additional machinery is necessary. Costs of process engineering do not only refer to constructions and machines, but also include costs related with safety regulations, and waste as well as waste water management issues. Depending on the type of natural dyeing process, engineering costs will be low if the process is designed in a simple manner. However, in terms of technical performance and quality control it will be more costly in any case if two process engineering streams have to be maintained. At present, two process engineering streams (natural colorants and synthetic colorants) are required, because specific colours such as pure black cannot be provided with natural colorants in connection with an environmentally friendly dyeing process. However, additional costs are justified because natural colorants broaden the colour range and thus also enlarge opportunities for product innovations.

Precondition 3: acceptable costs of natural colorants and dyeing processes. Natural colorants can be produced from primary agricultural products such as madder root, from herbs collected in natural parks and from residues being left over in other production processes, such as timber bark or onion peels [6, 7]. Production costs of dyestuff differ depending on the source of the raw materials. The overall cost structure of plant dyes differs depending on the type of commercially available formulations. Both the distribution of the standardized raw material packed in bags and dyestuff concentrated in solid or liquid form is feasible. In both cases the colouring matter has to be extracted from the plant material, but the time and location of the extraction process varies. In the latter, a 'plant dyestuff supplier' should extract the dye, minimize the volume or produce a colorant powder. Therefore packaging size would be minimized and handling becomes similar to conventional dyestuff handling. The sale of standardized plant material bags needs modification of the process. Due to the similarities of boiling tea, the extraction takes place in the dyehouse right before dyeing. If natural colorants are used in the form of dry and finely grinded plant powder, dyestuff extraction can be carried out during the dyeing process, thus saving energy costs for evaporating and up-heating, as opposed to liquid colorants. Therefore cost assessment has to refer to both the dyestuff product as well as the dyeing process as a whole. With regard to the dyestuff product, cost assessment has to be carried out for the whole range of requested colours. Colours that are produced from residues might be cheaper than colours produced from plants that were cultivated exclusively for dyestuff production; however, the total amount of costs for the entire colour portfolio might be acceptable compared to synthetic ones.

Precondition 4: competitive cost of naturally dyed products. Whether increased costs of natural colorants are acceptable or not depends on the share of expenditures for dyestuff in comparison with the revenue generated when selling the dyed product. Depending on the type of dyed product (consumer product or semi-manufactured product), acceptable costs for natural dyestuffs vary: For example, regarding the price of dyed wool, the share of dyestuff cost is rather high, while the share of dyestuff cost is low regarding the price at which premium clothes are sold. As a consequence, a wool-producing company providing a semi-manufactured product will hardly be able to accept increased dyestuff costs, while end-consumer-oriented companies selling their own products on the premium clothes market might be prepared to pay more.

Precondition 5: security of natural colorant supply. For industrial processes it is a prerequisite that materials are supplied within a defined period of time and with a guaranteed quality. In fact, natural colorants have to be traded the same way as synthetic colorants: a shade card has to be presented displaying the available colours and showing the dyeing result that can be achieved with the described dyeing process. Apart from the procurement procedure, clients expect complimentary service in addition to the shade cards: engineers need to be available to assist in the implementation of the natural dyeing process.

In the past few years natural colorants based on dry powder and the respective dyeing process were developed for industrial use in Austria. Natural dyestuff and the dyeing process were developed in close cooperation with an Austrian company interested in the application of these dyes [2, 3]. So far, project activities succeeded in generating positive results concerning preconditions 1, 2 and 3. Whether precondition 4 is met or not depends on the company's strategy (business-to-business orientation or consumer orientation). Precondition 5 is practically solved, and preparations for setting up a dye production and trading company have started in Austria [4].

21.3 Challenges for the Industrial Use of Natural Colorants

If the basic prerequisites mentioned above are positively concluded, details will have to be assessed. There are two important aspects that are connected with the design and marketing strategy of the company planning to use natural colorants: the range of available colours and the colour appearance in terms of reproducibility.

21.3.1 Quality of Raw Material and Reproducibility of Colours

Dyes from plants are natural products with all their positive and seemingly negative properties. On the one hand, selected natural colorants are produced by nature without hazardous by-products, which are generated in production processes of organic chemistry [7]. On the other hand, naturally grown plants never deliver exactly the same quality of dyes as they did the year before: even if it is the same species, even race in terms of plant taxonomy, the composition of dyes will differ, depending on the location (soil, nutrient conditions, climate and weather conditions) where the plant was growing. This applies to naturally bred and harvested raw materials. It is definitely possible to produce exactly the same quality of raw material under controlled laboratory conditions with a defined nutrient solution, temperature, humidity and lighting. This procedure is well known from food production; however, this type of production process is questionable from the sustainability perspective. Although the quality of raw material is not stable, it is necessary to guarantee a defined reproducibility of colours. This can be achieved through measures such as mixing dyestuff and varying the colorant's concentration of the dyeing bath.

21.3.2 Range of Available Colours

Depending on the requirements concerning the origin of dyestuff and the dyeing process itself, the full range of colours might not yet be available. In Austria, natural dyes and the respective dyeing process were developed together as one, to find an optimized solution in terms of resource use, energy and water consumption, as well as water pollution. This optimized dyeing process is mainly based on solid, finely ground plant powder and water, which is used to extract the dyestuff and to prepare the dye bath at the same time. Due to environmental reasons the use of mordant was limited to alum and iron mordant. In fact, there are environmental and economical benefits from doing without any mordant altogether. However, alum and iron mordant were accepted for their improved fastness properties and broadening the colour range.

The process saves energy, compared to the dyeing process with liquid natural colorant products, and the residues can be returned to agriculture for composting or gasification for biogas production. From the environmental point of view it is a condition that dyestuff is available locally and that a long transportation distance is avoided. In this case, environmental demands impose restrictions on the availability of colours: with the procedure described above it is not possible yet to produce black-coloured textiles, nor green nor blue, other than indigo.

It depends on the marketing and design strategy of a company whether the whole range of colours and defined tones are required, or whether limitations and variations are accepted in order to go for specific and unique copies.

Table 21.1 *Available range of colours (solid natural colorant) in 2006. Reproduced by permission of Bundesministerium für Verkehr, Innovation und Technologie, Wien, 2007, from Reference [4]*

Colour	*Fastness*	*Raw material*	*Botanical systematic*	*Source of origin*	*Price*[a]
Yellow	Good	Onion peels	*Allium cepa L.*	Food industry	O
Yellow	Good	Mayweed	*Anthemis tinctoria L.*	Agriculture	€
Yellow	Good	Mignonette	*Reseda luteola L.*	Agriculture	€
Yellow	Fair	Canada golden rod	*Solidago canadensis L.*	Agriculture	€
Brown	Good	Ash-tree bark	*Fraxinus excelsior L.*	Timber industry	O
Brown	Good	Alder bark	*Alnus glutinosa L.*	Timber industry	O
Blue	Fair	Wine grape pomace	*Vitis vinifera L.*	Wine farmer	O
Grey	Good	Black tea pomace	*Camellia sinensis L.*	Food industry	O
Red	Good	Madder root	*Rubia tinctorum L.*	Agriculture	€

[a] Rough estimation: O, low price of raw material; €, high price of raw material.

21.4 Consumer Expectations

The economic success of new developments, technologies and products depends on whether consumer expectations are met. Market research supports decision makers to answer the question whether consumer demand for the product will justify market introduction. In market research, experts use methods from different scientific disciplines such as psychology, social sciences and statistics in order to analyse product concepts, product properties, brand images, trends, consumer needs, consumer attitudes and consumer behaviour. Potential customers as well as the acceptance or rejection of innovations can be analysed by means of market research. However, the potential of market research is clearly limited when it comes to analysing radical innovations, or inventions, respectively: it is impossible to ask consumers about their attitude towards something they are not familiar with. However, as regards natural colorants the innovation is an incremental one, and in this field market research can provide valuable information about potential markets.

Market research is divided into two sections: qualitative and quantitative market research [8]. While quantitative market research delivers exact information, e.g. which consumer group buys which amount of a specific product, qualitative market research outlines motives, attitudes, rational and emotional reasons and thus provides information on how to interpret the results of quantitative market research. Qualitative market research contributes to a better understanding of consumer needs and arguments (consumer insights) and, finally, to a better understanding of the reasons why a product is purchased or not. This is the basis for the enhanced consumer orientation that has increasingly gained importance in product development during the past few years [9–11].

Product values are not limited to the functional properties of a product or a technology, but include additional advantages of the product as well. With respect to the buying decision, the emotional component clearly dominates. Thanks to these emotional components, the consumer is not only interested in buying the product because of its functional properties but will have the desire to buy it due to the emotional benefit. The emotional component is the reason why the price/performance ratio is perceived differently: a basis commodity turns to a luxury product (see, for example, developments in the automobile industry and fashion industry: compared to the basic service transport, cars provide status and adventure, and these benefits are much more desirable than the basic service; trendy dresses have become much more than just clothes: the service of protection is taking a back seat to the additional benefits of status and self-confidence). Many products today work in this way and hardly sell without emotional surplus value [9]. These findings point out the problems of products that are marketed based on the motives ‘sustainable/environmentally friendly’. These motives aim at ethical values and a sense of responsibility. Results of market research show that the target group which responds to this type of message is rather small, whereas marketing concepts that enhance personal emotional advantages are widely accepted. Consumer interviews demonstrate that in most cases aspects of sustainable development are considered ‘very important’, but that the purchase decision is made based on the following motives: price, brand, the individually perceived product value and the emotional surplus value. If the strategy sets out to address a large segment of consumers, the emphasis on the emotional surplus value is crucial in marketing sustainable products [11].

21.4.1 Market Research for Naturally Dyed Products

At the University of Applied Sciences FHWN Wieselburg, market research activities were conducted to investigate promising marketing routes for naturally dyed products. The market research was divided into two sections: qualitative and quantitative market research. Quantitative market research (e.g. method interviews) delivers exact information, but does not deliver any information about the motives to explain consumer behaviour and attitude. Therefore also qualitative market research (e.g. a methods focus group discussion, qualitative interviews) is required, outlining motives and attitudes and thus providing valuable information on how to interpret the results of quantitative market research. Qualitative market research contributes to a better understanding of the reasons why a product is purchased or not [12].

In this project, methods applied included focus group discussions and qualitative personal interviews, based on a questionnaire guideline. The objective was to learn more about consumer behaviour, rather than creating statistics on purchase decisions. Two focus group discussions served to find out about consumers' attitudes: potential consumers with a broad range of different attitudes (represented by age, gender, income and values) were invited to discuss several product concepts (various descriptions of the product) and real pilot products; in this case, they were women's tights dyed with natural colorants. An experienced moderator was in charge of the focus group discussions. Based on the findings a specific questionnaire for conducting qualitative personal interviews was developed in order to investigate important aspects in more detail [13].

The results of market research are as follows [12–14]:

- On the whole, consumers accept naturally dyed textiles.
- Consumers associate dyes from plants with 'health' and 'colours'.
- Natural dyes make a difference; they are special and exclusive.
- Natural dyes represent nature, the cycle of nature, and they take part in the cycle of nature.
- Natural dyes represent ecological awareness, responsibility and fairness.

Lack of acceptance concerns the following aspects:

- Consumers believe that dyes from plants are less stable and that naturally dyed textiles will lose colour during wearing and washing.
- Consumers are concerned about natural dyes causing allergic reactions.

These results represent social and personal ideas and values, rational and emotional motives. Arguments against natural dyes are characterized by a stronger personal concern, due to the close relationship between the textile and the consumer's body. Even if these concerns are not justified, they do exist, and producers of naturally dyed textiles have to consider them in their marketing concepts in order to ensure the success of products on the market.

Figures 21.1 and 21.2 show the interview results of two basic questions (open questions without given answers to choose) in detail: 'What do you associate with natural dyes from plants in connection with textiles?' and 'Why are you interested in a naturally dyed product?' Of the interviewed consumers 38 % associated natural dyes with 'colours', while 30 % associated it with 'health', while only 6.5 % referred to the environment. Of the interviewed consumers 54 % responded that 'health' is the most important motive for being interested in a naturally dyed product, while 18 % were interested for environmental reasons.

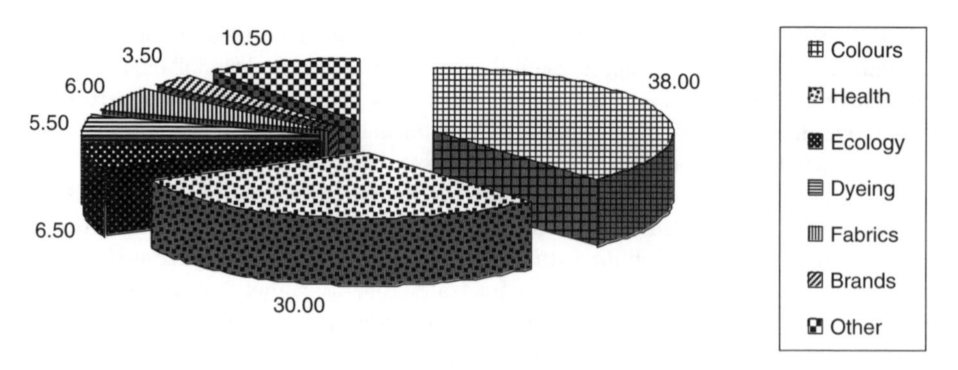

Figure 21.1 *'What do you associate with natural dyes from plants in connection with textiles?' (Number of interviewed persons: 49, numbers in the figure in percentage) [12, 13]*

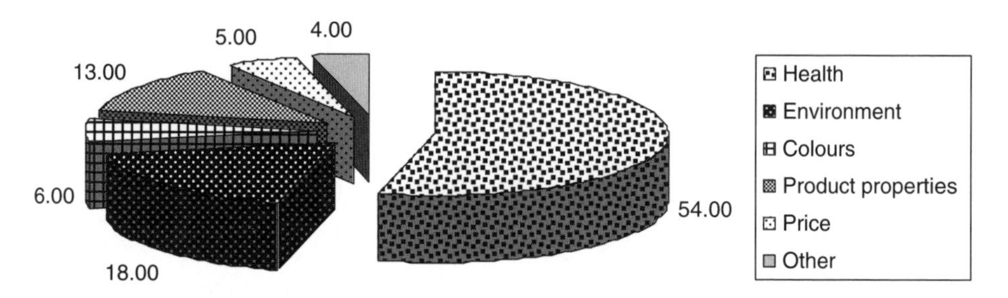

Figure 21.2 *'Why are you interested in a naturally dyed product?' (Number of interviewed persons: 49, numbers in the figure in percentage) [12, 13]*

Analyses showed that the acceptance of marketing concepts which enhance personal emotional advantages is higher than that of concepts which focus on responsibility and ethic-related factors. This might lead to the conclusion that the factor of personal advantage sells best. However, analyses also showed that consumers do accept sustainability as a buying motive, as long as no personal disadvantage is connected with the product. For example, 'Naturally dyed textiles are good for the environment/the ecosystem/the producer…, but at the same time the product is less attractive/more expensive/less reliable … for me.'

The widespread notion that 'sustainable products are less attractive and are only bought by tree huggers and freaks', is based on a purely emotional evaluation and on a historical background. Nevertheless, this association exists, and producers of sustainable products are well advised to take this into account when developing marketing concepts [12, 14].

The market research carried out at the University of Applied Sciences FHWN Wieselburg, resulted in the following suggestions for marketing naturally dyed textiles [12, 13]:

- Emphasize the origin of the plants the dyes are produced of. (Message: with this product I capture nature's colours and can take them home with me.)

- Emphasize the exceptional position of naturally dyed textiles. (Message: this product is special and exclusive and it helps me to differentiate myself from others.)
- Emphasize the natural origin and strong context with nature. (Message: I participate in the cycles of nature, I can take a piece of nature home and I 'm good to myself.)

These suggestions support the ongoing trends of individualism, wellness and exclusiveness. Consumers are generally prepared to pay more for these products.

21.5 Production Costs of Natural Colorant Products

In Austria, interviews with the demand side showed that the textile industry clearly requires comparable prices. Prices of natural colorants have to be within the range of synthetic dyes, although results from market research indicate that consumers might accept higher prices for naturally dyed products. Based on this information, studies were carried out to determine production costs of natural colorants from agricultural raw materials (primary production) and from residues from other production processes.

21.5.1 Cost Categories

Production of natural colorants in the form of a dry plant powder and in liquid form includes several cost categories, which are displayed in Table 21.2.

Table 21.2 *Cost categories of natural colorant production according to type of colorant. Reproduced by permission of Bundesministerium für Verkehr, Innovation und Technologie, from Reference [2]*

Solid colorant (plant powder)		*Liquid colorant*	
Agricultural primary production	*Residues from other processes*	*Agricultural primary production*	*Residues from other processes*
Agricultural production		Agricultural production	
Harvesting	Collection	Harvesting	Collection
Drying	Drying	Drying	Drying
Cutting and grinding	Cutting and grinding	Cutting and grinding	Cutting and grinding
		Extraction and evaporating	Extraction and evaporating
Standardization and quality control	Standardization and quality control	Standardization and quality control	Standardization and quality control
Packaging	Packaging	Packaging	Packaging

Compared with the production of a liquid colorant product, the cost category 'extraction and evaporation' does not exist when producing a solid plant powder. Therefore, in terms of production cost, liquid colorant products tend to be more expensive than solid ones. On the other hand, handling a liquid concentrate might be more comfortable for customers and worldwide distribution might be easier. Which type of natural colorant to produce, the solid one or the liquid one or both, depends on the business plan of the company producing and distributing the natural colouring products [4].

21.5.2 Aspects Influencing Production Costs

Production costs of solid colorants (powdered plant material) are composed of three main factors: production costs of the raw material, drying costs and standardization costs. These cost categories are influenced by specific aspects, which are described in the following paragraphs.

Costs for production of raw material. In the case of residue utilization from other production processes, raw material is purchased at a low cost. If there is already a competitive utilization (as there is, for example, for timber bark and berry pomace), prices for them can be taken as a guideline. If there is no competitive utilization yet, costs only occur for collecting the material. If agricultural primary production is needed, the decrease in production costs is connected with the increase in yield, the reason being that there is a specific share of fixed costs independent from the yield, such as costs for machines and fuel for tillage. Thus, as yield is increasing, the share of these costs per kg yield is decreasing. The increase in yield from 1 up to 2 ton/ha leads to the most significant decrease in production costs per kg yield.

Costs for drying process. Drying costs occur depending on the amount of material that has to be processed. Furthermore, drying costs are connected with the required treatment and drying temperature (depending on the valuable substances, which might be of use as by-products, e.g. for the pharmaceutical industry). Another important factor is the energy carrier used for the drying process. In 2006 and 2007, increasing oil prices caused the shutdown of oil-fired drying plants in Austria due to lack of profit [15].

Costs for standardization of natural colorant and quality control. These costs do not depend on the amount of material that needs to be standardized. The most important factor is the time needed to find the correct mixture of dry powder that produces a defined colour with the dyeing process described in the shade card used by the sales agents. Shade cards display a specific colour, which can be achieved when using the described colorant with the described dyeing process. Clients order colorants based on the shade cards they receive from the sales agent of the company selling colorants. It does not influence costs in any way whether standardization has to be carried out for 100 kg of natural colorant or for 2000 tons. The larger the amount of natural colorant product, the lower the share of standardization costs. In our research project, costs for standardizing different amounts of a natural colorant were calculated based on a defined procedure: costs for standardizing 1 ton of the natural colorant product was 10 % of costs, which amount for standardizing 100 kg natural colorant product [2].

Table 21.3 gives an overview of the production costs of raw materials from residues and from agricultural primary production. Cost categories considered are agricultural production and harvesting (if primary production material is processed) or collection (if organic residues are processed), drying, cutting and grinding, standardization and quality control, and packaging. Costs do not include the process engineering support in the textile companies. Calculations for agricultural primary production are based on data for mechanical production, average labour costs and a yield of 2 tons/ha (dry matter). These basic data were collected in the course of our research project [2].

Table 21.3 Overview of production costs for standardized natural colorants (€/kg). Reproduced by permission of Bundesministerium für Verkehr, Innovation und Technologie, from Reference [2]

Raw material (origin)	Raw material production	Treatment and drying (drying: 30% thermal solar energy)	Cutting, grinding/ packaging	Standardized colorant (1000 kg)	Standardized colorant (500 kg)	Standardized colorant (100 kg)
Mignonette (agriculture)	2.31	1.28	0.26	4.08	4.31	6.13
Madder root (agriculture)	4.88	2.45	0.21	7.77	8.00	9.82
Berry pomace (food industry, wet)	0.57	2.29	0.13	3.22	3.45	5.27
Tea pomace (food industry, wet)	0.57	2.29	0.13	3.22	3.45	5.27
Vegetable (food industry, wet)	0.57	2.29	0.13	3.22	3.45	5.27
Timber bark (decoration material)	0.17	1.21	0.24	1.85	2.08	3.90
Timber bark (produced manually)	0.57	1.21	0.24	2.25	2.48	4.30
Timber bark (heating plant)	0.12	1.21	0.24	1.80	2.03	3.85
Onion peels (food industry)	0.07	0	0.57	0.87	1.10	2.92
Nutshell peels (agricultural product[a])	6.27	0	0.24	6.74	6.97	8.79
Berry pomace (food industry, dry)	1.57	0	0.13	1.93	2.16	3.98

[a] The use for food production is hindered.

Costs displayed in the table are based on the following assumptions:

- For mignonette and madder root, the costs of raw material production are based on a yield of 2 tons/ha of dry matter. If yield is only 1 ton/ha, costs for raw material production will be estimated at 4.62 €/kg (mignonette) and 9.76 €/kg (madder root).
- The costs for treatment and drying are based on the assumption that a solar thermal plant is utilized, thus saving 30 % of energy costs for fuel (fossil oil).
- The three columns on the right-hand side contain the total costs for the standardized product, taking into account different standardization costs and depending on the amount of material that undergoes standardization. Costs for standardization amount to 2.28 €/kg (total amount of material is 100 kg), 0.46 €/kg (total amount of material is 500 kg), 0.23 €/kg (total amount of material is 1000 kg) and 0.11 €/kg (total amount of material is 2000 kg).

A rough comparison with the prices of a plant trading company (Alfred Galke GmbH, Germany) showed that in most cases production costs for standardized colorants were far below catalogue prices for nonstandardized plant products. Only costs of madder root exceed catalogue prices. [2]

This rough comparison leads to the conclusion that natural colorants may be introduced to the market at competitive prices, compared with companies trading similar products. However, the basic question is: Is it possible to offer natural colorants at prices that are in the range of prices of synthetic dyes?

21.5.3 Prices of Synthetic Dyes – How Much Are Textile Companies Prepared to Pay for Dyes?

An important factor influencing the textile industry's willingness to pay is the share of dye costs compared with the profit gained from selling the dyed product. If the share of dye costs is very low, the company might be ready to accept a higher price for natural colorants (e.g. an Austrian company who produces knitted fabric in the premium segment). If, however, the share of dye costs is already very high the tolerance towards increased prices of dyestuff will be low (e.g. an Austrian wool producer).

One of the Austrian projects focused on the price structure according to different market segments/products. Therefore an evaluation of required colour shades, colour depth and an estimated amount of plant dyes was performed. On the basis of two representative Austrian companies, this will be discussed in the following.

Company A is a producer of knitted polyamide, specializing in dyeing of knitted polyamide, for the project focusing on polyamide tights and polyamide stockings. Company B is a producer of wool yarn and a wool yarn dyeing company, here focusing on the dyeing of wool yarn. The price level is given for Austrian production costs. Table 21.4 presents the results of a survey on dyestuff costs carried out at Company B.

Table 21.4 Dyestuff costs of Company B for three exemplary colour gradations. Reproduced by permission of Bundesministerium für Verkehr, Innovation und Technologie, from Reference [2]

Colour gradation	% Dyestuff of total weight	Dyestuff costs: average dyestuff price is 25 €/kg
Light colours	0.3 % dyestuff of total weight	0.075 €/kg wool
Medium colours	2 % dyestuff of total weight	0.5 €/kg wool
Dark colours	8 % dyestuff of total weight	2.0 €/kg wool

At Company A, which produces knitted fabric in the premium segment, research showed high dyestuff costs. For tights, the upper limit was determined at € 7 per kg of textile. However, due to the fact that prices for black colour are far below this limit, and due to the fact that black is used for most products, prices of natural colorants should not exceed € 5 per kg of dyed product [2].

While it would be very difficult to compete with the synthetic dyestuff price at Company B (€ 2.0 per kg of wool), competition with synthetic dyestuff prices at Company A seems to be realistic under the following conditions:

- The minimum yield of raw materials from agricultural primary production is 2 tons/ha of dry matter.
- The drying process has to be optimized concerning energy costs and depending on valuable substances that could be of use as a by-product for the pharmaceutical or cosmetic industry.
- At least 1 ton of standardized natural colorant has to be sold per year.

21.5.4 Acceptable Production Costs through a Mixed Portfolio (Agricultural Primary Production and Residues from Other Production Processes)

Production costs of dyes from agricultural production are high compared with production costs for dyes from residues. However, agricultural primary production provides colours that cannot be produced from residues such as yellow colour. Therefore, production costs are not assessed in an isolated way, but altogether: the percentage of specific colours was determined and compared to the total colour portfolio of Company A and Company B, respectively (see Table 21.5). Based on this investigation, the estimated demand for specific plant raw material is presented in Table 21.5. The combination of different plant sources allows the total amount of expenses to be calculated for the natural colorants, and as an average number the amount might be within the range of synthetic colorants.

Table 21.5 *Demand for natural colorant according to type of colour. Reproduced by permission of Bundesministerium für Verkehr, Innovation and Technologie, from Reference [2]*

Raw material	*Company A*		*Company B*	
	% plant	*% colour compared with total colour portfolio*	*% plant*	*% colour compared with total colour portfolio*
Wine (*Vitis vinifera* L.)	20	Blue/purple	22	Blue/purple
Raspberry (*Rubus idaeus* L.)	10	Grey	6	Grey
Black tea (*Camellia sinensis* L.)	10	Total share brown	6	Total share brown
Nutshell peels (*Juglans regia* L.)	8	colours: 39 %	6	colours: 24 %
Ash-tree bark (*Fraxinus excelsior* L.)	11		6	
Alder bark (*Alnus glutinosa* L.)	10		6	
Mignonette (*Reseda luteola* L.)	3	Total share yellow	5	Total share yellow
Mayweed (*Anthemis tinctoria*)	3	colours: 15 %	5	colours: 25 %
Tansy (*Tanacetum vulgare* L.)	3		5	
Saw-wort (*Serratula tinctoria* L.)	3		5	
Canadian golden rod (*Solidago canadensis* L.)	3		5	
Onion peels (*Allium cepa* L.)	5	Orange	7	Orange
Dyer's woodruff (*Galium tinctorium* L.)	3	Total share red colours: 11 %	5	Total share red colours: 16 %
Cleavers (*Galium aparine* L.)	3		5	
Madder root (*Rubia tinctorum* L.)	5		6	
	100		**100**	

Assessed in a separate way, the red colorant produced from madder root is beyond the range of accepted prices for colorants, for instance.

Assuming that there are three amounts of natural colorant demand, being 1000, 3000 and 5000 kg, the corresponding amounts of raw materials result in differing costs for the total of requested colorants. Table 21.6 shows the required amounts of raw material depending on the share and type of colours needed for the colour portfolio requested by Company A and Company B; it also displays the corresponding costs of the total number of natural colorants.

Table 21.6 *Production costs in €/kg standardized natural colorants[a] considering different amounts of demand and colour portfolio requested by Company A and Company B. Reproduced with permission of Bundesministerium für Verkehr, Innovation und Technologie, Wien, 2003, from Reference [2]*

Total demand in kg		*1000*	*3000*	*5000*		*1000*	*3000*	*5000*
Requested colours Company A		€	€	€	*Requested colours Company B*	€	€	€
Berry pomace (raw material dry)	30 %	738	1758	2778	28 %	704	1656	2608
Tea pomace (raw material wet)	10 %	527	1125	1723	6 %	407	766	1125
Timber bark (manually)	29 %	813	1985	3157	18 %	591	1318	2046
Onion peels	5 %	260	324	388	7 %	272	362	452
Mignonette	15 %	766	1843	2920	25 %	1125	2920	4715
Madder root	11 %	1057	2716	4375	16 %	1434	3847	6260
	100 %	4160	9749	15339	100 %	4533	10868	17204
Production costs for total of natural colorants per kg		4.16	3.25	3.07		4.53	4.53	3.62

[a]Assumptions for raw material production: 2 tons/ha yield of dry matter; drying with 30% thermal solar energy; standardisation cost is 227.5 € per batch, meaning that standardisation costs per kg differ depending on the standardised amount (see also Table 21.3). Overview of production costs for standardised natural colorants (€/kg). Calculations are based on data published in Reference [2].

Average production costs for standardized natural colorants add up to € 4.53 and € 3.07 per kg respectively, the reasons for variations being the amount of demand and type of requested colours (see Table 21.6).

Table 21.6 show that the demand for natural colorants has to amount to 3000 kg at least, in order to ensure profit for the natural colorant-producing company, considering the upper price limit of € 5 per kg of dyed product, as discussed previously. As 1 kg of solid dyestuff is required to dye 1 kg of textile, therefore, in the case of the colour portfolio of Company A, production costs amount to 3.25 € per kg of dyed product, if the demand for natural colorants is 3000 kg.

For the market this is a realistic amount, even with the potential to be increased: 3000 kg of natural colorant serve to dye 150 000 tights, representing 1.9 % of the annual tights turnover of Company A.

21.6 Closed-Loop Economy: Towards a Zero-Emission and Zero-Waste Society

Industrial production systems process materials by using energy and water, usually causing airborne emissions, waste water and solid waste. Concepts such as 'end-of-pipe environmental protection' operate through adding on filter technologies at the end of the production process in order to gather airborne emissions as well as substances from waste water; they do not cause any changes in the process engineering system. Filters have to be disposed of in the same way as solid waste. It is not economically sound to dispose of something that was bought as a raw material at high expense before. The occurrence of solid waste, airborne emissions and waste water strongly indicates that there is optimization potential regarding the technical process engineering system. In addition, disposal has become more and more expensive, and so have materials, energy and water. Even if the cost for disposal sites is not relevant in some countries because of abundant unused space, it has to be taken into account that disposal is connected with transport, and transportation costs are increasing due to rising energy prices and cost emerging from damages caused by carbon dioxide emissions. Therefore, starting from the 'end-of-pipe technologies', new 'cleaner production systems' have been developed, focusing on a precautionary approach and aiming at changes in process engineering to avoid airborne emissions, solid waste and waste water instead of disposing of them at the end of the production process [16]. The application of 'cleaner production systems' implies also the use of new materials. As a consequence, the next step towards unloading the environment was the development of the 'ecodesign' concept, meaning that reduction of environmental loadings is not limited to the production process but is extended towards the whole life cycle of the product [17]. Products are designed in such a way that hazardous substances are avoided, energy and water consumption during production and use is low and re-use or recycling is made possible. The last step in this development is the concept of 'sustainable consumption', addressing the fact that in many cases customers are not interested in the specific product but in the service provided by the product. Therefore, after having dealt with end-of-pipe technologies, cleaner production concepts and ecodesign, service concepts have been developed [18].

The driving force behind this development is the vision of a 'zero-emission and zero-waste society', recently stated in the 6th Environmental Action Plan of the European Commission [19] and in the Austrian Program on Technologies for Sustainable Development, subprogram 'Factory of Tomorrow' [20]. The subprogram 'Factory of Tomorrow' addresses the trade, industry and service enterprises that produce and provide products of tomorrow using materials of tomorrow to meet tomorrow's needs. The achievement of zero-waste and zero-emission technologies and methods of production implies the increased use of renewable raw materials for materials and products as well as the increased use of renewable sources of energy in the production process and in the enterprise as a whole.

Considering the vision of a 'zero-emission and zero-waste society', it is therefore important to use a dyeing process with natural colorants based on locally available resources. Furthermore, it is imperative that natural raw materials are treated to gain the optimum output: the whole plant has to be made use of, with each part processed for a specific utilization. Thus, natural colorants are gained from what remains from processing agricultural products or by-products from processing natural colorants are used for another purpose, and in both cases residues from the colouring procedure are processed to be used as raw materials or for energy production.

From this point of view, the zero-emission and zero-waste society can also be described as a closed-loop economy, combining economic success with unloading the environment. Due to the fact of externalization of costs this sustainable concept has not yet been implemented: harmful industrial processes and products are still in use because it is not the one causing negative impact who pays but the general public, who compensates for the damage being represented by external costs [21]. Therefore, the term 'economical competitiveness' refers to existing legal conditions which shape the so-called 'free market'. If legal conditions change, competitiveness of sustainable concepts will change, too.

21.7 Conclusion: Aspects Influencing Market Development for Natural Colorants

Meeting technical requirements is only one factor for the economical success of natural colorants. The other very important aspect is to raise demand in order to reduce production costs. In Austria, a research project completed in 2003 showed that costs for plant dyestuff will amount to the price of expensive synthetic dyes and thus be acceptable for one company involved in the above-mentioned project. A precondition for this estimation is a substantial demand for dyestuff in order to keep production costs low [2].

There are three fundamental prerequisites that need to be fulfilled in order to increase the demand for natural colorants considerably:

- Farmers need a commercial structure to launch their products on the technical market. A professional supplier organization must collect raw materials for natural dyes from different agricultural sources and provide a group of dyes available at high standardized quality, deliverable to the dye house within a short period of time, all year long.
- Dyers require technical support because a rapid introduction of natural dyes into full-scale production processes is of particular importance to react efficiently to the ever-changing fashion demands. The technique of application of the dyes must be described and presented in comparably high technical quality, as done at present by the manufacturers of synthetic dyes. Defined recipes and profound technical support are needed.
- Depending on the corporate strategy, textile companies ask for a full range of colours, which cannot be provided at this point. To provide a pure black colour with natural colorants and the respective environmentally friendly dyeing process needs further investigations. There is a big demand for dark colours; therefore research needs to focus on how to provide natural dark colours in an acceptable and feasible way.

Another important factor for the economic success of natural colorants in the textile industry is the added value – the emotional benefit attached to the dyed product. The use of new dye products and the application of new dyeing processes create additional challenges for the company, in terms of infrastructure and quality control, because synthetic dyes are not likely to be completely replaced; rather, natural colorants will be applied in addition. Thus, even if costs for natural colorants are reasonable, natural dyes might be only applied if they are incorporated in a marketing strategy that leads to a better positioning in the market.

References

1. T. Bechtold, A. Turcanu, E. Ganglberger and S. Geissler, Natural dyes in modern textile dyehouses – How to combine experiences of two centuries to meet the demands of the future? Presentation in the course of *The Seventh European Round table on Cleaner Production*, Lund, Sweden, 2–4 May 2001.
2. S. Geissler, E. Ganglberger, T. Bechtold, A. Mahmut, A. Hartl and O. Schütz, FARB und STOFF – Sustainable Development durch neue Kooperationen und Prozesse, *Berichte aus der Energie- und Umweltforschung 25/2003*, Bundesministerium für Verkehr, Innovation und Technologie, Wien, 2003.
3. B. Rappl, Ch. Pladerer, M. Meissner, N. Prauhart, G. Roiszer-Bezan, B. Friedrich, E. Egger-Rollig, E. Ganglberger, S. Geissler, Th. Bechtold, R. Mussak, A. Mahmud-Ali, A. Grimm, G. Jasch and E. Freudenthaler, TRADEMARKFarb&Stoff. Von der Idee zum marktfähigen Handelsprodukt: Pflanzenfarben für die Textilindustrie, *Berichte aus der Energie- und Umweltforschung 21/2006*, Bundesministerium für Verkehr, Innovation und Technologie, Wien, 2006.
4. A. Wenisch and C. Pladerer, Risikominimierung entlang der Wertschöpfungskette vom pflanzlichen Rohstoff bis zum Farbstoff, *Berichte aus der Energie- und Umweltforschung 08/2007*, Bundesministerium für Verkehr, Innovation und Technologie, Wien, 2007.
5. S. Geissler, E. Gangberger, T. Bechtold, S. Sandberg, O. Schütz, A. Hartl and R. Reiterer, Potenzial an nachwachsenden Rohstoffen unter Aspekten der Nachhaltigkeit: Produktion von farbstoffliefernden Pflanzen in Österreich und ihre Nutzung in der Textilindustrie, Bundesministerium für Verkehr, Innovation und Technologie, Wien, 2001.
6. T. Bechthold, R. Mussak, A. Mahmud-Ali, E. Ganglberger and S. Geissler, Extraction of natural dyes for textile dyeing from coloured plant wastes released from the food and beverage industry, *Journal of the Science of Food and Agriculture*, **86**(2), 233–242(10) (30 January 2006).
7. S. Geissler, E. Ganglberger, T. Bechtold, A. Hartl and O. Schütz, Ashen's bark, raspberries' mash and Canadian golden rod – natural dyes for the textile industry from residues and agricultural primary production: broad range of colours and acceptable prices by combining agricultural products and residues from food and wood processing industry, Contribution to *Green-Tech 2002* , Floriade, Niederlande, 24–26 April 2002.
8. L. Berekoven, W. Eckert and P. Ellenrieder, Marktforschung. Methodische Grundlagen und praktische Anwendung, 11. Überarbeitete Auflage, Gabler, Wiesbaden, 2006.
9. H. Karmasin, Produkte als Botschaften, Carl Ueberreuter, Wien, 2003.
10. M. Horx, Konsument 2010, VNR Verlag für die Deutsche Wirtschaft AG, Bonn, 2001.
11. S. Geissler, R. Mussak, T. Bechtold, S. Klug and T. Vogel-Lahner, Nachwachsend = nachhaltig? Eine Analyse am Beispiel pflanzlicher Textilfärbung, in *GAIA 15/1*, 2006, pp. 44–53.
12. S. Geissler, S. Klug, T. Bechthold and R. Mussak, Marketing products from renewable resources: the example of naturally dyed textiles produced by industry, in *Schriftenreihe Nachwachsende Rohstoffe'*, Nachwachsende Rohstoffe für die Chemie 9, Symposium 2005 and 4th International Green-Tech Conference, pp. 275–286.
13. B. Fux, E. Grill, E. Pfeiffer and E. Rohrschach, Endbericht der Projektgruppe Trademark ,Teil I, Marketing und Marktforschung, Fachhochschule Wiener Neustadt FH-Diplomstudiengang Wieselburg für Produkt- und Projektmanagement, Wieselburg, 2004.
14. S. Geissler, S. Klug, T. Bechthold and R. Mussak, Marketing products from renewable resources: the example of naturally dyed textiles produced by industry, *Green-Tech Newsletter*, **8** (1), pp. 2–3 (2005).

15. U. Pock, F. Glechner, M. Wattaul and D. Kharchenko, Investkredit Bank AG – Research: Modellrechnungen zum Ölpreis – September 2006, 19.09.2006, Volksbank Gruppe, Investkredit.
16. ISO 14040 ff, Life Cycle Assessment; ISO 14040, Goal and Scope (1997); ISO 14041, Life Cycle Inventory Analysis (1998); ISO 14042, Life Cycle Impact Assessment (2000); ISO 14043, Life Cycle Interpretation (2000), International Standardisation Organization, Geneva.
17. The United Nations Conference on Environment and Development, Rio de Janeiro, 3–14 June 1992, Rio Declaration on Environment and Development, Article 15.
18. *Approaching Zero Emissions – A Special Issue of the Journal of Cleaner Production*, Elsevier, 2007.
19. Commission of the European Communities, Communication from the Commission to the Council, the European Parliament, the European Economic and Social Committee and the Committee of the Regions, Thematic Strategy on the Sustainable Use of Natural Resources, Brussels, 21.12.2005 COM(2005) 670 final.
20. www.fabrikderzukunft.at.
21. P. Bickel and R. Friedrich, ExternE Externalities of Energy Methodology 2005 Update, Stuttgart: Institut für Energiewirtschaft und Rationelle Energieanwendung IER/Universität Stuttgart, 2005.

Index

Note: Page numbers in *italic* type refer to figures; those in **bold** type relate to tables

Handbook of Natural Colorants Edited by Thomas Bechtold and Rita Mussak